Biotransformation of Waste Biomass into High Value Biochemicals

Satinder Kaur Brar • Gurpreet Singh Dhillon
Carlos Ricardo Soccol
Editors

Biotransformation of Waste Biomass into High Value Biochemicals

Springer

Editors
Satinder Kaur Brar
Institut national de la recherche scientifique
Centre - Eau Terre Environment (INRS-ETE)
University of Québec
Québec, QC, Canada

Gurpreet Singh Dhillon
Institut national de la recherche scientifique
Centre - Eau Terre Environment (INRS-ETE)
University of Québec
Québec, QC, Canada

Carlos Ricardo Soccol
Bioprocess Engineering and Biotechnology
Department of Chemical Engineering
Federal University of Parana
Curitiba, Paraná, Brazil

ISBN 978-1-4614-8004-4 ISBN 978-1-4614-8005-1 (eBook)
DOI 10.1007/978-1-4614-8005-1
Springer New York Heidelberg Dordrecht London

Library of Congress Control Number: 2013945156

Printed on acid-free paper

Springer is part of Springer Science+Business Media (www.springer.com)

Preface

World War II saw the onslaught of the chemical industries on a global scale. In fact, the chemical industry was perceived as a panacea to improve the everyday life and bring in new comforts and styles for the human beings. Meanwhile, over this decades-long period, the chemicals have resulted in all possible environmental and human health damages which are beyond repair even in the time to come. Most of the impacts, such as carcinogenicity, mutagenicity and teratogenicity are proliferating to such an extent that the countries have to spend billions as a health cost to repair these damages. Later on, "green chemistry," also known as clean chemistry or benign and sustainable chemistry came on the chemical industry scene. Green chemistry refers to "the design of chemicals and formulation of processes that reduce the risk to humans and minimize environment pollution." Green chemistry traces back several decades and can be linked to impactful environmental activists, such as Rachel Carson. Her 1962 publication, "Silent Spring," helped direct the public's awareness to pesticides and their ties to environmental pollution. As reflected in the previous decade spanning to today, there has been a shift in the emergence of green chemistry trends. As eco-awareness spreads to the consumer market and as the hazards of certain materials and chemicals become better known, companies and manufacturers are working to revamp the way they use and/or produce chemicals in their products. Eventually, this trend has been also parallelly taken over by the interest in biochemicals, that is, to produce chemicals using biological processes derived from plants, microorganisms and other living organisms. Biochemicals derived from biomass are generally obtained through industrial fermentation processes that make efficient use of a broad range of microorganisms to produce high-value fine chemicals, bulk chemicals, enzymes for use in pharmaceuticals through bio-catalysis, and a broad range of industrial chemicals (e.g., insect repellants, solvents, plastics, antibiotics, vitamins, and food additives, among others).

Another facet of the chemical industry has been its large-scale dependence on petroleum-based chemicals, and it is one of the many pressing challenges facing our planet. We can free ourselves from the dependency of petroleum industry-based feedstock though the bio-revolution. In this pursuit, opportunities to develop new and innovative bio-products and bio-processes are expanding every day. According to the Future Bio-Pathways Report released in February 2011, there are dynamic growth opportunities to develop new products (e.g., cellulose-based) that can be converted into bio-chemicals and used in novel ways, such as making bulletproof vests, food additives, and greener tires. Opportunities also exist to convert older, smaller-scale pulp mills to produce a range of bio-chemicals to serve niche markets. The potential market size for bio-chemicals over the next 10 years is confounding and exceeds that of the traditional products industry. According to the Organization for Economic Co-operation and Development (OECD), the bio-economy will contribute 1–14 new drugs per year, and will be responsible for 10 % of chemical production by 2030 (http://www.biotech.ca/en/policy-matters/beyondmoose-and-mountains/bioeconomyfacts.aspx). Worldwide demand for industrial enzymes is expected to reach U.S. $4.4 billion by 2015, with a compound annual growth rate of 6 %.

The biochemicals offer a range of environmental and economic advantages. The population at large benefits from biochemicals through reduced pollution. The biochemicals are derived from materials abundant in nature which dramatically reduces the amount of pollution generated from the extraction and processing of crude oil. Products derived from plant/microbial matter is highly biodegradable and in most cases can be disposed of safely and inexpensively. The private sector also accrues benefits from biochemicals in several ways. Biochemicals provide an environmental compliance tool for manufacturers. Increasingly stringent regulations regarding the use and disposal of petroleum-based chemicals are raising the administrative cost of these materials for manufacturers. Earlier, the manufacturers have often substituted a petrochemical not on the Toxic Release Inventory (TRI) to avoid environmental regulations only to discover that a later version of the TRI included the substitute petrochemical. Substituting biochemicals can be a permanent solution to the regulatory problem. The economics of replacing petrochemicals with biochemicals are increasingly favorable. Not only can a manufacturer save money by avoiding costly permits and compliance penalties, there is also a dramatic reduction in hazardous waste disposal costs. Companies using biochemicals in their manufacturing processes also can appeal to "green" consumers, an increasing portion of the market.

By 2025, biochemical is expected to make up 18 billion gallons, or $50 billion worth, with 9–14 billion gallons coming from the biotechnological conversion of biomass. Through industrial biotechnology, carbon emissions can be cut and companies can work towards becoming greener and it would be good to tap into this potential. According to a recent life cycle assessment analysis by a research group at Institute for Energy and Environmental Research, Heidelberg, Germany, biomass as a sustainable carbon source is very important for the organic chemistry and hence biochemicals production.

During the last decade there has been a tangible shift from incremental innovation in the biochemicals field, where some specific steps in existing technologies were replaced to save energy or to make the overall production process more efficient, towards more radical innovation with the development of new production processes entirely based on renewable resources. This makes production of biochemicals today one of the rare technologies which might help to radically change production chains in terms of sustainability along the whole product life cycle. Although currently the market share of biotechnology-produced bio-based chemicals is relatively small, the importance could grow very quickly depending on the substitution potential that bio-based materials have compared with their petrochemical counterparts. Some industrial biotechnology processes create the same molecules that are produced petrochemically. In these instances, the substitution could be complete, given that they are price-competitive. Other biotechnological processes lead to the creation of different compounds that have similar functionalities to petrochemical products, and in this case the biotechnology-derived molecules may not be suitable for a complete substitution of their petrochemical counterpart. They may nevertheless occupy or even create a niche within the overall market.

There are many publications which discuss about the production of compounds on large scale, such as biofuels, enzymes and organic acids from agro-industrial wastes. However, there is dearth of literature which discusses about the biovalorization of agro-industrial wastes into specialty biochemicals. The agro-industrial wastes have enormous potential to be converted to high-cost biochemicals, such as pharmaceuticals compounds (antibiotics, small peptides), antioxidants, aroma compounds, pigments, active food ingredients, exopolysaccharides and biosurfactants among others.

This book presents current and extensive information on biovalorization approach for the agro-industrial waste residues. Under each section, the chapters presents up to date and detailed information on agro-industrial waste residues and bioconversion technology to obtain biochemicals of economic importance. Implementation of biovalorization approach for agro-industrial wastes aims to mitigate pollution and other environmental problems and will promote economic benefits, such as production of biochemicals and reduction of waste management costs. This book covers the gap and will contain contributions from experts in field of processing of agro-industrial waste residues to various valuable biochemical products. This book will provide valuable information for academic researchers, graduate students and industry scientists working in industrial microbiology, biotechnology, bioprocess technology, waste management and the food industry.

We do hope that the book will provide a complete compilation of different biochemicals obtained from bioprocessing of residual biomass converging towards a bio-economy.

Québec, QC, Canada — Satinder Kaur Brar
Québec, QC, Canada — Gurpreet Singh Dhillon
Curitiba, Paraná, Brazil — Carlos Ricardo Soccol

Contents

Contributors

Ganesh Kumar Arumugam Marine Biotechnology, National Institute of Ocean Technology (Ministry of Earth Sciences, Govt. of India), Chennai, Tamil Nadu, India

Neha Babbar Central Institute of Post-Harvest Engineering and Technology, Ludhiana, Punjab, India

Vandana Bali Biotechnology Research Laboratory, Department of Food Engineering & Technology, Sant Longowal Institute of Engineering & Technology, Longowal, Punjab, India

Satinder Kaur Brar INRS-ETE, Université du Québec, Québec, QC, Canada

Lígia C. Cardoso Department of Biotechnology and Bioprocess Engineering, Federal University of Paraná—UFPR, Curitiba, PR, Brazil

Júlio C. De Carvalho Department of Biotechnology and Bioprocess Engineering, Federal University of Paraná–UFPR, Curitiba, PR, Brazil

Emna Chaabouni INRS-ETE, Université du Québec, Québec, QC, Canada

Juliana de Oliveira Department of Biotechnology and Bioprocess Engineering, Federal University of Paraná—UFPR, Curitiba, PR, Brazil

Deeplina Das Department of Biotechnology, Indian Institute of Technology Guwahati, Guwahati, Assam, India

Pradeep Dewapriya Marine Biochemistry Laboratory, Department of Chemistry, Pukyong National University, Busan, South Korea

Estera Szwajcer Dey Pure and Applied Biochemistry, Chemical Centre, Lund University, Lund, Sweden

Gurpreet Singh Dhillon Biorefining Conversions Network (BCN), Department of Agricultural, Food and Nutritional Sciences (AFNS), University of Alberta, Edmonton, AB, Canada

INRS-ETE, Université du Québec, Québec, QC, Canada

Zeineb Dkhili Institut National Agronomique de Tunisie (INAT), Tunis, Tunisia

Leandro F. dos Santos Department of Biotechnology and Bioprocess Engineering, Federal University of Paraná–UFPR, Curitiba, PR, Brazil

Fatma Gassara INRS-ETE, Université du Québec, Québec, QC, Canada

Vanessa Ghiggi Department of Biotechnology and Bioprocess Engineering, Federal University of Paraná—UFPR, Curitiba, PR, Brazil

Dharani Gopal Marine Biotechnology, National Institute of Ocean Technology (Ministry of Earth Sciences, Govt. of India), Chennai, Tamil Nadu, India

Arun Goyal Department of Biotechnology, Indian Institute of Technology Guwahati, Guwahati, Assam, India

Krishnamoorthy Hegde Department of Biotechnology, IIT Guwahati, Assam, India

Susan Grace Karp Biotechnology and Bioprocess Engineering Department, Centro Politecnico, Federal University of Parana, Curitiba, PR, Brazil

Surinder Kaur Department of Mycology and Plant Pathology, Institute of Agricultural Sciences, Banaras Hindu University (BHU), Varanasi, India

Se-Kwon Kim Marine Biochemistry Laboratory, Department of Chemistry, Pukyong National University, Busan, South Korea

Marine Bio-Process Research Center, Pukyong National University, Busan, South Korea

Elena Kiseleva Collection of Microorganisms, Institute of Microbiology, Belarus National Academy of Sciences, Minsk, Belarus

Emily Kolomiets Collection of Microorganisms, Institute of Microbiology, Belarus National Academy of Sciences, Minsk, Belarus

Shweta Kumari Biotechnology Research Laboratory, Department of Food Engineering & Technology, Sant Longowal Institute of Engineering & Technology, Longowal, Punjab, India

Luiz Alberto Junior Letti Biotechnology and Bioprocess Engineering Department, Centro Politecnico, Federal University of Parana, Curitiba, PR, Brazil

Naceur M'hamdi Institut National Agronomique de Tunisie (INAT), Cité Mahrajène Le Belvédère, Tunis, Tunisia

Mamta Defence Institute of Physiology & Allied Sciences, Timarpur, Delhi, India

Kshipra Misra Defence Institute of Physiology & Allied Sciences, Timarpur, Delhi, India

Sangeeta Negi Department of Biotechnology, Motilal Nehru National Institute of Technology, Allahabad, India

Alessandra Cristine Novak Department of Biotechnology and Bioprocess Engineering, Federal University of Paraná -UFPR, Curitiba, PR, Brazil

Galina Novik Collection of Microorganisms, Institute of Microbiology, Belarus National Academy of Sciences, Minsk, Belarus

Harinder Singh Oberoi Central Institute of Post Harvest Engineering and Technology, Ludhiana, Punjab, India

P.S. Panesar Biotechnology Research Laboratory, Department of Food Engineering & Technology, Sant Longowal Institute of Engineering & Technology, Longowal, Punjab, India

Kirubagaran Ramalingam Marine Biotechnology, National Institute of Ocean Technology (Ministry of Earth Sciences, Govt. of India), Chennai, Tamil Nadu, India

Cristine Rodrigues Department of Biotechnology and Bioprocess Engineering, Federal University of Paraná—UFPR, Curitiba, PR, Brazil

Rosália Rubel Department of Biotechnology and Bioprocess Engineering, Federal University of Paraná–UFPR, Curitiba, PR, Brazil

Harvinder Singh Saini Department of Microbiology, Guru Nanak Dev University, Amritsar, India

Saurabh Jyoti Sarma INRS-ETE, Université du Québec, Québec, QC, Canada

Venkatesh Selvaraj Marine Biotechnology, National Institute of Ocean Technology (Ministry of Earth Sciences, Govt. of India), Chennai, Tamil Nadu, India

Anastasiya Sidarenka Collection of Microorganisms, Institute of Microbiology, Belarus National Academy of Sciences, Minsk, Belarus

Partap Bir Singh Department of Microbiology, Guru Nanak Dev University, Amritsar, India

Keli Sobral Biotechnology and Bioprocess Engineering Department, Centro Politecnico, Federal University of Parana, Curitiba, PR, Brazil

Vanete Tomaz Soccol Biotechnology and Bioprocess Engineering Department, Centro Politecnico, Federal University of Parana, Curitiba, PR, Brazil

Carlos Ricardo Soccol Department of Biotechnology and Bioprocess Engineering, Federal University of Parana—UFPR, Curitiba, PR, Brazil

Eduardo Bittencourt Sydney Department of Biotechnology and Bioprocess Engineering, Federal University of Paraná—UFPR, Curitiba, PR, Brazil

Luciana Porto de Souza Vandenberghe Department of Biotechnology and Bioprocess Engineering, Federal University of Paraná—UFPR, Curitiba, PR, Brazil

Mausam Verma CO2 Solutions, Québec, QC, Canada

Adenise Lorenci Woiciechowski Biotechnology and Bioprocess Engineering Department, Centro Politecnico, Federal University of Parana, Curitiba, PR, Brazil

Part I
General Concepts

Chapter 1
Waste Biomass: A Prospective Renewable Resource for Development of Bio-Based Economy/Processes

Surinder Kaur, Gurpreet Singh Dhillon, Saurabh Jyoti Sarma, Satinder Kaur Brar, Kshipra Misra, and Harinder Singh Oberoi

1.1 Introduction

The sustainable socioeconomic development owing to the continued pace of world economic growth heavily relies upon a secure supply of raw material inputs for agriculture, industry, energy, and related sectors. The development heavily depends on energy, its applications ranging from home appliances, transportation, and industrial processes to supply commodities for our daily needs. To fulfill the energy needs, we consume nearly 500 Quadrillion Btu (QBtu) of energy, and majority of it (92 %) comes from nonrenewable resources, such as petroleum, coal, and nuclear and natural gas (Khanal and Lamsal 2010). Energy demand is expected to escalate by around 44 % by 2030 mostly due to the increasing demand from developing countries, such as India and China. However, today's heavy reliance on nonrenewable resources, especially fossil fuels, is increasingly constrained by economic, political, and

S. Kaur
Department of Mycology and Plant Pathology, Institute of Agricultural Sciences, Banaras Hindu University (BHU), Varanasi 221005, India

G.S. Dhillon (✉)
Biorefining Conversions Network (BCN), Department of Agricultural, Food and Nutritional Sciences (AFNS), University of Alberta, Edmonton, AB, Canada

INRS-ETE, Université du Québec, 490 Rue de la Couronne, Québec, QC, Canada G1K 9A9
e-mail: garry_dhillons@yahoo.com; gdhillon@ualberta.ca

S.J. Sarma • S.K. Brar
INRS-ETE, Université du Québec, 490 Rue de la Couronne, Québec, QC, Canada G1K 9A9

K. Misra
Defence Institute of Physiology & Allied Sciences, Timarpur, Lucknow Road, Delhi 110054, India

H.S. Oberoi
Central Institute of Post-Harvest Engineering and Technology, P.O. PAU, Ludhiana 141004, Punjab, India

S.K. Brar et al. (eds.), *Biotransformation of Waste Biomass into High Value Biochemicals*,
DOI 10.1007/978-1-4614-8005-1_1, © Springer Science+Business Media New York 2014

environmental factors. The dependence on these conventional resources is also accompanied by a heavy reliance on chemical and thermochemical processes. However, due to the continuous and fast depletion of the conventional energy resources and the growing awareness and concern regarding the environmental effects of their utilization, there has been a major challenge in the recent past to identify and develop alternate energy sources. In this regard, the bio-based processes are growing at a faster pace, although currently their role in the global economy is trivial. There are increasing initiatives from both public and private sector interests that support the supply of our energy needs and other industrial products through biological processes and/or biomass resources.

Rapid increase in volume and types of agricultural and industrial waste biomass, as a result of intensive agriculture in the wake of population growth, food processing, and improved living standards, is becoming a burgeoning problem. The waste biomass being rich in carbon and other vital nutrients is highly amenable to biological degradation and emits methane and leachate. Moreover, the open burning of agricultural wastes, such as rice stubble by the farmers to clear the lands, generates CO_2 and other pollutants. Hence, improper management of agricultural and agro-industrial waste biomass is contributing towards climate change, water and soil contamination, and local air pollution which jeopardizes the health of flora and fauna. Furthermore, this waste biomass is of high value with respect to material and energy recovery.

In the context of bio-based economy, the current chapter discusses the different sources, types, and nature of waste biomass. The overview of the different management strategies applied for the value addition of different types of waste biomass is discussed. This chapter also provide insights into the role of biotransformation of waste biomass resources for developing bio-based economy/processes. Finally, the chapter gives a brief summary of directly extractable high-value biochemicals from waste biomass.

1.2 Waste Biomass

Biomass is a renewable resource and refers to any material having recent biological origin, such as plant materials, agricultural crops, and even animal manure. According to National Renewable Energy Laboratory (NREL), biomass can be defined as any plant-derived organic matter. Biomass available for energy on a sustainable basis includes herbaceous and woody energy crops, agricultural food and feed crops, agricultural crop wastes and residues, wood wastes and residues, aquatic plants, and other waste materials including some municipal wastes. Biomass is a very heterogeneous and chemically complex renewable resource. Owing to its natural abundance, sustainability, and often low cost, biomass is a potential alternative to nonrenewable energy sources for production of chemicals. Biomass has a chemical composition comprised of C, H, O, and N, similar to fossil feedstocks which contain C and H. Currently, the annual worldwide production of biomass is estimated to exceed 100 trillion kilograms (Xu et al. 2008). However, presently, only 5 % of chemicals are derived from renewable resources (Lucia et al. 2006). Hence, there is an enormous potential for production of bio-based chemicals to compete with their fossil-derived counterparts.

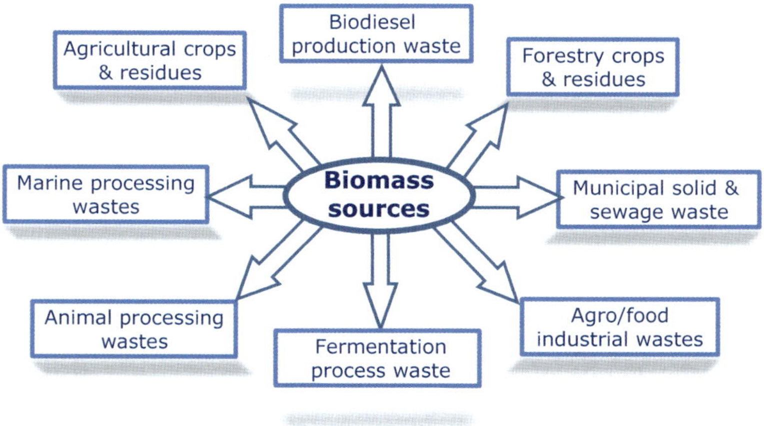

Fig. 1.1 Different types of waste biomass

1.2.1 Types of Waste Biomass/Potential Waste Biomass Resources

Globally, 140 billion metric tons of biomass is generated every year from agriculture. The main sources of biomass waste are given below (Fig. 1.1).

- *Agricultural and agro-industrial wastes*: Agricultural biomass generally comprises of leftovers after grain separation, such as residual stalks, straw, leaves, roots, husk, nut or seed shells, waste wood, and animal husbandry waste. Some common examples are coconut (fronds, husk, shell), coffee (hull, husk, ground), corn (cob, stover, stalks, leaves), cotton (stalks), nuts (hulls), peanuts (shells), rice (hull/husk, straw, stalks), sugarcane (leavings, bagasse, molasses), vegetable wastes, etc.
- *Animal husbandry wastes*: Manure from cattle, poultry, and hogs.
- *Food processing wastes*: Include by-products and leftovers processing, such as fruit pomace wastes (peels, seeds, and pulp) and wastewater sludge, brewery wastes (brewer's spent grain, spent hops, wastewaters, and surplus yeast), winery wastes (solid by-products include marcs, pomace, and stems and may account on average for almost 30 % (w/w) of the grapes and liquid sludge from organic wastewater treatment plants), starch industry wastes, dairy industry wastes (whey), and meat processing wastes.
- *Forestry residues*: Wood chips, bark, sawdust, timber slash, and mill scrap.
- *Municipal waste*: Solid household wastes, wastewater sludge, waste paper, and yard clippings.
- *Marine processing wastes*: fish industry waste (scales, skin, visceral mass (viscera, air bladder, gonads, and other organs), head, fins, and visceral mass) and crustacean shell and shell fish waste (head and body carapace).

- *Biotechnological industry wastes*: Waste fungal/bacterial/yeast/microalgae biomass.
- *Biodiesel industry wastes*: Crude glycerol from biodiesel production.

This high volume of biomass can be converted to an enormous amount of energy and raw materials. Agricultural waste biomass converted to energy can substantially displace nonrenewable-based fossil fuels, reduce emissions of greenhouse gases (GHG), and provide renewable energy. Biomass is a renewable resource that has a steady and abundant supply, especially those biomass resources that are by-products of agricultural activity. With the increasing global concerns to combat climate change, countries are now looking for alternative sources of energy to minimize GHG emissions. Apart from being carbon neutral, the utilization of biomass for energy decreases reliance on the consumption of fossil fuel, hence, contributing to energy security and climate change mitigation while closing the carbon cycle loop. Currently, as the debate on food versus fuel gets intensified, the biomass can provide extra income to farmers without compromising the production of main food and even nonfood crops.

Although there is an increasing trend on the utilization of biomass for energy and other industrial products, biomass is still largely underutilized and left to rot or openly burned in the fields, especially in developing countries. Mostly, these countries do not have strong regulatory laws to control such environmentally unfriendly practices or either fail to implement them. As a common practice, the burning of agricultural residue (e.g., open field burning of rice stubble) results in air pollution which poses risk to human and ecological health. Biomass is a renewable resource that causes problems when not used. The challenge, therefore, is to convert biomass as a resource for energy and other productive uses.

1.2.2 Nature of Biomass Feedstock

Agricultural crops can be roughly divided according to the composition of their (main) economic products, such as sugar, starch (grains, tubers), oilseed, protein, or fiber crop and crops for specialty products (pharmaceutics, cosmetics, dyes, fragrance, and flowers). Besides the main harvested product, all crop processing systems yield more or less secondary products/by-products and residues. These products may find an application depending on demand and possibilities for economic conversion. Biomass residues can be categorized into three main groups: (1) primary biomass residues, available at the farm; (2) secondary biomass residues, released in the agro-food industry; and (3) tertiary biomass, which is remaining after use of products. The characteristics that influence availability and suitability of the waste biomass as feedstocks are the nature of biomass, moisture content, the density, and the seasonality of supply.

Lignocellulosic biomass comprising forestry, agricultural, and agro-industrial wastes are abundant, renewable, and inexpensive energy sources. Such wastes include a variety of materials, such as sawdust, poplar trees, sugarcane bagasse,

Table 1.1 The main components of different lignocellulose wastes (refs. Nigam et al. 2009; Dhillon et al. 2011b)

Lignocellulose waste	Cellulose (wt %)	Hemicellulose (wt %)	Lignin (wt %)
Sugarcane bagasse	40.0	27.0	10.0
Rice straw	36.2	19.0	9.9
Wheat straw	32.9	24.0	8.9
Barley straw	33.8	21.9	13.8
Rye straw	37.6	30.5	19.0
Oat straw	39.4	27.1	17.5
Corn cobs	33.7	31.9	6.1
Corn stalks	35.0	16.8	7.0
Cotton stalks	58.5	14.4	21.5
Soya stalks	34.5	24.8	19.8
Sunflower stalks	42.1	29.7	13.4
Apple pomace	7.2	–	23.5
Brewer's spent grain	13.1–25.4	28.4–29.96	11.9–27.8
Citrus waste	8.8	4.4	3.7

waste paper, brewer's spent grains, coconut coir and shell, fruit pomace and liquid sludge, switch grass, and straws, hull, stems, stalks, leaves, husks, shells, and peels from cereals like rice, wheat, corn, sorghum, and barley, among others. Lignocellulosic biomass is chemically composed of three main fractions: cellulose, hemicellulose, and lignin in varied concentrations (Table 1.1) with smaller amounts of proteins, lipids, and ash (Fig. 1.2). Cellulose is a polymer of glucose (a C6 sugar), which can be used to produce glucose monomers for fermentation to produce a variety of products, such as renewable fuels, platform chemicals, organic acids and biopolymers among others. Hemicellulose is a copolymer of different C5 and C6 sugars including xylose, mannose, and glucose, depending on the type of biomass. Lignin is a branched polymer of aromatic compounds. Both the C5 sugars and the lignin fragments can be used as feedstock for the production of various value-added products including high-value biochemicals in a biorefinery.

These polymers are closely associated with each other constituting the cellular complex of the vegetal biomass. Basically, cellulose forms a skeleton which is surrounded by hemicellulose and lignin (Fig. 1.2). The pretreatment of lignocellulosic biomass helps to disrupt the 3D network structure of lignin, cellulose, and hemicellulose, allowing high yields of fermentable sugars to be produced in subsequent enzymatic hydrolysis. Different pretreatments used for the separation of different polymers in lignocellulosic waste are given in Fig. 1.3. The pretreatments help the enzymes for easy excess for the biomass hydrolysis to simple sugars.

Currently, biomass pretreatment is still a necessary step to establish a cheap sugar platform for bioethanol and other biochemicals. An ideal pretreatment technology should target the three basic requirements: simple process, cost-effective, and high sugar recovery.

Cellulose and hemicellulose are sugar-rich fractions of interest for use in fermentation processes, since microorganisms may use the sugars for growth and

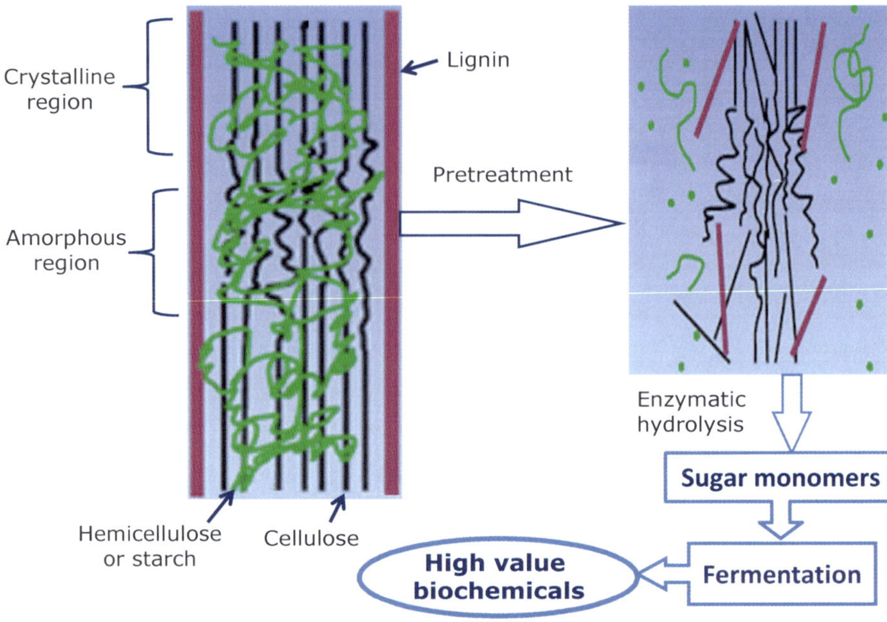

Fig. 1.2 The structure of lignocellulosic material and routes for its biotransformation to high-value biochemicals

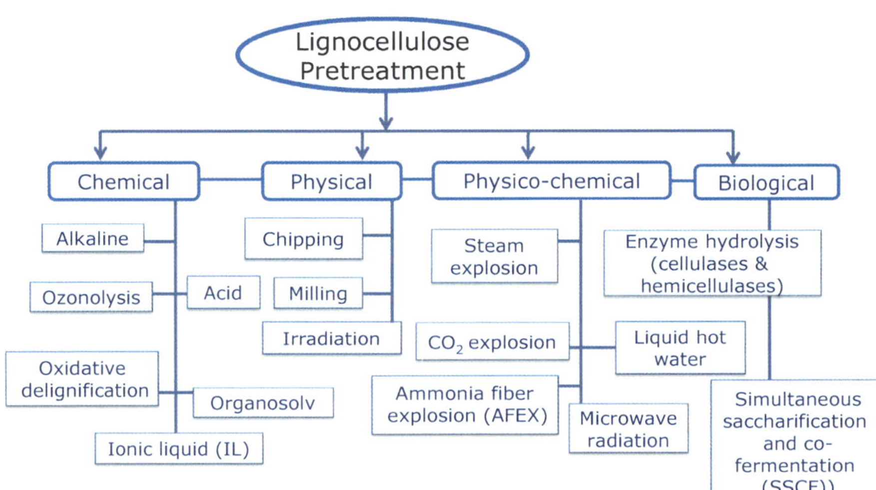

Fig. 1.3 Different pretreatments for the hydrolysis of lignocellulosic biomass

Table 1.2 Examples of biomass residues for different crops

Type of crop	Primary residues	Secondary residues	Residue ratio[a]
Fruits and nuts	Seeds		–
		Fruit pulp, peelings, fruit pomace	0.2–0.4
Vegetables	Leaves, stems, etc.		0.2–0.5
		Peelings, skin	0.1–0.2
Grains (wheat, corn, rice, barley, millet)	Straw (stover)	–	1.0–2.0
	Chaff (hulls, husks)	Bran, cobs	0.2–0.4
Sugarcane	Leaves and tops	–	0.3–0.6
	–	Bagasse	0.3–0.4
		Molasses	–
Tubers, roots (potato, cassava, beet)	Foliage, tops	–	0.2–0.5
		Peels	0.1–0.2
Oil seeds	Hulls		0.2–1.2
		Press cake	0.1–0.2
Sunflower, olive	Foliage, stems		0.2–0.5
Cocos, palm oil	Husks, fronts	Shells	0.3–0.4
Soy, rape, peanut	Foliage	Seed coat, shells	0.3–0.5

Sources: UNDP (2007), Van Dam (2002, Rosillo-Calle et al. (2007)
[a]Residue ratio refers to ratio of dry matter weight to crop produced

production of value-added compounds, such as ethanol, food additives, organic acids, enzymes, among others. Submerged and solid-state fermentation systems have been used to produce compounds of industrial interest from lignocellulosic wastes, as an alternative for valorization of these wastes and also to solve environmental problems caused by their disposal. When submerged fermentation systems are used, a previous stage of hydrolysis for separation of the lignocellulose constituents is required.

Few common primary and secondary residues from agricultural crops are given in Table 1.2. There is significant variation in the quantities available. For instance, in some cases, residues amount to only about 10–20 % of the crop by weight, whereas in other cases, the residues might actually be greater than the original crop. As evident from the Table 1.2, grain crops tend to have the highest overall residue ratio, amounting to as much as double the crop weight. For this reason, utilization of straw from grains should be a much higher priority for utilization of this largely untapped reservoir of biomass resources.

Lignocellulose wastes are accumulated every year in large quantities, causing environmental problems. However, due to their chemical composition based on sugars and other compounds of interest, they could be utilized for the production of a number of value-added products. Therefore, besides the environmental problems caused by their accumulation in the nature, the nonuse of these materials constitutes a loss of potentially valuable sources. .

The underutilized biomass resources from different possible sources, such as primary agricultural production, agro-industries, and municipal waste, are generally available in abundant quantities at negligible costs. Agriculture-based wastes, such

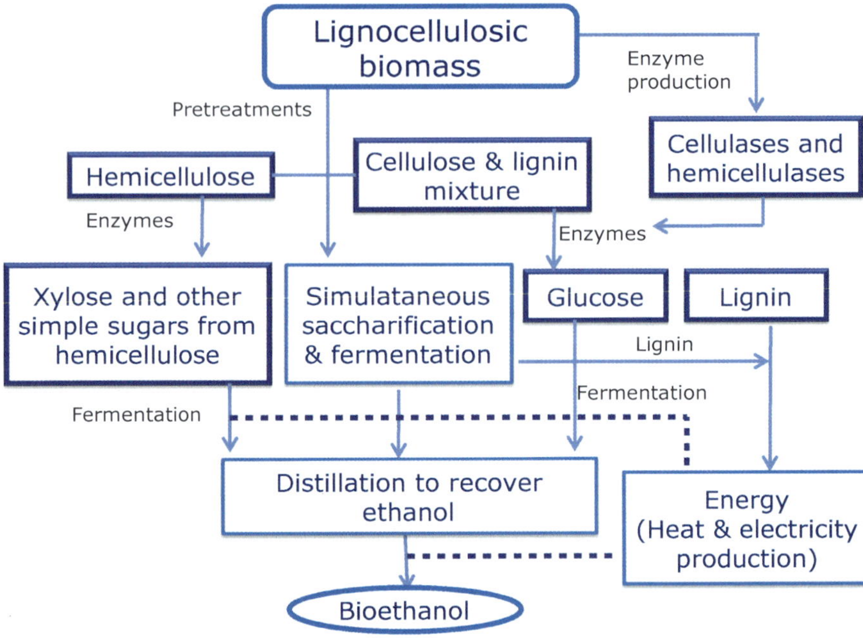

Fig. 1.4 Process showing the pretreatment and enzymatic hydrolysis of lignocellulosic biomass to produce sugar syrups for production of bioethanol

as straws or seed hulls, can be harvested and collected at the farm or at central processing units. Others wastes, such as food industry wastes, are only available in dispersed/diluted forms and need collection systems to be installed at particular industries.

Earlier, agricultural residues were promoted mainly for energy (e.g., bioethanol production) use, often at low efficiency (Fig. 1.4). However, it is now more widely recognized that there are in fact other possible routes that may provide higher-value-added products or could serve as complementary products via coproduction schemes alongside energy applications. The sugar-rich syrups produced after pretreatment and enzymatic hydrolysis of lignocellulosic biomass can be used for the production of high-value products. Currently, such integrated processes are recurring theme in industrial biotechnology development (van Dam et al. 2005). For instance, microalgae/fungal/yeast cultivations involving production of various products result in waste microalgae/fungal/yeast biomass as a by-product. The fungal biomass is rich in chitin which can be extracted and transformed to its deacetylated derivative, chitosan (Dhillon et al. 2012a; Kaur and Dhillon 2013a). Similarly, microalgae biomass resulting after lipid extraction for biodiesel is also rich in carbohydrates, proteins and other products, such as pigments.

Crude glycerol (CG) is a waste by-product of biodiesel production process. For every 100 kg of biodiesel produced by transesterification of vegetable oil/animal fat/microalgae-derived lipids, 10 kg of CG is produced. CG is a carbon-rich source and an emerging and less expensive feedstock for bioprocess technology. CG can be used for the production of a wide range of products, such as ethanol and biohydrogen. More recently, it has been evaluated for the production of high-value biochemical, such as eicosapentaenoic acid, docosahexaenoic acid, glycolipid, biosurfactant, 1,3-propanediol, and antibiotics, such as cephalosporin C (Pyle et al. 2008; Athalye et al. 2009; Liu et al. 2011; Shin et al. 2011; Ferreira et al. 2012).

1.3 Bio-Based Economy/Processes

Recently, a great deal of research is being devoted to the area of sustainable processes (Bruggink et al. 2003). The need for such processes stems from the burgeoning human population and the accompanied required growth in availability of materials and energy (Song 2006). A significant part of the developments is dedicated to bio-based sustainable processes, which make use of renewable feedstocks, such as agro-industrial wastes and industrial by-products, to decrease the use of nonrenewable fossil resources which are depleting very quickly. Owing to the higher efficiency in terms of energy and materials and the reduction of environmentally unfriendly wastes, the bio-based processes are clearly advantageous. In view of bio-based green processes, the identification and assessment of environmentally sound technologies that promote the use of biomass, i.e., conversion of lignocellulosic biomass into energy and raw materials, is highly desired.

The bio-based economy can be defined as consisting of those sectors that derive a majority of their market value from biological processes and/or products derived from natural materials, as compared to products/processes allied with nonrenewable resources and/or purely based on chemical processes. The industrial portion of the bio-economy is somewhat distinct from agricultural, forestry, and other sectors, in the sense that raw materials are utilized to make industrial feedstocks or products or to drive industrial processes. Sustainable feedstock supply is one of the key issues for the evolution towards the bio-based economy. Therefore, the resource base needs to be identified from the perspective of supply and demand. The waste biomass derived from crop residues of food and feed production, forestry residues, fermentation process wastes, food/beverage processing wastes, marine crops and processing wastes, municipal waste, manure and animal products, and biological process-derived wastes are potential candidates/resources towards the realization of a bio-based economy.

The deliberate importance of the bio-economy is linked to those areas in which bio-based products and processes can provide alternative for fossil- or mineral-based products and/or chemical processes. Since the vast majority of industrial products and processes are currently centered on nonrenewable resources and minerals, such substitution has considerable potential to make various industry sectors more sustainable in the long run while also reducing environmental impacts in the near future, especially with regard to the Kyoto protocol aiming towards reducing GHG emissions and land disposal requirements. The utilization of bio-based renewable resources holds great potential value for industries in various sectors, such as energy, platform chemicals, biopolymers, and health/personal care products. In general, a bio-based economy offers many benefits and opportunities:

- New areas of economic growth and development for the many regions especially rural areas that have abundant biomass resources.
- Creation of new innovative business sectors and entrepreneurial skills.
- Improved energy security, by reducing dependence on nonrenewable resources, such as fossil fuels.
- Enhanced economic and environmental linkages between the agricultural sector and a more prosperous and sustainable industrial sector.
- Mitigation of GHG emissions.
- Improved health by alleviating exposure to harmful substances through substitution of natural bio-based materials for chemical and synthetic materials.
- Employment creation and rural development.
- Avoid the competition of land used as raw material for industry with other land uses, especially in relation to food and animal feed (competition for other uses of biomass, especially food, feed, and fiber).

1.4 Value Addition of Waste Biomass

Transformation of waste biomass to various biotechnological products and bioenergy is carried out through different routes. The major routes comprise biological, chemical, and thermal processes and are depicted in Fig. 1.5. The conversion of biomass either can result in final products or may provide building blocks for further processing.

1.4.1 Biotransformation of Biomass

Biological transformation involves the utilization of living organisms or enzymes (biocatalysts) to catalyze the conversion of biomass into specialty and commodity chemicals. Generally, it is considered to be the most flexible mode for conversion of biomass into various industrial products (Dale 2003). Compared to chemical transformations, where high temperatures and pressures are involved, operating conditions for biological transformations are relatively mild. Fermentation is the primogenital and the most fundamental and mature area of biotechnology for biological transformation. For centuries, fermentation was used for preserving and processing food and beverages. Only in the last several decades due to current advancements in biotechnology, it has been used to bring to market a wide variety of fermentation-based products, including platform chemicals, renewable fuels, biopolymers, antibiotics, amino acids, organic acids, and pharmaceuticals using various agro-industrial feedstocks. Some commercial bulk chemicals, such as ethanol, lactic acid, citric acid, acetone, and butanol, have been produced via yeast,

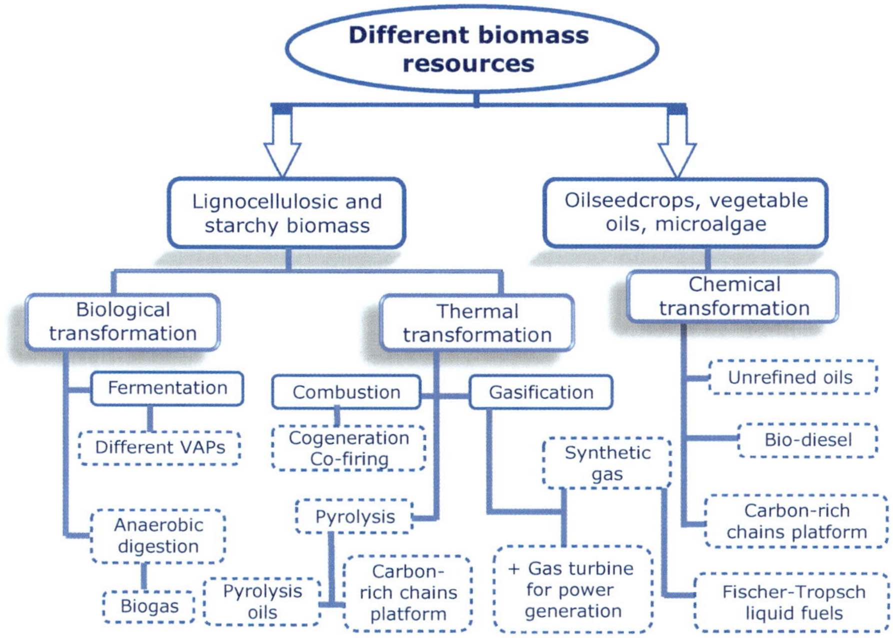

Fig. 1.5 Conversion routes of biomass to bioenergy and other biotechnological products

fungal and bacterial fermentation processes (Atsushi et al. 1996; Huang et al. 2005; Ezeji et al. 2007; Dhillon et al. 2011c).

Recently, there has been increasing interest in the utilization of biocatalysts to transform renewable resources into biochemicals, owing to high yield and selectivity, and fewer by-products as compared to chemical synthesis. Table 1.3 shows the biotransformation of different wastes to high-value biochemicals through different processes. However, due to the metabolic restriction in microorganisms, only a few bulk products currently are produced via fermentation (Danner and Braun 1999). Therefore, development of new technologies to broaden the product range is necessary. Advances in genetic engineering have been viewed as a powerful tool for genetic manipulation of multistep catalytic systems involved in cell metabolism (Zha et al. 2004). Recombinant DNA technology has been used to clone and manipulate gene encoding enzymes in organisms. Recombinant microorganisms, with altered sugar metabolism, are able to ferment sugar to few specialty biochemicals, which cannot be produced by the corresponding wild strain (Danner and Braun 1999). For instance, catechol and adipic acid were produced from glucose using genetically modified *Escherichia coli*. Both glucose and xylose, in cellulosic biomass, have been converted into ethanol by recombinant *Saccharomyces* strains (Anastas and Kirchhoff 2002). Hence, it is imperative that the recombinant strains can be used for the efficient utilization of pentose and hexose sugars from the abundant lignocellulosic biomass. Moreover, immobilized enzyme systems and whole cells have been used to produce various biochemicals from biomass.

Table 1.3 Biotransformation of different wastes to high-value biochemicals through different processes

Waste biomass	High-value product/microorganisms	Remarks	References
Agricultural wastes (vegetable and fruit processing wastes)			
Rice husk and straw	Antibiotic, neomycin by *Streptomyces marinensis*	SSF	Ellaiah et al. (2004)
Sugarcane bagasse/molasses and corn steep waste	Antibiotic, cephalosporin C- by *Acremonium chrysogenum*	SSF	Cuadra et al. (2008)
Wheat bran flour and coconut oil cake	Antibiotic, cyclosporin A by *Tolypocladium inflatum*	SSF	Survase et al. (2009)
Peanut shells, corn pomace/husk/cob, wheat bran, cassava peels, coconut oil cake, groundnut oil cake, groundnut shell, and rice husk	Antibiotics—tetracycline *and* oxytetracycline (*Streptomyces strains*); rifamycin B (*Amycolatopsis sp.*)	SSF	Asagbra et al. (2005a, b); Mahalaxmi et al. (2010); Vastrad and Neelagund (2012)
Apple pomace and sludge	Natural antioxidants, biopolymers, organic acids	SSF and SmF	Dhillon et al. (2011c); Ajila et al. (2011); Gassara et al. (2012)
Apple pomace	*Antibiotic*, mevastatin by *Streptomyces fradiae*	SSF	Vastrad and Neelagund (2011)
Grape pomace	Various antioxidant compounds	Extraction	Knoblich et al. (2005)
Tomato peel and seed by-products	Carotenoids from peel: lycopene, lutein, β-carotene, and *cis*-β-carotene. Carotenoids from seeds: Lycopene and other carotenoids	Extraction	Strati and Oreopoulou (2011); Spatafora and Tringali (2012)
Date palm juice by-products	Xanthan exopolysaccharides (EPS)—*Xanthomonas campestris*	43.35 g/l	Ben Salah et al. (2010)
Cassava residues	Astaxanthin by *Phaffia rhodozyma* (*yeast*)	0.060 mg/g	Yang et al. (2011)
Cassava bagasse	Polyketide mix (pigment)—*Monascus sp.*	SSF	Carvalho et al. (2007)
Citrus peel, mango kernel, banana peel, litchi pericarp and seeds, olive pomace, pomegranate peels and seeds	Different phenolic compounds	Antimicrobial compounds	Arogba (2000); Puravankara et al. (2000); Someya and Okubo (2002); Obied et al. (2005); Tehranifara et al. (2011); Duan et al. (2007)

Strawberry/blackberry/raspberry pomace	Anthocyanins, tannins, starches, saponins, polypeptides and lectins, polyphenols, lactones, flavones, and phenons	Antimicrobial compounds	Krisch et al. (2009)
Beet root pomace	Phenolic, flavonoid betalaine	Antimicrobial compounds	Canadanovic et al. (2011)
Citrus peels	Flavonoids, saponins, steroids, terpenoids, tannins, and alkaloids	Antimicrobial compounds	Ashok et al. (2011)
Guava bagasse	Epicatechin, quercetin and caffeic	Antimicrobial compounds	Martin et al. (2012)
Beet molasses, sugarcane molasses, waste residue of rice bran oil	2-phenylethanol, acetoin, vanillin	Aroma compounds	Etschmann et al. (2003); Xiao et al. (2007); Zheng et al. (2007)
Food processing industry waste			
Starchy wastewater	Poly(β-hydroxybutyric) (PHB)—*Alcaligenes latus*	55 % (g/g)	Yu (2001)
Potato peels	Gallic acid, caffeic acid, vanillic acid	–	Zeyada et al. (2008)
Carrot peels	Phenols, β-carotene	–	Chantaro et al. (2008)
Cucumber peels	Chlorophyll, pheophytin, phellandrene, caryophyllene	–	Zeyada et al. (2008)
Dairy waste			
Whey	β-carotene by *Blakeslea trispora (fungus)*	170 mg/g	Varzakakou et al. (2010)
Reconstituted whey	β-carotene by *Sporobolomyces roseus* (yeast)	0.55 mg/g	Marova et al. (2012)
Whey	Pigment canthaxanthin—*Dietzia natronolimnaea* (bacteria)	0.020 mg/g	Khodaiyan et al. (2008)
Beverage production wastes			
Brewer's spent grain—requires no preliminary detoxification steps and overall production is favored by high initial xylose concentrations	Xylitol (sweetener), a rare sugar that exists in low amounts in nature—used to combat dental caries, diabetes, disorders in lipid metabolism, and parenteral and renal lesions and to prevent lung infection, otitis, and osteoporosis	Acid hydrolysis and yeast fermentation	Mussatto and Roberto (2005, 2008)

(continued)

Table 1.3 (continued)

Waste biomass	High-value product/microorganisms	Remarks	References
Brewer's spent grain	Ferulic acid, hydroxycinnamic acid	(1) alkaline hydrolysis—0.3 % (2) Esterase from *A. niger*—3.3 %	Bartolomè et al. (1997, 2002, 2003)
Brewer's spent grain	Pullulan—an extracellular water-soluble microbial polysaccharide produced by strains of *Aureobasidium pullulans*	Maximum conc. (6.0 g/l) after 72 h of fermentation	Roukas (1999)
Wine industry waste (pomace—grape seeds, skins, stems)	Polyphenolic antioxidants—e.g., gallic acid, anthocyanins, proanthocyanidins, flavanols, and hydroxycinnamates	Extraction	Guendez et al. (2005); Pinelo et al. (2005); Makrisa et al. (2007)
Marine processing wastes			
Fish industry waste	Fish oil	Extraction	
Crustacean shell wastes	Carotenoid pigments—astaxanthin, canthaxanthin, 4-hydroxyechinenone, 3-hydroxycanthaxanthin, echinenone, isocryptoxanthin, β-carotene	Extraction	Pokorny et al. (2001)
Biotechnology industry wastes			
Fungal-based processes waste	Biopolymers—chitin/chitosan, proteins, pigment, and minerals	Extraction	Dhillon et al. (2012a)
Microalgae-based processing waste	Pigments, proteins, antioxidants, polysaccharides, vitamins, triglycerides, polyunsaturated fatty acids	Extraction	Mata et al. (2010)

Currently, research efforts are ongoing to isolate, identify, characterize, and even tailor microorganisms and enzymes in order to better utilize renewable resources to produce structurally diverse and complex chemicals. Biotransformation of biomass to higher-value chemicals provides advantages of high yield and selectivity, as well as minimum waste streams. However, there are still problems with current biological transformation technologies including both upstream and downstream processes. The capital costs related to energy requirements, such as pretreatment, sterilization, production, agitation, aeration, temperature control, and finally recovery of target products from aqueous systems with low product concentration, result in high-cost processes (Danner and Braun 1999). Further, considerable investment is required to make processes highly efficient and continuous (Dodds and Gross 2007). Therefore, there are research opportunities in the development of new economic biological transformation technologies which could effectively transform biomass into high-value biochemicals.

Biological conversion or biotransformation is a well-established process and comprises of fermentation and anaerobic digestion. Sugar and starchy crops provide the main feedstocks for the process of fermentation in which a microorganism converts the sugars into bioethanol. As an economic alternative to costly sugars, lignocellulosic biomass can be used as feedstock after pretreatment which helps to break it down into simple sugars. The pretreatment can be carried out by enzymes or acids. Although acid hydrolysis offers the more mature conversion platform, enzymatic hydrolysis appears to offer the best long-term option in terms of technical efficiency. Besides its recalcitrant structure, the efficient hydrolysis of lignocellulosic wastes and subsequent conversion of sugar syrups to various value-added products also depends upon various other factors, such as crystalline structure of cellulose, amount and nature of lignin present, and production of various inhibitory compounds during acid hydrolysis (Fig. 1.6). Lignocellulosic conversion would greatly increase the supply of raw materials available for production of various high-value products. The lignin residues could be used as fuel for the energy required and even providing surplus energy, resulting in significantly improved energy balances and resulting potential reductions in GHG emissions. The following sections discusses the potential of two underutilized wastes (marine processing and biotechnological process wastes) for the production of high-value biochemicals.

1.4.1.1 Biotransformation of Marine Processing Wastes

Large quantities of marine processing by-products are accumulated as aquaculture waste and shells of crustaceans and shellfish. Generally, the fishery by-products find applications for production of low-economic-value products, such as fish oil, fishmeal, fertilizer, pet food, and fish silage (Choudhury and Gogoi 1995). Currently, studies have identified a number of high-value bioactive compounds from fish wastes, such as fish muscle-derived peptides, collagen and gelatin, fish oil (source of omega-3 fatty acids), fish bone (consists of 60–70 % of inorganic substances, mainly composed of calcium phosphate and hydroxyapatite), and other

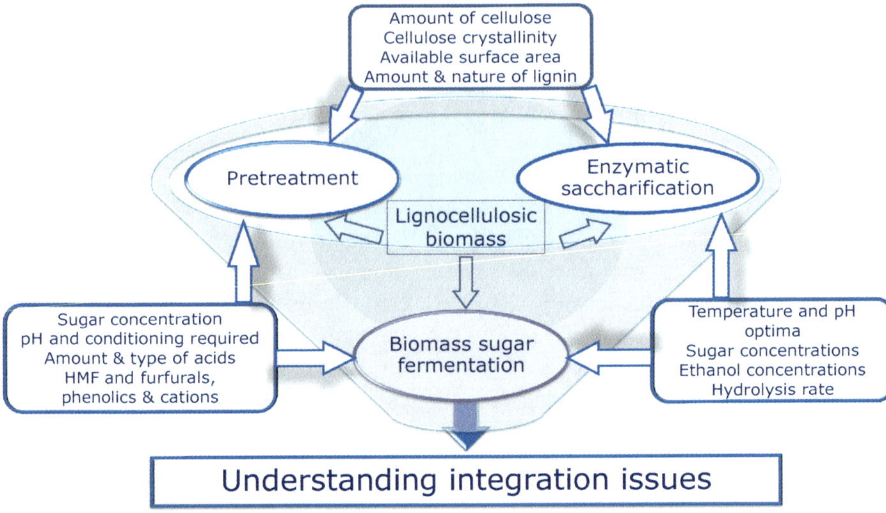

Fig. 1.6 Schematic diagram showing different aspects of lignocellulosic hydrolysis and its value addition

visceral organs (rich in a range of proteolytic enzymes including pepsin, trypsin, chymotrypsin, and collagenases) (Kim et al. 2001; Je et al. 2005). Lipid-based compounds that can be recovered from fish waste include fish oil, omega-3 fatty acids, phospholipids, squalene, vitamins, and cholesterol. Recovery of oil or lipids from fish industry waste offers not only the revenue generation but makes it suitable for other applications, such as spreading on land as a fertilizer or feedstuffs in swine diets to meet the protein requirements and as a substitute for common protein sources (i.e., soybean meal and commercial fishmeal) (Esteban et al. 2006).

Similarly, the other important class of by-products from marine bioprocessing plants includes crustacean shells and shellfish wastes mainly in the form of head and body carapace. These body parts comprise 48–56 % depending on the species. The efficient utilization of shellfish and crustacean shell by-products also becomes an environmental priority due to increased quantity of accumulation from processing plants as well as slow natural degradation of these materials. Shellfish and crustacean shells are a potential source of high-value biochemicals, such as biopolymers (chitin, chitosan), pigments (a carotenoid, astaxanthin), minerals, and proteins (Kaur and Dhillon 2013a, b) (Fig. 1.7). Most crustaceans, such as shrimp, lobsters, and crabs, are important reservoirs of natural carotenoids, such as astaxanthin and its esters (Sachindra et al. 2005).

However, the recovery of shell waste products, such as pigments and proteins, through chemical methods is complicated and the biological value of chemically extracted compounds is low. Additionally, these methods generate large quantities of hazardous chemical wastes. This has led to amplified interest in biotechnology research regarding the identification and extraction of high-grade, low-volume bioactive compounds produced from crustacean shell wastes. Recently, fermentation

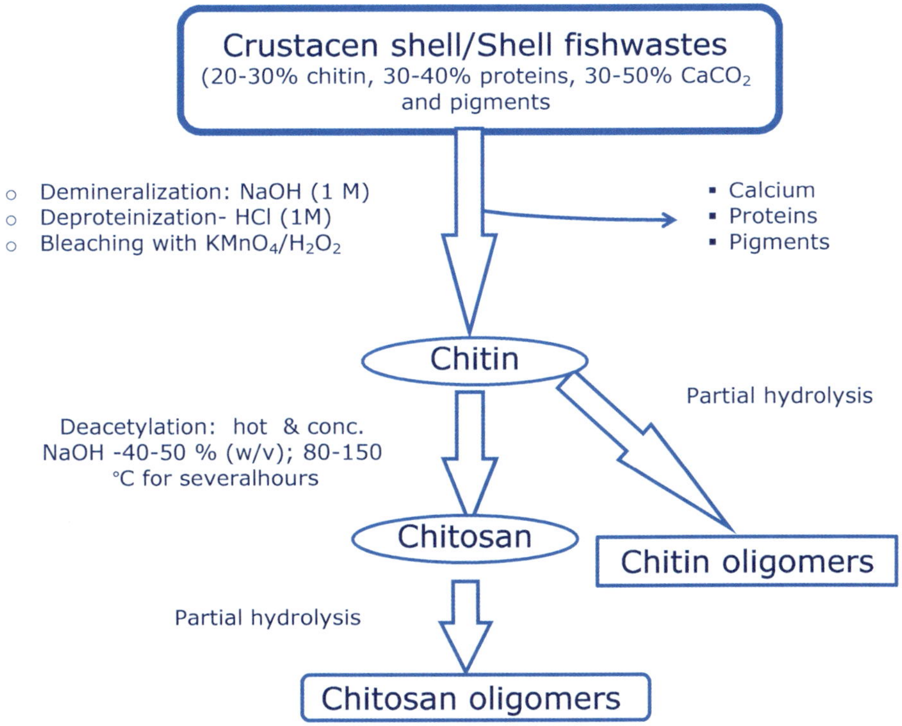

Fig. 1.7 Schematic diagram for preparation of proteins, pigments, chitin, chitosan, and their oligomers from marine wastes

has also been reported as a suitable and economic method to extract carotenoid pigments from crustacean shell wastes. These bioactive compounds can be extracted and purified with technologies varying from simple to complex. Furthermore, some of these bioactive compounds have been identified to possess nutraceutical potentials that are beneficial in human health promotion. Therefore, development of new technologies in exploration of new bioactive compounds from marine processing wastes will alleviate costs associated with its safe disposal. The bioactive compounds from marine processing wastes will add high value to marine waste and represent unique challenges and opportunities for the seafood industry.

The commercial applications of marine fish processing by-products are expanded every year. However, their applicability as bioactive compounds and their nutraceutical properties are not well described. High-value profit can be achieved by identifying bioactive compounds and exploring their nutraceutical properties and pharmaceutical and personal care applications. Identification of nutraceutical potential of natural compounds is a growing field and marine processing by-products represent potential feedstocks for this purpose. To date, only a limited number of bioactivities have been identified from isolated compounds and mandate future research developments to apply them for the human health promotion.

1.4.1.2 Biotransformation of Fermentation/Biotechnological Process Wastes

The advancements in bioprocess technology led to commercialization of various biotechnological/fermentation processes for the production of various bioproducts, such as food and beverages, organic acids, antibodies, pharmaceutical products, and renewable fuels among others. These microorganism-mediated processes result in thousands of tons of waste biomass, such as of yeast, bacteria, fungi, and algae. These waste are rich in various kinds of bioactive compounds, such as biopolymers, proteins, lipids, and pigments, among others.

Chitin and chitosan occur naturally in some fungi (*Mucoraceae*). Fungal cell walls are composed of polysaccharides and glycoproteins. Polysaccharides, such as chitin and glucan, are the structural components, whereas the glycoproteins, namely, mannoproteins, galactoproteins, xylomannoproteins, and glucuronoproteins, form the interstitial components of fungal cell walls (Bowman and Free 2006; Dhillon et al. 2012a). Commercially, chitin and chitosan are mainly derived from the marine processing wastes, such as shrimp, crabs, squids, and lobsters shell by chemical deacetylation, using a hot concentrated base solution (30–50 % w/v) at high temperatures (<100 °C) for a prolonged time (Dhillon et al. 2012a). However, the chitosan obtained by such treatments suffers some inconsistencies, such as protein contamination, inconsistent levels of deacetylation, and high molecular weight (MW), which results in variable physicochemical characteristics (No et al. 2000). There are some additional problems, such as environmental issues, due to the large amount of waste (concentrated alkaline solution), seasonal limitation of seafood shell supply, and high cost (Wu et al. 2005). In this context, production and purification of chitin and chitosan from the cell walls of waste fungal mycelium (Fig. 1.8) offers the advantage of being environmentally friendly and provides greater potential for a consistent product (Dhillon et al. 2012a; Kaur and Dhillon 2013a). Additionally, β-glucan can also be isolated from the mycelia chitosan–glucan complex and has important applications in biomedicine (Pomeroy et al. 2001).

Edible mushrooms are produced and consumed on a large scale. The amount of waste remaining after removing the edible part mainly consists of stalks and mushrooms with irregular dimensions and shapes and accounts for 5–20 % of the total production volume. In the USA alone, mushroom production results in nearly 50,000 metric tons of mushroom waste material per year with no suitable commercial application (Wu et al. 2005). The huge amount of wastes of edible mushrooms, such as *Agaricus bisporus*, *Lentinus edodes*, *Pleurotus* species, and *Volvariella volvacea*, among others, can be potentially used for the extraction of the high-value-added product chitosan, which nowadays finds promising applications in various fields (Dhillon et al. 2012a; Kaur and Dhillon 2013a; Dhillon et al. 2013).

Aspergillus niger strains are extensively used for the bioproduction of citric acid (CA) (Dhillon et al. 2011a, b, c, 2012b, c). The annual worldwide production of CA is estimated to be 1.7 million tons, which results in 0.34 million tons of *A. niger* mycelium waste per year, and furthermore, the industry continues to expand with an annual growth rate of 5 % (Wu et al. 2005; Dhillon et al. 2011c). *A. niger* strains contain approximately 15 % chitin, which can be separated and transformed into chitosan (Dhillon et al. 2012a).

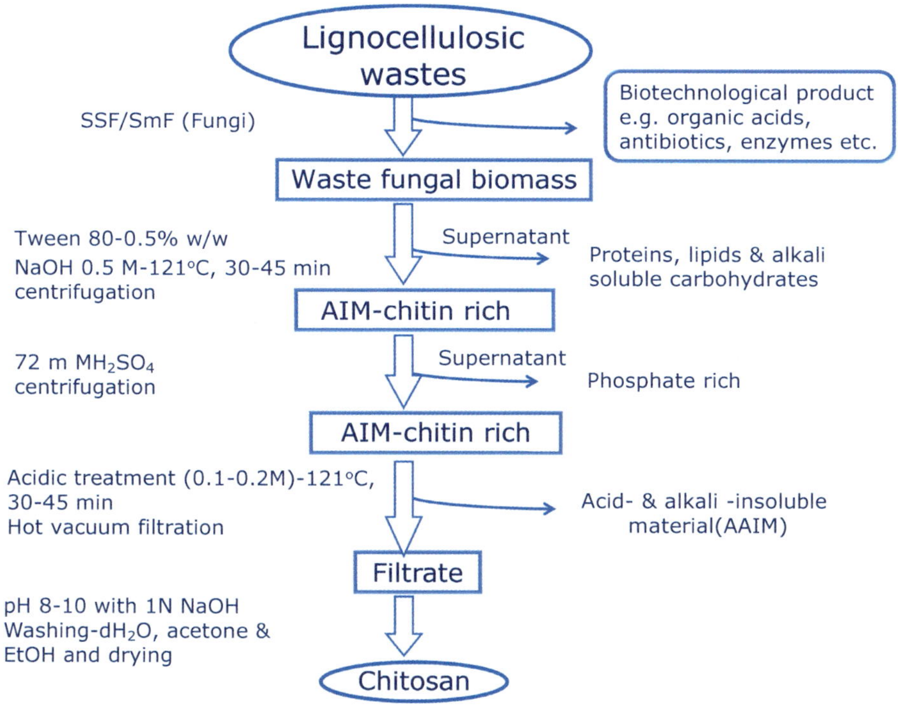

Fig. 1.8 Extraction of chitosan and other products from waste fungal mycelium (*SmF* submerged fermentation, *SSF* solid-state fermentation)

Penicillium chrysogenum is widely used for the large-scale production of antibiotics. As a by-product of the antibiotic industry, a large amount of *P. chrysogenum* mycelia waste is managed by incineration. Only a small percentage is used as an additive for cattle feed and in agriculture as fertilizers. Similarly, another important microbial strain, *Rhizopus oryzae*, is widely used in the food industry. Some yeast strains, such as *Saccharomyces cerevisiae*, find commercial applications in the brewery and bioethanol production. These yeasts strains are rich source of proteins and biopolymers. The development of bio-based economy mandates need to develop some integrative technology to utilize the unlimited waste mycelium resulting from fermentation industries which has not only a commercial advantage but also an ecological benefit.

1.4.2 Direct Extraction of Biochemicals from Biomass

Generally, the waste biomass contains many extractable compounds of high value which can be extracted directly from waste biomass. Fruit industry wastes, such as apple pomace (AP), are rich source of natural antioxidant compounds. Due to health and environmental awareness, sustainable food production and value addition of agro-industrial wastes is the principal issue in the agro- and food processing industry. AP is an excellent source of natural antioxidants, such as catechins, procyanidins, caffeic acid, phloridzin, phloretin glycosides, quercetin glycosides, chlorogenic acid, among others. Apple pomace, including seeds, contains polyphenolics with the strong antioxidant activity of quercetin glycosides, phloridzin, and its oxidative products (Schieber et al. 2003; Sanchez-Rabaneda et al. 2004; Guyot et al. 2007; Cetkovic et al. 2008). Similarly, many other fruit wastes are rich source of various natural oxidants and hence can be viewed as potential sources of bioactive phenolics.

Ferulic acid, a precursor for vanillin, occurs in a relatively high concentration in the form of xylan polysaccharide ester in corn fiber. The ferulic acid was extracted from corn fibers using novel fungal and bacterial feruloyl esterases (Shin et al. 1978). Vanillin is commonly used in the flavor and fragrance industries and it can be recovered by alkaline oxidation of lignin in the presence of a copper catalyst (Azadbakht et al. 2004). Ecket et al. (2007) described a more benign and cost-efficient method to extract vanillin from lignin using a gas-expanded liquid. Arabinogalactan and quercetin dihydrate were isolated from larch wood (Kuznetsova et al. 2008). Direct extraction of high-value biochemicals is a promising pathway for utilizing renewable resources, irrespective of scale. From an economic point of view, the extraction of high-value-added chemicals from biomass can be the most profitable, though the availability and variety of chemicals are limited.

Brewer's spent grains are the by-products of mashing process in brewery which is carried out in order to solubilize the malt and cereal grains to ensure adequate extraction of the wort (water with extracted matter). Following different separation strategies, the amount of BSG generated could be about 85 % of the total by-products (Tang et al. 2009). BSG is a readily available, high-volume, low-cost by-product of brewing and is a potentially valuable resource for industrial exploitation (Dhillon et al. 2012d). According to Stojceska et al. (2008) about 3.4 million tons of BSG from the brewing industry is produced in the EU every year. Ferulic (4-hydroxy-3-methoxy-cinnamic acid) and *p*-coumaric acid (4-hydroxycinnamic acid) are the most abundant phenolic acids in BSG (Bartolomè et al. 1997, 2002, 2003). The extraction of these high-value biochemicals opens up new possibilities for the use of BSG. Ferulic acid exhibits a number of potential applications, such as natural antioxidant, food preservative/antimicrobial agent, anti-inflammatory agent, photoprotectant, and as a food flavor precursor, while *p*-coumaric exhibits chemoprotectant and antioxidant properties (Bartolomè et al. 2002; Mussatto et al. 2007). Similarly, winery waste which is composed of solid pomace is rich in antioxidant polyphenols which can be extracted by various mild hydrolysis methods.

The food processing industry produces large volumes of both solid and liquid wastes. These wastes pose increasing dumping and severe pollution problems and represent a loss of valuable biomass and nutrients. In the past they often have been dumped or utilized without treatment for low-value applications, such as for animal feed or as fertilizers. However, due to the increasing environmental awareness as well as for economic motives and the need to conserve energy and new materials, recently new methods and policies for waste handling and treatment have been introduced in the recovery, bioconversion, and utilization of valuable constituents from these wastes. With the advancement in technology, food processing wastes might have a potential for recycling raw materials or for conversion into useful products of higher value as a by-product, or even as raw material for other industries, or for the use as food or feed/fodder after biological treatment. The biotransformation of food processing residues is receiving increased attention regarding the fact that these residual matters represent a possible and utilizable resource for transformation to high-value products.

1.5 Conclusions and Future Prospects

Interest in waste biomass utilization has increased dramatically over the last few years as a renewable resource alternative for fossil fuels as well as an input into other industrial processes. The environmental impacts of waste biomass utilization for energy and other commodity products are quite significant and are arguably greater in scale and scope than any other class of energy resources, viz., renewable or nonrenewable, due to the intensive use of land, water, and other resources. Developing countries have a vast agricultural resource base for alternatives for bioenergy and industrial biotechnology. The majority of biomass is found in rural areas resulting from agriculture processes. Therefore, the bio-economy has the potential to provide much needed diversification of the rural economy. The biomass has been viewed as an alternative to energy; while it has potential it can also be used as input into various biotechnological processes for production of various consumer products. Production of value-added by-products serves to expand a bio-based economy or sustainable development, offering alternatives for fossil fuel-based products and facilitating a lower overall cost of production. A greater reliance on bio-based resources and biological processes is an inevitable part of an overall sustainability transition, and thus the main questions for technical innovation and policy development relate to how to positively impact the nature and pace of such changes. In many developing countries, biomass is currently a significant source of energy and materials only for local and traditional uses. The biomass is generally used inefficiently with very few higher-value-added product markets. Bio-based renewable resources can provide raw materials for many new and growing biotechnological industries while also stimulating rural development, job creation, and GHG reduction. In assessing the options and strategies of a bio-based economy, economic, environmental, and social issues need to be addressed to ensure that sustainable development objectives can be met.

References

Ajila CM, Brar SK, Verma M, Tyagi RD, Valéro JR (2011) Solid-state fermentation of apple pomace using Phanerochaete chrysosporium: Liberation and extraction of phenolic antioxidants. Food Chem 126:1071–1080. doi:10.1016/j.foodchem.2010.11.129
Anastas PT, Kirchhoff MM (2002) Origins, current status, and future challenges of green chemistry. Acc Chem Res 35:686–694
Arogba SS (2000) Mango (*Mangifera indica*) kernel: chromatographic analysis of the tannin, and stability study of the associated polyphenol oxidase activity. J Food Comp Anal 13:149–156
Asagbra AE, Sanni AI, Oyewole OB (2005a) Solid-state fermentation production of tetracycline by *Streptomyces* strains using some agricultural wastes as substrate. World J Microbiol Biotechnol 21:107–114
Asagbra AE, Oyewole OB, Odunfa SA (2005b) Production of oxytetracycline from agricultural wastes using *Streptomyces* species. Niger Food J 23:174–182
Ashok K, Narayani M, Subanthini A, Jayakumar M (2011) Antimicrobial activity and phytochemical analysis of citrus fruit peels—utilization of fruit waste. Int J Eng Sci Technol 3(6):5414–5421
Athalye SK, Garcia RA, Wen Z (2009) Use of biodiesel-derived crude glycerol for producing eicosapentaenoic acid (EPA) by the fungus *Pythium irregulare*. J Agric Food Chem 57:2739
Atsushi S, Somsak S, Kohtaro K, Shoji U (1996) Direct production of citric acid from Starch by a 2-deoxygluxose-resistant mutant strain of *Aspergillus niger*. J Ferment Bioeng 81:320–323
Azadbakht M, Ebrahimzadeh MA, Koolayan S (2004) Preparation of lignin from wood dust as vanillin source and comparison of different extraction methods. Int J Biol Biotechnol 1:535–537
Bartolomè B, Faulds CB, Williamson G (1997) Enzymic release of ferulic acid from barley spent grain. J Cereal Sci 25:285–288
Bartolomè B, Faulds CB, Sancho AI (2002) Mono- and dimeric ferulic acid release from brewer's spent grain by fungal feruloyl esterases. Appl Microbiol Biotechnol 60:489–493
Bartolomè B, Gómez-Cordovés C, Sancho AI, Díez N, Ferreira P, Soliveri J, Copa-Patiño JL (2003) Growth and release of hydroxycinnamic acids from Brewer's spent grain by *Streptomyces avermitilis* CECT 3339. Enzyme Microb Technol 32:140–144
Ben Salah R, Chaari K, Besbes S et al (2010) Optimisation of xanthan gum production by palm date (*Phoenix dactylifera* L.) juice by-products using response surface methodology. Food Chem 121:627–633
Bowman SM, Free SJ (2006) The structure and synthesis of the fungal cell wall. Bioessays 28:799–808
Bruggink A, Straathof AJJ, van der Wielen LAM (2003) A fine chemical industry for life science products: green solutions to chemical challenges. In: von Stockar U, van der Wielen LAM (eds) Process Integration in biochemical engineering. Springer, Berlin, Heidelberg/NY, USA, pp 70–113
Canadanovic BJM, Savatovic SS, Cetkovic GS, Vulic JJ, Djilas SM, Markov SL, Cvetkovic DD (2011) Antioxidant and antimicrobial activities of beet root pomace extracts. Czech J Food Sci 29:575–585
Carvalho JC, Oishi BO, Woiciechowski AL, Pandey A, Soccol CR (2007) Effect of substrates on the production of *Monascus* biopigments by solid-substrate fermentation and pigment extraction using different solvents. Indian J Biotechnol 6:194–199
Cetkovic G, Canadanovic-Brunet J, Djilas S, Savatovic S, Mandic A, Tumbas V (2008) Assessment of polyphenolic content and in vitro antiradical characteristics of apple pomace. Food Chem 109:340–347
Chantaro P, Devahastin S, Chiewchan N (2008) Production of antioxidant high dietary fiber from carrot peel. LWT- Food Sci Technol 41:1987–1994
Choudhury GS, Gogoi BK (1995) Extrusion processing of fish muscle. J Aquat Food Prod Tech 4:37–67

Cuadra T, Fernandez FJ, Tomasini A, Barrios-Gonzalez J (2008) Influence of pH regulation and nutrient content on cephalosporin C production in solid-state fermentation by *Acremonium chrysogenum* C10. Lett Appl Microbiol 46:216–220

Dale BE (2003) "Greening" the chemical industry: research and development priorities for bio-based industrial products. J Chem Technol Biotechnol 78:1093–1103

Danner H, Braun R (1999) Biotechnology for the production of commodity chemicals from biomass. Chem Soc Rev 28:395–405

Dhillon GS, Brar SK, Verma M, Tyagi RD (2011a) Utilization of different agro-industrial wastes for sustainable bioproduction of citric acid by *Aspergillus niger*. Biochem Eng J 54:83–92. http://www.sciencedirect.com/science/journal/1369703X

Dhillon GS, Brar SK, Verma M, Tyagi RD (2011b) Recent trends in citric acid bioproduction and recovery. Food Bioproc Technol 4:505–529

Dhillon GS, Brar SK, Verma M, Tyagi RD (2011c) Enhanced solid-state citric acid bioproduction using apple pomace waste through response surface methodology. J Appl Microbiol 110: 1045–1055

Dhillon GS, Brar SK, Kaur S (2012a) Green synthesis approach: extraction of chitosan from fungus Mycelia. Crit Rev Biotechnol. doi:10.3109/07388551.2012.717217

Dhillon GS, Brar SK, Kaur S, Verma M (2012b) Rheological studies during submerged citric acid fermentation by *Aspergillus niger* in stirred fermenter using apple pomace ultrafiltration sludge. Food Bioproc Technol. doi:10.1007/s11947-011-0771-8

Dhillon GS, Brar SK, Verma M (2012c) Biotechnological potential of industrial wastes for economical citric acid bioproduction by *Aspergillus niger* through submerged fermentation. Int J Food Sci Technol 47:542–548

Dhillon GS, Kaur S, Brar SK (2012d) Flocculation and haze removal from crude fermented beer using in-house produced laccase via koji fermentation with *Trametes versicolor* using brewery spent grain. J Agric Food Chem 60(32):7895–904

Dhillon GS, Kaur S, Sarma SJ, Brar SK, Surampalli RY (2013) Recent development in applications of important biopolymer chitosan in biomedicine, pharmaceuticals and personal care products. Curr Tissue Eng 2(3):20–40

Dodds DR, Gross RA (2007) Chemicals from biomass. Science 318:250–1

Duan X, Jiang Y, Su X, Zhang Z, Shi J (2007) Antioxidant properties of anthocyanins extracted from litchi (*Litchi chinensis Sonn.*) fruit pericarp tissues in relation to their role in the pericarp browning. Food Chem 101:1365–1371

Ecket C, Liotta C, Ragauskas A et al (2007) Tunable solvents for fine chemicals from the biorefinery. Green Chem 9:545–548

Ellaiah P, Shrinivasulu B, Adinarayana K (2004) Optimization studies on neomycin production by a mutant strain of *Streptomyces marinensis* in solid-state fermentation. Process Biochem 39:529–534

Esteban MB, Garcia AJ, Ramos P, Marquez MC (2006) Evaluation of fruit–vegetable and fish wastes as alternative feedstuffs in pig diets. Waste Manag 27:193–200

Etschmann MMW, Sell D, Schrader J (2003) Screening of yeasts for the production of the aroma compound 2-phenylethanol in a molasses-based medium. Biotechnol Lett 25:531–536

Ezeji T, Qureshi N, Blaschek H (2007) Production of acetone butanol from liquefied corn starch, a commercial substrate, using *Clostridium beijerinckii* coupled with product recovery by gas stripping. J Ind Microbiol Biotechnol 34:771–777

Ferreira TF, Ribeiroa RR, Ribeirob CMS, Freirec DMG, Coelhoa MAZ (2012) Evaluation of 1, 3-propanediol production from crude glycerol by *Citrobacter freundii* ATCC 8090. Chem Eng Trans 27:157–162

Gassara F, Ajila CM, Brar SK, Verma M, Tyagi RD, Valéro JR (2012) Liquid state fermentation of apple pomace sludge for the production of ligninolytic enzymes and liberation of polyphenolic compounds. Process Biochem 47:999–1004. doi:10.1016/j.procbio.2012.03.001

Guendez R, Kallithraka S, Makris DP, Kefalas P (2005) An analytical survey of the polyphenols of seeds of varieties of grape (*Vitis vinifera* sp.) cultivated in Greece: implications for exploitation as a source of value-added phytochemicals. Phytochem Anal 16:17–23

Guyot S, Serrand S, Querre JML, Sanoner P, Renard CMGC (2007) Enzymatic synthesis and physicochemical characterization of phloridzin oxidation products, a new water soluble yellow dye deriving from apple. Innov Food Sci Emerg Technol 8:443–450

Huang LP, Jin B, Lant P, Zhou J (2005) Simultaneous saccharification and fermentation of potato starch wastewater to lactic acid by *Rhizopus oryzae* and *Rhizopus arrhizus*. Biochem Eng J 23:265–276

Je JY, Park PJ, Kwon JY, Kim SK (2005) A novel angiotensin I converting enzyme inhibitory peptide from Allaska Pollack (Theragra chalcogramma) frame protein hydrolysate. J Agric Food Chem 52:7842–7845

Kaur S, Dhillon GS (2013a) The versatile biopolymer chitosan: potential sources, evaluation of extraction methods and applications. *Critical reviews in Microbiology*. DOI:10.3109/10408 41X.2013.770385

Kaur S, Dhillon GS (2013b) Recent trends in biological extraction of chitin from marine shell wastes: A review. *Critical Reviews in Biotechnology*. DOI:10.3109/07388551.2013.798256

Khanal SK, Lamsal BP (2010) Bioenergy and biofuels production: some perspectives. In: Khanal SK, Surampalli RY, Zhang TC, Lamsal BP, Tyagi RD, Kao CM (eds) Bioenergy and biofuel from biowastes and biomass. ASCE, New York, pp 1–22

Khodaiyan F, Razavi SH, Mousavi SM (2008) Optimization of canthaxanthin production by *Dietzia natronolimnaea* HS-1 from cheese whey using statistical experimental methods. Biochemical Engineering J 40:415–422

Kim SK, Kim YT, Byun HG, Nam KS, Joo DS, Shahidi F (2001) Isolation and characterization of antioxidative peptides from gelatin hydrolysate of Allaska Pollack skin. J Agric Food Chem 49:1984–1989

Knoblich M, Anderson B, Latshaw D (2005) Analyses of tomato peel and seed byproducts, and their use as a source of carotenoids. J Sci Food Agr 85:1166–1170

Krisch J, Galgoczy L, Papp T, Csaba Vagvolgyi C (2009) Antimicrobial and antioxidant potential of waste products remaining after juice pressing. J Eng Ann Fac Eng tome vii(year. Fascicule 4):131–134

Kuznetsova SA, Danilov VG, Kuznetsov BN, Taraban'Ko VE, Pervyshina EP, Alexaandrova NB (2008) Fine chemicals from larch wood biomass. http://www.brdisolutions.com/pdfs/bcota/abstracts/26/z120.pdf

Liu Y, Koh CMJ, Ji L (2011) Bioconversion of crude glycerol to glycolipids in *Ustilago maydis*. Bioresour Technol 102:3927

Lucia LA, Argyropoulos DS, Adamopoulos L, Gaspar AR (2006) Chemicals and energy from biomass. Can J Chem 84:960–970

Mahalaxmi Y, Sathish T, Subba Rao C, Prakasham RS (2010) Corn husk as a novel substrate for the production of rifamycin B by isolated *Amycolatopsis sp*. RSP3 under SSF. Process Biochem 45(1):47–53

Makrisa DP, Boskoub G, Andrikopoulos NK (2007) Polyphenolic content and in vitro antioxidant characteristics of wine industry and other agri-food solid waste extracts. J Food Compos Anal 20(2007):125–132

Marova I, Carnecka M, Halienova A, Certik M, Dvorakova T, Haronikova A (2012) Use of several waste substrates for carotenoid-rich yeast biomass production. Environ Manag 95:338–342

Martin JGP, Porto E, Correa GB, Matias de Alencar S, Micotti da Gloria E, Cabral ISR, Maria L, de Aquino L (2012) Antimicrobial potential and chemical composition of agro-industrial wastes. J Nat Prod 5:27–36

Mata TM, Martins AA, Caetano NS (2010) Microalgae for biodiesel production and other applications: a review. Renew Sustain Energy Rev 14:217–232

Mussatto SI, Roberto IC (2005) Acid hydrolysis and fermentation of brewer's spent grain to produce xylitol. J Sci Food Agric 85:2453–2460

Mussatto SI, Roberto IC (2008) Establishment of the optimum initial xylose concentration and nutritional supplementation of brewer's spent grain hydrolysate for xylitol production by *Candida guilliermondii*. Process Biochem 43:540–546

Mussatto SI, Dragone G, Roberto IC (2007) Ferulic and *p*-coumaric acids extraction by alkaline hydrolysis of brewer's spent grain. Ind Crop Prod 25:231–237

Nigam PS, Gupta N, Anthwal A (2009) Pre-treatment of agro-industrial residues. In: Nigam PS, Pandey A (eds) Biotechnology for agro-industrial residues utilization, 1st edn. Springer, Netherlands, pp 13–33

No HK, Lee KS, Meyers SP (2000) Correlation between physicochemical characteristics and binding capacities of chitosan products. J Food Sci 65:1134–1137

Obied HK, Allen MS, Bedgood DR, Prenzler PD, Robards K, Stockmann R (2005) Bioactivity and analysis of biophenols recovered from olive mill waste. J Agric Food Chem 53:823–837

Pinelo M, Rubilar M, Jerez M, Sineiro J, Núñez MJ (2005) Effect of solvent, temperature, and solvent-to-solid ratio on the total phenolic content and antiradical activity of extracts from different components of grape pomace. J Agric Food Chem 53:2111–2117

Pokorny J, Yanishlieva N, Gordon MH (2001) Antioxidants in food: practical applications. CRC Press, Boca Raton, Boston, New York Washington, DC

Pomeroy S, Tupper R, Cehun-Aders M, Nestel P (2001) Oat betaglucan lowers total and LDL-cholesterol. Aust J Nutr Diet 58:51–55

Puravankara D, Boghra V, Sharma RS (2000) Effect of antioxidant principles isolated from mango (*Mangifera indica L.*) seed kernels on oxidative stability of buffalo ghee (butter-fat). J Sci Food Agric 80:522–526

Pyle DJ, Garcia RA, Wen Z (2008) Producing docosahexaenoic acid (DHA)-rich algae from biodiesel-derived crude glycerol: effects of impurities on DHA production and algal biomass composition. J Agric Food Chem 56:3933

Rosillo-Calle F, de Groot P, Hemstock SL, Woods J (2007) Non-woody biomass and secondary fuels. In: Rosillo-Calle F, de Groot P, Hemstock SL, Woods J (eds) The biomass assessment handbook. Earthscan, London, UK

Roukas T (1999) Pullulan production from brewery wastes by *Aureobasidium pullulans*. World J Microbiol Biotechnol 15:447–450

Sachindra NM et al (2005) Carotenoids in different body components of Indian shrimps. J Sci Food Agric 85:167–172

Sanchez-Rabaneda F, Jauregui O, Lamuela-Raveentos RM, Viladomat F, Bastida J, Codina C (2004) Qualitative analysis of phenolic compounds in apple pomace using liquid chromatography coupled to mass spectrometry in tandem mode. Rapid Commun Mass Spectrom 18:553–563

Schieber A, Hilt P, Streker P, Endress HU, Rentschler C, Carle R (2003) A new process for the combined recovery of pectin and phenolic compounds from apple pomace. Innovat Food Sci Emerg Tech 4(1):99–107

Shin HD, McClendon S, Taylor F, Chen RR (1978) Enzymatic extraction of ferulic acid from agriculture waste for high-valued products. Paper presented at AIChE annual meeting, Cincinnati, OH. Oct 30–Nov 4 2005

Shin HY, Lee JY, Choi HS, Lee JH, Kim SW (2011) Production of cephalosporin C using crude glycerol in fed-batch culture of *Acremonium chrysogenum* M35. J Microbiol 49:753

Someya SYY, Okubo K (2002) Antioxidant compounds from bananas (*Musa cavendish*). Food Chem 99:351–354

Song C (2006) Global challenges and strategies for control, conversion and utilization of CO2 for sustainable development involving energy, catalysis, adsorption and chemical processing. Catal Today 115:2–32

Spatafora C, Tringali C (2012) Valorization of vegetable waste: identification of bioactive compounds and their chemo-enzymatic optimization. Open Agr J 6:9–16

Stojceska V, Ainsworth P, Plunkett A, Ibanoglu S (2008) The recycling of brewer's processing by-product into ready-to-eat snacks using extrusion technology. J Cereal Sci 47:469–479

Strati IF, Oreopoulou V (2011) Effect of extraction parameters on the carotenoid recovery from tomato waste. Int J Food Sci Technol 46:23–29

Survase SA, Shaligram NS, Pansuriya RC, Annapure US, Singhal RS (2009) A novel medium for the enhanced production of cyclosporin A by *Tolypocladium inflatum* MTCC 557 using solid state fermentation. J Microbiol Biotechnol 19(5):462–467

Tang D, Yin G, He Y, Hu S, Li B, Li L, Liang H, Borthakur D (2009) Recovery of protein from brewer's spent grain by ultrafiltration. Biochem Eng J 48:1–5
Tehranifara A, Selahvarzia Y, Kharrazia M, Bakhshb VJ (2011) High potential of agro-industrial by-products of pomegranate (*Punica granatum* L.) as the powerful antifungal and antioxidant substances. Ind Crop Prod 34(3):1523–1527
United Nations Industrial development organization (2007) Industrial biotechnology and biomass utilisation: Prospects and challenges for the developing world. vienna, pp 1–186
Van Dam JEG (2002) "Wet processing of coir—drying, bleaching, dyeing, softening and printing," CFC/FAO Techno-economic manual No 6
Van Dam JEG, de Klerk-Engels B, Struik PC, Rabbinge R (2005) "Securing renewable resources supplies for changing market demands in a biobased economy"- Industrial Crops and Products 21:129–144
Varzakakou M, Roukas T, Kotzekidou P (2010) Effect of the ratio of (+) and (−) mating type of *Blakeslea trispora* on carotene production from cheese whey in submerged fermentation. World J Microbiol Biotechnol 26:2151–2156
Vastrad BM, Neelagund SE (2011) Optimization and Production of Neomycin from Different Agro Industrial Wastes in Solid State Fermentation. Int J Pharm Sci Drug Res 3(2):104–111
Vastrad BM, Neelagund SE (2012) Optimization of process parameters for rifamycin b production under solid state fermentation from *Amycolatopsis mediterranean* MTCC14. Int J Curr Pharmaceut Res 4(2):101–108
Wu T, Zivanovic S, Draughon FA, Conway WS, Sams CE (2005) Physicochemical properties and bioactivity of fungal chitin and chitosan. J Agric Food Chem 53:3888–3894
Xiao Z, Liu P, Qin JY, Xu P (2007) Statistical optimization of medium components for enhanced acetoin production from molasses and soybean meal hydrolysate. Appl Microbiol Biotechnol 74(1):61–68
Xu Y, Hanna MA, Isom L (2008) "Green" chemicals from renewable agricultural biomass—a mini review. Open Agr J 2008(2):54–61
Yang J, Tan H, Yang R, Sun X, Zhai H, Li K (2011) Astaxanthin production by *Phaffia rhodozyma* fermentation of cassava residues substrate. Agri Eng Int 13:1–6
Yu J (2001) Production of PHA from starchy wastewater via organic acids. J Biotechnol 86:105–112
Zeyada NN, Zeitoum MAM, Barbary OM (2008) Utilization of some vegetables and fruit waste as natural antioxidants alex. J Food Sci Tech 5:1–11
Zha W, Shao Z, Frost JW (2004) Rational pathway engineering of type I fatty acid synthase allows the biosynthesis of triacetic acid lactone from D-glucose in vivo. J Am Chem Soc 126: 4534–4535
Zheng L, Zheng P, Sun Z, Bai Y, Wang J, Guo X (2007) Production of vanillin from waste residue of rice bran oil by *Aspergillus niger* and *Pycnoporus cinnabarinus*. Bioresour Technol 98(5): 1115–1119

Chapter 2
Pretreatment Strategies to Enhance Value Addition of Agro-industrial Wastes

Adenise Lorenci Woiciechowski, Susan Grace Karp, Keli Sobral, Júlio Cesar de Carvalho, Luiz Alberto Junior Letti, Vanete Tomaz Soccol, and Carlos Ricardo Soccol

2.1 Introduction

The utilization of agro-residues as substrate to generate new products of commercial interest through a bioprocess is considered an important strategy for the development of sustainable technologies that are essential, nowadays, for many reasons: (1) the use of a residue of an industrial process will demand less efforts and resources to treat and dispose the residue through conventional techniques, and (2) value addition of these wastes besides saving money on conventional treatment will provide a raw material at a relatively low cost. The proposal is ideally based on integrated processes, where the residue of one process will not be only treated and disposed but used as raw material for a new process mainly through biotechnological pathways. The use of agro-industrial residues as raw material for new bioprocesses usually deserves a study of many key points to evaluate the feasibility of the proposal. Among the points to be carefully analyzed is the residue nature, involving a complete characterization to determine the carbon source type, the presence of the appropriate nutrients and/or toxic compounds, and pretreatment steps to avoid deterioration of the material and/or to ensure the material accessible to chemical or biochemical transformation and the accessibility of the carbon source and nutrients to the microorganism metabolism. An appropriate pretreatment step may also be necessary for the detoxification of the residues that contains anti-metabolites. These pretreatments include (1) drying or concentration, (2) grinding and size classification, (3) improvement of the accessibility to the carbon sources through thermal or enzymatic treatment, (4) balancing the nutrient contents, (5) reduction of the concentration

A.L. Woiciechowski • S.G. Karp • K. Sobral • J.C. de Carvalho • L.A.J. Letti
V.T. Soccol • C.R. Soccol (✉)
Biotechnology and Bioprocess Engineering Department, Centro Politecnico,
Federal University of Parana, 81 Jardim das Américas, Curitiba, PR 531-990, Brazil
e-mail: soccol@ufpr.br

S.K. Brar et al. (eds.), *Biotransformation of Waste Biomass into High Value Biochemicals*,
DOI 10.1007/978-1-4614-8005-1_2,

of toxic compounds, and (6) transformation of recalcitrant compounds. Consideration about transportation, drying, grinding and sieving, thermochemical and/or enzymatic hydrolysis, detoxification cases, analysis, and composition considerations, including technical and economical aspects, will be discussed. The characterization of the residue and the choice of suitable pretreatment strategies are essential as these steps directly affect the whole process.

2.2 Residue Characterization

The appropriateness of a residue to a particular application depends on its effective composition. Hence, the first step for a suitable choice of any alternative substrate is its composition characterization. Some important points to be evaluated are:

2.2.1 Physical State

It is practical to use the raw material in its original physical state. This will eliminate at least one operation to prepare the material for the new industrial processing. Among the alternatives available to use a residue as substrate, submerged fermentation (SmF) for liquid media and solid state fermentation (SSF) for solid media without free water and a range of fermentation alternatives between both strategies are used. Solid substrates, for example, can be used preferably through SSF. When it is not possible or viable, if the raw material is at solid state, to dissolve or hydrolyze the nutrients present at the solid matrix is necessary to develop a process through submerged fermentation. Glucose syrup, for example, to be used in SmF can be obtained from acid or enzymatic hydrolysis of potato starch residue or starch cassava bagasse. On the other hand, a liquid residue should be used preferably in submerged fermentation or while searching a solid support to be impregnated with the liquid medium if the target process is through SSF. Hence, to improve a process using a residue as raw material, the first alternative is to use the residue in its original physical state. In any case, technical and technological aspects and alternatives of the process must be studied.

2.2.2 Nature of Carbon Source

Each plant species has a particular composition in terms of structural molecules that depends on the variety, cultivar, climate, and soil where it is produced and time of the harvest. Each industrial process suitable to extract from the plant the main industrial product (starch, oil, protein) produces a residue besides the main product. The residue composition of each process varies depending on the processing itself, the technology used, the raw material processed, and the kind of main product recovered. Depending on the plant species, the carbon composition can also vary.

In relation to the nature of the carbon source, biomass organic matter can be classified into various types:

2.2.2.1 Phytobiomass

The biomass is produced through the photosynthesis using the solar energy and inorganic molecules, such as carbon dioxide and water, in the presence of chlorophyll. The plant species synthesize its biomass and acquire its energetic resources through photosynthesis. With a structural matrix composed of cellulose, hemicelluloses, and lignin, according to Carioca and Arora (1984) depending on the nature of the additional energetic molecule synthesized, phytobiomass can be classified into:

(a) Saccharide crops: Besides the cellulose, hemicelluloses, and lignin structure, this phytobiomass is characterized by the accumulation of energetic molecules in the form of a simple sugar. This directly fermentable sugar is the main energetic source of the species. The accumulated sugar depends on the species and can be a monosaccharide, such as glucose, fructose, and galactose; a disaccharide, such as sucrose; a polysaccharide, such as pectin, inulin, xylans, and galactoxylans; or a mixture of them in different proportions but at least one of them. Examples of saccharide species biomass are sugarcane juice, grape juice and grape skin, citric juice and citric pulp, apple juice and pomace, and sweet sorghum juice, among others. This kind of phytobiomass has a great importance due to a considerable part of the biomass being composed of directly fermentable or easily hydrolyzed sugar.

(b) Starchy crops: This phytobiomass species is characterized by an accumulation of energetic molecules in the amylaceous form at a certain part of the plant usually the root, but not only there. Crude starch is the main energetic source of the vegetal specie, such as cassava, potato, and corn, among others, and its processing residues, cassava bagasse and corn steep liquor, for example. This polymeric molecule (starch) can be hydrolyzed in acid or enzymatic medium to release mainly glucose from the starchy grain, producing a broth with large sugar concentration. Another strategy to use this phytobiomass or the residue of its industrial processing is to select an amylase-producing microorganism that will hydrolyze the starch during its growth and metabolism. Normally, this plant species is a substrate to produce and recover amylases. Among the microorganisms, amylase producers are filamentous fungi, such as *Aspergillus*, *Pleurotus*, *Rhizopus*, some yeasts, and bacteria.

(c) Oilseed biomass: Soya, sunflower, palm, and olive are examples of this phytobiomass species characterized by an accumulation of an oil energetic molecule, usually in its seed, grain, or pulp. The plant species is industrially processed in order to obtain the oil, for commercial aims, and the processing residue can be reprocessed to recover other molecules, such as protein, sugar, bioactive products, and others. For fermentative processes, specifically, these residues are of special interest due to the presence of large amounts of carbohydrates, nitrogen, and minerals besides residual oil.

(d) Lignocellulosic biomass: The most important phytobiomass available all over the world. It has a complex composition formed by cellulose fibers, hemicelluloses, protected by a lignin matrix in a very typical structure responsible for its mechanical, chemical, and biological resistance. Depending on the species, a specific extract and ashes complete its composition. These species include hardwood; wood from trees, such as broadleaves (oak, eucalyptus); and softwood, such as wood from coniferous (*Pinus*). It also includes agricultural residues and industrial residues, such as husks, bagasse, grass, straw, leaves, branches, shavings, and all kinds of trimmings and prunings. The amount, diversity, and presence all over the world of this source demonstrate its importance. Nowadays, the industrial exploration of this species is spotted mainly to produce cellulose pulp from the paper industry through some known thermochemical processes. Hemicellulose residues and lignin form the black liquor, residue of the process. Many processing alternatives are being studied to efficiently recover these fractions, mainly the sugars from the hemicellulosic fraction, due to its potentiality to be used as substrate for fermentative processes, producing a large number of biomolecules. Alternatives including mild thermochemical processes in acid medium are being studied to firstly hydrolyze the hemicelluloses followed by alkaline wash to remove the lignin and lastly recover and hydrolyze the cellulose. Nowadays, the search of processes to generate biofuels to substitute fossil fuels has urged the research groups around the world to improve technical and economic knowledge for a viable second-generation biofuel production.

Hemicelluloses are amorphous polysaccharides non-starchy and noncellulosic constituents, depending on the species, comprising hexoses (glucose, mannose, galactose), pentoses (xylose, arabinose), uronic acids (glucuronic, methylglucuronic, galacturonic), and deoxyhexoses (rhamnose, fucose). The hemicellulosic hydrolysate contains the simple sugar that constitutes the hemicelluloses of the plant species. This syrup can be potentially fermented by a selected microorganism to produce many biomolecules of industrial importance, including ethanol.

Cellulose is a crystalline homopolymer, constituted by glucose units linked by β-(1–4) linkage, generating a linear and flat structure, allowing the packing and superposition of the cellulose molecules, producing a compact and stable structure, which characterize the cellulose fiber. Acidic or enzymatic hydrolysis will furnish directly and easily fermentable glucose syrup.

Lignins are complex polymers, specific for each plant species; it is a three-dimensional network constituted by units derivated from phenyl propane with different degrees of methoxylation and acetylation linked by carbon–carbon or ether linkage. Lignin is linked to the hemicellulose structure through molecular bond. The physicochemical nature of the lignin and diversity of components are responsible for its recalcitrant characteristics.

2.2.2.2 Microbial Biomass

A complex carbon source due to its particular composition has been used as nitrogen source for many large-scale processes. For economic reasons, nitrogen source compounds should be chosen among the low-price materials, for example, ammonium salts, urea, and nitrate salts, but depending on the microorganism, an organic nitrogen source must be tested. In these cases, some organic residues rich in nitrogen can be tested to compose the fermentative medium. Yeast extract is a nitrogen source and substrate for many microorganisms, containing many amino acids (alanine, arginine, cystine, glutamic acid, glycine, histidine, leucine, isoleucine, valine, serine, threonine, proline) and peptides, vitamins (thiamine, riboflavin, niacinamide, pantothenic acid), and carbohydrates (glycogen and trehalose) that can be hydrolyzed to glucose. Yeast extract can be produced through autolysis at 45–55 °C or plasmolysis by using high concentrations of a soluble salt, mainly NaCl. Both differ in quality depending on the process used, but in general, the first one possesses high preservative properties.

2.2.2.3 Animal Biomass

Other protein hydrolysates, peptones from meat, casein, gelatin, and keratin can also be used as organic nitrogen source to supply complex nitrogen components. They largely vary in composition depending on the origin; thus, the amino acid and oligopeptide composition must be carefully analyzed, and it is a good option when a specific substance is necessary to make up the fermentation medium. To supply proline, peptone from gelatin can be used, but it lacks sulfur-containing amino acids. Keratin peptone can supply proline and cystine, but has no lysine. Even being an option to specific amino acid supplementation, the utilization of peptones usually is limited by the price, due to its relatively high production cost, mainly for industrial fermentation, involving high substrate volumes.

2.2.3 Nutrient Sources

Media used as substrate in fermentative process must contain all components in a proper form and concentration to address the microorganism physiology. When properly balanced, it gives all conditions for the cell growth, energetic metabolism, and metabolic products production. At lab scale, pure substances can be used for a medium composition, but at industrial scale, less expensive yet balanced materials must be tested. Normally complex and undefined substances, but properly characterized, can be used.

2.2.4 Toxic Compounds

The presence of toxic compounds in the residue composition may demand some kind of special treatment in order to remove or previously degrade these toxic compounds such as solvent extraction, lixiviation, sparging for volatile toxic compound release, and chemical or enzymatic degradation to generate two or more nontoxic substances or at least less toxic chemical products. If these strategies were not technical/economic possible, the selection of a resistant strain must be tried to run the fermentative process.

2.3 Transport and Storage Considerations

To use any agro-industrial residue, the place where it is generated, the place where it will be used, and its stability may be a concern, since agro-industrial residues, besides their organic nature, may have a high microbial load and usually are easily degraded. Transportation costs can be prohibitive to the whole process, due to distance and/or temperature control needs. In general, it is recommended to place the unit where the residue will be used near by the factory where it is generated, to be reused within a short period of time. If not possible, the logistic of the process must be studied in order to preserve the residue characteristics and consider all transportations costs.

Marrison and Larson (1995) developed the standard expression for transport costs as given in (2.1):

$$\text{Cost}(\text{in } \$/\text{ton}) = \text{fixed costs} + (\text{UTC} \times \text{distance}) \tag{2.1}$$

where UTC is the unitary transport cost, in $ per ton × km, and distance is in km. Most transportation of residual biomass is done by trucks, for which fixed costs may vary from 3.7 to 5.7 U$/t and variable costs may range from 0.085 to 0.146 U$/t.km for straw and stover or for wood chips, respectively, in Canada, data corrected for 2012 dollars (Searcy et al. 2007). Rail and barrage transportation have fixed costs of 17.1 and 34.1 and variable costs of 0.03 and 0.01, respectively (Montross and Crofcheck 2010). In Brazil, the fixed costs are around 7.5 U$/t and variable costs are around 0.06 U$/t.km for transportation by trucks.

Storage is usually cheap, but it should be as short as possible, because of the risk of degradation and infestation, and care must be taken about fire and for exposed heaps of wetting by rain.

2.4 Pretreatment Strategies

2.4.1 Drying and Concentration

Drying is the process of removal of a volatile substance, usually but not exclusively water, from a solid material through the evaporation. Humidity is the amount of water present in a material, and during the material drying, the humidity is reduced to a final acceptable or an equilibrium value. In most cases, drying is a finishing product operation, but for agro-industrial residues, it is necessary to reduce the water activity, in order to prevent deterioration reactions, microbial growth, chemical redox reaction, and the enzymatic reactions (Pessoa and Kilikian 2005).

According to Coulson and Richardson (1991), drying process is essential for (1) reduction of volume and weight of the material, (2) improvement of storage and handling characteristics of the product, and (3) reduction of transportation, packing, and storage costs.

The liquid content of a residue may be reduced by mechanical processes, such as centrifugation or pressing, but the final solid still contains high humidity and water activity (a_w). Drying must be more effective and reduce the humidity to suitable values. Hence, thermal methods are more adequate and will be discussed in this chapter.

2.4.1.1 Drying Process

During the drying process, the liquid aggregated on the material is removed by evaporation, passing from the liquid to the gaseous phase. The evaporation happens at a temperature below the liquid boiling temperature at the internal system pressure. Thus, drying is a complex process that involves simultaneously heat and mass transfer, resulting in significant changes in the physical, chemical, and structural properties. Water loss can cause cellular structural stress, microstructure alteration, increasing porosity, and material shrinkage (Laopoolkit and Suwannaporn 2011).

When the material is put in contact with hot air, heat is transferred from the air to the material under the effect of temperature gradient between them. Simultaneously, the partial pressure difference of water vapor between the air and the material surface determines the mass transfer of water vapor from the material to the air (Perry et al. 1997).

According to Mc Cabe et al (1993) and Coulson and Richardson (1991), the material can be in different forms such as flakes, granules, crystals, powder, plates, or leaves showing different properties. Also, the liquid to be removed can be at the solid surface or at its interior, or partially outside and inside, in two basic forms: (1) free moisture, the water in excess relative to the equilibrium humidity content, and (2) bound moisture, the water retained in such way that it has a vapor pressure below the free water at the same temperature. It can be inside small capillaries, adsorbed on the surface, or inside the material cells. Aspects of the raw material, characteristics, and quality of the final product must be considered during the drying process.

2.4.1.2 Drying Rate

The drying rate or the variation of the humidity of each material along time is an important parameter for the process. The humidity variation depends on the differences of the moisture inside each material (free/bounded). The curve obtained during

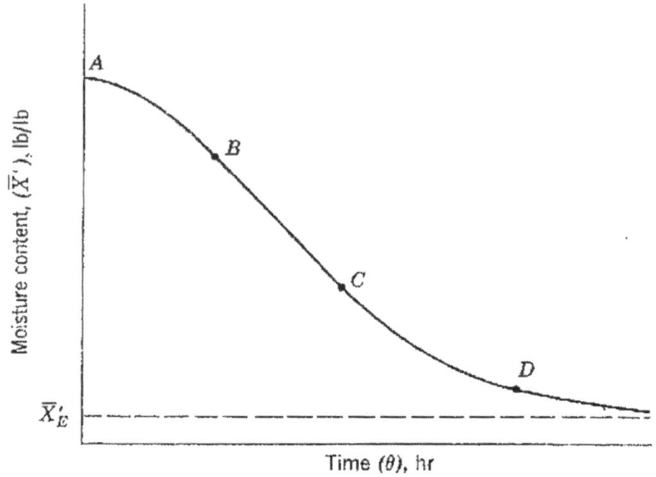

Fig. 2.1 Typical drying curve for constant drying conditions, moisture content as a function of time (Adapted from Foust et al. 1960)

monitoring of the variation of the moisture content with time allows the determination of the drying velocity for moisture content, and the curve shape varies with the structure, kind and granulometry of the material, thickness of the material layer, and kind of dryer (Coulson and Richardson 1991).

The main factors to be evaluated to correctly choose the drying method/equipment are directly related to the mass and heat transfer mechanism, to the physical material characteristics, and to the desirable final quality of the product.

Figure 2.1 shows a typical drying curve. AB segment represents the unsteady-state period. BC segment is the constant-rate drying part, where the entire exposed material surface is saturated with water and drying proceeds from a free liquid surface without the influence of the solid. CD segment is the period with a decreasing drying rate, where there is a lack of free liquid at the material surface as the liquid movement to the surface is slower than the movement of mass from the surface to the drying air. At point D and at lower moisture values, there is no significant amount of liquid at the material surface. The wet part of the surface dries by convective transfer of heat and mass to the drying gas stream, and vapor from inside of the material diffuses to the dry places of the surface and then into the gas stream. This mechanism is slower than the convective transfer from the saturated surface. All evaporation occurs from the interior of the solid, and the material moisture contents continue to fall until the equilibrium moisture content (EMC) is reached (X_E) and the drying process stops.

Figure 2.2 shows simulated drying curves for a 1,000 kg of *Miscanthus* grass. The initial biomass humidity is 30 %, reasonable for grasses and straw partially air-dried in the field. The drying temperature and the initial air moisture determine the drying rate and the equilibrium moisture content.

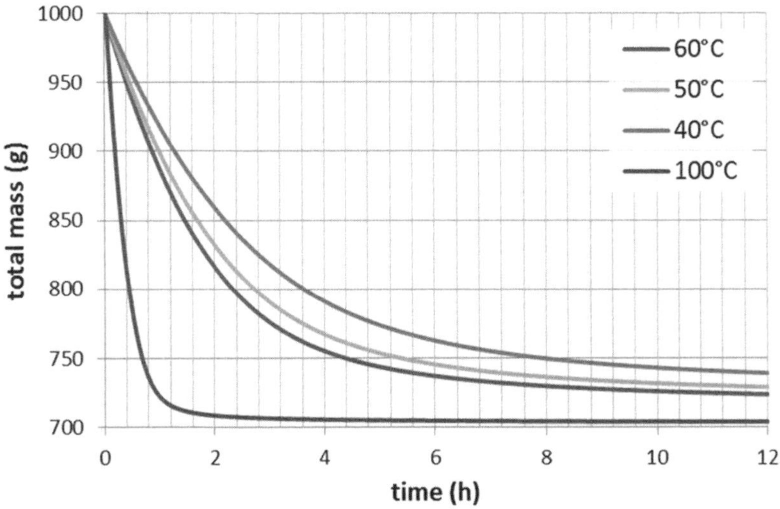

Fig. 2.2 Simulated drying curves for *Miscanthus* at several temperatures. Inlet air is at 20 °C and 50 % relative humidity. Data based on the equilibrium curve of Fig. 2.3 and literature thermodynamic data for water

Materials with free water, such as fruit residues, may have an initial lower drying rate than a constant drying rate period and finally an exponential decay of the drying rate as shown in Fig. 2.1; materials without free water (as is the case) show only the falling rate period.

The EMC of a material depends on the ambient temperature and relative humidity, being inversely proportional to the temperature and directly proportional to the relative humidity (RH). To know the behavior of a certain material is useful to determine the suitable humidity for storage, considering the usual weather.

Figure 2.3 shows EMC isotherms for selected biomasses. While *Jatropha* seeds, for example, show relatively stable moisture content for a wide range of RHs, the EMC of a grass such as *Miscanthus* grows from 10 to 20 % with RH going from 40 to 80 %. The RH in the isotherm may be used to predict the a_w of the solid, which should be below 0.8 to prevent the growth of most fungi.

2.4.1.3 Equipment

The classification of the drying equipment depends on the heat transfer method and the properties and characteristics of the material to be dried. In general, drying equipment can be divided into two groups: (1) direct dryers where the material is put in direct contact with the drying gas heat and (2) indirect dryers where the heat is supplied through other way, such as radiation, conduction, high-frequency electric field, and microwave (Mc Cabe et al. 1993; Perry et al. 1997). Both types can be used for residue drying.

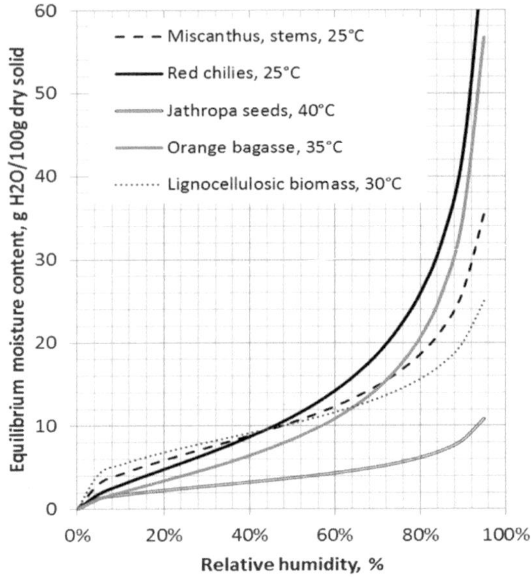

Fig. 2.3 Equilibrium moisture curves for 5 biomasses. Regression models using the Oswin modified equation and data from Arabhosseini et al. (2010), Kallemullah and Kailappan (2004), Kartikaa et al. (2012), Fioretin et al. (2010), Achargee et al. (2011)

There is a large variety of equipment and drying processes. The first criterion to be analyzed is the volume and characteristic of the material to be dried. Later, the heating method, material feeding mode, and cost must be evaluated (Al-Kassir et al. 2005).

For industrial scale, tray dryers, conveyors and tunnel dryers, rotatory dryers, and fluidized and fixed bed dryers can be used. According to Perry et al. (1997), the main dryers are:

Conveyors and tunnel dryers: The material is transported through a drying tunnel, where hot air circulates, transversal, countercurrent, or parallel to the material. The advantage of this equipment is the operational flexibility, and it can be used to dry materials of various sizes and forms, as the hot air velocity can be adjusted and the residence time does not depend on the particle characteristics, although it will determine the material final humidity.

Rotatory dryers: Consists of a big slightly inclined cylinder, rotating around an axis, with internal paddles that enhance the thermal exchange between the hot air and the material. Beyond the cylinder movement and the gravity action, the material is constantly rebutted favoring the drying and driven to the dryer discharge while the volatile mass is transported by the gas flow. Normally, dryers operate in countercurrent, where the hot air enters opposite to the material discharge, enhancing the process thermal yield. In this case, parameters such as density, form, particle size, and the equipment inclination significatively influence the biomass velocity along the dryer and the retention time.

Tray dryers: It consists of a thermal isolated camera, with heating systems and forced ventilation through trays placed in shelves. The air movement allows heat conservation and improves drying efficiency. It is the simplest equipment, used mainly for discontinuous operation and in low scale.

2.4.2 Grinding and Size Classification

When searching a substrate for a fermentative process from a solid material or residue, the medium size of the particles must be defined, taking in view some particularities of the process; mainly about size, the substrate particles should be neither big nor small. Smaller particles provide bigger surface areas and consequently bigger transformation grade as far as its enhanced surface contact favors microbial growth but on the other hand can result in clumping and clogging of the fermentation medium, negatively affecting respiration/aeration, microbial growth, and process yield. Solid medium must have a proper granulometry to avoid medium compaction, preserving empty space among particles to allow air circulation through the mass, an efficient oxygen supply necessary, and an efficient dissipation of gas and heat produced during the microorganism activity, harmful to the process (Del Bianchi et al. 2001; Souza et al. 2007; Santos et al. 2005; Ruiz et al. 2012).

Particle material size and porosity will directly affect the superficial surface and the microorganism accessibility to the substrate, the enzyme activity, and the microorganism metabolism which are surface phenomena (Santos et al. 2005). In general, it is considered that the size reduction of a substrate is better for the fermentation process. However, there is a limit to decrease the particle size.

The ideal particle size is arbitrary and varies according to the substrate, microorganism, and process used. The particle uniformity in terms of size will provide a uniform process even for chemical and biochemical reaction (Izumi et al. 2010; Hendriks and Zeeman 2009).

Pandey et al. (2001) studied wheat bran and corn flour particles at the proportion of 9:1 with diameters between 425 and 500 μm and 500 and 600 μm, respectively, which reached the highest amyloglucosidase production, although they showed that diameter between 180 μm and 1.4 mm had presented similar yields.

Kumar et al. (2003) studied four different sugarcane bagasse granulometry varying between 0.64 and 2.0 mm for citric acid production. The best results were obtained with particle size between 1.2 and 1.6 mm. This particle size provided solid medium with high porosity, resulting in better heat and mass transfer. On the other hand, bigger particles (1.6–2.0 mm) provided smaller surface area to the microorganisms and presented lower citric acid production. For the test using smaller particles (0.64–1.2 mm), the production was affected by the low heat and mass transfer.

Studies of neomycin production using the *Streptomyces marinensis* mutant strain by Bapiraju et al. (2004), with three particles sizes of wheat flour (small, intermediate, and big), found that the best yield was obtained with the larger granulometry.

Yuan et al. (2011) investigated the influence of the wheat stem particle size (1–5 and 10 mm) to produce hydrogen, acetate, and butyrate using a microflora from a biogas plant; they obtained the best results with the smaller particle size.

2.4.2.1 Grinding Process

During grinding or particle size reduction, the average size of the solid particles is reduced by the application of an impact force, compression, or abrasion. Advantages of the particle size reduction are (1) increase of the ratio surface/volume of the material and (2) uniformity of the material particle size, helping the homogeneity of the substrate (Mc Cabe et al. 1993; Gauto and Rosa 2011).

The different methods for particle size reduction are classified according to the range of the particle size produced; crushers are used for the production of bigger and medium particles or mills that produce smaller particles and fine dust (Gauto and Rosa 2011; Coulson and Richardson 1991).

2.4.2.2 Equipment Used for Fragmentation

There is a large variety of equipment for solid size reduction that must be evaluated according to the finality, initial investment, and the operational cost. The proper equipment choice must take into account the nature, amount, and dimensions of the material to be treated. The main properties to be evaluated are hardness, structure, humidity, crushing strength, tendency to slip or impaste, dust production, and explosion risks (Coulson and Richardson 1991).

Large and medium particle sizes are produced using crushers or shredders, such as mandibles, swivels, hammer, rolls, discs, and conicals. They have little application to fragment industrial residues to biotechnological use, where smaller particle dimensions are necessary. In this case, mills are the best choice, which produce medium to fine particle size. The main types of mills are:

Disc mill: They are generally used to reach a fine granulation. They are constructed with two steel discs placed on horizontal axis, one or both movable. One of them is mounted on to an eccentric support, allowing that the two milling faces of the disc are continuously approaching and moving away. The material is fed at the center of the discs and discharged by centrifugal force. Fine material granulometry is obtained (Perry et al. 1997).

Roll mills: They are used for cereal milling; it provides a product with uniform texture. There are basically two or more heavy cylinders with the same diameter that spin in opposite directions with equal or different velocities. Fed particles are submitted to compression and cutting forces. The distance between the rolls is regulated and must be adjusted to the raw material conditions and desired granulometry (Coulson and Richardson 1991).

Knives and hammer mills: They are used to produce a material thinner than the rolls mill. This equipment is appropriate for cereal milling destined to extract any cereal component and/or to produce a fine powder. It is a mill of impact, constructed with a rotor that spins at high velocity inside a cylindrical camera. At the edge of the rotor is fixed a series of hammers or knives. The material is broken or cut under the impact of the hammers or knives and pulverized by the blades of the mill while

Fig. 2.4 Sieving

forced to pass through a screen placed at the opening between the hammer or knives and the camera. They are used for milling fibrous or brittle materials (Coulson and Richardson 1991).

Ball mills: It is basically a cylinder that spins supported on a horizontal or slightly inclined axis, with the internal space charged with balls of steel or porcelain. The size reduction takes place by the impact and friction of the material with the balls when the mill spins, producing a very homogeneous material even in liquid medium. They are recommended to obtain a very uniform liquid suspension mainly for the ceramic industry. The final size of the particles depends on the material properties, weight and diameter of the balls, spin cylinder velocity, time of the processing, and level of the material inside the cylinder (Coulson and Richardson 1991).

2.4.2.3 Granulometric Separation

Grinded material classification in different granulometries is required for better material homogenization and to assure a uniform process. Figure 2.4 shows the simplest and most common method in the granulometric separation that consists of passing the material through a series of sieves with meshes progressively smaller during vibration. Fractions are separated according to the meshes of the screens. The average particle size of each fraction and the amount of material of each fraction are determined. Later, it is possible to build a granulometric distribution curve of each milling process (Gomide 1983). In lab scale, a set of sieves with screens progressively smaller can separate efficiently the size fractions of a milled material. However, at industrial scale, a continuous sieving process must be used.

For large-scale processes the available equipment are:

- *Slug sieve*: They are constituted of a long horizontal cylinder perforated (screen) with slight inclination, spinning at a low velocity. The screen opening along the cylinder increases progressively in the exit direction and allows the separation of the various material size fractions.
- *Shaking sieves*: In this case, the particle movement in the sieving surface allows the size separation. Normally, these sieves are horizontal but sometimes inclined to provoke the material displacement.
- *Vibrating sieves*: They are equipment of high capacity and efficiency, mainly for obtaining fine particles. They differ from the shaking sieves at the frequency and vibration amplitude and are bigger than the vibrating ones.

2.4.3 Thermochemical Hydrolysis

Many natural substrates, such as lignocellulosic biomass, have a complex microstructure that makes the material hard to digest. One approach to enhance further processing and/or to recover each constituent of the material as a partial hydrolysis or fiber expansion can be tried. Some thermochemical methods have been proposed for pretreating and hydrolyzing agro-industrial lignocellulosic wastes, such as dilute acid hydrolysis (Hernández-Salas et al. 2009; Balat et al. 2008; Zhang et al. 2007), steam explosion (Hernández-Salas et al. 2009; Hendriks and Zeeman 2009; Balat et al. 2008; Ramos et al. 1992; Ramos et al. 2000; Glasser and Wright 1997), alkali washing (Hernández-Salas et al. 2009; Hendriks and Zeeman 2009; Balat et al. 2008), and ammonia fiber expansion (Hendriks and Zeeman 2009; Balat et al. 2008), among others.

2.4.3.1 Acid Hydrolysis

Dilute acid hydrolysis can be an effective pretreatment to recover the sugars from the hemicellulosic part of the material, to improve further lignin separation, and to produce partially pure cellulose. Sulfuric acid can be used for pretreatment and hydrolysis (Lavarack and Griffin 2002), but other reagents, such as hydrochloric, nitric, and other acids, can also be used (Gámez et al. 2006; Rodríguez-Chong et al. 2004). Chen et al. (2012) found that the pretreatment using dilute sulfuric acid has been considered as one of the most cost-effective methods. The biomass into dilute acid solution is submitted to controlled and moderate temperature by means of conventional heating that hydrolyzes the sugars from the hemicelluloses and causes the softening of the lignin, facilitating its alkaline removal from the residual material, generating relatively pure cellulose. Microwave-assisted heating can also be used to pretreat biomass. The microwave electromagnetic field may create nonthermal effects which accelerate the destruction of crystal structures. Binod et al. (2012) developed a process using biomass microwave treatment in alkali solution (1 % NaOH) followed by acid pretreatment (1 % H_2SO_4) and enzymatic hydrolysis, achieving an overall reduction of sugar yield of 0.83 g/g dry sugarcane bagasse.

Hot acid pretreatment has been used to hydrolyze the simple sugars derived from the biomass of polysaccharides, mostly hemicelluloses. The resulting sugars can degrade to furfural (from pentoses) and to 5-hydroxy-methyl-furfural or HMF (from hexoses). These compounds are inhibitory for microorganisms, and their production means loss of fermentable sugars. Organic acids, such as maleic and fumaric, have been suggested as alternatives to avoid HMF formation (Kootstra et al. 2009), but mild thermal conditions can prevent formation of furfural and HMF.

2.4.3.2 Steam Explosion

Steam explosion is a promising method for the pretreatment of lignocellulosic biomass (Soccol et al. 2011) and can be performed in the presence or absence of an acid or alkali catalyst. The grinded biomass is treated with high-pressure saturated steam, at temperatures varying from 160 to 260 °C, and then the pressurized reactor is quickly decompressed, releasing the material to the normal pressure, undergoing an explosion. The process causes the disrupting of the material's structure, degradation of hemicellulose, and lignin partial disruption due to the high temperature, thus facilitating the subsequent obtention of cellulose for further hydrolysis (Öhgren et al. 2007).

Rocha et al. (2012) used steam explosion process to treat sugarcane bagasse (50 % of moisture) at a pressure of almost 1.3 MPa (190 °C) during 15 min, depressurizing suddenly the reactor after this time. The treatment recovered an average of 82.7 ± 4.3 % of the hemicelluloses, and cellulose was hydrolyzed at the ratio of 11.8 ± 3.7 %, probably from the polymer amorphous region. Lignin was solubilized at the proportion of 7.9 ± 9.1 %. Furfural and HMF production is more pronounced at the steam explosion process due to the hard thermal conditions used.

2.4.3.3 Alkaline Pretreatment

The biomass alkaline pretreatment has being studied a lot because it can remove and modify the lignin from the biomass crystalline structure, improving hydrolysis of the remaining polysaccharides and removing acetyl groups and various uronic acid substitutions on hemicellulose. Depending on the temperature and alkali concentration, most of the hemicellulose sugars are degraded; thus, treatment conditions must be carefully studied. Alkaline pretreatment probably causes saponification and hydrolysis of intermolecular ester bonds cross-linking of xylan hemicelluloses and lignin or lignin and other hemicelluloses. Dilute NaOH treatment of lignocellulosic material causes rupture of structural linkages between lignin and carbohydrates of hemicelluloses and disruption of the lignin structure, leading to a decrease in the lignin polymerization degree, a decrease in cellulose crystallinity, and a separation of the hemicellulose sugars (Soccol et al. 2011; Fan et al. 1987).

Rocha et al. (2012) reported a process where sugarcane bagasse was submitted to the steam explosion and then reacted with a NaOH solution 1.0 % (w/v), for 1 h at 100 °C, using a solid–liquid ratio of 1:10 (w/v). They got 92.7 ± 3.9 % of lignin removal from the biomass, hydrolyzed 31.1 ± 3.5 % of the cellulose, and achieved 94.7 ± 0.9 % of the hemicellulose hydrolysis.

2.4.3.4 Ammonia Fiber Expansion

During ammonia fiber expansion (AFEX) process, biomass is treated with liquid ammonia under pressure (100–400 psi) and temperature (70–200 °C) before rapidly releasing to the atmospheric pressure (Bals et al. 2010). This process causes the decrease in the crystalline cellulose, increasing the rate of enzymatic hydrolysis; hydrolyzes hemicellulose; solubilizes and depolymerizes lignin; and increases the size and number of micropores in the plant cell wall (Mosier et al. 2005).

Krishnan et al. (2010) working with sugarcane bagasse and cane leaf residues reported that AFEX pretreatment improved the accessibility of cellulose and hemicelluloses to the enzymatic hydrolysis by breaking ester linkages and other lignin–carbohydrate complex bonds. The enzymatic hydrolysis efficiency of the AFEX pretreated bagasse and cane leaf residue by cellulases was approximately 85 %, and the use of hemicellulases during enzymatic hydrolysis promoted the xylan hydrolysis to 95–98 %.

2.4.3.5 Organosolv

Organosolv is the biomass pretreatment with an aqueous solution of an organic solvent with or without an acid or alkali catalyst. The process efficiently solubilizes lignin from lignocellulosic materials promoting the partial hydrolysis of lignin bonds and solubilization of the most of the hemicelluloses sugars, resulting in a liquid phase containing the lignin and hemicellulosic sugars and a solid phase composed by a cellulose-enriched pulp. The addition of a catalyst can enhance the lignin removal efficiency (Mesa et al. 2011; Sun and Cheng 2002). The possibility of lignin and polyoses recovering from the liquid phase by distillation with the simultaneous recycling of solvents is an important advantage of this technique when compared with other aqueous-based processes (Novo et al. 2011).

Novo et al. (2011) developed a process using glycerol–water mixtures and obtained a pulp with a residual lignin amount lower than 8 %, extent of delignification close to 80 %, and residual cellulose content higher than 80 %.

Mesa et al. (2011) processed sugarcane bagasse with the combination of a dilute acid pretreatment followed by the organosolv pretreatment with ethanol and NaOH under optimized conditions (60 min, 195 °C, 30 % (v/v) ethanol), yielding a residual solid material containing 67.3 % (w/w) glucose, which was easily recovered by enzymatic hydrolysis.

2.4.4 Enzymatic Hydrolysis

Enzymes can be used to hydrolyze lignin, cellulose, and hemicellulose which are the main components of lignocellulosic agro-industrial wastes. The advantages of enzymatic hydrolysis over the previous methods may include mild reaction conditions, higher product yields and fewer side reactions, less energy demand, and less reactor resistance to pressure and corrosion (Lee 1997).

White-rot basidiomycetes produce several ligninolytic enzymes that catalyze one-electron oxidation of lignin units, producing aromatic radicals (Giardina et al. 2000). Lignin is normally not degraded as sole carbon and energy sources, requiring additional co-substrates such as cellulose, hemicellulose, or glucose (Silva et al. 2010). There are four major groups of ligninolytic enzymes produced by the white-rot fungi: lignin peroxidase (LiP; 1,2-bis(3,4-dimethoxyphenyl)propane-1,3-diol:hydrogen-peroxide oxidoreductase; EC 1.11.1.14), manganese-dependent peroxidase (MnP; Mn(II):hydrogen-peroxide oxidoreductase or manganese peroxidase; EC 1.11.1.13), versatile peroxidase (VP; EC 1.11.1.16), and laccase (benzenediol: oxygen oxidoreductase; EC 1.10.3.2). However, the process of lignin biodegradation can be further enhanced by the action of other enzymes, such as glyoxal oxidase (EC 1.2.3.5), aryl-alcohol oxidase (veratryl alcohol oxidase; EC 1.1.3.7), pyranose 2-oxidase (glucose 1-oxidase; EC 1.1.3.4), cellobiose/quinone oxidoreductase (EC 1.1.5.1), and cellobiose dehydrogenase (EC 1.1.99.18) (Wong 2009).

Both LiP and MnP belong to the class of peroxidases that oxidize their substrates by two consecutive one-electron oxidation steps with intermediate cation radical formation. Due to its high redox potential, the preferred substrates for LiP are non-phenolicmethoxyl-substituted lignin subunits, and the oxidation occurs in the presence of H_2O_2 (Tuor et al. 1995; Wong 2009), whereas MnP acts exclusively as a phenoloxidase on phenolic substrates using Mn^{2+}/Mn^{3+} as an intermediate redox couple (Tuor et al. 1995). Versatile peroxidases are a group of enzymes, primarily recognized as manganese peroxidases, which exhibit activities on aromatic substrates similar to that of LiP. These enzymes not only are specific for Mn (II) but also oxidize phenolic and non-phenolic substrates that are typical for LiP, including veratryl alcohol, methoxybenzenes, and lignin model compounds in the absence of manganese (Wong 2009).

Laccases are blue multicopper oxidases able to oxidize a variety of phenolic compounds including polyphenols, methoxy-substituted phenols, diamines, and a considerable range of other compounds, with concomitant reduction of molecular oxygen to water (Autore et al. 2009; Dwivedi et al. 2011). They oxidize phenols and phenolic lignin substructures by one-electron abstraction with formation of radicals that can repolymerize or lead to depolymerization (Higushi 1989). These enzymes have been found to oxidize also non-phenolic compounds in the presence of a mediator (e.g., 2,2′-azinobis-3-ethylbenzthiazoline-6-sulfonate or ABTS) (Wong 2009). Laccases are more readily available and easier to manipulate than both LiP and MnP. Moreover, these enzymes find many industrial applications in the areas of food products, pulp and paper, textiles, nanobiotechnology, soil bioremediation, synthetic chemistry, and cosmetics (Couto and Herrera 2006).

Enzymatic hydrolysis can also be employed to produce reducing sugars from cellulose and hemicellulose. Utility cost of enzymatic hydrolysis is low compared to acid or alkaline hydrolysis because it is usually conducted at mild conditions (pH 4.8 and temperature 45–50 °C) and does not cause corrosion problems mainly for cellulose hydrolysis (Duff and Murray 1996). Both bacteria and fungi can produce cellulases and hemicellulases for hydrolysis of lignocellulosic materials.

The factors that affect the enzymatic hydrolysis of cellulose and hemicellulose include substrates, enzymatic activity, and reaction conditions (Sun and Cheng 2002). Substrate concentration is one of the main factors that affect the yield and initial rate of enzymatic hydrolysis. At low substrate levels, an increase of substrate concentration normally results in an increase of yield and reaction rate of the hydrolysis (Cheung and Anderson 1997).

When cellulose is used as raw material, the cellulase complex is responsible for enzymatic hydrolysis of pretreated cellulosic biomass. There are three major types of cellulases involved in the hydrolysis of cellulose: endo-β-1,4-glucanase (EG), which acts randomically at the molecule producing reducing and nonreducing ends in the cellulose polymer; cellobiohydrolase (CBH) or exoglucanase, which liberates cellooligosaccharides and cellobiose from these reducing and nonreducing ends; and β-glucosidase (BGL) that cleaves cellobiose and liberates glucose (Mathew et al. 2008).

The enzymes of the cellulase complex are strongly inhibited by their hydrolysis products: glucose and short cellulose chains. Several methods have been developed to reduce the inhibition of hydrolysis, including the use of high concentrations of enzymes, the supplementation of β-glucosidases during hydrolysis, and the removal of sugars during hydrolysis by ultrafiltration or simultaneous saccharification and fermentation (Lin and Tanaka 2006).

2.5 Conclusion

The search of economical and more efficient industrial processes is the objective of many research groups around the world, some of them supported by important companies and some from researcher's institutes. Moreover, the development of environmentally friendly processes, with fewer losses, less wastage, generating fewer residues, or reusing the residues in a new process, appears as an alternative to a sustainable world. In addition, use of the energy of the green carbon, phytobiomass produced by photosynthesis will make the world independent of the fossil carbon energy, seems to be the reasonable way to reach this goal. Many steps and various technologies must be developed to harness the phytobiomass energy. It requires the establishment of some operations in sequence to set a proper and particular industrial process.

References

Achargee TC, Coronella CJ, Vasquez VR (2011) Effect of thermal pretreatment on equilibrium moisture content of lignocellulosic biomass. Bioresour Technol 102(7):4849–4854

Al-Kassir A, Gañan J, Tinaut FV (2005) Theoretical and experimental study of a direct contact thermal screw dryer for biomass residues. Appl Therm Eng 25(17–18):2816–2826

Autore F, Del Vecchio C, Fraternali F, Giardina P, Sannia G, Faraco V (2009) Molecular determinants of peculiar properties of a *Pleurotus ostreatus* laccase: analysis by site-directed mutagenesis. Enzym Microb Tech 45:507–513

Arabhosseini A, Huisman W, Müller J (2010) Modeling of the equilibrium moisture content (EMC) of Miscanthus (*Miscanthus* × *giganteus*). Biomass Bioenergy 34(4):411–416

Balat M, Balat H, Cahide OZ (2008) Progress in bioethanol processing. Progr Energ Combust Sci 34:551–573

Bals B, Rogers C, Jin M, Balan V, Dale B (2010) Evaluation of ammonia fiber expansion (AFEX) pretreatment for enzymatic hydrolysis of switchgrass harvested in different seasons and locations. Biotechnol Biofuels. doi:10.1186/1754-6834-3-1

Binod P, Satyanagalakshmi K, Sindhu R, Janu KU, Sukumaran RK, Pandey A (2012) Short duration microwave assisted pretreatment enhances the enzymatic saccharification and fermentable sugar yield from sugarcane bagasse. Renew Energy 37:109–116

Bapiraju KVSN, Sujatha P, Ellaiah P, Ramana T (2004) Mutation induced enhanced biosynthesis of lipase. Afr J Biotechnol 3(11):618–621

Carioca JOB, Arora HL (1984) Biomassa: fundamento e aplicações tecnológicas. UFC, Fortaleza

Chen WH, Ye SC, Sheen HK (2012) Hydrolysis characteristics of sugarcane bagasse pretreated by dilute acid solution in a microwave irradiation environment. Appl Energ 93:237–244. doi:10.1016/j.apenergy.2011.12.014

Cheung SW, Anderson BC (1997) Laboratory investigation of ethanol production from municipal primary wastewater. Bioresour Technol 59:81–96

Coulson JM, Richardson JF (1991) Chemical engineering, vol II. Oxford, Pergamon, Londres

Couto SR, Herrera JLT (2006) Industrial and biotechnological applications of laccases: a review. Biotechnol Adv 24:500–513

Del Bianchi VL, Moraes IO, Capalbo DMF (2001) Fermentação em estado sólido. In: Schmidell W, Lima UA, Aquarone E, Borzani W (eds) Biotecnologia industrial: engenharia bioquímica. Edgard Blücher Ltda, São Paulo, pp 247–276

Duff SJB, Murray WD (1996) Bioconversion of forest products industry waste cellulosics to fuel ethanol: a review. Bioresour Technnol 55:1–33

Dwivedi P, Vivekanand V, Pareek N, Sharma A, Singh RP (2011) Co-cultivation of mutant *Penicillium oxalicum* SAU_E-3.510 and *Pleurotus ostreatus* for simultaneous biosynthesis of xylanase and laccase under solid-state. N Biotechnol 28:616–626. doi:10.1016/j.nbt.2011.05.006

Fan LT, Gharpuray MM, Lee YH (1987) Cellulose hydrolysis, 1st edn. Springer, New York

Fioretin LD, Menon BT, Barros STD, Pereira NC, Lima OC, Modenes AO (2010) Isotermas de sorção do resíduo agroindustrial do bagaço de laranja. Rev Brasileira de engenharia agrícola e ambiental 14(6):653–659

Foust AS et al (1960) Principles of unit operations. Wiley, New York

Gámez S, González JJ, Ramírez JA, Garrote G, Vázquez M (2006) Study of the sugarcane bagasse hydrolysis by using phosphoric acid. J Food Eng 74:78–88

Gauto MA, Rosa GR (2011) Processos e operações unitárias da indústria Química. Ciência Moderna Ltda, Rio de Janeiro

Giardina P, Palmieri G, Fontanella B, Rivieccio V, Sannia G (2000) Manganese peroxidase isoenzymes produced by Pleurotusostreatus grown on wood sawdust. Arch Biochem Biophys 376(1):171–179

Glasser WG, Wright RS (1997) Steam-assisted biomass fractionation. II. Fractionation behavior of various biomass resources. Biomass Bioenergy 14:219–235

Gomide R (1983) Operações unitárias: operações com sistemas sólidos granulares (1). Cempro, São Paulo

Hendriks ATWM, Zeeman G (2009) Pretreatments to enhance the digestibility of lignocellulosic biomass. Bioresour Technol 100:10–18

Hernández-Salas JM, Villa-Ramírez MS, Veloz-Rendón JS, Rivera-Hernández KN, González-César RA, Plascencia-Espinosa MA, Trejo-Estrada SR (2009) Comparative hydrolysis and fermentation of sugarcane and agave bagasse. Bioresour Technol 100:1238–1245

Higushi T (1989) Mechanisms of lignin degradation by lignin peroxidase and laccase of white-rot fungi. In: Lewis NG, Paice MG (eds) Plant cell wall polymers, biogenesis and biodegradation, vol 399. ACS Symposium Series, Washington, pp 482–502

Izumi K, Okishio Y, Nagao N, Niwa C, Yamamoto S, Toda T (2010) Effects of particle size on anaerobic digestion of food waste. Int Biodeter Biodegr 64(7):601–608

Kallemullah S, Kailappan R (2004) Moisture sorption isotherm of red chillies. Biosystems Eng 88(1):95–104

Kartikaa IA, Yulianib S, Kailakub SI, Rigalc L (2012) Moisture sorption behaviour of jatropha seed (*Jatropha curcas*) as a source of vegetable oil for biodiesel production. Biomass Bioenergy 36:226–233

Kootstra AMJ, Beeftink HH, Scott EL, Sanders JPM (2010) Comparison of dilute mineral and organic acid pretreatment for enzymatic hydrolysis of wheat straw. Biochem Eng J 46:126–131

Krishnan C, Sousa LC, Jin M, Chang L, Dale BE, Balan V (2010) Alkali-based AFEX pretreatment for the conversion of sugarcane bagasse and cane leaf residues to ethanol. Biotechnol Bioeng 107(3):441–450

Kumar D, Jain VK, Shanker G, Srivastava A (2003) Citric acid production by solid state fermentation using sugarcane bagasse. Process Biochem 38(12):1731–1738

Laopoolkit P, Suwannaporn P (2011) Effect of pretreatments and vacuum drying on instant dried pork process optimization. Meat Sci 88:553–558

Lavarack BP, Griffin GJ (2002) The acid hydrolysis of sugarcane bagasse hemicellulose to produce xylose, arabinose, glucose and other products. Biomass Bioenergy 23:367–380

Lee J (1997) Biological conversion of lignocellulosic biomass to ethanol. J Biotechnol 56:1–24

Lin Y, Tanaka S (2006) Ethanol fermentation from biomass resources: current state and prospects. Appl Microbiol Biotechnol 69:627–642

Marrison CI, Larson ED (1995) Cost versus scale for advanced plantation-based biomass energy systems in the US. EPA symposium on greenhouse emissions and mitigation research, Washington

Mathew GM, Sukumaran RK, Singhania RR, Pandey A (2008) Progress in research on fungal cellulases for lignocellulose degradation. J Sci Ind Res 67:898–907

Mc Cabe WL, Smith JC, Harriot P (1993) Unit operations in chemical engineering, 5th edn. Book Company, New York

Mesa L, González E, Cara C, González M, Castro E, Mussatto SI (2011) The effect of organosolv pretreatment variables on enzymatic hydrolysis of sugarcane bagasse. Chem Eng J 168:1157–1162

Montross M, Crofcheck C (2010) Energy crops for the production of biofuels. In: Thermochemical conversion of biomass to liquid fuels and chemicals. Crocker M (ed). RSC, London pp 26–45

Mosier N, Wyman C, Dale BE, Elander R, Lee YY, Holtzapple M, Ladisch M (2005) Features of promising technologies for pretreatment of lignocellulosic biomass. Bioresour Technol 96:673–686

Novo LP, Gurgel LVA, Marabezi K, Aprigio A, Curvelo S (2011) Delignification of sugarcane bagasse using glycerol-water mixtures to produce pulps for saccharification. Bioresour Technol 102:10040–10046

Öhgren K, Vehmaanperä J, Siika-Aho M, Galbe M, Viikari L, Zacchi G (2007) High temperature enzymatic prehydrolysis prior to simultaneous saccharification and fermentation of steam pretreated corn stover for ethanol production. Enzym Microb Tech 40(4):607–613

Pandey A (1991) Effect of particle size of substrate on enzyme production in solid-state fermentation. Bioresour Technol 37(2):169–172

Perry RH, Green DV, Maloney JO (1997) Chemical engineers' handbook, 7th edn. McGraw-Hill, Malasia

Pessoa A Jr, Kilikian BV (2005) Purificação de Produtos Biotecnológicos. Manole, São Paulo

Ramos LP, Breuil C, Kushner DJ, Saddler JN (1992) Steam pretreatment conditions for effective enzymatic hydrolysis and recovery yields of *Eucalyptus viminalis* wood chips. Holzforschung 46:149–154

Ramos LP, Carpes ST, Silva FT, Ganter JLMS (2000) Comparison of the susceptibility of two hardwood species, *Mimosa scabrella*Benth and *Eucalyptus viminalis*Labill, to steam explosion and enzymatic hydrolysis. Braz Arch Biol Tech 43:185–206

Rocha GJM, Gonçalves AR, Oliveira BR, Olivares EG, Rossell CEV (2012) Steam explosion pretreatment reproduction and alkaline delignification reactions performed on a pilot scale with sugarcane bagasse for bioethanol production. Ind Crop Prod 35:274–279

Rodríguez-Chong A, Ramírez JA, Garrote G, Vázquez M (2004) Hydrolysis of sugarcane bagasse using nitric acid: a kinetic assessment. J Food Eng 61:143–152

Ruiz HA, Rodríguez-Jasso RM, Rodríguez R, Contreras-Esquivel JC, Aguilar CN (2012) Pectinase production from lemon peel pomace as support and carbon source in solid-state fermentation column-tray bioreactor. Biochem Eng J 65:90–95

Santos SFM, Wanderley LR, Souza RLA, Pinto GAS, Silva FLH, Macedo GR (2005) Caracterização físico-química do pedúnculo de caju in natura e do resíduo seco. In: 1th Simpósio Brasileiro de Pós-colheita de Frutos Tropicais. SBF, João Pessoa-PB, 29 November – 02 December 2005 (CD Rom)

Searcy E, Flynn P, Ghafoori E, Kumar A (2007) The relative cost of biomass energy transport. Appl Biochem Biotechnol 140:639–652

Silva I.S, Menezes CR, Franciscon E, Santos EC, Durrant LR (2010) Degradation of lignosulfonic and tannic acids by ligninolytic soil fungi cultivated under icroaerobic conditions. Brazilian Archives of Biology and Tech 53(3) http://dx.doi.org/10.1590/S1516-89132010000300026

Soccol CR, Faraco V, Karp S, Vandenberghe LPS, Thomaz-Soccol V, Woiciechowski AL, Pandey A (2011) Lignocellulosic bioethanol: current status and future perspectives. In: Pandey A, Larroche C, Ricke SC, Dussap CG, Gnansounou E (eds) Biofuels: alternative feedstocks and conversion processes. Academic, San Diego, pp 101–122

Souza RA, Amorim BC, Silva FLH, Conrado L (2007) Caracterização do resíduo seco do maracujá para utilização em fermentação semi-sólida. In: 16th Simpósio Nacional de Bioprocessos. Federal University of Paraná, Curitiba, 1–5 August 2007 (CD Rom)

Sun Y, Cheng J (2002) Hydrolysis of lignocellulosic materials for ethanol production: a review. Bioresour Technol 83:1–11

Tuor U, Winterhalter K, Fiechter A (1995) Enzymes of white-rot fungi involved in lignin degradation and ecological determinants for wood decay. J Biotechnol 41:1–17

Wong DWS (2009) Structure and action mechanism of ligninolytic enzymes. Appl Biochem Biotechnol 157:174–209

Yuan X, Shi X, Zhang P, Wei Y, Guo R, Wang L (2011) Anaerobic biohydrogen production from wheat stalk by mixed microflora: kinetic model and particle size influence. Bioresour Technol 102(19):9007–9012

Zhang YP, Ding S, Mielenz JR, Cui J (2007) Fractionating recalcitrant lignocellulose at modest reaction conditions. Biotechnol Bioeng 97:214–223

Chapter 3
Thermochemical Transformation of Agro-biomass into Biochar: Simultaneous Carbon Sequestration and Soil Amendment

Mausam Verma, Naceur M'hamdi, Zeineb Dkhili, Satinder Kaur Brar, and Kshipra Misra

3.1 Introduction

"Biochar" might be a relatively new term; however, it is not a new product. Biochar (BC) is the thermochemical degradation product of organic materials in the absence of oxygen. It can be distinguished from charcoal by its reaction conditions for production as well as its application potential in soil amendments. Soils throughout the world contain biochar deposited through natural events, such as forest and grassland fires (Krull et al. 2008). In fact, areas high in naturally occurring biochar, such as the west Mississippi River and east Rocky Mountains in North America, are some of the most fertile soils in the world. Historical use of biochar dates back at least 2,000 years (O'Neill et al. 2009). In the Amazon Basin, evidence of extensive use of biochar can be found in the unusually fertile soils known as Terra Preta and Terra Mulata by ancient indigenous civilizations (O'Neill et al. 2009). Due to the large amounts of biochar incorporated into its soils, this region still remains highly fertile despite centuries of leaching from heavy tropical rains. In parts of Asia, notably Japan and Korea, the use of biochar in agriculture also has a long history. Recently, heightened interest in more sustainable farming systems, such as Korean Natural

M. Verma (✉)
CO2 Solutions, 2300, rue Jean-Perrin, Québec, QC, Canada G2C 1T9

N. M'hamdi • Z. Dkhili
Institut National Agronomique de Tunisie (INAT), 43 Rue Charles Nicole, Cité Mahrajène Le Belvédère, Tunis 1082, Tunisia

S.K. Brar
INRS-ETE, Universite du Quebec, 490, Rue de la Couronne, Quebec, QC, Canada G1K 9A9

K. Misra
Defence Institute of Physiology & Allied Sciences, Timarpur, Lucknow Road, Delhi 110054, India

S.K. Brar et al. (eds.), *Biotransformation of Waste Biomass into High Value Biochemicals*, DOI 10.1007/978-1-4614-8005-1_3,

Farming, has revived the use of biochar in Western agriculture. As of now, the possible connections between biochar properties and the soil biota and their implications for soil processes have yet to be studied systematically. Nevertheless, the comparative estimation of effects of biochar on overall soil properties has mostly concluded biochar as an advantageous agent for agriculture.

3.1.1 What Is Biochar?

Biochar is char derived from the thermal conversion of biomass that is used for nonenergy purposes. It may however have an alternative use as an energy carrier (Mašek and Brownsort 2010). Syngas or pyrolysis gas is the noncondensable product of pyrolysis containing CO, CO_2, H_2, CH_4, and higher hydrocarbons. Biochar is the charred by-product of biomass pyrolysis, the heating of plant-derived material in the absence of oxygen in order to capture combustible gases. The term biochar is a relatively recent development, emerging in conjunction with soil management and carbon sequestration issues (Lehmann et al. 2006). Biochar is a name for charcoal when it is used for particular purpose, such as soil amendment. Similar to all charcoal, biochar is created by pyrolysis of biomass.

It has previously been used in connection with charcoal production (Karaosmanoglu et al. 2000; Demirbas 2004a, b). The rationale for avoiding the term "charcoal" when discussing fuel may stem from the intent to distinguish it from coal. Indeed, coal is formed very differently from charcoal and has separate chemical and physical properties, although in very specific cases the differences in properties can become blurred. The establishment of the term "agrichar" is closely related to that of biochar, with the desire to apply charred organic matter to soil. "Biochar" is preferred here as it includes the application of charred organic matter in settings outside of agriculture, such as promoting soil remediation or other environmental services. And the term emphasizes biological origin, distinguishing it from charred plastics or other nonbiological material. "Char" is a term that is often used interchangeably with charcoal, but is sometimes applied to refer to a material that is charred to a lesser extent than charcoal, typically as a product of fire (Schmidt and Noack 2000). Biochar is charcoal created by pyrolysis of biomass (Laird et al. 2009). The resulting charcoal-like material is a form of carbon capture and storage. Charcoal is a stable solid and rich in carbon content and thus can be used to lock carbon in the soil. Biochar is of increasing interest because of concerns about climate change caused by emissions of carbon dioxide (CO_2) and other greenhouse gases (GHG). Biochar is a way for carbon to be drawn from the atmosphere and is a solution to reducing the global impact of farming (and in reducing the impact from all agricultural waste). Since biochar can sequester carbon in the soil for hundreds to thousands of years, it has received considerable interest as a potential tool to slow global warming. The burning and natural decomposition of trees and agricultural matter contributes a large amount of CO_2 released to the atmosphere. Biochar can store this carbon in the ground, potentially making a significant reduction in atmospheric GHG levels; at the same time its presence in the earth can improve water quality, increase soil fertility, raise agricultural productivity, and reduce pressure on old growth forests.

Biomass is defined as a renewable source of fixed carbon in the short term. This includes agricultural residues such as crop residues, manures, industrial wastes such as paper mill sludge, and residues from sugar mills and waste wood products (Roberts et al. 2010).

Shackley and Sohi (2010) defined biochar as the porous carbonaceous solid produced by thermochemical conversion of organic materials in an oxygen-depleted atmosphere. Likewise, according to Lehmann and Joseph (2009), biochar is commonly defined as charred organic matter, produced with the intent to deliberately apply to soils to sequester carbon and improve soil properties.

3.1.2 The Origin of Biochar Management and Research

The scientific attention towards biochar originates from the research in charcoal amended anthropogenic soils situated close to current or historical settlements in the Amazon region and estimated to cover an area of 500 km^2 (Smith 1980). The term biochar originated in the bioenergy literature in the late 1990s to distinguish grain-derived activated carbon from similar coal-derived materials. The two concepts of using charcoal as a soil improver and as a greenhouse gas mitigation strategy arose separately in the early 1990s. Biochar is an ancient practice that converts agricultural waste into a soil enhancer that can hold carbon, boost food security, and discourage deforestation. The process creates a fine-grained, highly porous charcoal that helps soils retain nutrients and water. Although the practice of charcoal application for soil fertility building is thought to be several thousand years old, the research in biochar for carbon sequestration is relatively new, emerging with the rising scientific and political awareness of climate change (Bruun 2011). Biochar is found in soils around the world as a result of vegetation fires and historic soil management practices. It seems that can be an important tool to increase food security and cropland diversity in areas with severely depleted soils, scarce organic resources, and inadequate water and chemical fertilizer supplies. Biochar also improves water quality and quantity by increasing soil retention of nutrients and agrochemicals for plant and crop utilization (Sohi et al. 2009). All over the world, scientists are investigating the incredible properties of Biochar. If revitalizing soil and capturing carbon were not enough, the production of charcoal produces gas, called syngas that can be used as fuel, providing clean, renewable energy. When the biochar is buried in the ground as a soil enhancer, the system can become carbon negative. The matching of the term "biochar" with the climate change mitigation concept did not occur until 2005 (Lehmann et al. 2005). While both research and development of biochar for environmental management at a global scale are a somewhat recent development, it is by no means new in certain regions and has even been the subject of scientific research for quite some time. For example, Trimble (1851) shared observations of evidence of the effect of charcoal dust in increasing and quickening vegetation. Early research on the effects of biochar on seedling growth (Retan 1915) and soil chemistry (Tryon 1948) yielded detailed scientific information. The use of biochar has, for some time, been recommended in various horticultural contexts as a substrate for potting mix (Santiago and Santiago 1989). Morley (1927) considers that charcoal acts as a sponge in the soil, absorbing and retaining water, gases, and solutions. He even remarks that "as a soil's purifier and an of moisture's absorber, charcoal has no equal" (Morley 1929). Liebig (1878) describes a practice in China where waste biomass was mixed and covered with soil and set on fire to burn over several days until a black earth is produced, which reportedly improved plant vigor.

3.1.3 Production of Biochar

Biochar production can only be estimated based on certain assumptions as the commercial scale processes for biochar are still underway to establish their market. In other words, biochar might require some more time to be added as a mainstream product in economy. Nevertheless, the research till date on biochar as a valuable product for agriculture and environment always suggested its vast potential (Denyes et al. 2012; Anderson et al. 2011; Barrow 2011; Cross and Sohi 2011; Fabbri et al. 2011; Galinato et al. 2011). Technically, biochar production is a feasible and sustainable process both energetically and economically; however, in order to overcome initial hurdles such as social and cultural acceptance, a persistent political will might be needed to facilitate its commercialization. Until date, many estimates of biochar production have been carried out by several researchers, which forecasted biochar production for future years (Gaunt and Lehmann 2008). According to Lehmann et al. (2006), the global estimated potential of biochar production was already in the order of gigatonnes (approximately, $0.6 \pm 0.1 \times 10^9$ tonnes per year by 2006). Recently, the authors also extrapolated their calculations for future potential of biochar production, for example, the biochar production could reach about 5.5–9.5 gigatonnes per year by 2100. However, such approximation seems to be too susceptible to even minor variations in parameters of estimation (up to about 1,000 %); therefore, Woolf et al. (2010) came up with a reasonably conservative assumption that if all existing agricultural crop residues were used to produce biochar, it would contribute to about 1 gigatonne per year at present. The gigatonne scale per year of carbon emission is equivalent to about 12 % of current anthropogenic carbon emission. Moreover, at this pace, it is fairly possible to offset global atmospheric carbon emissions by year 2050, and if a concerted effort is made by countries worldwide, this goal is possible even by year 2030. The authors also marked that these goals are achievable in a sustainable manner, without endangering food security, habitat, or soil conservation. Interestingly, the authors did not consider about municipal, forestry, and industrial biomass for biochar production, which can be substantial resources for biochar; therefore, the estimation made by the authors is easily achievable as far as biochar feedstock is concerned, if a strategic plan for carbon sequestration through biochar production is in place. As also explained earlier, definition of biochar is inclusive of products, such as wood charcoal. Therefore, if wood charcoal production is analyzed from its utility as biochar, about 47.2×10^6 tonnes of wood charcoal was produced in the year 2011, which is mostly derived from forestry biomass (FAO 2012). This quantity is about 5 % of annual gigatonne scale biochar production. The chemical composition and proportions of these constituents can vary greatly with the type of biomass; therefore, the suitability of a particular biomass as a potential feedstock for biochar production could depend upon various factors such as moisture content, calorific value, carbon, hydrogen, nitrogen, oxygen, volatile compounds, and ash content. In addition, compatibility with existing equipments for pretreatment and/or handling is important for mass scale processing. Therefore, several other possible biochar feedstocks are plausible. Moreover, application of suitable biochar production techniques such as pyrolysis can also make the biochar production economically feasible and energetically autonomous. Therefore, biochar production goal can be achieved at commercial scale.

3.1.4 Biomass Resources for Biochar Production

Biochar can be produced from various types of feedstock, such as agricultural crop residues, forestry residues, wood waste, organic portion of municipal solid waste, and animal manures. Nevertheless, the applicability of each type of biomass as feedstock depends upon the physical characteristics, chemical composition, environmental, as well as social, economic, and logistical factors (Duku et al. 2011; Verheijen et al. 2010a, b). The feedstock materials as biomass resources for biochar can be categorized as shown in the flow diagram below (Fig. 3.1).

As a matter of fact, biochar can be sourced from almost everything that contains organic matter rich in carbon and hydrogen. Therefore, there is a very wide range of biomass that can be a potential candidate; nonetheless, production process requirement will be a decisive factor in its usability as feedstock. As also mentioned earlier, that even by moderately conservative estimates, a significant portion of biomass can be available for biochar production. In particular, Canada has major biomass resource (approximately, 15 million tonnes per year) in the form of municipal solid waste and forestry residues (Verma et al. 2012). In addition, Canada is also a major supplier of wood pellets for fuel in world market (approximately, 10 % of 10 million tonnes per year wood pellets). Likewise, many biomass-based countries such as Ghana have annual agricultural biomass residues production capacity of about four million tonnes per year, in addition to about one million tonnes per year from forestry, municipal, and miscellaneous sectors. Therefore, biochar production can be commercially viable if a government-supported strategic plan is put in place as the feedstock for this product can be obtained from diverse sources in a sustainable manner. The support from government will be required mainly to overcome initial barrier of social acceptance and to demonstrate its viability.

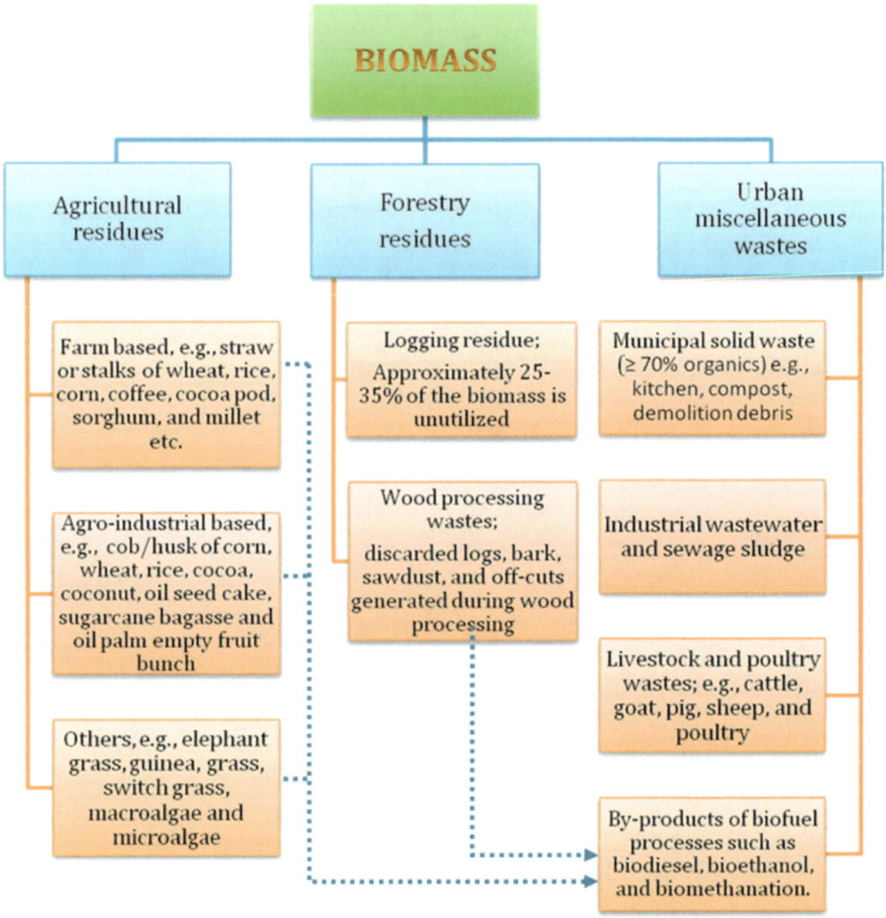

Fig. 3.1 Potential and concurrent resources for biochar production

3.2 Biochar Production from Biomass by Thermochemical Conversion Technologies

Thermochemical conversion technologies for biomass into biochar include carbonization, pyrolysis, and gasification processes (Masek et al. 2011; Shackley et al. 2011; Yoder et al. 2011; Mahinpey et al. 2009). The difference between carbonization and pyrolysis processes is subtle, as both processes have similar principle of operation; however, carbonization is aimed at biochar as primary product, whereas pyrolysis produces biochar as a by-product. Different biomass conversion processes for biochar production are presented in Fig. 3.2. The biochar produced from

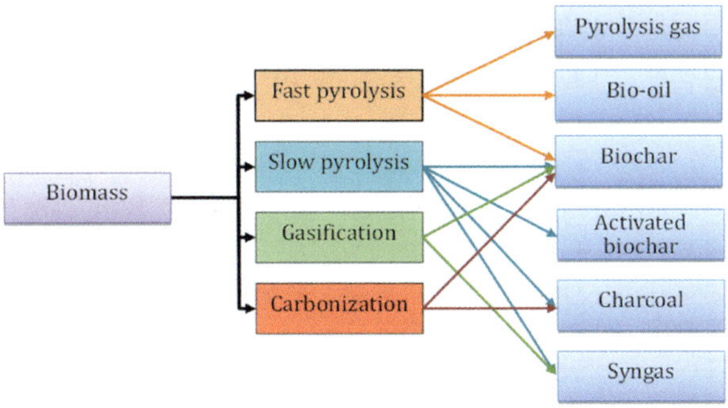

Fig. 3.2 Different thermochemical processes for biochar production

these processes varies greatly in terms of C, H, N, S, O, and ash content depending not only upon feedstock source and pretreatments such as particle sizing and drying but also upon the operating conditions of production (e.g., temperature, heating rate, residence time, and pressure). Likewise, the characteristics of biochar such as density, particle size distribution, moisture content, and pH depend on the feedstock type as well as pyrolysis reaction conditions. For example, biochar produced from wood is reported to be coarse and highly resilient with carbon content up to 80 %. These properties of biochar were reported to be due to the high lignin content (e.g., olive husks) which also yields high biochar content in the pyrolysis products as a result of the stability of lignin to thermal degradation. On the other hand, biochar produced from crop residues (e.g., maize, wheat, rye) and animal manures is generally finer and less robust with low mechanical strength due to lower lignin content (Demirbas 2004a, b). Similarly, particle size distribution of biochar produced from sawdust and woodchips under different slow pyrolysis conditions showed that particle size generally decreased with increase in temperature in the range of 450–700 °C (Abdullah and Wu 2009). In general, the process conditions in the slow pyrolysis are long vapor residence times (≥10 s), reaction temperatures between 450 and 650 °C at atmospheric pressure, and heating rates in the range of 0.01–2.0 °C s^{-1}. At these conditions, the rate of increased cracking reactions increases that shifts the liquid organic (bio-oil) yield towards the biochar yield (Sohi et al. 2009). From the existing literature, it can be concluded that slow pyrolysis and intermediate pyrolysis both are suitable for higher biochar yields, whereas fast pyrolysis provides higher liquid yields, which can be suitable for biofuel production for energy. Therefore, the production of biochar can be optimized via slow pyrolysis and intermediate pyrolysis processes. The operating conditions for high biochar yields are (a) high lignin, ash, and nitrogen contents in the biomass, (b) low pyrolysis temperature (<400 °C), (c) reaction at high pressure, (d) higher residence time, (e) extended vapor/solid interaction, (f) low heating rate, (g) large biomass particle size, and (h) optimized heat transfer. These processes for the production of biochar can be carried out in batch or continuous modes depending upon production scale.

3.2.1 Physical and Thermal Characteristics and Chemical Composition of Biochar

As also mentioned earlier, the elemental composition of biochar is mainly C, H, N, S, and O. In addition, biochar also has ash content. These biochar components vary with feedstock source as well as operating conditions of the production process. An example of biochar chemical composition and calorific values at different operating pressure is shown in Table 3.1.

It is evident from this table that the elemental composition of biochar substantially changes for any change in operating parameter (e.g., pressure) for a given biomass type. This change was also found to be influential in affecting the HHV (high heating value). Moreover, Table 1 also presents biochar elemental composition due to variation in operating temperature.

3.2.2 Environmental Impact of Biochar Application to Soils

Biochar has substantial potential for soil improvement because of its unique physical, chemical, and biological properties and their interactions with soil and plant communities. Biochar characteristics (chemical composition, surface chemistry, particle and pore size distribution) as well as physical and chemical stabilization mechanisms of biochar in soils regulate the effects of biochar on soil functions (Verheijen et al. 2010a, b). Biochar addition to the soil improves soil fertility by improving the soil carbon storage, reducing the loss of N through percolating water, and fixing atmospheric N through biological nitrogen fixation. Biochar incorporation into soil is expected to enhance overall sorption capacity of soils towards anthropogenic organic contaminants (polycyclic aromatic hydrocarbons (PAHs), pesticides, and herbicides), in a mechanistically different (and stronger) way than amorphous organic matter.

Understanding and improving environmental quality by reducing soil nutrient leaching losses, reducing bioavailability of environmental contaminants, sequestering C, reducing greenhouse gas emissions, and enhancing crop productivity in highly weathered or degraded soils has been the goal of agroecosystem researchers and producers for years (Novak and Busscher 2012). The review by Spokas et al. (2012) presents a compelling introduction to the environmental impacts of biochar. In addition to holding considerable carbon sequestration potential, biochar has been shown to improve soil quality and crop yields. The positive effects have mainly been attributed to biochar's ability to absorb plant nutrients (Chan et al. 2007; Steiner et al. 2008) and to increase soil water holding capacity (Brockhoff et al. 2010). Besides, biochar can be a simple yet powerful tool to combat climate change. As organic materials decay, greenhouse gases such as carbon dioxide and methane (which is 21 times more potent as a greenhouse gas than CO_2) are released into the atmosphere. By charring the organic material, much of the carbon becomes fixed

Table 3.1 Effect of operating temperature on biochar yield and elemental composition from different wood-based biochar samples (Kim et al. 2012; Abdullah and Wu 2009, and Mahinpey et al. 2009)

Sample	Yield	M	Ash	FC	VM	C	H	N	S	O	BET (m^2/g)	HHV (kJ/kg)
Wood char—control	–	4.5	1.3	14.8	79.4	49.1	6.1	0.13	0.02	44.7	–	–
Wood char 300 °C	–	4.6	0.7	35.3	59.5	60.3	5.3	0.18	0.02	34.14	–	–
Wood char 320 °C	–	2.9	0.9	56.9	39.3	72.9	4.6	0.24	0.02	22.21	–	–
Wood char 330 °C	–	2.7	1.1	59.2	37	73.6	4.5	0.27	0.03	21.52	–	–
Wood char 400 °C	–	4.5	1.2	69.9	24.4	79.1	3.7	0.29	0.04	16.8	–	–
Wood char 450 °C	–	4	1.4	75.3	19.3	82.9	3.3	0.32	0.03	13.38	–	–
Wood char 500 °C	–	4.5	1.3	79.8	14.4	85.5	3	0.34	0.03	11.14	–	–
Collie coal	–	5.5	8.5	51.1	34.9	74	4.3	1.3	0.6	19.8	–	–
Biochar—control	–	–	4.5	–	–	48.8	6	0.2	–	45	–	–
Biochar 300 °C	60.7	–	7.9	–	–	63.9	5.4	0.3	–	30.4	2.9	–
Biochar 400 °C	33.5	–	7.7	–	–	70.7	3.4	0.4	–	25.5	4.8	–
Biochar 500 °C	14.4	–	0.3	–	–	90.5	2.5	0.3	–	6.7	175.4	–
Wheat straw—control	–	–	–	–	–	40.8	5.8	0.2	0.3	52.9	–	16,298
Wheat straw char at 10 psi	–	–	–	–	–	71.1	3.0	0.3	0.0	25.6	–	28,049
Wheat straw char at 20 psi	–	–	–	–	–	71.2	3.0	0.3	0.0	25.5	–	28,135
Wheat straw char at 30 psi	–	–	–	–	–	69.6	3.0	0.5	0.0	27.0	–	27,235
Wheat straw char at 40 psi	–	–	–	–	–	69.6	2.9	0.4	0.0	27.2	–	27,152

Yields ≡ wt%, wet basis; elemental analysis (C,H,N,S,O) ≡ wt%, dry basis; ash analysis ≡ wt%, dry basis; BET-N_2 ≡ surface area (m^2/g, dry basis) into a more stable form, and when the resulting biochar is applied to soils, the carbon is effectively sequestered (Liang et al. 2008). It is estimated that use of this method to tie up carbon has the potential to reduce current global carbon emissions by as much as 10 % (Woolf et al. 2010).

There is a large imbalance between carbon release to the atmosphere and carbon uptake by other compartments that leads to a continued increase in atmospheric CO_2. Biochar was of great importance in increasing soil carbon storage, improving soil fertility, as well as maintaining the balance of soil ecosystems, and it could act as a kind of soil fertilizer or amendment to increase crop yield and plant growth by supplying and retaining nutrients. Biochar is the carbon-rich product obtained when biomass, such as wood, manure, or leaves, is heated in a closed container with little or no available air pyrolysis. In more technical terms, biochar is produced by the so-called thermal decomposition of organic material under limited supply of oxygen (O_2) and at relatively low temperatures (<700 °C). When plant tissues are used as raw materials for biochar production, heat produced during combustion volatilizes a significant portion of the hydrogen and oxygen, along with some of the carbon contained within the plant's tissues. The remaining carbonaceous materials contain many poly-aromatic (cyclic) hydrocarbons, some of which may contain functional groups with oxygen or hydrogen. The pyrolysis process greatly affects the qualities of biochar and its potential value to agriculture in terms of agronomic performance or in carbon sequestration. Because of its macromolecular structure dominated by aromatic C, biochar is more recalcitrant to microbial decomposition than uncharred organic matter. Biochar is believed to have long mean residence times in soil, ranging from 1,000 to 10,000 years. "Biochar" is the appropriate term where charred organic matter is applied to soil in a deliberate manner, with the intent to improve soil properties. This distinguishes biochar from charcoal that is used as fuel for heat, as a filter, as a reductant in iron making, or as a coloring agent in industry.

3.2.2.1 Increases Soil Carbon Storage

Most carbon in the soil is lost as greenhouse gas (CO_2) into the atmosphere if natural ecosystems are converted to agricultural land. Soils contain 3.3 times more carbon than the atmosphere and 4.5 times more than plants and animals on earth (Goudriaan 1995). This makes soil an important source of greenhouse gases but also a potential sink if right management is applied. The use of biochar can permanently increase the soil's carbon content and establish a carbon sink for atmospheric CO_2. Biochars are stable forms of charcoal produced through a high-temperature, low-oxygen combustion process known as pyrolysis. Biochars have the potential to store carbon for very long periods. However, the amount of carbon stored in the soil is the balance between the rate at which organic matter is added and the rate at which it decomposes and returned to the atmosphere as CO_2. There is a large imbalance between carbon release to the atmosphere and carbon uptake by other compartments which leads to a continued increase in atmospheric CO_2 equivalent to a rate of 4.1×10^9 tons of carbon per year (IPCC 2007).

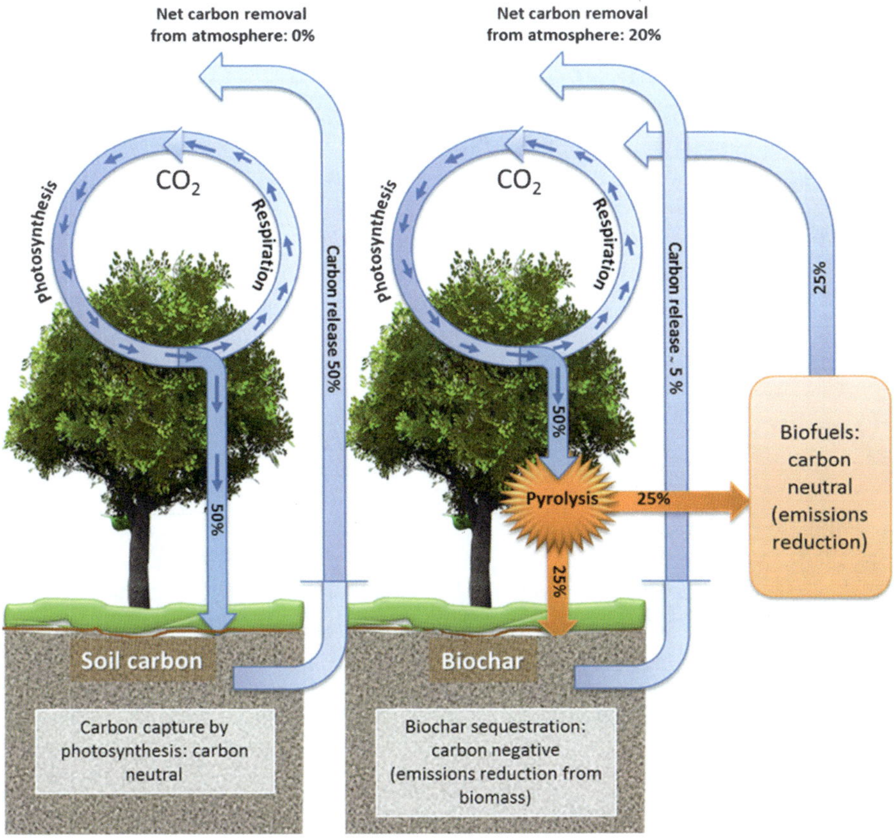

Fig. 3.3 Biochar sequestration—a carbon-negative technology (adapted from Lehmann 2007)

Biochar has twofold higher carbon content than ordinary biomass. Moreover, biochar locks up rapidly decomposing carbon in plant biomass in a much more durable form. Once biochar is incorporated into soil, it is difficult to imagine any change in practice that would cause a sudden loss of stored carbon. Plant biomass decomposes in a relatively short period of time, whereas biochar is orders of more stable magnitudes. Lehmann (2007) investigated the behavior of biochars in arable and forest soil in a greenhouse experiment in order to prove that these amendments can increase carbon storage in soils (Fig. 3.3).

3.2.2.2 Reduces Leaching of N into Groundwater

The incorporation of biochar into biosolids-amended soil mitigates nitrate leaching over the short term. This delay should be beneficial both to the environment, which receives lower nitrate loadings, and to plants, which have N held in the root zone for longer periods. Biochar can play a key role in nutrient cycling, potentially affecting nitrogen retention when applied to soils. Knowles et al. (2011) indicate in their study that by including biochar in biosolids, soil amendment can mitigate nitrate leaching from over the short term. Major et al. (2012) studied nutrient leaching in a Columbian Oxisol following a 20 Mg ha^{-1} biochar application. In general, nutrient leaching with biochar applications relative to unamended soils was greater at 0.6 m than at the 1.2 m in depth. Leaching differences were evident even though no differences in net water flux were present between the two treatments. The authors suggested that biochar may have influenced nutrient retention throughout the root zone. Schomberg et al. (2012) found that some biochars reduced N leaching losses, but soil N fractions were not increased with biochar application. Much of the apparent reductions in leaching were due to NH_3 volatilization loss from high ash biochars.

3.2.2.3 Reduction in Emissions of Greenhouse Gases (GHGs)

Soil biochar application is promoted as a climate change mitigation tool, due to its potential to increase long-term soil carbon pools and reduce greenhouse emissions. Biochars are reputed to affect soil N transformation processes, but only a few studies have tested in detail the influence of biochars on soil N_2O emissions and inorganic N leaching (Pal Singh et al. 2010; Major et al. 2009). Biochar is potentially a mitigation option for reducing the world's elevated carbon dioxide emissions, since the embodied carbon can be sequestered in the soil. Biochar also has the potential to beneficially alter soil nitrogen transformations. Adding a charred biomass material called biochar to glacial soils can help reduce emissions of the greenhouse gases carbon dioxide and nitrous oxide, according to Yanai et al. (2007) and Pal Singh et al. (2010). Biochar application to soil could affect N_2O emissions by altering soil properties (pH, aggregation, CEC) and the availability and distribution of key electron acceptors (O_2, NO_3^-) and donors (NH_4^+, dissolved organic matter), inducing catalytic reduction of N_2O to N_2 following oxidation and subsequent reactions of biochars with soil minerals and influencing microbial community structures, and microbial enzymes and processes (N mineralization–immobilization turnover, nitrification, denitrification) involved in N cycling in soil (Šimek and Cooper 2002; Yanai et al. 2007; Van Zwieten et al. 2009).

3.2.2.4 Increases Cation Exchange Capacity

The summation of values from the compulsive exchange of cations (isomorphic substitution of calcium, sodium, potassium, and magnesium) is often used as a quantitative method for determining the ability of a substrate to retain positively charged ions on its surface cation exchange capacity (CEC). This value can then give an indication of the surface charge of the substrate, depending on pH and the size of the exchanging ions (Waters et al. 2010). However, charge in soils can be both permanent and pH dependent (Bohn et al. 2001). The surface charge of substrates can also be determined by the difference between pH_{KCl} and pH_{H2O} (Black and Waring 1976). Hence, a net negative number would suggest a negative surface charge, and vice versa.

3.2.2.5 Moderating Soil Acidity

Soil acidity is a serious constraint for crop production in many regions of the world. According to Klau Berek et al. (2011), the addition of biochar to agricultural soils has recently received much attention due to the apparent benefits to soil quality and enhanced crop yields, as well as the potential of gaining carbon credits by carbon sequestration. Biochar reduces soil acidity which decreases liming needs, but in most cases does not actually add nutrients in any appreciable amount.

3.2.2.6 Increases Water Retention

Water retention of soil is determined by the distribution and connectivity of pores in the soil matrix, which is largely affected by soil texture, aggregation, and soil organic matter content (Brady and Weill 2004). Biochar has a higher surface area and greater porosity relative to other types of soil organic matter and can therefore improve soil texture and aggregation, which improves water retention in soil. These starting physical properties in biochar occur at a range of scales and affect the proportion of water that can be retained. Soil moisture retention improvement is an indirect result of alterations in soil aggregation and structure after biochar application (Brodowski et al. 2006). Tryon (1948) studied the effect of charcoal on the percentage of available moisture in soils of different textures and found different response among soils. In sandy soil, the addition of charcoal increased available moisture by 18 % after adding 45 % biochar by volume, while no changes were observed in loamy soil, and soil available moisture decreased in the clayey soil. The high surface area of biochar can lead to increased water retention, although the effect seems to depend on the initial texture of the soil. Improved water holding capacity with biochar additions is most commonly observed in coarse-textured or sandy soils (Gaskin et al. 2007; Glaser et al. 2002). The impact of biochar additions on moisture content may be due to increased surface area relative to that found in coarse-textured soils (Glaser et al. 2002). Therefore, improvements in soil water retention by biochar additions may only be expected in coarse-textured soils or soils with large amounts of macropores. Additionally, a large amount of biochar may need to be applied to the soil before it increases water retention.

3.2.2.7 Increases Number of Beneficial Soil Microbes

Biochar properties may enhance soil microbial communities and create microenvironments that encourage microbial colonization. Biochar pores and its high internal surface area and increased ability to absorb organic matter provide a suitable habitat to support soil microbiota that catalyzes processes that reduce N loss and increase nutrient availability for plants (Winsley 2007). The pores are suggested to serve as a refuge by protecting microbes from predation and desiccation, while the organic matter adsorbed to biochar provides C energy and mineral nutrient requirements (Warnock et al. 2007; Saito and Marumoto 2002).

3.2.3 As a Valuable Soil Amendment

Biochar is a solid material obtained from the carbonization of biomass. However, it may be added to soils to improve soil functions and to reduce greenhouse gas emissions from biomass. Biochar also has appreciable carbon sequestration value. These properties are measurable and verifiable in a characterization scheme, or in a carbon emission offset protocol. The application of biochar in soils is based on its agricultural value from enhanced soil nutrient retention and water holding capacity, its ability to permanent carbon sequestration, and reduced greenhouse gas emissions and methane (CH_4) release (Lehmann et al. 2006; McLaughlin 2010). Biochar is found in soils around the world as a result of vegetation fires and historic soil management practices. It can be an important tool to increase cropland diversity in areas with severely depleted soils, scarce organic resources, and inadequate water and chemical fertilizer supplies. Further, biochar improves water quality and quantity by increasing soil retention of nutrients and agrochemicals for plant and crop utilization.

3.3 The Biochar Option to Improve Plant Yields and Crop Productivity

Several papers in this special collection describe plant growth responses to biochar. Research consistently reveals that poor soils enriched with biochar grow bigger, stronger plants that yield higher crop quantity and quality. Currently, a large number of studies have been conducted where biochar application has shown significant agronomic benefits with a minor number of studies showing no significant effects of biochar application on crop productivity and some studies reporting adverse effects. Biochar increases crop productivity in many tropical soils. Yield increases have frequently been reported that are directly attributable to the addition of biochar (Lehmann et al. 2003). The reasons probably include improved water retention, reduced leaching, and better availability of nutrients to plant roots. The majority of currently published studies assessing the effect of biochar on crop yield is generally small scale, almost all short term, and sometimes conducted in pots where environmental fluctuation is removed.

Field trials with biochar application have also shown increased yields of many plants, especially where they are added with mineral fertilizers or with organic fertilizers, such as manure.

Field experiments carried out by Islami et al. (2011) reported the beneficial effects of biochar on the productivity of cassava (*Manihot esculenta Crantz*)-based cropping system in the degraded uplands of East Java. Many studies revealed that biochar seems to increase crop yields (Yamato et al. 2006; Chan et al. 2008). Although increases in cowpea and maize yield (Yamato et al. 2006), soybean (Tagoe et al. 2008), upland rice (Asai et al. 2009), and lowland rice (Masulili et al. 2010) following biochar application have been reported.

Unlike fertilizers, biochar has an extremely long life in soils. Charcoal is carbon-rich and gives it the ability to persist in the soil indefinitely by not being susceptible to biological decay. Biochar also attracts microbes and beneficial fungi and holds on to nutrients that are put into the soil. One of the major challenges in agriculture is to make the nutrients in the soil available to the plant when the plant can benefit from them. Furthermore, biochar helps conserve plant nutrients. Biochar in soil also has the ability to hold moisture and save on irrigation costs. Besides, it modifies the soil's performance by retaining moisture and making it available during periods of low precipitation and dry soil conditions (Trujillo 2002; Skjemstad et al. 2002; Lehmann 2007; Yu et al. 2009). Lehmann (2007) may think that biochar being black in color would heat up in the sun, but biochar helps the soil stay moist even in full sunlight (Lehmann et al. 2006; Lehmann 2007). Biochar also has significant impacts on soil drainage, and its addition compensates for the native soil deficiency. Biochar also makes a significant contribution to mycorrhiza by promoting microbe populations (Read et al. 2004; Rillig and Mummey 2006). It increases the availability of mycorrhiza by detoxifying soil water by adsorbing compounds that inhibit microbe growth, providing a protective habitat for microbes and improving soil moisture management in which mycorrhiza thrives.

3.3.1 Biochar and Climate Change

Climate change and global warming are two terms used for the predicted and observed increase in temperature. While the current temperature increase is caused by human influence on the earth's carbon cycle, the greenhouse effect is a naturally occurring process. Besides global warming mitigation, biochar can also be viewed from the perspective of adaptation to climate change. In a world dependent on fossil energy, it is easy to see the carbon capture benefits of biochar as offsets against current and future fossil fuel emissions. As we face the catastrophic impacts of climate change, efforts to wangle the climate are proliferating. Among these is a proposal to use soils as a medium for addressing climate change by scaling up the use of biochar. Biochar application to soils, therefore, may play both a global warming mitigation and a climate change adaptation role. Biochar is a soil supplement that sequesters carbon in the soil and thus may help to mitigate global climate change. It has the potential to curtail greenhouse gas emissions and other environmental hazards in the near term and to benefit agricultural producers as a soil amendment and source of renewable energy (Bracmort 2010). Many scientists believe there is already an unsafe excess of carbon dioxide in the atmosphere; this obligates the nations that caused the excess to abate it (Schils et al. 2008). The carbon in biochar resists degradation and can hold carbon in soils for hundreds to thousands of years. Biochar is produced through pyrolysis or gasification processes that heat biomass in the absence (or under reduction) of oxygen. In addition to creating a soil enhancer, sustainable biochar practices can produce oil and gas by-products that can be used as fuel, providing clean, renewable energy (Denman et al. 2007). When the biochar is buried in the ground as a soil enhancer, the system can become carbon negative. Biochar and bioenergy coproduction can help combat global climate change by displacing fossil fuel use and by sequestering carbon in stable soil carbon pools. It may also reduce emissions of nitrous oxide (Van Zwieten et al. 2010; Yanai et al. 2007).

3.4 Conclusions and Future Outlook Required

This paper reviewed a number of conclusions that can have practical importance for future experiments and real world application of biochar in agriculture and environment. Biochar production and application in soils has a very promising potential for the development of sustainable agriculture and climate change mitigation. It is clear that biochar has different effects in different soils and climates. However, many other benefits are attached with biochar such as reduction of greenhouse gas emissions, increase in nutrient and water retention of soils, stabilization of native soil organic matter, and decrease in the bioavailability of a range of contaminants within soils. Though a number of benefits have been identified within the literature, biochar has been found to improve agriculturally significant soil parameters such as soil pH, cation exchange capacity, and soil water holding capacity. Researchers have found the increase in these performance parameters has improved nitrogen use efficiency and therefore crop productivity in limited field trials. Further, biochar has the potential to reduce greenhouse gas emissions through carbon sequestration, as well as potentially decreasing methane and nitrous oxide emissions from the soil. The positive effects of biochar application and its potential for C-sequestration are accompanied by the production of thermal energy during the pyrolysis. Thermochemical processes have advantage over "biological only" processes where lignocellulosic biomass needs various pretreatment steps, time, and investments. At present, thermochemical processes have a foothold at commercial scale for heat and electricity. Due to the lack of research about their effect on sustainable development, future research should strive to delineate and identify what characteristics can be attributed to different impacts and outcomes. And since there are both agronomic and environmental benefits of biochar production and application in soil, implementation of agricultural schemes involving the application of biochar should first be critically evaluated.

References

Abdullah H, Wu H (2009) Biochar as a fuel: 1. Properties and grindability of biochars produced from the pyrolysis of mallee wood under slow-heating conditions. Energ Fuel 23(8): 4174–4181

Anderson CR, Condron LM, Clough TJ et al (2011) Biochar induced soil microbial community change: implications for biogeochemical cycling of carbon, nitrogen and phosphorus. Pedobiologia 54(5–6):309–20

Asai H, Samson BK, Stephan HM, Songyikhangsuthor K, Homma K, Kiyono Y, Inoue Y, Shiraiwa T, Horie T (2009) Biochar amendment techniques for upland rice production in Northern Laos 1. Soil physical properties, leaf SPAD and grain yield. Field Crop Res 111:81–84

Barrow CJ (2011) Biochar: potential for countering land degradation and for improving agriculture. Appl Geogr 34:21–8

Black AS, Waring SA (1976) Nitrate leaching and adsorption in a krasnozem from Redland Bay, QLD. II. Soil factors influencing adsorption. Aust J Soil Res 14(2):181–188

Bohn HL, McNeal BL, O'Connor GA (2001) Soil chemistry, 3rd edn. Wiley, New York

Bracmort K (2010) Biochar: examination of an Emerging concept to mitigate climate change. Congressional research service. CRS report for Congress, 24 May 2010. http://www.nationalaglawcenter.org/assets/crs/R40186.pdf

Brady NC, Weill RR (2004) Elements of the nature and properties of soils, 2nd edn. Pearson Prentice Hall, Upper Saddle River, NJ, pp 111–112

Brockhoff SR, Christians NE, Killorn RJ, Horton R, Davis DD (2010) Physical and mineral-nutrition properties of sand-based turf grass root zones amended with biochar. Agro J 102: 1627–1631

Brodowski S, John B, Flessa H, Amelung W (2006) Aggregate-occluded black carbon in soil. Eur J Soil Sci 57:539–546

Bruun EW (2011) Application of fast pyrolysis biochar to a loamy soil: effects on carbon and nitrogen dynamics and potential for carbon sequestration. PhD thesis Risø-PhD-78 (EN)—Biosystems division-Risø-DTU Marts

Chan KY, Van Zwieten L, Meszaros I, Downie A, Joseph S (2007) Agronomic values of greenwaste biochar as a soil amendment. Aust J Soil Res 45:629–634

Chan KY, Van Zwieten BL, Meszaros I, Downie D, Joseph S (2008) Using poultry litter biochars as soil amendments. Aust J Soil Res 46:437–444

Cross A, Sohi SP (2011) The priming potential of biochar products in relation to labile carbon contents and soil organic matter status. Soil Biol Biochem 43(10):2127–2134

Demirbas A (2004a) Effect of temperature and particle size on biochar yield from pyrolysis of agricultural residues. J Anal Appl Pyrolysis 721:243–8

Demirbas A (2004b) Determination of calorific values of bio-chars and pyro-oils from pyrolysis of beech trunk barks. J Anal Appl Pyrolysis 72:215–219

Denman KL, Brasseur G, Chidthaisong A, Ciais P, Cox PM, Dickinson RE, Hauglustaine D, Heinze C, Holland E, Jacob D, Lohmann U, Ramachandran S, da Silva Dias PL, Wofsy SC, Zhang X (2007) Couplings between changes in the climate system and bio geochemistry. In: Solomon S, Qin D, Manning M, Chen Z, Marquis M, Averyt KB, Tignor M, Miller HL (eds) Climate change 2007: the physical science basis. Contribution of working group I to the fourth assessment report of the intergovernmental panel on climate change. Cambridge University Press, Cambridge, United Kingdom and New York, NY, USA

Denyes MJ, Langlois VS, Rutter A, Zeeb BA (2012) The use of biochar to reduce soil PCB bioavailability to Cucurbita pepo and Eisenia fetida. Sci Total Environ 437:76–82

Duku MH, Gu S, Hagan EB (2011) Biochar production potential in Ghana—a review. Renew Sustain Energy Rev 15(8):3539–51

Fabbri D, Torri C, Spokas KA (2011) Analytical pyrolysis of synthetic chars derived from biomass with potential agronomic application (biochar). Relationships with impacts on microbial carbon dioxide production. J Anal Appl Pyrolysis 93:77–84

FAO (2012) FAOSTAT–Forestry data. http://faostat.fao.org/site/626/DesktopDefault.aspx?PageID=626#ancor. Accessed 10 Oct 2012

Galinato SP, Yoder JK, Granatstein D (2011) The economic value of biochar in crop production and carbon sequestration. Energy Policy 39(10):6344–50

Gaskin JW, Speir A, Morris LM, Ogden L, Harris K, Lee D, Das KC (2007) Potential for pyrolysis char to affect soil moisture and nutrient status of a loamy sand soil. In: Proceedings of the Georgia Water Resources Conference, 27–29 March 2007. University of Georgia, Athens, GA

Gaunt JL, Lehmann J (2008) Energy balance and emissions associated with biochar sequestration and pyrolysis bioenergy production. Environ Sci Technol 42(11):4152–8

Glaser B, Lehmann J, Zech W (2002) Ameliorating physical and chemical properties of highly weathered soils in the tropics with charcoal—a review. Biol Fertil Soils 35:219–230
Goudriaan (1995) In: Beran MA (ed) Carbon sequestration in the biosphere. Processes and prospects, vol 33, Springer, Heidelberg, Berlin, p 3–18
IPCC (2007) Climate change 2007: the physical science basis. Contribution of working group I to the fourth assessment report of the intergovernmental panel on climate change. Cambridge University Press, Cambridge, 996 pp
Islami T, Guritno B, Basuki N, Suryanto A (2011) Biochar for sustaining productivity of cassava based cropping systems in the degraded lands of East Java, Indonesia. J Trop Agr 49(1–2): 40–46
Karaosmanoglu F, Isigigur-Ergundenler A, Server A (2000) Biochar from the straw stalk of rapeseed plant. Energ Fuel 14:336–339
Kim KH, Kim J-Y, Cho T-S, Choi JW (2012) Influence of pyrolysis temperature on physicochemical properties of biochar obtained from the fast pyrolysis of pitch pine (Pinus rigida). Bioresour Technol 118:158–62
Klau Berek A, Hue N, Ahmad A (2011) Beneficial use of biochar to correct soil acidity. Hānai'Ai/ the food provider. University of Hawaii (Sept-Oct-Nov 2011)
Knowles OA, Robinson BH, Contangelo A, Clucas L (2011) Biochar for the mitigation of nitrate leaching from soil amended with biosolids. Sci Total Environ 409(2011):3206–3210
Krull ES, Lehmann J, Skjemstad J, Baldock J (2008) The global extent of black C in soils; is it everywhere? In: Schroder HG (ed) Grasslands; ecology, management and restoration. Nova Science Publishers, Inc, New York, pp 13–17
Laird DA, Brown RC, Amonette JE, Lehmann J (2009) Review of the pyrolysis platform for coproducing bio-oil and biochar. Biofuels Bioproducts Biorefining-Biofpr 3:547–562
Lehmann J (2007) Bio-energy in the black. Front Ecol Environ 5:381–387
Lehmann J, Joseph S (2009) Biochar for environmental management: science and technology. Earthscan, London
Lehmann J et al (2003) Nutrient availability and leaching in an archaeological Anthrosol and a Ferralsol of the Central Amazon basin: fertilizer, manure and charcoal amendments. Plant Soil 249:343–357
Lehmann J, Gaunt J, Rondon M (2005) Bio-char sequestration in soil: a new frontier. http://soilcarboncenter.kstate.edu/conference/PowerPoint_files/Lehmann_Baltimore_05_files/frame.htm
Lehmann J, Gaunt J, Rondon M (2006) Biochar sequestration in terrestrial ecosystems. A review. Mit Adapt Strat Global Chang 11:403–427
Lehmann J, Rillig MC, Thies J, Masiello CA, Hockaday WC, Crowley D (2011) Biochar effects on soil biota—a review. Soil Biol Biochem 43(9):1812–1836
Liang BJ, Lehmann D, Solomon S, Sohi JE, Thies JO, Skjemstad FJ, Luizao F, Engelhard MH, Neves EG, Wirick S (2008) Stability of biomass derived black carbon in soils. Geochimicaet Cosmochimica Acta 72:6096–6078
Mahinpey N, Murugan P, Mani T, Raina R (2009) Analysis of bio-oil, biogas, and biochar from pressurized pyrolysis of wheat straw using a tubular reactor. doi:10.1021/ef8010959
Major J, Steiner C, Downie A, Lehmann J (2009) Biochar effects on nutrient leaching. In: Lehmann J, Joseph S (eds) Biochar for environmental management: science and technology. Earthscan, London, UK, pp 271–287
Major J, Rondon M, Molina D, Riha SJ, Lehmann J (2012) Nutrient leaching in a Colombian savanna Oxisol amended with biochar. J Environ Qual 41:1076–1086
Mašek O, Brownsort PA (2010) Research on production of bespoke biochar. Poster presented in the 2nd UKBRC conference, Rothamsted, UK
Masek O, Brownsort P, Cross A, Sohi S (2011) Influence of production conditions on the yield and environmental stability of biochar. Fuel. doi:10.1016/j.fuel.2011.08.044
Masulili A, Utomo WH, Ms S (2010) Rice husk biochar for rice based cropping system in acid soil. The characteristics of rice husk biochar and its influence on the properties of acid sulfate soils and rice growth in West Kalimantan, Indonesia. J Agr Sci (Canada) 3:25–33

McLaughlin H (2010) Biochar and energy linkages. In: Biochar and energy co-products, assessment of biochar's benefits for the United States of America, June 2010, U.S.-focused biochar report, www.biochar-us.org/pdf%20files/biocharreportlowres.pdf
Morley J (1927) Following through with grass seeds. Natl Greenkeeper 1(1):p15
Morley J (1929) Compost and charcoal. Natl Greenkeeper 3(9):8–26
Novak JM, Busscher WJ (2012) Selection and use of designer biochars to improve characteristics of southeastern USA Coastal Plain degraded soils. In: Lee JE (ed) Advanced biofuels and bio-products. Springer Science, New York
O'Neill B, Grossman J, Tsai MT, Gomes JE, Lehmann J, Peterson J, Neves E, Thies JE (2009) Bacterialcommunity composition in Brazilian Anthrosolsand adjacent soils characterized using culturing andmolecular identification. Microb Ecol 58:23–35
Read DJ, Leake JR, Perez-Moreno J (2004) Mycorrhizal fungi as drivers of ecosystem processes in heath land and boreal forest biomes. Can J Bot 82:1243–1263
Retan GA (1915) Charcoal as a means of solving some nursery problems. Forest Q 13:25–30
Rillig MC, Mummey DL (2006) Mycorrhizas and soil structure. New Phytol 171:41–53
Roberts KG, Gloy BA, Joseph S, Scott NR, Lehmann J (2010) Life cycle assessment of biochar systems: estimating the energetic, economic and climate change potential. Environ Sci Technol 44:827–833
Saito M, Marumoto T (2002) Inoculation with arbuscular mycorrhizal fungi: the status quo in Japan and the future prospects. Plant Soil 244:273–279
Santiago A, Santiago L (1989) Charcoal chips as a practical substrate for container horticulture in the humid tropics. Acta Horticulturae 238:141–147
Schils et al. (2008). Review of existing information on the interrelationships between soil and climate change. The CLIMSOIL report. http://ec.europa.eu/environment/soil/review_en.htm
Schmidt MWI, Noack AG (2000) Blackcarbon in soils and sediments: analysis, distribution, implications, and current challenges. Global Biogeochem Cy 14:777–794
Schomberg HH, Gaskin JW, Harris K, Das KC, Novak JM, Busscher WJ, Watts DW, Woodroof RH, Lima IM, Ahmedna M, Rehrah D, Xing B (2012) Influence of biochar on nitrogen fractions in a coastal plain soil. J Environ Qual 41:1087–1095
Shackley S, Sohi SP (eds) (2010) An assessment of the benefits and issues associated with the application of biochar to soil. report to the department for environment. Food and Rural Affairs and the Department of Energy and Climate Change, London
Shackley S, Carter S, Knowles T et al (2011) Sustainable gasification-"biochar systems? A case-study of rice-husk gasification in Cambodia, Part I: context, chemical properties, environmental and health and safety issues. Energy Policy 42:49–58
Šimek M, Cooper JE (2002) The influence of soil pH on denitrification: progress towards the understanding of this interaction over the last 50 years. Eur J Soil Sci 53:345–354
Singh BP, Hatton BJ, Singh B, Cowie AL (2010) The role of biochar in reducing nitrous oxide emissions and nitrogen leaching from soil. 19th World Congress of soil science, soil solutions for a changing World, Brisbane, Australia, 1–6 Aug 2010
Singhal SK, Sharma VK, Pandey RN (2011) Biochar: a charred organic matter and its importance in soil on January 2 2011. Soil Sci. Admin Division of Soil Science and Agricultural Chemistry, IARI, Pusa, New Delhi
Skjemstad JO et al (2002) Charcoal carbon in U.S. agricultural soils. Soil Sci Soc Am J 66: 1249–1255
Smith NJH (1980) Anthrosols and human carrying capacity in Amazonia. Ann Assoc Am Geogr 70:553–566
Sohi S, Loez-Capel SE, Krull E, Bol R (2009) Biochar's roles in soil and climate change: a review of research needs. CSIRO, land and water science report 05/09, February; 2009. p 64. www.ias.ac.in/currsci/10nov2010/1218.pdf
Spokas KA, Cantrell KB, Novak JM, Archer DA, Ippolito JA, Collins HP, Boateng AA, Lima IM, Lamb MC, McAloon AJ, Lentz RD, Nichols KA (2012) Biochar: a synthesis of its agronomic impact beyond carbon sequestration. J Environ Qual 41:973–989

Steiner C, Das K, Garcia M, Forster B, Zech W (2008) Charcoal and smoke extract stimulate the soil microbial community in a highly weathered xanthic ferralsol. Pedobiologia 51:359–366

Tagoe SO, Takatsugu Horiuchi T, Matsui T (2008) Effects of carbonized and dried chicken manures on the growth, yield, and N content of soybean. Plant Soil 306:211–220

Trimble WH (1851) On charring wood. Plough Loom Anvil 3:513–516

Trujillo L (2002) Fluxos de nutrientes em solo de pastagem abandonada sob adubacao organica e mineral na Amazonia central. M.Sc. thesis, INPA and University of Amazonas, Brazil

Tryon EH (1948) Effect of charcoal on certain physical, chemical, and biological properties of forest soils. Ecol Monogr 18:81–115

Van Zwieten L, Singh B, Joseph S, Kimber S, Cowie A, Chan KY (2009) Biochar and emissions of non-CO_2 greenhouse gases from soil. In: Lehmann J, Joseph S (eds) Biochar for environmental management: science and technology. Earthscan, London, UK, pp 227–249

Van Zwieten L, Kimber S, Morris S, Downie A, Berger E, Rust J, Scheer C (2010) Influence of biochars on flux of N_2O and CO_2 from Ferrosol. Aust J Soil Res 48:555–568

Verheijen F, Jeffery S, Bastos AC, van der Velde M, Diafas I (2010) Biochar application to soils: a critical scientific review of effects on soil properties, processes and functions. European Commission. http://eusoils.jrc.ec.europa.eu/esdbarchive/eusoilsdocs/other/EUR24099.pdf

Verheijen F, Jeffery S, Bastos AC, Van Der Velde M, Diafas I (2010) Biochar application to soils—a critical scientific reviewof effects on soil properties, processes and functions. JRC Scientific and technical reports. http://publications.jrc.ec.europa.eu/repository/bitstream/111111111/13558/1/jrc_biochar_soils.pdf

Verma M, Godbout S, Brar SK, Solomatnikova O, Lemay SP, Larouche JP (2012) Biofuels production from biomass by thermochemical conversion technologies. Int J Chem Eng 542426:18. doi:10.1155/2012/542426

von Liebig J (1878) Chemische briefe. C. F. Winter'sche Verlagshandlung, Leipzig and Heidelberg, Germany

Warnock DD, Lehmann J, Kuyper TW, Rillig MC (2007) Mycorrhizal responses to biochar in soil—concepts and mechanisms. Plant Soil 300:9–20

Waters D, Condon J, Van Zwieten L, Moroni S (2010) Biochar-ion interactions: an investigation of biochar charge and its effect on ion retention. 19th World Congress of soil science, soil solutions for a changing World, Brisbane, Australia, 1–6 Aug 2010

Winsley P (2007) Biochar and bioenergy production for climate change mitigation. New Zeal Sci Rev 64:5–10

Woolf D, Amonette JE, Street-Perrott FA, Lehmann J, Joseph S (2010) Sustainable biochar to mitigate global climate change. Nat Commun 1:1–9, www.nature.com/ncomms/journal/v1/n5/full/ncomms1053.html

Yamato M, Okimori Y, Wibowo IF, Anshori S, Ogawa M (2006) Effects of the application of charred bark of Acacia mangium o n the yield of maize, cowpea and peanut, and soil chemical properties in South Sumatra, Indonesia. J Soil Sci Plant Nutr 52:489–495

Yanai Y, Toyota K, Okazaki M (2007) Effects of charcoal addition on N2O emissions from soil resulting from rewetting air-dried soil in short-term laboratory experiments. Soil Sci Plant Nutr 53:181–188

Yoder J, Galinato S, Granatstein D, Garcia-Perez M (2011) Economic tradeoff between biochar and bio-oil production via pyrolysis. Biomass Bioenergy 35(5):1851–62

Yu XY, Ying GG, Kookana RS (2009) Reduced plant uptake of pesticides with biochar additions to soil. Chemosphere 76:665–671

Part II
Bioactive Secondary Metabolites

Chapter 4
Microbial Pigments

Júlio C. De Carvalho, Lígia C. Cardoso, Vanessa Ghiggi, Adenise Lorenci Woiciechowski, Luciana Porto de Souza Vandenberghe, and Carlos Ricardo Soccol

4.1 Introduction

There are three main sources of color additives for foods, drugs, and cosmetics: (1) synthetic colors, (2) plant-derived pigments, and (3) microbial pigments. In chemical terms, soluble colored substances are colorants and insoluble are pigments; however, in biological context, the colored substances are called pigments irrespective of its solubility. Although the term "biopigment" has a bit of redundancy, it is used to refer pigments of natural origin.

Microbial pigments or biopigments are multitude of chemical structures capable of absorbing light in the visible range (400–700 nm). There is an ever-growing number of biopigments. These molecules may possess other properties, which may or may not be compatible with the industrial use: vitamins riboflavin or β-carotene, antioxidants as most carotenoids and xanthophylls, and antimicrobial activity of some fungal polyketides. This chapter presents an overview of microbial pigments as potential food and drug color additives, presenting a brief description of the origin of their color and the physiological role of pigments in microorganisms, followed by a prospect of their use and a final section with representative classes of biopigments which may be produced using agro-industrial wastes.

J.C. De Carvalho (✉) • V. Ghiggi • A.L. Woiciechowski
L.P. de S. Vandenberghe • C.R. Soccol
Department of Biotechnology and Bioprocess Engineering,
Federal University of Paraná—UFPR, Rua Cel. Francisco H dos Santos,
100, Caixa Postal 19060, ACF Centro Politécnico, 81531-980 Curitiba, PR, Brazil
e-mail: jccarvalho@ufpr.br

L.C. Cardoso
Positivo University, Curitiba, PR, Brazil

S.K. Brar et al. (eds.), *Biotransformation of Waste Biomass into High Value Biochemicals*,
DOI 10.1007/978-1-4614-8005-1_4,

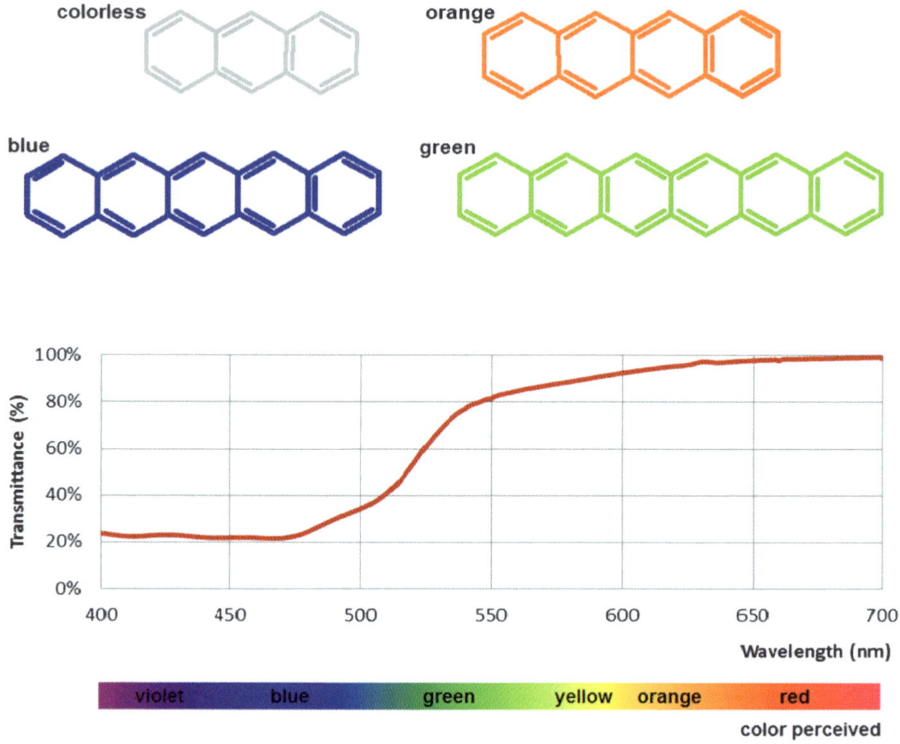

Fig. 4.1 The number of conjugated bonds (molecular structures on *top*) affects the absorption band and the color of the pigments; the *graphic* shows the transmittance spectrum for *Monascus* pigments, which appear *red*. *Sources*: Chemical structures from The Merck index, 2006; transmittance curve obtained at laboratory

4.1.1 The Origin of Color

When light interacts with matter, there may be absorption, reflection, refraction, and even reemission depending on the wavelength of incident light, chemical composition, and physical structure of the material giving rise to the multitude of colors that we see. A material may absorb the incident light unspecifically or selectively: if the absorption is unspecific, we perceive the color of the material as white, gray, or black, while if the absorption of one or more wavelengths is more pronounced, we perceive the material as having the complimentary color of that absorbed. Figure 4.1 illustrates the relationship between structure and color: molecular orbitals absorb and reemit light, and substances with multiple conjugated double bonds (a common trait in organic colored substances) tend to do so in the visible range (Nassau 2003; Meléndez-Martínez et al. 2007). Color may also arise or be modified on interaction of transition metal ions in complexes, as in porphyrin rings in hemoglobin and chlorophyll.

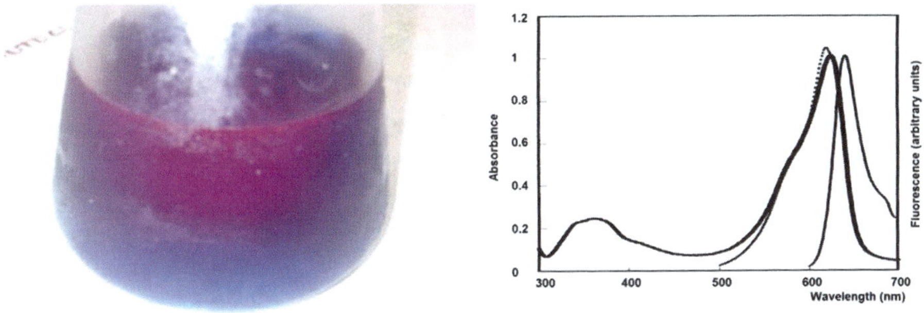

Fig. 4.2 Aqueous solution of phycocyanin from cyanobacteria and its absorption and emission spectra (*Sources*: photograph from Walter et al. 2011; spectra from Tooley et al. 2001)

Electrons in molecular orbital's absorb photons, which leap to higher energetic states and revert towards its fundamental state by releasing the energy, perhaps with several smaller leaps and radiations of different wavelengths. The wavelengths reemitted are usually outside the visible range, the effect being the net and selective absorption of visible light. However, if the reemission occurs in the visible range, we have a fluorescent pigment: for example, phycocyanin absorb mostly green, yellow, and orange light and reemit a bit of red light as can be seen in Fig. 4.2.

The part of a molecule responsible for light absorption is called a chromophore. Despite the enormous variety of biopigments, some recurring structures appear in nature and are illustrated in Fig. 4.3. There are a number of biological roles for these molecules, the most important ones being (a) their antioxidant nature, (b) their use as antennae for energy absorption, and (c) reserve substances.

Several microbial pigments are powerful antioxidants because their conjugated systems are susceptible to electrophilic attack. That is the case for carotenoids and xanthophylls, which are generally several times more efficient than ascorbic acid or butyl hydroxyl toluene (BHT) as antioxidants. Colored substances may also act as a sunscreen, protecting the cell by absorbing UV radiation and thus reducing the formation of DNA-damaging free radicals. Photosynthetic microorganisms such as cyanobacteria and microalgae rely on pigments such as chlorophylls and phycobilins for transferring light energy to electrons which will be used for carbon reduction in photosynthesis—a mechanism which produced the oxygen in our atmosphere and is in the base of virtually all food chains. Some biopigments such as phycobilins, chlorophyll, and prodigiosin may also act as nitrogen reserve in microorganisms. Besides these functions, there are several other cases in which light-absorbing molecules play an important role, such as in eyespots of microorganisms and in light-activated response mechanisms, such as the circadian rhythms of *Neurospora* sp. or as UV-induced damage correction mechanisms. Whenever these light-absorbing molecules selectively absorb in visible range, the result is a colored substance.

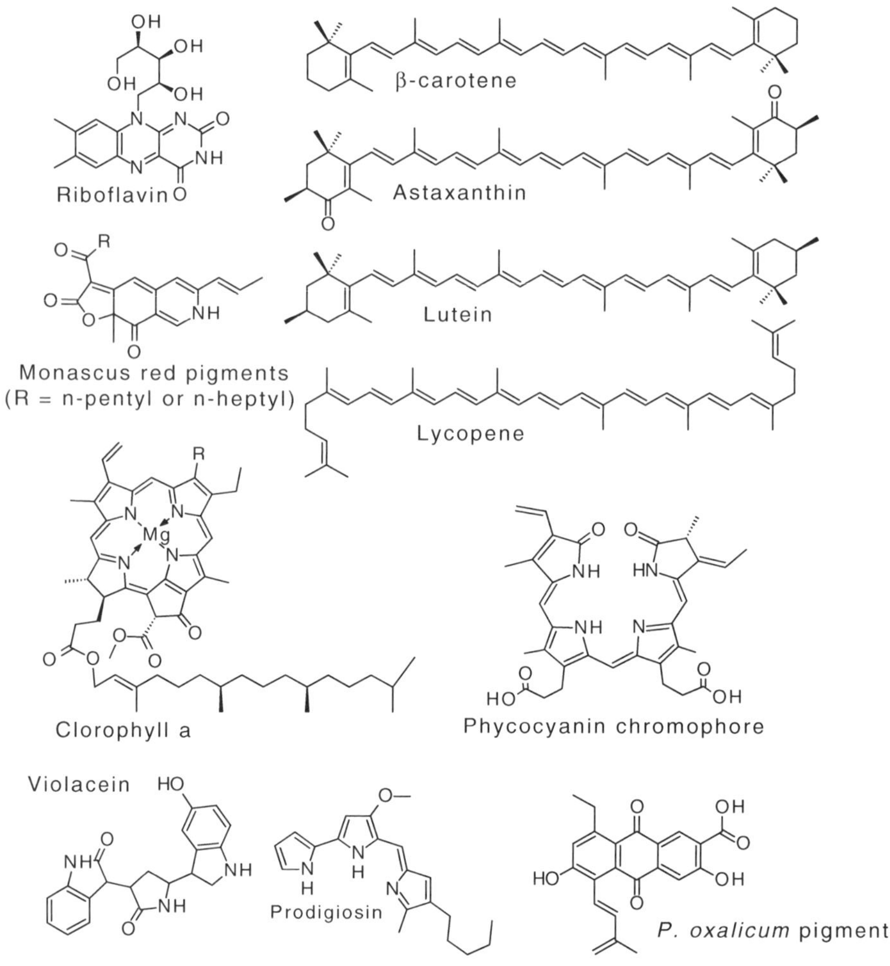

Fig. 4.3 Common microbial pigment core structures

4.1.2 Current Use of Biopigments

Several foods are naturally strongly colored and may be processed into extracts or powders used as food colors. That is the case of tomato-based lycopene, beet powder, paprika, and carrot oil. However, several nonfood plants are used for FD&C (food, drug, and cosmetic) colors, such as alfalfa for chlorophyll, marigold for lutein, annatto for bixin and norbixin, and cochineal (this, an insect) for carmine. Table 4.1 lists the natural color additives approved by the FDA (United States Food and Drug Administration) for food and feed use, excluding the mineral pigments. The pigments in the list are exempt from certification and may be used according to specific legislation for each type of food product (in general, *ad quantum satis* or as much as needed according to current good manufacturing practices).

Table 4.1 Color additives approved by the FDA for use in human food

Straight color	Uses and restrictions
Algae meal, dried	Chicken feed only
Annatto extract	GMP
Astaxanthin	Salmonid fish food only
Beet juice or powder	GMP
Canthaxanthin	Foods, salmonid fish feed, broiler chicken feed
Caramel	GMP
β-Apo-8′-carotenal	Foods and feeds
β-Carotene, natural and synthetic	GMP
Carmine or cochineal extract	GMP
Carrot oil	GMP
Corn endosperm oil	Chicken feed only
Copper chlorophyllin, sodic	Citrus-based dry beverage mixes
Cottonseed flour, toasted partially defatted cooked	GMP
Ferrous gluconate or lactate	Ripe olives
Fruit juice	GMP
Grape color extract	Non-beverage food only
Grape skin extract (enocianina)	Beverages and beverage bases
Haematococcus algae meal	Salmonid fish feed only
Lycopene, tomato extract, or concentrate	GMP
Paprika and paprika oleoresin	GMP
Paracoccus pigment	Salmonid fish feed only
Xanthophyllomyces dendrorhous (*Phaffia*) yeast	Salmonid fish feed only
Riboflavin	GMP
Saffron	GMP
Tagetes (Aztec marigold) meal and extract	Chicken feed only
Turmeric and turmeric oleoresin	GMP
Vegetable juice	GMP

From: FDA (2012). GMP, "good manufacturing practices," vary with the class of food

Biopigments are usually more susceptible to chemical attack than their synthetic counterparts (Aberoumand 2011) and may not resist the processing or the intended shelf life in some formulated products. However, this very limitation may be a reflex of a desirable trait at least in some cases: for example, the molecules would be easily degraded in the body.

There is a high degree of homology among eukaryotic metabolism, and animals are expected to manage diverse metabolites from fungi, yeast, and algae (with some notable exceptions such as mycotoxins). Most of the natural pigments are easily oxidized (e.g., carotenoids), hydrolyzed (e.g., phycobilins), or excreted due to its aqueous solubility (e.g., riboflavin). Table 4.2 shows the most important natural food colors derived from microorganisms, its origin, and possible metabolic roles. While some pigments such as astaxanthin are merely healthy, others such as those with provitamin A activity (β-carotene and β-cryptoxanthin) are actually essential for human nutrition.

Table 4.2 Microbial production of pigments in use as natural food colorants or with technology well developed and possible metabolic roles

Molecule (color)	Microorganism	Metabolic role
Lutein (yellow)	*Spongiococcum excentricum*	Antioxidant, may help slow macular degeneration
Ankaflavin (yellow)	*Monascus* sp.	Antimicrobial
Anthraquinoid (red)	*Penicillium oxalicum*	
Astaxanthin (salmon)	*Haematococcus pluvialis*	Antioxidant
	Xanthophyllomyces dendrorhous	
	Paracoccus carotinifaciens	
β-carotene (orange)	*Blakeslea trispora*	*Provitamin-A activity, antioxidant*
	Dunaliella salina	
Monascorubramin (red)	*Monascus* sp.	Antimicrobial
Phycocyanin (blue)	*Arthrospira platensis*	*Antioxidant*
Riboflavin (yellow)	*Ashbya gossypii*	*Vitamin B2*
Rubropunctatin (orange)	*Monascus* sp.	Antimicrobial
Lycopene (red)	*Blakeslea trispora*	Antioxidant

Sources: adapted from Nelis and De Leenheer (1989); Margalith (1992); Soni (2007)

For application in processed products, a biopigment must be formulated (which will be discussed at the end of the chapter) to be stable enough in its life cycle. For products with low water activity such as liquors, drugs, and several cosmetics, stability is hardly an issue, however in moist formulations two problems arise: first, the pigment must be adequately dispersed—and lipophilic pigments require a suitable vehicle for dispersion—and, second, the chromophores may behave differently in different pHs, just as happens with fruit anthocyanins. Finally, some pigments may require the addition of an antioxidant in the formulation, or a protective barrier (such as specific packaging) in the processed food, but this is an issue already addressed because of the other sensitive chemical components of foods and cosmetics.

The market for food biopigments grows steadily despite of economic turmoil. Estimated as a 35 million USD market in the late 1980s, 250 million in 2000, and 600 million in 2011 (Yarnell 2012), biopigments grew from an 11 % niche of food colors in 1987 to a 27 % share in 2000. When dried foods (vegetables and microalgae) and nature-identical carotenoids used for fish and poultry feeds are included, the market surpasses 1.2 billion, with an annual growth of 2.3 % (BBC Research 2010). The steep rise in the biopigment market may be attributed to a mix of more stringent regulations regarding synthetic colors, the development of new pigment formulations, and the increase in fish farming. Considering the industrialization and consumption of processed food by countries like China, India, and Brazil, it would not be an over estimation that the biopigment market will grow to over 1.5 billion USD by 2020.

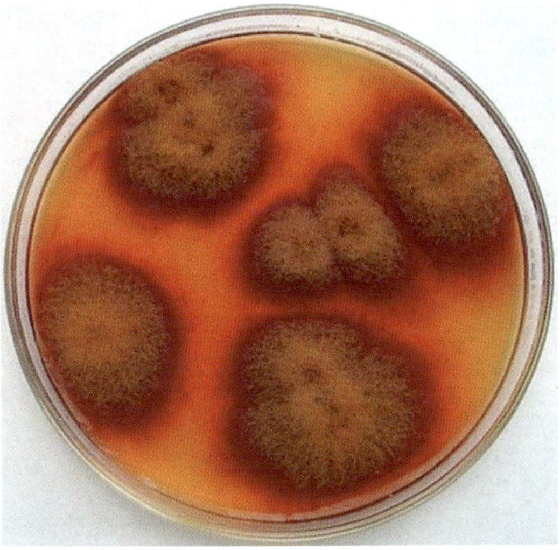

Fig. 4.4 Monascus colonies in an YM agar plate. Mycelium is colored and pigments diffuse through the medium

4.1.3 Why Microbial Pigments?

There are several natural pigments derived from vegetables and animals, in some cases from agro-industrial wastes such as tomato (lycopene), grape (anthocyanins), and palm (carotenoids) processing residues. However, source variability and presence in low concentration of pigment in those target fruits require processing of large amounts of agro-industrial waste. On the other hand, the use of selected microorganisms which are able to synthesize specific pigments has the advantages of consistent batches and high concentration and quality. Besides these advantages, microorganisms may be selected or modified in search of suitable color additives.

The selection of color-producing microorganisms is straightforward: the observation of colored colonies in agar plates. A diffusion halo is formed if the pigment is liberated to the medium and is at least partially water soluble (Fig. 4.4). The isolation of new microorganisms may be directed towards an acid stable pigment by controlling the pH of the isolation medium.

Further mutagenesis may lead to enhanced production. After initial isolation, the microorganism must be cultivated in a suitable medium so that the pigment may subsequently be isolated in order to determine eventual biological activity and finally be identified by LC-MS. In order to enhance the production of pigments, the careful evaluation of the effect of several substrates, macro- and micronutrients, as well as pH and temperature is done. The comprehension of the metabolic pathways leading to the target molecule is also important, for these pathways may be manipulated either by using inductors or repressors in the culture media or by gene knockout or promotion.

Despite of possible usefulness as bioactive compounds, the antibiotic pigments have much restricted use, and the route to market will be far more complicated. For example, *Monascus* pigments have been used for centuries in the Western countries but are still not regulated (therefore not permitted) in the USA and Europe (Puttananjaiah et al. 2011). Antioxidant pigments, at the other side, are generally welcome and even if the molecule is new, there is the possibility of its use as a nutraceutical ingredient, although petitions to regulatory agencies still will have to be done for FD&C compliance. This is easier with pigments derived from GRAS microorganisms, such as phycocyanin from *Spirulina* sp. Table 4.2 shows the most important microbial biopigment either in production or in late development state.

4.2 Biopigment Production Cases

With hundreds of microbial genera capable of producing biopigments, there are only a handful of substances which are industrially produced, as shown in Table 4.2. For these biopigments, technological and regulatory barriers have been transposed, which does not mean that there is no space for further development. Actually, these are the most likely platforms to be used at agro-industrial valorization initiatives, from which new or improved strains and products may progressively be developed. This section describes some cases selected by microorganism type, i.e., the production using yeasts, microalgae and cyanobacteria, filamentous fungi, and bacteria, for selected pigments.

4.2.1 Carotenoids from Yeast and Fungi

The color and nutraceutical properties of carotenoids have attracted the attention of food, cosmetic, and pharmaceutical industries. The food industry uses this kind of molecule as natural food colorants, as dietary supplements, and fortified foods (Vílchez et al. 2011). For example, β-carotene, the carotenoid with the largest market share can be found in margarine, cheeses, and fruit juices. The carotenoids are also used in pharmaceutical and cosmetic due to their nutraceutical properties.

Carotenoids are naturally found in bacteria and fungi (Table 4.3) and microalgae (Table 4.4). These important natural pigments have colors ranging from yellow to red (Perez-Fons et al. 2011). More than 600 different structures of these biomolecules, synthesized in vegetables and microorganisms, have been characterized.

It is clear, from Table 4.3, that the productivity of biopigments is quite diverse. For comparison, the vegetative cycle of carrots is around 100 days, with 70 mg/kg of β-carotene and for watercress 50–70 days, with 60 mg/kg of β-carotene. The slowest growing microorganism from Table 4.3 has a 5-day cycle (considering 10 % inoculum), with a final carotenoid concentration of 250 mg β-carotene/kg of biomass, on average.

Table 4.3 Comparison of microbial carotenoid production using different waste substrates as carbon source

Microorganism	Molecule	Culture medium	X_{max} (g/L)	P_{max} (mg/L)	Conc. (mg/g)	$\sim\mu_x$ (h^{-1})	References
Blakeslea trispora (fungus)	β-carotene	Corn steep liquor	20	800	40	0.022	Papaioannou and Liakopoulou-Kyriakides (2010)
Blakeslea trispora (fungus)	β-carotene	Whey	8	1,360	170	0.023	Varzakakou et al. (2010)
Sporobolomyces roseus (yeast)	β-carotene	Reconstituted whey	4.71	2.58	0.55	–	Marova et al. (2012)
Rhodopseudomonas palustris (bacterium)	β-carotene	Sodium succinate	2.58	1.78	0.69	–	Kuo et al. (2012)
Rhodotorula glutinis (yeast)	β-carotene	Potato extract	5.70	1.08	0.19	–	Marova et al. (2012)
Dietzia natronolimnaea (bacteria)	Canthaxanthin	Whey	3.29	2.87	0.87	0.020	Khodaiyan et al. (2008)
Phaffia rhodozyma (yeast)	Astaxanthin	Cassava residues	8.6	2.98	0.35	0.060	Yang et al. (2011)
Sporobolomyces ruberrimus (yeast)	Torularhodine	Technical glycerol	30	3.7	0.12	0.040	Razavi and Marc (2006)

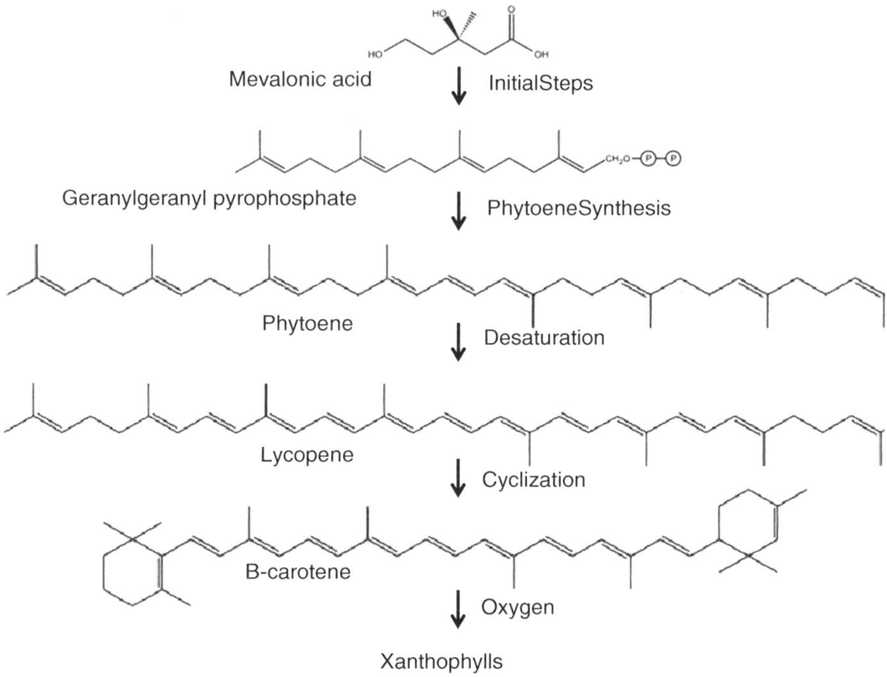

Fig. 4.5 Carotenoid biosynthesis pathway (Silva 2004)

The sequence of carotenoid biosynthesis in microorganisms is summarized in Fig. 4.5. The biosynthesis starts with the mevalonic acid that by different reactions (initial steps) produces the geranylgeranyl pyrophosphate (GGPP). Two molecules of GGPP are condensed to synthesize phytoene, which is transformed into lycopene through few steps of desaturation. β-carotene is finally formed from lycopene through cyclization. Oxygenation of carotenoids gives the xanthophylls such as astaxanthin (Wang et al. 2007).

There are several ways to improve carotenoid synthesis in fungi and yeasts, such as addition of inducing substances and light stimulation. But since these organisms are heterotrophs, culture media optimization is by far the most important aspect for a production system. Nowadays, the challenge is to reduce the production costs of carotenoids from bioprocesses. The use of cheap industrial by-products as nutrient sources and use of a microorganism with high carotenoid yield can contribute to the minimization of production cost (Tinoi et al. 2005). As showed on Table 4.3, carotenoids can be produced by microorganisms able to use different kinds of waste substrates as carbon sources. Grape must, glucose syrup, beet molasses, sugar cane molasses, soybean flour extract, technical glycerol, and whey are examples of by-products that have been reported in the literature as low-cost substrates for carotenoid production (Buzzini and Martini 1999; Razavi and Marc 2006; Khodaiyan et al. 2008; Papaioannou and Liakopoulou-Kyriakides 2010; Yang et al. 2011).

Marova et al. (2012) have shown in their work the ability of the yeast *Rhodotorula glutinis* to use industrial waste (whey) for high-value β-carotene production.

Table 4.4 Selected carotenoid-producing algae

Microalga	Carotenoid	Medium[a]	*X* (g/L)	Conc. (mg/g)	Maximum specific growth rate	References
Chlorella zofingiensis	Astaxanthin	BBM with glucose	10.2	1	0.031 h^{-1}	Ip and Chen (2005)
Coelastrella striolata	Canthaxanthin Astaxanthin *β*-carotene	BBM	2.7	47.5 1.5 7	0.30 day^{-1}	Abe et al. (2007)
Coccomyxa onubensis	*β*-carotene Lutein	K9	1.6	2.88 6.48	0.50 day^{-1}	Vaquero et al. (2012)
Haematococcus pluvialis	Astaxanthin	BBM	2.2	13.5	–	Harker et al. (1996)
Chlorella zofingiensis	Astaxanthin	Bristol, modified	10	1.25	0.043 h^{-1}	Ip et al. (2004)
Dunaliella salina	*β*-carotene	f2	–	14[b]	0.55 day^{-1}	Kleinegris et al. (2011)
Haematococcus pluvialis	Astaxanthin	Standard	3	12–15	0.56 day^{-1}	Garcıa-Malea et al. (2005)
Muriellopsis sp.	Lutein	Arnon, modified	5.37	6.51	0.17–0.23 h^{-1}	Del Campo et al. (2000)
H. pluvialis (wild type)	Astaxanthin	NIES medium	1.6	47.62	0.07	Hong et al. (2012)
H. pluvialis (mutant)			2.25	54.78	0.08 h^{-1}	

[a]Except where specified, these are mineral-based media. Recipes may be found at UTEX, SAG, or CCMP collections web sites
[b]Estimated. The original reference reports 28.1 mg/L carotenoids

Another source of carotenoids from yeast may be as by-products of nutraceutical oil production, although the fractionation of the mixture may be challenging.

The production of β-carotene by *Blakeslea trispora*, which is one of the best known processes today, was at a time elusive: carotenoids after conversion into pheromones are used as basis for communication in this fungus, and the use of two mating types is the key for the production of the large amounts of carotenoids (Papaioannou and Liakopoulou-Kyriakides 2010). However, microalgal carotenoid productivity may surpass that of fungi, as is discussed in the next section.

4.2.2 *Carotenoids from Microalgae*

Carotenoids are synthesized by microalgae for photoprotection, oxidation-protective agent, and as part of the light-harvesting complexes for photosynthesis. Being ubiquitous in microalgae, carotenoids are even used as a primary classification key for genera. Table 4.4 presents cases of carotenoid-producing microalgal cultures.

Although β-carotene holds the largest market for carotenoids, astaxanthin—a high-value keto-carotenoid pigment—is increasingly being used as feed additive in aquaculture. Farmed salmonid production was almost nonexistent in 1980 but reached 2.41 million tons in 2010 and had an average increment of 3.8 % per year in the previous 8 years (data from FAO 2012). This represents a huge market for fish feed, in which astaxanthin is included in order to confer the attractive coloration of these fish, and helps maintain their normal growth and survival (Shen et al. 2009). In addition, the strong antioxidative activity of astaxanthin over other carotenoids such as β-carotene, zeaxanthin, and lutein has attracted tremendous commercial interest for medicinal and nutraceutical uses (Miki 1991).

The occurrence of astaxanthin in the freshwater microalga *Haematococcus pluvialis* led to the development of two-stage cultures of this alga: the production intracellular pigment involves changes in the metabolism associated with a morphological transformation from green vegetative cells to deep-red, astaxanthin-rich, immobile aplanospores (Elliot 1934) that take place when the culture is subjected to stress conditions such as high irradiances (Kobayashi et al. 1992) usually in combination with nutrient deprivation (Boussiba et al. 1992; Margalith 1999; Orosa et al. 2001).

The incorporation of well-developed *Haematococcus* and *Dunaliella* production systems into conventional (agro industrial) wastewater treatment reduces production costs of algal biomass, which in turn can be applied to production of bioactive substances, bioenergy, or valuable chemicals (Hoffmann 1998). Municipal wastewaters and piggery wastes are very rich in nutrients but must not be used to feed microalgae until effluent preprocessing and biomass post-processing guarantee pathogen destruction. But agro-industrial wastes such as *manipueira* (cassava processing wastewater) and vinasse (ethanol production wastewater) may be conveniently used as culture media for mixotrophic growth of microalgae (Soccol et al. 2012). Table 4.5 shows the main components of autotrophic of culture media and of two residues.

In a conventional biological wastewater treatment process, external carbon sources such as methanol or acetate are usually needed to convert nitrate into nitrogen gas (Tchobanoglous et al. 2003), and excess biomass generated needs to be treated and disposed of in a safe and cost-effective way—which leads to high operating costs (Yang et al. 2003). However, the assimilation of nitrate by microalgae has two advantages over conventional biological nutrient removal: (1) nitrate can be converted into biomass without any external carbon source, and (2) high-value products such as astaxanthin can be extracted from excess biomass. Because of the low rates of growth and nitrate uptake in microalgae, it may be difficult for this microalgal treatment process to be used in a mainstream treatment process, but it may have potential application as a subsidiary process in biological nutrient removal (Kang et al. 2006). Actually, the final step in wastewater treatment (stabilization) usually has a healthy population of microalgae.

At the other side, *direct* (i.e., untreated) use of agro-industrial wastewaters, even with a high organic load is possible, for suitably adapted species. Direct cultivation in these residues may even work as a selective trait: extreme pH or high osmolality, coupled to cultivation in shallow ponds, and high inoculum concentration may be

Table 4.5 Main components of culture media (in mg/L) for microalgae with potential for production of value-added biomass and production (in g/L) for *Chlorella* and *Spirulina*

	Seawater	*Vinasse*	*Manipueira*	*Spirulina*	*Chlorella* and *Scenedesmus*	*Porphyridium*	*Dunaliella*
HCO_3^-	145			12,200		28.6	1,235
Na^+	10,768	51.6		5,670	300	10,626	46,465
K^+	399	1,689	1,863	672	313	406.1	199
Ca^{2+}	412	368	227.5	10.9	4.2	408	11.9
Mg^{2+}	1,292	135	405	19.5	24.4	1,306	131.4
Fe^{2+}	0.002	17.6	15.4	2.0	0.9	4. 4	0.1
Cl^-	19,353	1,219		626		19,068	71,050
Br^-	66						
$H_2BO_3^-$	27		26.8		2.5	0.6	6
HPO_4^{3-}		92	498	276	385	49.9	10.0
SO_4^{2-}	2,712	1,538		633	98	2,576	468.3
NO_3^-	0.3			5,670	993	614	310
NH_4^+	0.03	13.2					
EDTA				0.2	3.1	15.6	0.5
Organic matter		9,300	55,000				
BOD		16,950	55,000				
COD		28,450	85,400				
Organic N		343.4	4,900				
Soluble carbohydrates			5,100				
Biomass production, g/L		Residue diluted to 30 %	Residue diluted to 30 %				
Chlorella	–	1.6	3.1	–	5–10	–	–
Spirulina	–	4.47	2.6	1–10	–	–	–

Source: adapted from Soccol et al. (2012), with permission. Empty cells indicate low or zero concentration, for media and seawater and low or unknown, for residues enough to guarantee the predominance of one algal species in open systems. In closed systems, the problem is reduced to a question of sterilization—to which bioindustries are well acquainted.

Comparing media in Table 4.5, one may note that the residues have high potassium and low nitrate concentration but carry organic nitrogen and phosphate, which explain why some algae thrive in these wastewaters. Kang et al. (2006) introduced the cultivation of *H. pluvialis* into secondary treatment of the primary-treated sewage (PTS) and primary-treated piggery wastewater (PTP), containing low and high concentrations of nitrate, respectively. The authors examined the characteristics of algal growth, nitrate assimilation, and astaxanthin biosynthesis by red cyst cells of *H. pluvialis* through subsequent strong photoautotrophic induction. The work showed that the inorganic wastes in the wastewater were removed successfully by *Haematococcus* cultivation, after which green vegetative cells were transformed by photoautotrophic induction to red aplanospores with substantial astaxanthin content of 39.7 mg L^{-1} and 83.9 mg L^{-1} on PTS and PTP-2 cultures, respectively.

Carotenoid bioavailability depend on post-processing of the biomass produced; intact astaxanthin-rich cysts of *Haematococcus* are poorly absorbed in salmonids (Sommer et al. 1991), and the biomass should be processed in order to enhance digestibility, e.g., via high-pressure homogenization.

4.2.3 Other Photosynthetic Pigments from Microalgae

Chlorophyll (which exists in several forms in microalgae and cyanobacteria) is usually produced from alfalfa, through solvent extraction and then purified and converted to cupric complexes. However, microalgae have much higher contents of chlorophylls (for example, 37.1 mg/g biomass in *Chlorella*, against 3.84 mg/g in alfalfa), and the prospective production of large amounts of microalgal biomass for biofuels or protein may provide raw material for microalgae-based chlorophyll.

In order to enhance the absorption of light, besides chlorophyll and carotenoids, several microorganisms have specialized phycobilin proteins. These antennae pigments are present in cyanobacteria and some algae (rhodophyta, cryptophyta, and glaucophyta) and have absorption spectra complementary to that of chlorophylls. Among phycobilins, phycocyanin is one of the most interesting pigments with a distinct blue color. Also, being a water-soluble pigment, it offers a large array of potential applications. Table 4.6 presents the production characteristics for representative photosynthetic microorganisms.

Phycocyanin production is done by cultivating the chosen microorganism and then processing the biomass. Since *Spirulina* supports very high pH for growth, large-scale open cultures of this cyanobacterium have been done even with mixotrophic cultures. However, without the total dependence of light, cells may reduce the concentration of photosynthetic pigments—for example, chlorophyll content drops in mixotrophic cultures, from approximately 23 mg/g in Chlorella to around 4 mg/g

Table 4.6 Selected phycocyanin (PC), phycoerythrin (PE) and chlorophyll (CHL), and allophycocyanin (APC)—producing microorganisms

Microorganism	Molecule	Culture medium	X_{max} (g/L)	P_{max} (mg/g)	μ (day^{-1})	References
S platensis	PC	ZRK+2.5 g/L glucose, 4klux	2.66	121	0.62	Chen et al. (1996)
S platensis	PC	ZRK		167		Yan et al. (2011)
	APC	ZRK		36.6		Yan et al. (2011)
S platensis		ZRK	10.24	54–125		Chen and Zhang (1997)
A platensis	PC	ZRK, 2.5 g/L glucose, 12 kLux	1.33		0.49	Ben et al. (2010) Better production in high light, low glu; not optimized
Synechocystis sp.	PC	Modified BG-11	0.2	120		
Galdieria sulphuraria	PC	Glucose and ammonia based, fed batch, high concn.	109	27	1.4	Graverholt and Eriksen (2007)
Pseudanabaena sp.	PE	ASN-III	0.89	44	0.1	Mishraa et al. (2012)
S. platensis	CHL	Mineral, nitrate based	1.9	11.6	0.15	Rangel-Yagui et al. (2004)

ZRK Zarrouk medium; *S* (*Spirulina*) and *A* (*Arthrospira*) refer to the same microorganism, but the original denomination given by the reference was maintained

(Cheirsilp and Torpee 2012); light intensity also affects the production, although cell concentration must be taken into account when analyzing irradiance.

While chlorophyll may be extracted from algal biomass just as in plant-based processes, phycocyanin processing requires cell disruption in a buffer, followed by filtration or centrifugation of the debris, concentration, and drying. Broiler additives may be produced by simply drying the biomass. Pure phycobilins may be obtained by fractional precipitation at 25 and 60 % ammonium sulfate, followed by a DEAE-Sepharose chromatography with a gradient from pH 5–3.6, reaching phycocyanin purity of 5.59 (A_{620}/A_{680}) and allophycocyanin purity of 5.19, with recoveries of 67 % and 80 %, respectively (Yan et al. 2011).

As it could be expected with a protein, phycocyanin is stable only in the pH range of 5.5–6, with a temperature below 47 °C (Chaiklahan et al. 2012). Outside this range, partial degradation of the pigment occurs through denaturation, insolubilization, and possibly due to hydrolysis (in extreme pHs). Several solutes may aid the stabilization, as well as the use of high-temperature, short-time (HTST) processing rather than long heat-incubation times.

Table 4.7 Pigments produced by fungi in agro-industrial residues

Microorganism	Molecule	Culture medium	X_{max}	P_{max}	References
Monascus sp.	Polyketide mix	Rice, SSF	330 mg/g substrate	1.87 mg/g substrate	Carvalho et al. (2006)
Monascus sp.	Polyketide mix	Cassava bagasse, SSF		0.3 mg/g substrate	Carvalho et al. (2007)
Monascus sp.	Polyketide mix	Jackfruit seed + minerals, SSF			Babitha et al. (2006)
Monascus purpureus	Polyketide mix	Gluten and bran-free wheat flour SmF	10.34 g/L	2.46 mg/g substrate	Dominguez-Espinosa and Webb (2003)
Monascus kaoliang	Polyketide mix	Wheat meal, SSF		60.64 mg/g substrate	Lin ad Iizuka (1982)
Penicillium oxalicum	Arpink red	Molasses, yeast autolysate		1.5 g/L	Dufossé (2006)
Ashbya gossypii	Riboflavin	Corn steep liquor, peptone, soybean oil		1–5 g/L	Lim et al. (2001)

SmF submerged fermentation, *SSF* solid substrate fermentation

4.2.4 Pigments from Fungi

There are hundreds of colored pigments produced by fungi. Several of these substances are bioactive, having, e.g., antibiotic, immunomodulatory, nephrotoxic, and hepatotoxic properties. Considering that fungi are usually unable to use light, the color of the pigment could be a consequence of interaction with light of a structure with other functionalities. Several accounts on the potential for biopigments production by fungi were done by Durán et al. (2002) and others, compiling data for several classes of biopigments mainly including melanins, flavins, carotenoids, quinones, and azaphilones. If at one side this large diversity represents an untapped source of promising molecules, at the other the bioactivity is a barrier to the market. Table 4.7 presents the most important fungal pigments which are used in foods (even if *Monascus*, as already stated, is limited to oriental countries).

Riboflavin (vitamin B2) is a well-known, permitted, stable water-soluble pigment added to a multitude of products to impart yellow color and as a nutritional ingredient. Although more than 75 % of the worldwide riboflavin production is synthetic, its industrial production by fungi is well established.

An anthraquinoid pigment, Arpink red, from *Penicillium oxalicum* species has a structure similar to that of cochineal carmine and may be an important substitute to the insect-derived pigment in the future. As a stable nontoxic pigment, its use for foods has been proposed (Dufossé 2006). A patent application claims that the compound and its derivatives have anticancer activity (Sardaryan 2006). The production by fermentation is straightforward, and its acidic functional group makes concentration very simple, via precipitation of the pigment by pH regulation.

Monascus pigments are sold either as raw fermented powders, concentrated or dried, or as fractionated extracts. Its production is done in submerged fermentation or, more usually, in solid substrate fermentation. There are dozens of substrates tested for its production, although rice and wheat meals (either integral or broken residual cereal) give the highest pigment production. After 7–10 days fermentation, the mass (*koji*, in the case of rice) may be dried or extracted with a suitable solvent. The solution may be concentrated and dried. *Monascus* pigments stability depend on pH and temperature; for pH 7–8 in aqueous media, the pigment resists processing for 30 min at 100 °C but may lose up to 20 % tinctorial strength at pH 4 (Carvalho et al. 2005). Alcoholic solutions are very stable. Monascus pigments may bind to amino groups, which may lead to increased stability in some formulations. Care must be taken for most strains produce citrinin, a yellow nephrotoxic mycotoxin; also, raw biomass may contain the anti-hypercholesteremic molecule lovastatin.

Another microbial pigment producer whose metabolites are bioactive is *Pycnoporus sanguineus*, a ubiquitous wood-growing fungus which produces phenoxazine analogs with antimicrobial activity.

4.2.5 Bacterial Pigments: Prodigiosin, Violacein, Pyocyanin, etc.

There are several well-studied bacterial pigments but these are not introduced in market because of its antimicrobial and eventual toxic activity, as is the case of prodigiosin and violacein. However, these pigments may have niche food uses (e.g., avoid fungal proliferation on the surface of meat products) or nonfood uses, as in textiles. At the other side, some bacteria produce harmless carotenoids. Examples of bacterial pigments produced using fermentation are presented in Table 4.8.

Prodigiosins are a class of tripyrrole antibiotic pigments produced by several microorganisms such as *Serratia marcescens* and *Hahella chejuensis*. These substances received recent renewed attention because of its reported immunosuppressant and anticancer properties (Williamson et al. 2006; Gulani et al. 2012) and potential involvement in the reduction of algal proliferation in algal blooms (Kwon et al. 2010). *Serratia* cultures produce almost 500 mg/L of prodigiosin in 2 days, at 30 °C. Giri et al. (2004) obtained excellent production in peanut seed broth (38.75 g/L), which indicated that peanut (and perhaps soy) processing residues could be an adequate substrate for the pigment.

Violacein is a purple diindole-pyrrole pigment derived from tryptophan. It is soluble in ethanol, and its biosynthesis and potential uses are still being studied. The same applies to the blue phenazine pigment pyocyanin, produced by *Pseudomonas aeruginosa*. Pyocyanin is a highly reactive metabolite which, being toxic to mammal

Table 4.8 Selected pigments produced by bacteria

Microorganism	Molecule	Culture medium	X_{max} (g/L)	P_{max} (mg/L)	References
Serratia marcescens	Prodigiosin	Maltose and peanut oil based	–	535	Gulani et al. (2012)
Serratia marcescens	Prodigiosin	Powdered peanut medium	–	39,000	Giri et al. (2004)
Hahella chejuensis, mutant	Prodigiosin	Sucrose and peptone based	–	2,600	Kim et al. (2008)
Chromobacterium violaceum	Violacein	Glucose and peptone based	21	430	Mendes et al. (2001)
Paracoccus carotinifaciens	Astaxanthin, canthaxanthin	Glucose and peptone based	–	25–40 (mg/g)	Hirschberg et al. (1999); Tanaka et al. (2011)

cells, cannot be used in foods; however, it is possible that its conjugation to proteins in leather and other materials may stabilize it, permitting its use as a textile pigment.

Paracoccus carotinifaciens is a bacterium which accumulates a mix of carotenoids. Recent patents cover proprietary isolates and mutant strains, and although the carotenoid content is not superior to that of selected microalgae, the specific growth rate is probably high, as common in bacteria. This biomass hit market relatively quickly and is already permitted as a fish feed supplement in the USA.

4.3 Biopigment Production and Formulation

Among the biopigments discussed in this chapter, few are produced in industrial scale; in some cases, the technology is relatively new and scaling up is being developed (as is the case for phycobilins), or there are regulatory issues (as with *Monascus* pigments in Europe and in the USA). The following section presents suitable, laboratory-tested conditions for producing and concentrating selected pigments.

Although the pigments discussed here vary widely, there are some general considerations which apply: we may be interested in the integral biomass or in a concentrated pigment. The substances of interest may be intra- or extracellular (or both), predominantly lipo- or hydrosoluble, and probably thermolabile and prone to oxidation. Finally, biomasses destined for feeds and supplements may have poor digestibility, requiring a cell disruption step. Four selected cases of biopigment production systems follow:

4.3.1 Production

β-carotene from *Blakeslea trispora*: inocula of spores of mating types + and − are used in the ratio 1:10 (e.g., 10^4 spores of type + and 10^5 spores of type −) (Varzakakou et al. 2010) in a suitable culture medium, such as (in g/L) D-glucose 50, olive oil 5.4, soybean oil 5.4, sunflower oil 5.4, corn steep liquor 80.0, span 20 10.0, Tween 80 0.1, casein hydrolysate 2.0, yeast extract 1.0, L-asparagine 2.0, KH_2PO_4 1.5, $MgSO_4{\cdot}7H_2O$ 0.5, BHT 0.02, and thiamine-HCl 0.005, as proposed by Papaioannou and Liakopoulou-Kyriakides (2010). After 8 days in aerated culture with moderate agitation at 26 °C, the cells are separated from the broth by filtration, partially dehydrated and slowly extracted with a suitable solvent such as chloroform. The solvent is separated from the biomass and evaporated; the dry raw extract is further processed. From 62 kg of corn steep liquor and 38 kg of glucose, there should be produced 20 kg of biomass which, extracted with 50 L chloroform, would give approximately and 8 kg extract containing 90 g of β-carotene.

Carotenoids from *Haematococcus*: This microalga may be cultivated in synthetic mineral media or in residues if sufficient nitrogen and phosphorus is present. An initial cell count of 10^3 cells/mL is a suitable inoculum to a basal medium containing (in g/L): KNO_3 2, K_2HPO_4 0.2, $MgSO_4{\cdot}7H_2O$ 0.2, soil extract 30 mL (according to SAG 2012), and micronutrients. Cultivation is done at 25 °C for 10–15 days, at moderate irradiation. Biomass should reach 2–3 g/L, and at this point the irradiation should be increased to 30–40 kLux with the addition of NaCl reaching a concentration of 0.4–0.6 %, conditions which induce carotenogenesis. After 5 days, carotenoid-rich biomass is separated by centrifugation. The biomass is thermally processed in order to enhance the digestibility of the algae meal (for instance, right before spray drying). From 100 kg of salts, there may be produced 125 kg of biomass with 1.65 kg of astaxanthin, requiring a production area (for a pond) of 200 m^2.

Phycocyanin from *Arthrospira* sp.: this cyanobacterium may be suitably cultivated in agro-industrial wastes, provided that there is enough nitrogen and phosphorus and that the pH is increased (if necessary) to around 9, which usually requires addition of bases. Growth in *manipueira*, for example, produces at least 2–3 g/L of biomass (or 10–15 g/L of residue, since it is diluted to 20 %). From 1,000 L of residue, diluted to 20 % and inoculated with a previous culture to an initial concentration of 0.1 g/L, approximately 12.5 kg of biomass may be obtained after 10–15 days, depending on irradiation and eventual contamination. This biomass is harvested by flocculation with a cationic copolymer such as a polyacrylamide-amine, followed by filtration and drying of the paste, in the case of algae meal for broilers. For phycocyanin production, the wet biomass is resuspended in 100 L of 0.1 M phosphate buffer and successively frozen and thawed (4 cycles). The phycobilins are liberated to the buffer, which is then centrifuged and desalinized by ultrafiltration and dialysis, giving a raw extract with 1.25–1.5 g of phycocyanin. Further purification is possible using fractional precipitation with ammonium sulfate. Such a process requires a production area of 25 m^2.

Pigments from *Monascus*: this fungus may be cultivated over residual cereals from mill processing, e.g., broken rice. The cereal is mixed with one part water and autoclaved, giving a product with approximately 56 % water. This cooked rice is inoculated with a spore suspension or with a fermented powder previously obtained, and is then cultivated for 7–10 days at 30 °C. In this solid substrate fermentation, it is important to maintain aeration through the medium. After cultivation, the material may be dried for production of a meal or extracted (without drying) with 95 % ethanol (2–3 L:kg of rice initially used). The extract is then filtered, concentrated by evaporation, and the viscous precipitate that forms may be processed into a powder or a liquid formulation. From 100 kg of broken rice, an extract or powder with approximately 200 g of mixed pigments may be produced. Further fractionation is uncommon but may be done by chromatography.

4.3.2 Formulation

The distribution (in the mass transfer sense) of a pigment in a product matrix is determined by its structure and formulation. Therefore, besides producing the molecule, it is usually desirable to have several physicochemically distinct presentable forms of the pigment, e.g., hydrosoluble and liposoluble concentrates, solid dispersible powders, and solid additives for feed. Microencapsulation and/or additives may be used to protect the pigment. Of the products to which biopigments may be applied, perhaps foods represent the most complex class, because of its naturally complex composition; however, the considerations below may be also applied to drugs and cosmetics.

Hydrosoluble pigments: Riboflavin, phycocyanin, cupric chlorophyllin, and other hydrosoluble pigments may be directly added to foods, in accordance with the *Codex Alimentarius* regulations. These pigments may be prepared and stored as dry powders or liquid concentrates, eventually with dispersing agents. Dry powders are preferred because of its low water activity and high stability.

Liposoluble pigments: Carotenoids, several xanthophylls, and chlorophyll dissolve poorly in water but may readily be dissolved in hot oils, fats, or highly concentrated alcoholic solutions (e.g., spirits). For lipid-rich foods, these pigments may be applied directly, but for foods with high water contents, the application must be done carefully to avoid phase separation. For example, application of β-carotene in emulsions such as sausages may lead to segregation of the pigment to lipid droplets and lead to a heterogeneous aspect. Prior assessment of color stability is necessary. Formulations for these pigments are usually solutions in oil (e.g., lycopene in soy oil) or dry powders.

There are commercial hydrosoluble preparations of these pigments: these are stable emulsions with edible oil micelles stabilized by USP/FDA/EC-approved emulsifiers. These are lipo- and hydrosoluble, mostly intended for use in beverages.

Microencapsulation consists of preparing a mixture of the pigment with a suitable support which surrounds or dissolves it: for example, hydrophobic carotenoids may be mixed with cyclodextrins forming water-soluble complexes; hydrophilic pigments such as phycocyanin may be emulsified in light oil with the aid of a lecithin, followed by spray drying. For example, one commercial astaxanthin formulation contains the extract dispersed into soy protein and contains alginate and hydroxypropyl cellulose, besides antioxidants to protect the material during processing and storage.

4.4 Future Developments

There are several ways in which biopigment research should focus: vegetable and animal cell cultures may lead to the production of already permitted pigments (for example, anthocyanins from grapes) using "cell reactors." Genetic tools are much more likely to be used in order to develop genetically modified hosts which will produce the desired pigment with very high productivities. For example, Das et al. (2007) report that yields of up to 18, 49, and 11.4 mg/g DCW of lycopene, β-carotene and astaxanthin were obtained in specific studies where *E. coli* was modified for production of carotenoids. The elucidation of the composition of structural pigments such as helically coiled cellulose in *Pollia* fruits (Vignolini et al. 2012) may lead to exciting new GMO structural pigment producers—today, pearlescent effects are obtained using mica, a mineral.

Molecular tools may be used for knockout of genes for toxin production in fungi, to enhance concentration through multiple gene copies, and regulating of pathways: mutagenesis as in *Ashbya gossypii* and further understanding of pheromone and carotene biosynthesis regulation in *Phycomyces blakesleeanus* (Tagua et al. 2012) is leading to enhanced production of carotenoids.

On the more traditional side, the fungal, bacterial, and microalgal diversity is yet to be explored, and new tools for molecule identification may be used for high throughput screening of ambiental samples. The fact that one of the newest biomasses permitted for feed use is a newly bacterial isolate (*Paracoccus* sp.) shows the potential of bioprospection. At the other side, species with known behavior may be mutated, with selective pressure, in order to develop better pigment producers—as it has been done successfully with *Haematococcus* algae.

References

Abe K, Hattori H, Hirano M (2007) Accumulation and antioxidant activity of secondary carotenoids in the aerial microalga *Coelastrella striolata* var *multistriata*. Food Chem 100:656–661

Aberoumand A (2011) A review article on edible pigments properties and sources as natural biocolorants in foodstuff and food industries. World J Dairy Food Sci 6:71–78

Babitha S, Soccol CR, Pandey A (2006) Jackfruit seed – a novel substrate for the production of *Monascus* pigments through solid-state fermentation. Food Technol Biotechnol 44:465–471

BBC Research (2010) Carotenoids global market folder. http://www.bbcresearch.com. Accessed 13 Oct 2012

Ben DR, Ghenim N, Trabelsi L, Yahia A, Challouf R, Ghozzi K, Ammar J, Omrane H, Ben OH (2010) Modeling growth and photosynthetic response in *Arthrospira platensis* as function of light intensity and glucose concentration using factorial design. J Appl Phycol 22:745–752

Boussiba S, Fan L, Vonshak A (1992) Enhancement and determination of astaxanthin accumulation in green alga *Haematococcus pluvialis*. Methods Enzymol 213:386–391
Buzzini P, Martini A (1999) Production of carotenoids by strains of *Rhodotorula glutinis* cultured in raw materials of agro-industrial origin. Bioresour Technol 71:41–44
Carvalho JC, Oishi BO, Pandey A, Soccol CR (2005) Biopigments from *Monascus*: strains selection, citrinin production and color stability. Braz Arch Biol Technol 48:885–894
Carvalho JC, Pandey A, Oishi BO, Brand D, Rodriguez-Leon JA, Soccol CR (2006) Relation between growth, respirometric analysis and biopigments production from *Monascus* by solid-state fermentation. Biochem Eng J 29:262–269
Carvalho JC, Oishi BO, Woiciechowski AL, Pandey A, Soccol CR (2007) Effect of substrates on the production of *Monascus* biopigments by solid-substrate fermentation and pigment extraction using different solvents. Ind J Biotechnol 6:194–199
Chaiklahan R, Chirasuwan N, Bunnag B (2012) Stability of phycocyanin extracted from *Spirulina* sp: influence of temperature, pH and preservatives. Process Biochem 47:659–664
Cheirsilp B, Torpee S (2012) Enhanced growth and lipid production of microalgae under mixotrophic culture condition: effect of light intensity, glucose concentration and fed-batch cultivation. Bioresour Technol 110:510–516
Chen F, Zhang Y, Guo S (1996) Growth and phycocyanin formation of *Spirulina platensis* in photoheterotrophic culture. Biotechnol Lett 18:603–608
Chen F, Zhang Y (1997) High cell density mixotrophic culture of *Spirulina platensis* on glucose for phycocyanin production using a fed-batch system. Enzyme Microb Technol 20:221–224
Das A, Yoon SH, Lee SH, Kim JY, Oh DK, Kim SW (2007) An update on microbial carotenoid production: application of recent metabolic engineering tools. Appl Microbiol Biotechnol 77:505–512
Del Campo JA, Moreno J, Rodrıguez H, Vargas MA, Rivas J, Guerrero MJ (2000) Carotenoid content of chlorophycean microalgae: factors determining lutein accumulation in *Muriellopsis* sp (*Chlorophyta*). J Biotechnol 76:51–59
Dufossé L (2006) Microbial production of food grade pigments. Food Technol Biotechnol 44:313–321
Durán D, Teixeira MFS, Conti R, Esposito E (2002) Ecological-friendly pigments from fungi. Crit Rev Food Sci Nutr 42:53–66
Elliot AM (1934) Morphology and life history of *Haematococcus pluvialis*. Archiv Protistenk 82:250–272
Dominguez-Espinosa RM, Webb C (2003) Submerged fermentation in wheat substrates for production of *Monascus* pigments. World J Microbiol Biotechnol 19:329–336
FDA (2012) Color additives inventories. http://www.fda.gov. Accessed 13 Oct 2012
FAO (2012) Yearbook of fishery statistics. http://www.fao.org/fishery/statistics/programme/publications/all/en. Accessed 15 Nov 2012
Garcıa-Malea MC, Brindley C, Del Río E, Acien FG, Fernandez JM, Molina E (2005) Modeling of growth and accumulation of carotenoids in *Haematococcus pluvialis* as a function of irradiance and nutrients supply. Biochem Eng J 26:107–114
Giri AV, Anandkumar N, Muthukumaran G, Pennathur G (2004) A novel medium for the enhanced cell growth and production of prodigiosin from *Serratia marcescens* isolated from soil. BMC Microbiol 4:11. Available at http://www.biomedcentral.com/content/pdf/1471-2180-4-11.pdf. Accessed 15 Nov 2012
Graverholt OS, Eriksen NT (2007) Heterotrophic high-cell-density fed-batch and continuous-flow cultures of *Galdieria sulphuraria* and production of phycocyanin. Appl Microbiol Biotechnol 77:69–75
Gulani C, Bhattacharya S, Das A (2012) Assessment of process parameters influencing the enhanced production of prodigiosin from *Serratia marcescens* and evaluation of its antimicrobial, antioxidant and dyeing potentials. Maln J Microbiol 8:116–122
Harker M, Tsavalos A, Young AJ (1996) Factors responsible for astaxanthin formation in the chlorophyte *Haematococcus pluvialis*. Bioresour Technol 55:207–217

Hirschberg J et al (1999) Carotenoid-producing bacterial species and process for production of carotenoids using same. United States Patent 5,935,808

Hoffmann JP (1998) Wastewater treatment with suspended and non-suspended algae. J Phycol 34:757–763

Hong ME, Choi SP, Park YI, Kim YK, Chang WS, Kim BW, Sim SJ (2012) Astaxanthin production by a highly photosensitive *Haematococcus* mutant. Process Biochem, 47:1972–1979. http://dx.doi.org/10.1016/j.procbio.2012.07.007

Ip PF, Wong KH, Chen F (2004) Enhanced production of astaxanthin by the green microalga *Chlorella zofingiensis* in mixotrophic culture. Process Biochem 39:1761–1766

Ip PF, Chen F (2005) Production of astaxanthin by the green microalga *Chlorella zofingiensis* in the dark. Process Biochem 40:733–738

Kang DK, An JY, Park TH, Sim SJ (2006) Astaxanthin biosynthesis from simultaneous N and P uptake by the green alga *Haematococcus pluvialis* in primary-treated wastewater. Biochem Eng J 31:234–238

Khodaiyan F, Razavi SH, Mousavi SM (2008) Optimization of canthaxanthin production by *Dietzia natronolimnaea* HS-1 from cheese whey using statistical experimental methods. Biochem Eng J 40:415–422

Kim SJ, Lee HK, Lee YK, Yim JH (2008) Mutant selection of *Hahella chejuensis* KCTC 2396 and statistical optimization of medium components for prodigiosin yield-up. J Microbiol 46:183–188

Kleinegris DMM, Janssen M, Brandenburg WA, Wijffels RH (2011) Continuous production of carotenoids from *Dunaliella salina*. Enzyme Microb Technol 48:253–259

Kobayashi M, Kakizono T, Nishio S, Nagai S (1992) Effects of light intensity, light quality and illumination cycle on astaxanthin formation in the green alga *Haematococcus pluvialis*. J Ferment Bioeng 74:61–63

Kuo FS, Chien YH, Chen CJ (2012) Effects of light sources on growth and carotenoid content of photosynthetic bacteria *Rhodopseudomonas palustris*. Bioresour Technol 113:315–318

Kwon SK, Park YK, Kim JF (2010) Genome-wide screening and identification of factors affecting the biosynthesis of prodigiosin by *Hahella chejuensis*, using *Escherichia coli* as a surrogate host. Appl Environ Microbiol 76:1661–1668

Lim SH, Choi JS, Park EY (2001) Microbial production of riboflavin using riboflavin overproducers, *Ashbya gossypii*, *Bacillus subtilis*, and *Candida famata*: an overview. Biotechnol Bioprocess Eng 6:75–88

Lin CF, Iizuka H (1982) Production of extracellular pigment by a mutant of *Monascus kaoliang* sp nov. Appl Environ Microbiol 43(3):671–676

Margalith PZ (1999) Production of ketocarotenoids by microalgae. Appl Microbiol Biotechnol 51:431–438

Marova I, Carnecka M, Halienova A, Certik M, Dvorakova T, Haronikova A (2012) Use of several waste substrates for carotenoid-rich yeast biomass production. J Environ Manag 95:338–342

Meléndez-Martínez AJ, Britton G, Vicario IM, Heredia FJ (2007) Relationship between the colour and the chemical structure of carotenoid pigments. Food Chem 101:1145–1150

Merck (2006) The Merck index. Merck &Co, Whitehouse Station, NJ

Margalith PZ (1992) Pigment microbiology. Chapman & Hall, Cambridge

Mendes AS, Carvalho JE, Duarte MCT, Durán N, Bruns RE (2001) Factorial design and response surface optimization of crude violacein for *Chromobacterium violaceum* production. Biotechnol Lett 23:1963–1969

Miki W (1991) Biological functions and activities of animal carotenoids. Pure Appl Chem 63:141–146

Mishraa SK, Shrivastava R, Mauryaa RR, Patidara SK, Haldarb S, Mishraa S (2012) Effect of light quality on the C-phycoerythrin production in marine cyanobacteria *Pseudanabaena* sp isolated from Gujarat coast, India. Protein Expr Purif 81:5–10

Nelis HJ, De Leenheer AP (1989) Microbial production of carotenoids other than β-carotene. In: Vandamme J (ed) Biotechnology of vitamins, pigments and growth factors. Elsevier, Essex

Nassau K (2003) The physics and chemistry of color: the 15 mechanisms. In: Shevell SK (ed) The science of color. Elsevier, New York, NY

Orosa M, Franqueira D, Cid A, Abalde J (2001) Carotenoid accumulation in *Haematococcus pluvialis* in mixotrophic growth. Biotechnol Lett 23:373–378

Papaioannou EH, Liakopoulou-Kyriakides M (2010) Substrate contribution on carotenoids production in *Blakeslea trispora* cultivations. Food Bioprod Process 8:305–311

Perez-Fons L, Steiger S, Khaneja R, Bramley PM, Cutting SM, Sandmann G, Fraser PD (2011) Identification and the developmental formation of carotenoid pigments in the yellow/orange *Bacillus* spore-formers. Biochim Biophys Acta 1811:177–185

Puttananjaiah MKH, Dhale MA, Govindaswamy V (2011) Non-toxic effect of *Monascus* purpureus extract on lactic acid bacteria suggested their application in fermented foods. Food Nutri Sci 2:837–843

Rangel-Yagui CO, Danesi EDG, Carvalho JCM, Sato S (2004) Chlorophyll production from *Spirulina platensis*: cultivation with urea addition by fed-batch process. Bioresour Technol 92:133–141

Razavi SH, Marc I (2006) Effect of temperature and pH on the growth kinetics and carotenoid production by *Sporobolomyces ruberrimus* H110 using technical glycerol as carbon source. Iran Chem Chem Eng 25:59–64

SAG (2012) List of media recipes. http://www.uni-goettingen.de/en/184982.html. Accessed 13 Oct 2012

Sardaryan E (2006) Food supplement. United States Patent application 20060247316. http://appft.uspto.gov/netacgi/nph-Parser?p=1&u=%2Fnetahtml%2FPTO%2Fsearch-adv.html&r=1&f=G&l=50&d=PG01&s1=20060247316.PN.&OS=PN/20060247316&RS=PN/20060247316. Accessed 15 Nov 2012

Shen H, Kuo CC, Chou J, Delvolve A, Jackson SN, Post J, Woods AS, Hoffer BJ, Wang Y, Harvey BK (2009) Astaxanthin reduces ischemic brain injury in adult rats. FASEB J 23(6): 1958–1968

Soccol CR, Sydney EB, de Carvalho JC, Dalmas Neto CJ, Coraucci Neto D, Assmann R, Thomaz-Soccol V (2012) Microalgae use in integrated processes for simultaneous carbon fixation, aqueous agroindustrial residues treatment and production of value-added biomolecules. Annals of the 5th Conference on Industrial Bioprocesses, IFIB, October 7–10, 2012, Taipei

Sommer TR, Pottsa WT, Morrisy NM (1991) Utilization of microalgal astaxanthin by rainbow trout (*Oncorhynchus mykiss*). Aquaculture 94:79–88

Soni SK (2007) Microbes: a source of energy for the 21st century. New India Publishing Agency, New Delhi

Tagua VG, Medina HR, Martín-Dominguez R, Eslava AP, Corrochano LM, Cerdá-Olmedo E, Idnurm A (2012) A gene for carotene cleavage required for pheromone biosynthesis and carotene regulation in the fungus *Phycomyces blakesleeanus*. Fungal Genet Biol 49:398–404

Tanaka T et al (2011) Microorganism and method for producing carotenoid using it. United States Patent 8,030,022

Tchobanoglous G, Burton FL, Stensel HD (2003) Wastewater engineering: treatment, disposal and reuse/Metcalf & Eddy Inc. McGraw-Hill, New York, NY

Tinoi J, Rakariyatham N, Deming RL (2005) Simplex optimization of carotenoid production by *Rhodotorula glutinis* using hydrolyzed mung bean waste flour as substrate. Process Biochem 40:2551–2557

Tooley AJ, Cai YA, Glazer AN (2001) Biosynthesis of a fluorescent cyanobacterial C-phycocyanin holo-α subunit in a heterologous host. Proc Natl Acad Sci 98:10560–10565

Silva MC (2004) Alterações na biossíntese de carotenoids em leveduras induzidas por agentes químicos. Thesis, University of Campinas

Vaquero I, Ruiz-Domínguez C, Márquez M, Vílchez C (2012) Cu-mediated biomass productivity enhancement and lutein enrichment of the novel microalga *Coccomyxa onubensis*. Process Biochem 47:694–700

Varzakakou M, Roukas T, Kotzekidou P (2010) Effect of the ratio of (+) and (-) mating type of *Blakeslea trispora* on carotene production from cheese whey in submerged fermentation. World J Microbiol Biotechnol 26:2151–2156

Vignolini S, Rudall PJ, Rowland AR, Moyroud E, Faden RB, Baumberg JJ, Glover BJ, Steiner Y (2012) Pointillist structural color in *Pollia* fruit. Proc Natl Acad Sci 109:15712–15715

Vílchez C, Forján E, Cuaresma M, Bédmar F, Garbayo I, Veja JM (2011) Marine carotenoids: biological functions and commercial applications. Mar Drugs 9:319–333

Walter A, de Carvalho JC, Thomaz-Soccol V, Faria ABB, Ghiggi V, Soccol CR (2011) Study of phycocyanin production from *Spirulina platensis* under different light spectra. Braz Arch Biol Technol 54:675–682

Wang F, Jiang JG, Chen Q (2007) Progress on molecular breeding and metabolic engineering of biosynthesis pathways of C30, C35, C40, C45, C50 carotenoids. Biotechnol Adv 25:211–222

Williamson NR, Fineran PC, Leeper FJ, Salmond GP (2006) The biosynthesis and regulation of bacterial prodiginines. Nat Rev Microbiol 4:887–899

Yan SG, Zhu LP, Su HN, Zhang XY, Chen XL, Zhou BC, Zhang YZ (2011) Single-step chromatography for simultaneous purification f C-phycocyanin and allophycocyanin with high purity and recovery from *Spirulina* (*Arthrospira*) *platensis*. J Appl Phycol 23:1–6

Yang XF, Xie ML, Liu Y (2003) Metabolic uncouplers reduce excess sludge production in an activated sludge process. Process Biochem 38:1373–1377

Yang J, Tan H, Yang R, Sun X, Zhai H, Li K (2011) Astaxanthin production by *Phaffia rhodozyma* fermentation of cassava residues substrate. Agricult Eng Int 13:1–6

Yarnell A (2012) Bringing blue to a plate near you. Chem Eng News 37:30–31

Chapter 5
Utilization of Agro-industrial Waste for the Production of Aroma Compounds and Fragrances

Saurabh Jyoti Sarma, Gurpreet Singh Dhillon, Krishnamoorthy Hegde, Satinder Kaur Brar, and Mausam Verma

5.1 Introduction

Aroma compounds or fragrances are the chemical compounds having pleasant smell. In general, they are volatile compounds which can easily reach the olfactory system. An aroma compound could also be a flavor which has a smell as well as a taste (e.g., diacetyl, 2-phenylethanol, acetoin, and vanillin). Aroma compounds are the integral parts of foods and beverages as well as they are also found in cosmetics and personal care products. Presently, flavors represent around 25 % of global food additive market (Longo and Sanromán 2006). Aroma/flavor compounds may naturally occur in foods and beverages or it could be externally included to enhance their taste/quality. Aroma/flavor compounds of foods and beverages may come from the raw materials or they may be produced during processing (e.g., fermentation) of such materials. Likewise, artificially added aroma/flavor compounds may be of natural origin, such as plant-derived materials, compounds produced by fermentation, or could be synthetic materials produced by chemical process. Most flavoring compounds presently available in the market are either synthetic compounds or products obtained by extraction from natural sources (Longo and Sanromán 2006). The foods and beverages containing aroma/flavor compounds of natural origin are

S.J. Sarma • S.K. Brar (✉)
INRS-ETE, Université du Québec, 490 Rue de la Couronne, Québec, QC, Canada G1K 9A9
e-mail: satinder.brar@ete.inrs.ca

G.S. Dhillon
Biorefining Conversions Network (BCN), Department of Agricultural, Food and Nutritional Sciences (AFNS), University of Alberta, Edmonton, AB, Canada

INRS-ETE, Université du Québec, 490 Rue de la Couronne, Québec, QC, Canada G1K 9A9

K. Hegde
Department of Biotechnology, IIT Guwahati, Assam, India 781039

M. Verma
CO2 Solutions, 2300, rue Jean-Perrin, Québec city, QC, Canada G2C 1T9

S.K. Brar et al. (eds.), *Biotransformation of Waste Biomass into High Value Biochemicals*, DOI 10.1007/978-1-4614-8005-1_5, © Springer Science+Business Media New York 2014

preferred by the customers and it could be considered as a marketing advantage (Krings and Berger 1998; Longo and Sanromán 2006). Actually, the value of naturally produced aroma/flavor compounds is significantly higher than their chemically synthesized counterparts. For example, vanillin produced by conventional chemical process has a market value of US $ 12/kg; however, at the same time natural vanillin could have a market value as high as US $ 1,200–4,000/kg (Sindhwani et al. 2012). Meanwhile, flavor compounds extracted from natural sources, such as botanical sources, are also considered natural and an environment/health friendly alternative to chemically synthesized products. However, the extraction methods may be time consuming and laborious and require more space and chemicals/organic solvents for feedstock pretreatment and product extraction. For example, prior to extraction of natural vanillin, vanilla plants should be cultivated in large scale and the flower should be manually pollinated. Therefore, it is a labor-intensive process and it is difficult to encourage the farmers for large-scale plantation of vanilla plants (Ramachandra Rao and Ravishankar 2000). Alternatively, production of natural aroma/flavor compounds by microorganisms through fermentation has great potential and this option has been largely investigated for these products (Feron et al. 1996). In fact, considerable advancements have been made in genetic engineering, enzyme, or bioprocess technology for flavor/aroma production. However, owing to economic constrains only a small fraction of commercially available natural flavor compounds are produced by microbial technology (Feron and Waché 2006). Requirement of expensive synthetic media components could be considered as the most important factor to limit the economic feasibility of such processes. Alternatively, utilization of agro-industrial wastes as the feedstock for biotechnological production of flavor/aroma compounds could bring economic feasibility for such commercial processes (Bicas et al. 2010). Thus, the present chapter discusses about a few commercially important aroma/flavor compounds for which biotechnological production has been widely explored. Furthermore, possible application of agro-industrial waste for the production of such compounds has been analyzed as well as recent advances on aroma compounds production using agro-industrial wastes has been summarized.

5.2 Types and Few Industrially Important Aroma Compounds

Based on physicochemical as well as sensorial properties, aroma compounds could be of different types. If only chemical properties are considered, they may be volatile fatty acids or esters, aldehydes, ketones, alcohols, and lactones. The detailed classification of different aroma compounds has been reviewed by Dastager (2009). Similarly, a detailed list of biotechnologically produced aroma/flavor compounds, their precursors, and market potential could be found in Feron and Waché (2006). Among these compounds some particular aroma compounds have been widely investigated for possible commercial production by bioprocess technology. In this section, few aroma compounds have been discussed to have a better understanding about their substrate specificity, microorganism involved in their production, productivity, present scale of production, and possible metabolic/genetic engineering strategy for improved product yield. Similarly, this discussion will be helpful to determine the suitability and nature/type of agro-industrial waste which could be possibly used for such processes.

5.2.1 Diacetyl

Diacetyl is a flavor compound having butter-like flavor which may be naturally present or could be artificially added as a flavoring agent to certain foods, such as dairy products (Duboff et al. 1996; Hugenholtz et al. 2000; Alvandi and Azar 2008; Quach et al. 2012). Diacetyl is a by-product resulting during beer and wine production process (Garc 1994; Fornachon and Lloyd 2006). The well-known chemical synthesis method for diacetyl production is through oxidative decarboxylation of α-acetolactate (Hugenholtz et al. 2000). In the case of fermentative diacetyl production, pyruvate is first converted to α-acetolactate which is further broken down to diacetyl by enzymatic reaction (Benson et al. 1996; Swindell et al. 1996). Actually, two molecules of pyruvate can produce one molecule of α-acetolactate by enzymatic condensation reaction (Hugenholtz et al. 2000). Lactic acid bacteria, such as *Lactococcus lactis and Lactobacillus casei,* are mostly investigated for fermentative diacetyl production (Benson et al. 1996; Swindell et al. 1996; Boumerdassi et al. 1997; Nadal et al. 2009). Alvandi and Azar (2008) have optimized different process parameters (temperature, agitation, and media components) for diacetyl production by lactic acid bacteria of *Lactococcus* and *Leuconostoc* genus. Based on the optimized parameters, a scale-up study using 10 L fermenter resulted in maximum diacetyl concentration of 945 mg/L (Alvandi and Azar 2008). Similarly, a patented study by Duboff et al. (1996) have demonstrated that diacetyl yield of above 500 ppm could be achieved by fermentation of a pectin substrate using lactic acid bacteria (Duboff et al. 1996). Meanwhile, in different investigation it was observed that in the case of citrate-positive *Lactococcus lactis* diacetyl production could be enhanced by the addition of Cu^{2+} and Fe^{3+} and simultaneously, citrate uptake activity could be inhibited probably due to the accumulation of diacetyl (Kaneko et al. 1990a). Redox potential of the culture media can play an important role in diacetyl production. According to Monnet et al. (1994), the redox potential of the culture drops towards the end of the process and for diacetyl production a high redox potential is preferable. Further, by constant agitation a relatively higher redox potential could be maintained as compared to an unagitated system and a higher diacetyl production could be obtained by a process with continuous agitation (Monnet et al. 1994). Likewise, initial oxygen concentration is another important factor in diacetyl production. Bassit et al. (1993) showed that diacetyl production could be improved by a factor of 18 by increasing the initial oxygen concentration from 0 to 100 % (Bassit et al. 1993). Furthermore, different genetic engineering strategies for improved diacetyl production are also known. According to Benson et al. (1996), the production of α-acetolactate, the intermediate of diacetyl production by pyruvate bioconversion, is catalyzed by two α-acetolactate synthases produced by *Lactococcus lacti*. The authors have used a plasmid-based gene expression strategy and 3.6-folds improvement in product formation has been reported (Benson et al. 1996). Boumerdassi et al. (1997) have used chemical mutagen *N*-methyl-*N'*-nitro-N-nitrosoguanidine to develop three mutant strains of *Lactococcus lactis*. The isolated mutants capable of diacetyl production were reported to have improved glucose utilization ability (Boumerdassi et al. 1997). Similarly, improved diacetyl production by genetic manipulation of *Lactococcus lactis* MG1363 has been reported by Swindell et al. (1996). According to the report, ilvBN – the genes encoding α-acetolactate synthase – was overexpressed in a mutant of *Lactococcus lactis* MG1363 and the resulting strain was able to produced higher level of the aroma compounds, such as acetoin and diacetyl (Swindell et al. 1996). Likewise, improved diacetyl and acetoin production by genetic manipulation of *Lactobacillus casei* has also been reported. Recently, Nadal et al. (2009) have mutated a *L. casei* strain having acetohydroxy acid synthase encoding ilvBN genes of *Lactococcus lactis*. According to the authors, the mutant was capable of improved production of diacetyl using whey permeate (Nadal et al. 2009). Similarly, application of yeast, such as *Debaryomyces hansenii,* is also known for diacetyl production (Deiana et al. 1990). From cheese, two strains of *Debaryomyces hansenii* have been isolated by Deiana et al. (1990). The production of the aroma compound by the isolated strains could vary depending upon the substrate as well as strain used (Deiana et al. 1990). Thus, from this discussion it could be concluded that considerable investigations on possible biotechnological production of diacetyl have been made and significant success have been achieved. Hence, diacetyl could be an ideal candidate for possible industrial production by using agro-industrial waste feed stocks. However, at the same time recent report on health hazards associated with vaporized diacetyl and its possible effect on diacetyl market should also be considered (Quach et al. 2012).

5.2.2 2-Phenylethanol

2-Phenylethanol is an aroma compound with rose-like fragrance which is widely used in cosmetics, food, and pharmaceutical industries (Hua and Xu 2011). Relatively small amount of it is also used in soft drinks and cookies for improving the flavor (Fabre et al. 1998; Etschmann et al. 2003). Biotransformation of L-phenylalanine by suitable yeast strain is mostly used for biotechnological production of 2-phenylethanol (Etschmann et al. 2003; Hua and Xu 2011; Achmon et al. 2011; Rong et al. 2011). Likewise, 2-phenylethanol can be extracted from the distillation residues of alcohol production (Savina et al. 1999). However, presently 2-phenylethanol is mostly produced by chemical synthesis. Benzene and styrene are the two raw materials mainly used for chemical synthesis of 2-phenylethanol; however, both of them are known as health/environmental hazards (Etschmann et al. 2003; Hua and Xu 2011). Flavors produced by such chemical processes are not considered as safe products and even their uses in food, beverages, and cosmetics have been

Table 5.1 Biotechnological production of 2-phenylethanol

Microorganism	Major substrate (s)	Production scale	References
Saccharomyces cerevisiae Ye9612	L-phenylalanine	25 mL	Achmon et al. 2011
Active dry yeast *Saccharomyces cerevisiae* powder	Ethanol (6–50 g/L) and L-phenylalanine (8 g/L)	500 mL	Rong et al. 2011
Kluyveromyces marxianus CBS 600	Beet molasses (60 g/L) and L-phenylalanine (7 g/L)	300 mL	Etschmann et al. 2003
Saccharomyces cerevisiae UCM Y-514 and UCM Y-524	Sucrose (9–18 %), yeast extract (0.75–1.5 %), and L-phenylalanine (0.1–0.2 mg/L)	750 mL	Mameeva et al. 2010
Kluyveromyces marxianus CBS 5670	Glucose.H_2O (77 g/L) and L-phenylalanine (7 g/L)	100 mL	Wittmann et al. 2002

restricted by European legislation (Xu et al. 2007). Similarly, botanical production of 2-phenylethanol is also known where the product is extracted from rose or essential oils; however, such processes are expensive (Etschmann et al. 2003). Thus, biotechnological production of food grade 2-phenylethanol has well prospective.

As shown in Table 5.1, *Saccharomyces cerevisiae* and *Kluyveromyces marxianus* are the two commonly used yeast strains for this purpose. Ehrlich pathway present in yeast, such as *S. cerevisiae,* can convert a range of amino acids to a variety of alcohols which are commonly known as fusel alcohols. Production of 2-phenylethanol by biotransformation of the amino acid L-phenylalanine is one common example of Ehrlich pathway. Moreover, recently a metabolically engineered *Escherichia coli* strain has been developed which can produce 2-phenylethanol along with five other aroma compounds (Koma et al. 2012). Further, from Table 5.1 it can be easily assumed that biotransformation of L-phenylalanine for 2-phenylethanol production has been extensively investigated by different researchers. Etschmann et al. (2003) have screened 14 different yeast strains for 2-phenylethanol production and according to the author, *Kluyveromyces marxianus* CBS 600 and CBS 397 were the two most productive strains (Etschmann et al. 2003). Similarly, Rong et al. (2011) have reported fed-batch production of 2-phenylethanol by using active dry yeast (*S. cerevisiae*). The authors also have investigated the catalytic effect of ethanol on the process and different parameters have been optimized (Rong et al. 2011). Similarly, Mameeva et al. (2010) have reported high (1.21 g/L) 2-phenylethanol production by *S. cerevisiae* UCM Y-514 and UCM Y-524 (Mameeva et al. 2010). In an interesting study by Wittmann et al. (2002), ^{13}C labeled L-phenylalanine has been used as the substrate for 2-phenylethanol production by *Kluyveromyces marxianus*. Finally, the authors have concluded that during the process, 73.3 % of the labeled L-phenylalanine was converted to the major products (2-phenylethanol and 2-phenylethylacetate), 22.4 % was catabolized via the cinnamate pathway, and 4.3 % was utilized for protein synthesis (Wittmann et al. 2002). Further, Eshkol et al. (2009) have screened different thermotolerant and multi-stress resistant *S. cerevisiae* strains for improved 2-phenylethanol production. Growth rate, biomass dry weight, and product yields were considered as the basic criteria for the selection and according to the authors high biomass production could be correlated with high product concentration. Moreover, based on the investigation thermotolerant strain, Ye9-612 has been identified as the most efficient strain for 2-phenylethanol production among different strains considered (Eshkol et al. 2009). Thus, a number of successful investigations could be found for 2-phenylethanol production; however, most of the investigations were carried out in relatively small scale and pilot scale efficiency of the process needs to be evaluated. Similarly, from Table 5.1 it could also be observed that apart from L-phenylalanine, the process should be supplemented with relatively high amount of carbon source, such as glucose, sucrose, or molasses. Likewise, in situ product recovery (ISPR) is a commonly used technique for biotechnological production of 2-phenylethanol. Actually, accumulation of 2-phenylethanol in the culture media is inhibitory for further product formation and hence, it needs to be continuously removed during fermentation. Liquid-liquid two-phase system, application of hydrophobic microsphere, and application of resin for preferential simultaneous adsorption/separation of the product from the culture broth are different techniques presently used for in situ recovery of 2-phenylethanol. In a two-phase system developed by Etschmann et al. (2003), oleyl alcohol was used as the second liquid phase for in situ product recovery. According to the author the strategy was successful to improve 2-phenylethanol production and product yield as high as 3 g/L has been reported (Etschmann et al. 2003). In a recent investigation, Wang et al. (2011) have used resin FD0816 (by weight 10 % of the medium) to obtain the overall 2-phenylethanol concentration as high as 13.7 g/L (Wang et al. 2011). Similarly, Hua et al. (2010) have screened a range of macroporous adsorbent resins and resin HZ818 has been selected for in situ recovery of 2-phenylethanol. The authors also have indicated that resin HZ818 can preferably adsorbed 2-phenylethanol with minimum adsorption of L-phenylalanine, the parent amino acid. Further, by adding the selected resin 2-phenylethanol concentration was found to be improved by 66.2 % (Hua et al. 2010). Similarly, application of hydrophobic polymethylmethacrylate microspheres for in situ recovery of 2-phenylethanol has been reported by Achmon et al. (2011). Microspheres of 1.53 ± 0.10 μm diameter were used for continuous removal of the product by the mechanism of swelling. The authors have reported tenfolds improvement in 2-phenylethanol productivity for an in situ product recovery system containing nearly 10 % (w/v) of the microspheres (Achmon et al. 2011). Likewise, improved production of 2-phenylethanol using ISPR technology has also been reported by Rong et al. (2011). Sendovski et al. (2010) have used immiscible ionic liquids as the second phase in a two-phase system used for in situ recovery of 2-phenylethanol. Nine different liquids were screened based on their biocompatibility with respect to yeast and anions [Tf2N] were reported to be the most biocompatible nonaqueous phase among them. Further, the authors have also reported that 3–5-folds improvement could be achieved by the application of ISPR technique using a biocompatible immiscible ionic liquid (Sendovski et al. 2010). Serp et al. (2003) have successfully developed a highly absorbent and mechanically stable resin to facilitate ISPR and to increase the volumetric productivity of 2-phenylethanol by a factor of 2. Moreover, according to the authors, the product could be easily back extracted from the resin and unlike a two-liquid phase system, no stable emulsion would be formed during the process. Therefore, the authors have concluded that easier downstream processing and possible commercial application of the proposed system could be expected (Serp et al. 2003). In order to study ISPR of 2-phenylethanol, Etschmann et al. (2005) have demonstrated improved 2-phenylethanol production by introducing an organophilic pervaporation unit (having a polyoctylmethylsiloxane membrane) for ISPR of 2-phenylethanol from the bioreactor (Etschmann et al. 2005). Thus, it can be concluded that 2-phenylethanol has significant prospect as biotechnological product and it has been widely explored by different researchers mostly in last decade. Further, possible application of agro-industrial waste as a substrate for 2-phenylethanol production could be considered. Most importantly, as ISPR is widely used for product recovery and the downstream processing is relatively simple, application of a complex feedstock, such as agro-industrial waste should not be a concern. Biotechnological production of 2-phenylethanol has been reviewed in detail by Etschmann et al. (2002) and Hua and Xu (2011) and potential readers can consult those work for further detail.

5.2.3 Acetoin

Acetoin is another flavor compound with typical butter-like flavor (Xu et al. 2011b). It can be found in wine, honey, coffee, and dairy products, such as butter, cheese, and fruits, such as strawberry and currants (Xu et al. 2011b; Liu et al. 2011). Owing to its flavor, it is mostly used as artificial flavor enhancer in food products. Moreover, it is also a precursor/intermediate for the synthesis of valuable fine chemicals and widely used as platform chemical (Xu et al. 2011b; Liu et al. 2011). Acetoin could be produced by traditional chemical processes; however, being a food additive its natural production has good commercial prospect. In recent years, biotechnological production of acetoin has been investigated by different researchers and considerable success has been achieved (Table 5.2).

In an interesting report, Kaneko et al. (1990b) have reported that hemin (an iron-containing porphyrin) or Cu^{2+} can stimulate the activity of certain enzymes (e.g., diacetyl synthase) of acetoin production pathway and can improve acetoin production from glucose by *Lactococci* sp. (Kaneko et al. 1990b). Similarly, Teixeira et al. (2002) have identified initial glucose concentration and temperature as the main parameters that can affect acetoin production by *Hanseniaspora guilliermondii*. According to the authors, 63 g/L initial glucose concentration and 28 °C incubation temperature were found to be optimum for acetoin production (Teixeira et al. 2002). Recently, the acetoin production by *Bacillus licheniformis* MEL09 has been investigated by Liu et al. (2011) and maximum acetoin concentration of 41.26 g/L was obtained (Liu et al. 2011).

Table 5.2 Biotechnological production of acetoin

Microorganism	Major substrate	Production scale	Maximum product concentration	Reference
Lactococcus lactis subsp. lactis 3022	Glucose	100 mL	1.054 mol/L	Kaneko et al. 1990b
Hanseniaspora guilliermondii	Glucose	–	367 mg/L	Teixeira et al. 2002
Bacillus subtilis CICC 10025	Molasses	5 L	35.4 g/L	Xiao et al. 2007
Bacillus licheniformis MEL09	Glucose	–	41.26 g/L	Liu et al. 2011
Serratia marcescens H32	Sucrose	3.7 L	60.5 g/L	Sun et al. 2012
Bacillus subtilis TH-49	Glucose	100 L	56.9 g/L	Xu et al. 2011a
Bacillus amyloliquefaciens FMME044	Glucose	7 L	51.2 g/L	Zhang et al. 2012a
Paenibacillus polymyxa CS107	Glucose	5 L	55.3 g/L	Zhang et al. 2012b

Similarly, Sun et al. (2012) have used sucrose and corn steep-based media and concluded that the concentration of these two components has significant effect on acetoin production. Further, the authors have proposed a two-stage agitation strategy for improved acetoin production and maximum concentration of 60.5 g/L has been reported (Sun et al. 2012). Two-stage agitation strategy has also been tested for acetoin production by *Bacillus amyloliquefaciens* FMME044 (Zhang et al. 2012a). The authors showed that upon depletion of glucose level in the media, 2,3-butanediol (produced during the process) could be transformed to acetoin. Interestingly, lower agitation is suitable for 2,3-butanediol production whereas, higher agitation speed could be attributed to further transformation of 2,3-butanediol to acetoin. Therefore, the authors have used 350 rpm in first 24 h and it was increased to 500 rpm for rest of the fermentation to record improved acetoin production (Zhang et al. 2012a). Likewise, Zhang et al. have investigated acetoin production by newly isolated *Paenibacillus polymyxa* CS107. According to the authors, fed-batch fermentation using a 5 L fermenter was capable of achieving a maximum concentration of 55.3 g/L of acetoin (Zhang et al. 2012b). Considering possible industrial production of acetoin by fermentation technology, feasibility of pilot scale production of acetoin has also been investigated. Xu et al. (2011a) have developed a mutant strain of *Bacillus subtilis* (TH-49) and investigated it for acetoin production by using a 100 L fermenter. The author have reported that *Bacillus subtilis* TH-49 was capable of producing acetoin as high as 56.9 g/L by using glucose as the main feedstock (Xu et al. 2011a). Thus, based on above discussion it could be concluded that acetoin is an industrially important aroma compound and considerable success has been achieved for its large-scale production by fermentation technology. However, techno-economic feasibility of the process should be evaluated and the possibility of reducing the process cost should be explored.

Table 5.3 Biotechnological production of Vanillin

Microorganism	Major substrate	Production scale	Maximum product concentration	Reference
Streptomyces sp. strain V-1	Glucose, ferulic acid	500 mL flask	19.2 g/L	Hua et al. 2007
Escherichia coli strain JM109	Ferulic acid	100 mL flask	2.52 g/L	Barghini et al. 2007
Engineered *Saccharamyces cerevisiae*	Glucose	2 L	500 mg/L	Brochado et al. 2010
Aspergillus niger CGMCC0774 and *Pycnoporus cinnabarinus* CGMCC1115	Ferulic acid (from waste residue of rice bran oil)	25 L	2.8 g/L	Zheng et al. 2007
Staphylococcus aureus	Ferulic acid and glucose	100 mL flask	45.7 mg/L	Sarangi et al. 2010
Bacillus species	Eugenol and dimethyl sulfoxide	500 mL flask	0.32 mg/L	Sindhwani et al. 2012
Strain KK-02	Ferulic acid	50 L	15 g/L	Yiyong and Hong 2011

5.2.4 Vanillin

It is one of the industrially produced aroma compounds, which is a crystalline powder in its isolated form (Priefert et al. 2001). Major application of vanillin is as flavoring agent in foods and beverages; however, a significant portion of vanillin produced in the world is also used in cosmetics and pharmaceutical products (Priefert et al. 2001). *Vanilla planifolia*, a vanillin-containing orchid is the major natural source of vanillin (Sindhwani et al. 2012). Similarly, other natural sources of vanillin are Peru balsam, essential oil of *Java citronella* and clove bud oil; however, it is mostly produced from guaiacol or lignin by using chemical processes (Priefert et al. 2001; Sindhwani et al. 2012). Alternatively, fermentative production of vanillin by biotransformation of ferulic acid, eugenol, and isoeugenol has been extensively investigated (Hua et al. 2007). Similarly, microbiological de novo synthesis of vanillin is also known (Hua et al. 2007). Due to high consumer perception, naturally prepared vanillin is considered more suitable as a food additive and hence, it has higher market value as compared to its synthetic version (Priefert et al. 2001; Hua et al. 2007; Sindhwani et al. 2012). A summary of the recent investigations on biotechnological production of vanillin is provided in Table 5.3.

For improved vanillin production by ferulic acid bioconversion, a genetically engineered *E. coli* strain (JM109) with genes from *Pseudomonas fluorescens* BF13 has been developed by Barghini et al. (2007). The recombinant strain was found to be capable of producing vanillin without accumulation of undesirable metabolites (Barghini et al. 2007).

In silico metabolic engineering strategy could be another interesting strategy for improved vanillin production. Recently, Brochado et al. (2010) have recently developed a metabolically engineered *Saccharomyces cerevisiae* strain and according to the authors, fivefold improvement in vanillin production has been achieved (Brochado et al. 2010). As already mentioned, similar to ferulic acid eugenol could also be used as a substrate for fermentative vanillin production. Sindhwani et al. (2012) have screened a range of eugenol-degrading bacteria for vanillin production. Based on their investigation the authors have identified *Bacillus species* strain BR as a novel eugenol-degrading bacterium capable of vanillin production (Sindhwani et al. 2012). Although, vanillin could be produced from more than one feedstock, all vanillin producing microorganisms are not capable of producing vanillin from different substrates. Bloem et al. (2007) have investigated vanillin production from different substrates by lactic acid bacteria. The authors have reported that tested lactic acid bacteria were not able to produce vanillin from the substrates, such as eugenol, isoeugenol, and vanillic acid. However, *Lactobacillus* sp. tested was capable of producing vanillin by using ferulic acid as the substrate (Bloem et al. 2007). Similarly, vanillin production by ferulic acid biotransformation has also been described by Sarangi et al. (2010). According to the authors, vanillin production by *Staphylococcus aureus* could be improved by nearly fourfold by the addition of glucose as the supplementary carbon source (Sarangi et al. 2010). Recently, Yiyong and Hong (2011) have demonstrated that final vanillin concentration of 15 g/L could be achieved by biotransformation of ferulic acid in a 50 L fermenter (Yiyong and Hong 2011).

Product inhibition is one common problem associated with fermentative production of vanillin. According to Hua et al. (2007), fed-batch production of vanillin by bioconversion of high concentration of ferulic acid could be problematic owing to substantial product inhibition (Hua et al. 2007). In situ product recovery could be considered as a potential technology to mitigate such problem for improved product yield. Hua et al. (2007) have screened a range of macroporous adsorbent resins for in situ recovery of vanillin produced during bioconversion of ferulic acid by *Streptomyces* sp. strain V-1. The results indicated that resin DM11 was the most efficient among tested adsorbents and was able to improve the productivity (Hua et al. 2007).

The compound isoeugenol could also be used as a substrate for vanillin production. Li et al. (2005) have investigated enzymatic conversion of isoeugenol into vanillin. According to the authors, the efficiency of the process could be improved by addition of an adsorbent to the process. Further, the authors have reported that a maximum vanillin concentration of 2.46 g/L could be achieved in 36 h using 10 g/L of powdered activated carbon as an adsorbent (Li et al. 2005). Less expensive downstream processing could be considered as a prerequisite for successful commercialization of biotechnological products. Vanillin is also not an exception and considerable success has been achieved in this regard. In a recent report, Badcock (2011) has described an organophilic pervaporation method for vanillin recovery from fermentation media. This method is a single step, sustainable, and solvent/adsorbent-free technique for efficient vanillin recovery (Badcock 2011).

Thus, it could be concluded that considerable success has been obtained for biotechnological vanillin production. However, possible application of agro-industrial waste as substrate could be considered as an option to alleviate the overall production cost.

5.3 Possible Application of Agro-Industrial Wastes for Aroma Compound Production

From above discussion it is evident that different investigations on aroma/flavor production by fermentation technology are remarkably successful. Further, it should be noted that in almost all such studies carbon sources, such as glucose, other synthetic supplementary media components, and precursor of the particular aroma compound, were unavoidably used (Tables 5.2 and 5.3). However, application of such synthetic media components can increase the process cost of flavor production. At the same time, it should be noted that aroma/flavor compounds produced by chemical processes are less expensive. Thus, biotechnologically produced products (aroma/flavor compounds) should at least have a reasonable price, if not lesser than the products produced by chemical processes. In this context, application of agro-industrial waste materials as the substrate for the production of aroma/flavor compounds could be considered. These are generally plant-based materials containing polysaccharides, proteins, and minerals (Gonçalves et al. 2011), and hence, they have great potential as the feedstock to be used for fermentation. Moreover, utilization of these materials may solve inherent environmental pollution and disposal problem associated with agro-industrial wastes (Chapla et al. 2010). Successful application of various agro-industrial wastes, such as sugarcane bagasse, sugar beet pulp, wheat stillage, apple pomace, and cassava bagasse for high-volume low-value products, such as ethanol, single cell protein, and mushrooms, could be found in the literature (Sasaki et al. 1991; Pandey et al. 2000; Davis et al. 2005; Ruanglek et al. 2006; Philippoussis 2009). However, relatively fewer reports on aroma/flavor compounds production using agro-industrial waste could be found in the literature and some of such reports are summarized in Table 5.4.

Fruity flavor production by *Ceratocystis fimbriata* has been studied by Soares et al. (2000). The authors have investigated possible application of coffee husk as the substrate for fruity aroma production using solid-state fermentation. Further, the report has also described the addition of glucose as additional carbon source as well as the effect of precursor (e.g., leucine, soya bean oil) and saline supplements on aroma production (Soares et al. 2000). According to the authors, acetaldehyde, ethanol, isopropanol, and ethyl acetate were the major volatile compounds identified in the headspace of the system. Similarly, production of other volatile compounds, such as ethyl isobutyrate, isobutyl acetate, isoamyl acetate, and ethyl-3-hexanoate, were also detected albeit at lower concentrations (Soares et al. 2000). Similarly, Christen et al. (1997) evaluated wheat bran, sugarcane bagasse, and cassava bagasse as possible substrates for aroma/flavor production by *Ceratocystis fimbriata*.

Table 5.4 Aroma compound production from agro-industrial waste-based substrate

Aroma/aroma compound	Waste-based substrate	Microorganism	Production scale	Reference
Pineapple aroma and banana odor	Coffee husk supplemented with glucose	*Ceratocystis fimbriata*	250 mL Erlenmeyer flasks	Soares et al. 2000
Fruity aroma and banana aroma	Wheat bran, cassava bagasse, and sugarcane bagasse with supplements	*Ceratocystis fimbriata*	–	Christen et al. 1997
Vanillin	Waste residue of rice bran oil	*Aspergillus niger* CGMCC0774 and *Pycnoporus cinnabarinus* CGMCC1115	25 L fermenter	Zheng et al. 2007
Acetaldehyde, ethyl propionate, and 3-methyl butanol	Amaranth grain	*Rhizopus oryzae*	250 mL Erlenmeyer flasks	Bramorski et al. 1998
Fruity aroma compounds	Cassava bagasse and giant palm bran	*Kluyveromyces marxianus*	250 mL Erlenmeyer flasks	Medeiros et al. 2000
Acetoin	Molasses, soybean meal hydrolysate, and supplements	*Bacillus subtilis* CICC 10025	5 L fermenter	Xiao et al. 2007
Fruity aroma	Citric pulp, soya bran, sugarcane molasses, and other supplements	*Ceratocystis fimbriata*	250 mL Erlenmeyer flasks	Rossi et al. 2009

According to the authors, sensory characteristics of the final product could vary with respect to the synthetic supplement added to the process and the aroma production process was found to be growth associated. Nearly 20 compounds including alcohols, aldehyde, esters, and ketones were detected in the headspace gas mixture of the process (Christen et al. 1997). In yet another report of aroma compound production by *Ceratocystis fimbriata* Rossi et al. (2009) have demonstrated that citric pulp, a waste generated during citric juice production could be successfully utilized for aroma compound production. The solid-state fermentation was carried out in 250 mL Erlenmeyer flasks and the results showed that application of supplementary carbon and nitrogen sources have a great role in improved aroma production using the waste feedstock. Soy molasses, soy bran, sugarcane molasses, and urea were used as supplementary nutrients, and it has been demonstrated that enrichment of the citric pulp with soy bran (50 %), sugarcane molasses (25 %), and mineral saline solution was found to have positive effect on the aroma production (Rossi et al. 2009). Likewise, a two-step process for vanillin production from waste residue of rice bran oil derived ferulic acid has been demonstrated by Zheng et al. (2007). According to the process, first vanillic acid was produced by *Aspergillus niger* CGMCC0774 and it was further bioconverted into vanillin by *Pycnoporus cinnabarinus* CGMCC1115 (Zheng et al. 2007). Edible fungus *Rhizopus oryzae* is known to produce volatile aroma compounds, such as acetaldehyde using solid agro-industrial waste as substrates. Bramorski et al. (1998) have demonstrated the production of volatile metabolites by *R. oryzae* using solid-state fermentation of various tropical agro-industrial waste substrates, such as amaranth grain and cassava bagasse (Bramorski et al. 1998).

The application of yeast strains for the production of aroma production from agro-industrial waste have been also attempted by different researchers. Medeiros et al. (2000) have studied aroma compound production from different agro-industrial wastes by the yeast *Kluyveromyces marxianus* and cassava bagasse and giant palm bran were found to be the potential substrates. The authors also have mentioned that addition of glucose as an additional carbon source was found to have significant effect on aroma compound production. Based on headspace gas analysis, the authors have reported that alcohols, aldehyde, and esters were present in the produced gas mixture and esters were responsible for its fruity aroma (Medeiros et al. 2000). From above discussion it could be concluded that fungi are the most investigated organism for biotechnological production of aroma compounds using agro-industrial wastes (Table 5.4). However, yeast strains could also be potentially used for this purpose. The application of bacteria for aroma compound production using agro-industrial by-products has also been researched. Xiao et al. (2007) have investigated agro-industrial by-products, such as molasses and soybean meal hydrolysate, as the feedstock for acetoin production by *Bacillus subtilis* CICC 10025. The authors have optimized the nutritional requirement by statistical techniques and concluded that molasses and soybean meal hydrolysate has remarkable influence on acetoin production with 99 % significant level (Xiao et al. 2007).

5.4 Concluding Remarks

Agro-industrial waste materials, such as cassava bagasse, giant palm bran, citric pulp, soya bran, and sugarcane molasses, have been successfully utilized for the production of a range of natural aroma/flavor compounds. Considering the nature of the feedstock, solid-state fermentation was the mostly used process for this purpose. Produced compounds were generally volatile; hence, complex nature of agro-industrial waste-based feedstock is not going to be a problem for downstream processing. Different fungal and bacteria as well as yeast strains are known to be used for aroma compound production using agro-industrial waste; however, fungi are the mostly used microorganisms. The fact that fungi can easily grow on solid feed stocks as they resemble their natural habitats could be a possible reason for such observation. Similarly, it is also observed that so far investigations were being carried out in relatively lab scale, using Erlenmeyer flasks. Therefore, scaling up of the process and an evaluation of its techno-economic feasibility are necessary. A range of volatile compounds may produce during the process and some of these compounds may be unwanted as an aroma compound. However, downstream processing of produced gas mixture is hardly dealt with in any reports and it will be interesting to explore it. Meanwhile, in the case of mostly studied aroma compounds such as, 2-phenylethanol and vanillin, product inhibition has been observed and therefore, in situ product recovery (ISPR) techniques are widely applied. Therefore, efficiency of such techniques should be demonstrated on industrial scale prior to their commercial application. Further, along with the precursor glucose is exclusively used as the supplementary carbon source for the production of the aroma compounds, such as 2-phenylethanol and vanillin. Hence, considering possible industrial production of such compounds by bioprocess technology, less expensive agro-industrial wastes could be evaluated as a replacement of glucose.

Acknowledgements The authors are sincerely thankful to the Natural Sciences and Engineering Research Council of Canada (Discovery Grants 355254) and INRS-ETE for financial support. The views or opinions expressed in this article are those of the authors.

References

Achmon Y, Goldshtein J, Margel S, Fishman A (2011) Hydrophobic microspheres for in situ removal of 2-phenylethanol from yeast fermentation. J Microencapsul 28(7):628–638. doi:10.3109/02652048.2011.599443

Alvandi H, Azar M (2008) Diacetyl production in batch fermentation process by lactic starter culture. Iran J Food Sci Technol 5(2):27–39

Badcock M (2011) Sustainable recovery of pure natural vanillin from fermentation media in one step. http://blogs.rsc.org/gc/2011/08/01/sustainable-recovery-ofpure-natural-vanillin-from-fermentation-media-in-one-step/. Accessed on 30 Oct 2012

Barghini P, Di Gioia D, Fava F, Ruzzi M (2007) Vanillin production using metabolically engineered *Escherichia coli* under non-growing conditions. Microb Cell Fact 6(1):13

Bassit N, Boquien C-Y, Picque D, Corrieu G (1993) Effect of initial oxygen concentration on diacetyl and acetoin production by *Lactococcus lactis* subsp. lactis biovar diacetylactis. Appl Environ Microbiol 59(6):1893–1897

Benson K, Godon J, Renault P, Griffin H, Gasson M (1996) Effect of<i>ilvBN -encoded α-acetolactate synthase expression on diacetyl production in Lactococcus lactis. Appl Microbiol Biotechnol 45(1):107–111. doi:10.1007/s00253005065

Bicas JL, Silva JC, Dionísio AP, Pastore GM (2010) Biotechnological production of bioflavors and functional sugars. Ciência e Tecnologia de Alimentos 30(1):07–18

Bloem A, Bertrand A, Lonvaud-Funel A, De Revel G (2007) Vanillin production from simple phenols by wine-associated lactic acid bacteria. Lett Appl Microbiol 44(1):62–67

Boumerdassi H, Monnet C, Desmazeaud M, Corrieu G (1997) Isolation and properties of *Lactococcus lactis* subsp. lactis biovar diacetylactis CNRZ 483 mutants producing diacetyl and acetoin from glucose. Appl Environ Microbiol 63(6):2293–2299

Bramorski A, Christen P, Ramirez M, Soccol CR, Revah S (1998) Production of volatile compounds by the edible fungus *Rhizopus oryzae* during solid state cultivation on tropical agro-industrial substrates. Biotechnol Lett 20(4):359–362

Brochado AR, Matos C, Møller BL, Hansen J, Mortensen UH, Patil KR (2010) Improved vanillin production in baker's yeast through in silico design. Microb Cell Fact 9(1):84

Chapla D, Divecha J, Madamwar D, Shah A (2010) Utilization of agro-industrial waste for xylanase production by *Aspergillus foetidus* MTCC 4898 under solid state fermentation and its application in saccharification. Biochem Eng J 49(3):361–369. doi:10.1016/j.bej.2010.01.012

Christen P, Meza J, Revah S (1997) Fruity aroma production in solid state fermentation by Ceratocystis fimbriata: influence of the substrate type and the presence of precursors. Mycol Res 101(8):911–919

Davis L, Jeon YJ, Svenson C, Rogers P, Pearce J, Peiris P (2005) Evaluation of wheat stillage for ethanol production by recombinant *Zymomonas mobilis*. Biomass Bioenergy 29(1):49–59

Deiana P, Cecchi L, Lodi R, Berardi E, Farris G, Fatichenti F (1990) Some aspects of diacetyl and acetoin production by *Debaryomyces hansenii*. Ital J Food Sci 2(1):35–42

Dastager SG (2009) Aroma compounds, biotechnology for agro-industrial residues utilisation. In: Pandey A, Nigam P (eds) Utilisation of Agro-Residues. Springer, Netherlands, pp 105–127. doi:10.1007/978-1-4020-9942-7_6

Duboff SA, Kwon SS, Vadehra DV (1996) Diacetyl production. EP Patent 0,564,770

Eshkol N, Sendovski M, Bahalul M, Katz-Ezov T, Kashi Y, Fishman A (2009) Production of 2-phenylethanol from L-phenylalanine by a stress tolerant *Saccharomyces cerevisiae* strain. J Appl Microbiol 106(2):534–542

Etschmann M, Bluemke W, Sell D, Schrader J (2002) Biotechnological production of 2-phenylethanol. Appl Microbiol Biotechnol 59(1):1–8

Etschmann MMW, Sell D, Schrader J (2003) Screening of yeasts for the production of the aroma compound 2-phenylethanol in a molasses-based medium. Biotechnol Lett 25(7):531–536. doi: 10.1023/a:1022890119847

Etschmann MMW, Sell D, Schrader J (2005) Production of 2-phenylethanol and 2-phenylethylacetate from L-phenylalanine by coupling whole-cell biocatalysis with organophilic pervaporation. Biotechnol Bioeng 92(5):624–634. doi:10.1002/bit.20655

Feron G, Bonnarme P, Durand A (1996) Prospects for the microbial production of food flavours. Trend Food Sci Technol 7(9):285–293. doi:10.1016/0924-2244(96)10032-7

Feron G, Waché Y (2006) Microbial biotechnology of food flavor production. Food Sci Technol 148:407, Marcel Dekker, New York

Fornachon J, Lloyd B (2006) Bacterial production of diacetyl and actoin in wine. J Sci Food Agric 16(12):710–716

Fabre C, Blanc P, Goma G (1998) 2-Phenylethyl alcohol: an aroma profile. Perfumer flavorist 23(3):43–45

Garc AI (1994) Modelling of diacetyl production during beer fermentation. J Inst Brew 100:179–183

Gonçalves FA, Sanjinez-Argandoña EJ, Fonseca GG (2011) Utilization of agro-industrial residues and municipal waste of plant origin for cellulosic ethanol production. J Environ Protect 2(10):1303–1309

Hua D, Ma C, Song L, Lin S, Zhang Z, Deng Z, Xu P (2007) Enhanced vanillin production from ferulic acid using adsorbent resin. Appl Microbiol Biotechnol 74(4):783–790. doi:10.1007/s00253-006-0735-5

Hua D, Lin S, Li Y, Chen H, Zhang Z, Du Y, Zhang X, Xu P (2010) Enhanced 2-phenylethanol production from L-phenylalanine via in situ product adsorption. Biocatal Biotransform 28(4):259–266. doi:10.3109/10242422.2010.500724

Hua D, Xu P (2011) Recent advances in biotechnological production of 2-phenylethanol. Biotechnol Adv 29(6):654–660. doi:10.1016/j.biotechadv.2011.05.001

Hugenholtz J, Kleerebezem M, Starrenburg M, Delcour J, De Vos W, Hols P (2000) *Lactococcus lactis* as a cell factory for high-level diacetyl production. Appl Environ Microbiol 66(9):4112–4114

Kaneko T, Watanabe Y, Suzuki H (1990a) Enhancement of diacetyl production by a diacetyl-resistant mutant of citrate-positive *Lactococcus lactis* ssp. lactis 3022 and by aerobic conditions of growth. J Dairy Sci 73(2):291–298. doi:10.3168/jds.S0022-0302(90)78672-9

Kaneko T, Takahashi M, Suzuki H (1990b) Acetoin fermentation by citrate-positive *Lactococcus lactis* subsp. lactis 3022 grown aerobically in the presence of hemin or Cu2+. Appl Environ Microbiol 56(9):2644–2649

Koma D, Yamanaka H, Moriyoshi K, Ohmoto T, Sakai K (2012) Production of aromatic compounds by metabolically engineered *Escherichia coli* with shikimate pathway expansion. Appl Environ Microbiol 78:6203–6216. doi:10.1128/aem.01148-12

Krings U, Berger R (1998) Biotechnological production of flavours and fragrances. Appl Microbiol Biotechnol 49:1–8

Li YH, Sun ZH, Zhao LQ, Xu Y (2005) Bioconversion of isoeugenol into vanillin by crude enzyme extracted from soybean. Appl Biochem Biotechnol 125(1):1–10

Liu Y, Zhang S, Yong YC, Ji Z, Ma X, Xu Z, Chen S (2011) Efficient production of acetoin by the newly isolated *Bacillus licheniformis* strain MEL09. Process Biochem 46(1):390–394

Longo MA, Sanromán MA (2006) Production of food aroma compounds: microbial and enzymatic methodologies. Food Technol Biotechnol 44(3):335–353

Mameeva O, Ostapchuk A, Podgorsky V (2010) The 2-phenylethanol and ethanol production by yeast *Saccharomyces cerevisiae*. http://www.nbuv.gov.ua/Portal/Chem_Biol/Mib/2010_1/2.pdf. Accessed on 17 Nov 2012

Medeiros ABP, Pandey A, Freitas RJS, Christen P, Soccol CR (2000) Optimization of the production of aroma compounds by *Kluyveromyces marxianus* in solid-state fermentation using factorial design and response surface methodology. Biochem Eng J 6(1):33–39. doi:10.1016/s1369-703x(00)00065-6

Monnet C, Schmilt P, Divies C (1994) Diacetyl production in milk by an α-acetolactic acid accumulating strain of *Lactococcus lactis* ssp. lactis biovar. diacetylactis. J Dairy Sci 77(10):2916–2924. doi:10.3168/jds.S0022-0302(94)77232-5

Nadal I, Rico J, Pérez-Martínez G, Yebra M, Monedero V (2009) Diacetyl and acetoin production from whey permeate using engineered *Lactobacillus casei*. J Ind Microbiol Biotechnol 36(9):1233–1237

Pandey A, Soccol CR, Nigam P, Soccol VT, Vandenberghe LPS, Mohan R (2000) Biotechnological potential of agro-industrial residues. II: cassava bagasse. Bioresour Technol 74(1):81–87

Philippoussis AN (2009) Production of mushrooms using agro-industrial residues as substrates. Biotechnology for agro-industrial residues utilisation. Springer, Netherlands, pp 163–196

Priefert H, Rabenhorst J, Steinbüchel A (2001) Biotechnological production of vanillin. Appl Microbiol Biotechnol 56(3):296–314. doi:10.1007/s002530100687

Quach AT, Liu C, Davies IY, Elston LD (2012) Toxicology report for Diacetyl. http://www.jakeheng.com/lulu.pdf. Accessed on 17 Nov 2012

Ramachandra Rao S, Ravishankar G (2000) Vanilla flavour: production by conventional and biotechnological routes. J Sci Food Agric 80:289–304

Rong S, Ding B, Zhang X, Zheng X, Wang Y (2011) Enhanced biotransformation of 2-phenylethanol with ethanol oxidation in a solid–liquid two-phase system by active dry yeast. Curr Microbiol 63(5):503–509. doi:10.1007/s00284-011-0008-0

Rossi S, Vandenberghe L, Pereira B, Gago F, Rizzolo J, Pandey A, Soccol C, Medeiros A (2009) Improving fruity aroma production by fungi in SSF using citric pulp. Food Res Int 42(4):484–486

Ruanglek V, Maneewatthana D, Tripetchkul S (2006) Evaluation of Thai agro-industrial wastes for bio-ethanol production by *Zymomonas mobilis*. Process Biochem 41(6):1432–1437

Sarangi PK, Nanda S, Sahoo H (2010) Maximization of vanillin production by standardizing different cultural conditions for ferulic acid degradation. NY Sci J 3(7):77–79

Sasaki K, Noparatnaraporn N, Nagai S, Martin A (1991) Use of photosynthetic bacteria for the production of SCP and chemicals from agroindustrial wastes. Bioconversion of waste materials to industrial products. Elsevier, New York, NY, pp 225–264

Savina JP, Kohler D, Brunerie P (1999) Method for extracting 2-phenylethanol. Google Patents

Sendovski M, Nir N, Fishman A (2010) Bioproduction of 2-phenylethanol in a biphasic ionic liquid aqueous system. J Agric Food Chem 58(4):2260–2265. doi:10.1021/jf903879x

Serp D, von Stockar U, Marison IW (2003) Enhancement of 2-phenylethanol productivity by *Saccharomyces cerevisiae* in two-phase fed-batch fermentations using solvent immobilization. Biotechnol Bioeng 82(1):103–110. doi:10.1002/bit.1054

Sindhwani G, Ilyas U, Aeri V (2012) Microbial transformation of eugenol to vanillin. J Microbiol Biotechnol Res 2(2):313–318

Soares M, Christen P, Pandey A, Soccol CR (2000) Fruity flavour production by *Ceratocystis fimbriata* grown on coffee husk in solid-state fermentation. Process Biochem 35(8):857–861

Sun J, Zhang L, Rao B, Han Y, Chu J, Zhu J, Shen Y, Wei D (2012) Enhanced acetoin production by *Serratia marcescens* H32 using statistical optimization and a two-stage agitation speed control strategy. Biotechnol Bioprocess Eng 17(3):598–605

Swindell SR, Benson KH, Griffin HG, Renault P, Ehrlich S, Gasson MJ (1996) Genetic manipulation of the pathway for diacetyl metabolism in *Lactococcus lactis*. Appl Environ Microbiol 62(7):2641–2643

Teixeira R, Cavalheiro D, Ninow J, Furigo A Jr (2002) Optimization of acetoin production by *Hanseniaspora guilliermondii* using experimental design. Braz J Chem Eng 19(2):181–186

Wang H, Dong Q, Guan A, Meng C, Xa S, Guo Y (2011) Synergistic inhibition effect of 2-phenylethanol and ethanol on bioproduction of natural 2-phenylethanol by *Saccharomyces cerevisiae* and process enhancement. J Biosci Bioeng 112(1):26–31. doi:10.1016/j.jbiosc.2011.03.006

Wittmann C, Hans M, Bluemke W (2002) Metabolic physiology of aroma-producing *Kluyveromyces marxianus*. Yeast 19(15):1351–1363

Xiao Z, Liu P, Qin JY, Xu P (2007) Statistical optimization of medium components for enhanced acetoin production from molasses and soybean meal hydrolysate. Appl Microbiol Biotechnol 74(1):61–68

Xu P, Hua D, Ma C (2007) Microbial transformation of propenylbenzenes for natural flavour production. Trends Biotechnol 25(12):571–576

Xu H, Jia S, Liu J (2011a) Production of acetoin by Bacillus subtilis TH-49. In: Consumer Electronics, Communications and Networks (CECNet), International Conference, 2011, IEEE, pp 1524–1527

Xu H, Jia S, Liu J (2011b) Development of a mutant strain of *Bacillus subtilis* showing enhanced production of acetoin. Afr J Biotechnol 10(5):779–788

Yiyong DUYZZZ, Hong C (2011) A new bioprocess to produce natural vanillin by microbial fermentation. Flavour Frag Cosmet 3:003

Zhang Y, Li S, Liu L, Wu J (2012a) Acetoin production enhanced by manipulating carbon flux in a newly isolated *Bacillus amyloliquefaciens*. Bioresource Technol 130:256–260. http://dx.doi.org/10.1016/j.biortech.2012.10.036

Zhang L, Chen S, Xie H, Tian Y, Hu K (2012b) Efficient acetoin production by optimization of medium components and oxygen supply control using a newly isolated *Paenibacillus polymyxa* CS107. J Chem Technol Biotechnol 87(11):1551–1557

Zheng L, Zheng P, Sun Z, Bai Y, Wang J, Guo X (2007) Production of vanillin from waste residue of rice bran oil by *Aspergillus niger* and *Pycnoporus cinnabarinus*. Bioresour Technol 98(5):1115–1119

Chapter 6
Antioxidants

Mamta, Kshipra Misra, Gurpreet Singh Dhillon, Satinder Kaur Brar, and Mausam Verma

6.1 Introduction

Antioxidants are the molecules that prevent cellular damage caused by oxidation of other molecules. Oxidation is a chemical reaction that transfers electrons from one molecule to an oxidizing agent. Oxidation reactions are known to produce free radicals. These free radicals are highly reactive species which contains one or more unpaired electrons in their outermost shell. Once they are formed, the chain reaction starts. Antioxidant reacts with these free radicals and terminates this chain reaction by removing free radical intermediates and inhibits other oxidation reactions by oxidizing themselves.

Though oxidation reactions are crucial for life, they can also be damaging. Plants and animals have a complex system of multiple types of antioxidants, such as vitamin C and vitamin E, as well as enzymes, such as catalase (CAT), superoxide dismutase (SOD), and various peroxidases (Hamid et al. 2010). Oxidative stress plays a key role in causing various human diseases, such as cellular necrosis, cardiovascular disease, cancer, neurological disorder, Parkinson's dementia, Alzheimer's disease, inflammatory disease, muscular dystrophy, liver disorder, and even aging (Amit and Priyadarsini

Mamta • K. Misra (✉)
Defence Institute of Physiology & Allied Sciences, Timarpur, Lucknow Road, Delhi 110054, India
e-mail: kmisra99@yahoo.com

G.S. Dhillon
Biorefining Conversions Network (BCN), Department of Agricultural, Food and Nutritional Sciences (AFNS), University of Alberta, Edmonton, AB, Canada

INRS-ETE, Université du Québec, 490 Rue de la Couronne, Québec, QC, Canada G1K 9A9

S.K. Brar
INRS-ETE, Université du Québec, 490, Rue de la Couronne, Québec, QC, Canada G1K 9A9

M. Verma
CO2 Solutions, 2300, rue Jean-Perrin, Québec, QC, Canada G2C 1T9

S.K. Brar et al. (eds.), *Biotransformation of Waste Biomass into High Value Biochemicals*, DOI 10.1007/978-1-4614-8005-1_6,

2011). Besides, there are some antioxidants in the form of micronutrients which cannot be manufactured by the body itself such as vitamin E, β-carotene, and vitamin C, and hence these must be supplemented in the normal diet (Teresa et al. 2011).

Antioxidants can also act as prooxidants when these are not present at the right place at the right concentration at the right time (Touriño et al. 2008). The relative importance of the antioxidant and prooxidant activities is not yet explored fully and needs further research.

In this chapter, authors have tried to discuss the various types, sources, synthesis, uses, and protective efficacy of antioxidant with examples.

6.2 Classification of Antioxidants

Antioxidants can be classified into two major types based on their source, i.e., natural and synthetic antioxidants (schematic representation of the classification of antioxidants is shown in Fig. 6.1).

6.2.1 Natural Antioxidants

Natural antioxidants either are synthesized in human body through metabolic process or are supplemented from other natural sources, and their activity very much depends upon their physical and chemical properties and mechanism of action. This can be further divided into two categories, i.e., enzymatic antioxidants and nonenzymatic antioxidants.

6.2.1.1 Enzymatic Antioxidants

Enzymatic antioxidants are uniquely produced in the human body and can be subdivided into primary and secondary antioxidant.

Primary Antioxidants

Primary antioxidants mainly include superoxide dismutase (SOD), catalase (CAT), and glutathione peroxidase (GPx) as described below.

Superoxide Dismutase Superoxide dismutase (SOD) enzyme is found in both the dermis and the epidermis. It removes the superoxide radical ($O_2^{\cdot-}$) and repairs the body cells damaged by free radical. SOD catalyzes the reduction of superoxide anions to hydrogen peroxide (6.1). SOD is also known to compete with nitric oxide (NO) for superoxide anion, which inactivates NO to form peroxynitrite. Therefore, by scavenging superoxide anions, it promotes the activity of NO (Chakraborty et al. 2009).

$$2O_2^{\cdot-} + 2H^+ \xrightarrow{SOD} H_2O_2 + O_2 \tag{6.1}$$

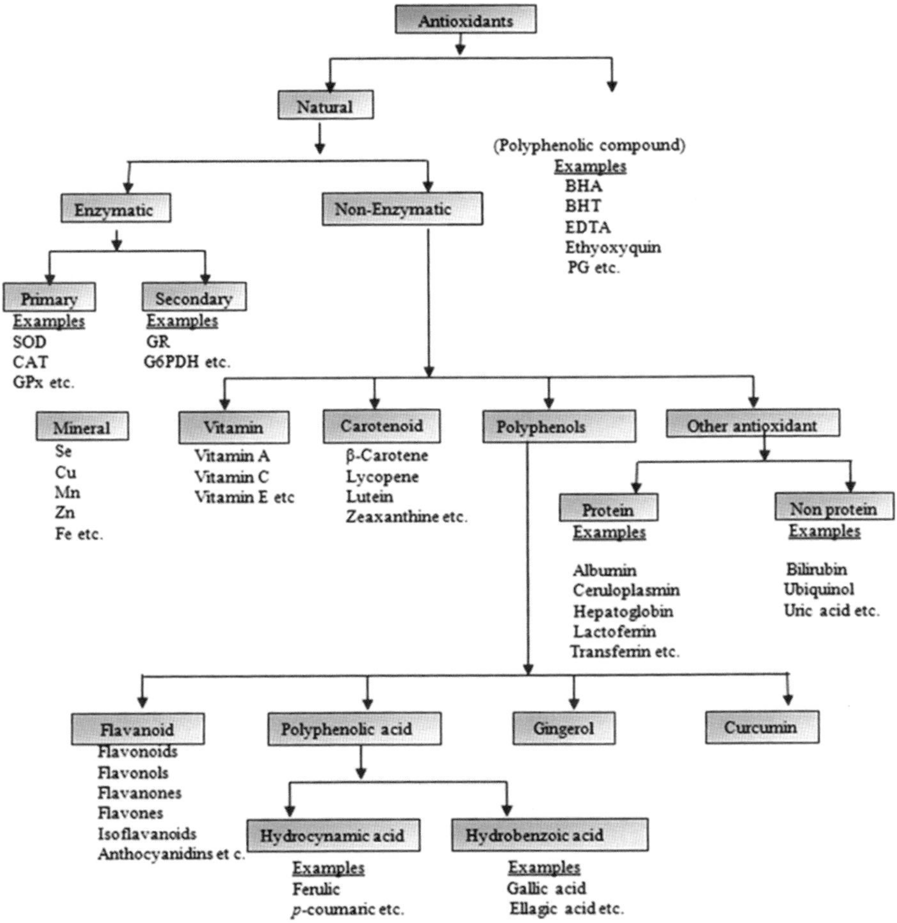

Fig. 6.1 Schematic representation of classification of antioxidants

Catalase Catalase enzyme (CAT) is found in the blood and most of the living cells and decomposes H_2O_2 into water and oxygen (6.2). Catalase with glucose peroxidase is also used commercially for the preservation of the fruit juices, cream consisting of egg yolk, and salad by removing the oxygen (Chakraborty et al. 2009).

$$2H_2O_2 \xrightarrow{CAT} 2H_2O + O_2 \tag{6.2}$$

Glutathione Peroxidase Glutathione peroxidase (GPx) is a group of selenium-dependent enzymes, and it consists of cytosolic, plasma, phospholipid hydroperoxide, and gastrointestinal glutathione peroxidase (Chakraborty et al. 2009). GPx (cellular and plasma) catalyzes the reaction of H_2O_2 by reduced glutathione (GSH); as a

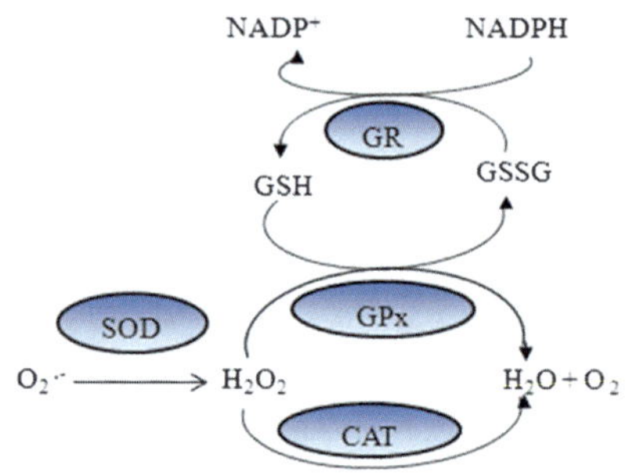

Fig. 6.2 Outline of the mechanism of enzymatic antioxidants in the removal of free radical

result, oxidized glutathione (GSSG) is produced (6.3) and it is again recycled to its reduced form by glutathione reductase (GR) and reduced nicotinamide adenine dinucleotide phosphate (NADPH).

$$2GSH + H_2O_2 \xrightarrow{GPx} GSSG + 2H_2O \tag{6.3}$$

Secondary Antioxidant

Secondary antioxidant includes glutathione reductase (GR) and glucose-6-phosphate dehydrogenase (G6PDH). G6PDH generates NADPH. GR is required to recycle the reduced glutathione (GSH) using secondary enzyme GR and NADPH (6.4).

$$GSSG + NADPH \xrightarrow{GR} NADP + 2GSH \tag{6.4}$$

Glutathione is a cysteine containing peptide-type antioxidant and is synthesized in the body cells. The thiol group in its cysteine moiety is a reducing agent and can be reversibly oxidized and reduced. A high level of glutathione is found in the cells (~3,100 μg/g of tissue) (Hissin and Hilf 1976), maintained in the reduced form (GSH) by the enzyme GR, and in turn reduces other metabolites and enzyme systems, such as ascorbate. Due to its high concentration and its role in maintaining redox state in the cells, it is considered one of the most important cellular antioxidants. (Outline of the mechanism of enzymatic antioxidants in the removal of free radical is shown in Fig. 6.2.)

6.2.1.2 Nonenzymatic Antioxidants

They are a class of the antioxidants which are not found in the body naturally but are required to be supplemented for the proper metabolism (Raygani et al. 2007). Some of the known nonenzymatic antioxidants are minerals, vitamins, carotenoids, polyphenols, and other antioxidants as listed below.

Minerals

Minerals are required in the body cells for the proper functioning of the enzymes. Their absence is known to affect the metabolism of many macromolecules. They include selenium, copper, iron, zinc, and manganese. They act as cofactors for the enzymatic antioxidants.

Iron (Fe) Iron is the most abundant trace metal found to bound with protein in the biological system. Normally the concentration of free iron is very low and the low concentrations of iron-binding proteins promote ROS production, lipid peroxidation, and oxidative stress (Dabbagh et al. 1984). Hence iron supplementation helps in reducing the oxidative stress.

Magnesium (Mg) Magnesium is a cofactor for glucose-6-phosphate dehydrogenase (G6PD) and 6-phosphogluconate dehydrogenase (6PGD) involved in pentose cycle which catalyzes the production of NADPH from NADP during the glucose metabolism and hence maintains the normal ratio of GSH to GSSG and a normal redox state in cells. Deficiency of magnesium reduces GR activity and GSSG does not reduce to GSH, hence causing oxidative damage to the cells (Fang et al. 2002).

Selenium (Se) Selenium is also a very important component of enzymatic antioxidant. In the presence of selenium (Se), glutathione peroxidase (GPx) plays a protective role against oxidation of lipid and protects the cell membrane and takes part in H_2O_2 and lipids' hydroxyperoxide metabolism. Hence, Se behaves like vitamin E and can be substituted in place of vitamin E and is used to prevent the risk of cancer and cardiovascular diseases (Sikora et al. 2008).

Copper (Cu), Zinc (Zn), and Manganese (Mn) SOD is a class of enzyme that consists of different types of SODs, depending upon their metal cofactor such as Cu–Zn and Mn. Cu–Zn SOD is found in the cytosol having Cu and Zn at their active sites which helps in proton conduction, whereas Mn-SOD is found in mitochondria and has Mn at its active site. These metals are responsible for SOD's antioxidant activities.

Vitamins

Vitamins form the class of micronutrients required for the proper functioning of the body's antioxidant enzyme system, such as vitamin A, vitamin C, vitamin E, and vitamin B. They cannot be synthesized in our body and hence need to be supplemented in the diet.

Vitamin A Vitamin A is helpful in night vision and in maintenance of epithelial cells in mucus membranes and skin. Because of its antioxidant properties, it assists immune system also and is found in three main forms: retinol, 3,4-didehydroretinol, and 3-hydroxyretinol. The main sources of this include sweet potatoes, carrots, milk, egg yolks, and mozzarella cheese.

Vitamin C Vitamin C is water soluble and is also called as ascorbic acid. It is found in fruits (mainly citrus), vegetables, cereals, beef, poultry, fish, etc. It is helpful in preventing some of the DNA damage caused by free radicals, which may contribute to the aging process and the development of diseases, such as cancer, heart disease, and arthritis.

Vitamin E Vitamin E is a lipid-soluble vitamin. This consists of eight different forms such as α-, β-, γ-, and δ-tocopherol and α-, β-, γ-, and δ-tocotrienol. Most abundantly found in almonds, safflower oil, soybean oils, oil of wheat germs, nuts, broccoli, fish oil, etc., α-tocopherol possesses highest bioavailability and is the most important lipid-soluble antioxidant which reacts with the lipid radical and protects the membranes from lipid peroxidation; as a result, oxidized α-tocopheroxyl radicals are produced that can be recycled to the reduced form through reduction by other antioxidants, such as ascorbate and retinol.

Carotenoid

Carotenoid consists of β-carotene, lycopene, lutein, and zeaxanthin. They are fat-soluble colored compounds found in fruits and vegetables. β-Carotene is found mostly in radish-orange-green color food items including carrots, sweet potatoes, apricots, pumpkin, mangoes, and cantaloupe along with some green and leafy vegetables, including collard greens, spinach, and kale. Lutein is abundant in green leafy vegetables such as collard greens, spinach, and kale (Hamid et al. 2010). Lutein is best known for its role in protection of retina against harmful action of free radicals and also prevents atherosclerosis (Sikora et al. 2008).

Although lycopene, lutein, canthaxanthin, and zeaxanthin do not possess provitamin A activity, β-carotene is known as a precursor for vitamin A (Fang et al. 2002). Tomato is a good source of lycopene and spinach is a good source of zeaxanthin. It has been shown that lycopene is a potent antioxidant and is the most effective compound in removing singlet oxygen found in tomatoes, watermelon, guava, papaya, apricots, pink grapefruit, and other foods.

Polyphenols

Polyphenols is a class of the phytochemicals that possess marked antioxidant activities. Their antioxidant activities depend on their chemical and physical properties which in turn regulates the metabolism depending on their molecular structures (Ajila et al. 2011). These consist of phenolic acids, flavonoids, gingerol, curcumin, etc. (Amit and Priyadarsini 2011).

Flavonoid is a major class of polyphenolic compound and is mostly found in vegetables, fruits, grains, seeds, leaves, flower, bark, etc. Some of the spices, such as ginger and turmeric, are also good sources of polyphenolic compound, e.g., gingerol is obtained from the rhizomes of ginger, whereas curcumin (diferuloylmethane) is the main bioactive component of turmeric and is known to possess good antioxidant activity. Curcumin is an excellent scavenger of ROS, such as $O_2^{\cdot -}$ radicals, lipid peroxyl radicals ($LO_2^{\cdot}$), OH radicals, and nitrogen dioxide ($NO_2^{\cdot}$)

radicals, which induced oxidative stress. Curcumin has been shown to inhibit lipid peroxidation and has been shown to increase GSH levels also in epithelial cells which lead to lower ROS production (Biswas et al. 2005).

Other Antioxidants

Transition Metal-Binding Proteins Albumin, ceruloplasmin, hepatoglobin, and transferrin are the transition metal-binding proteins found in human plasma, bind with transition metals, and control the production of metal catalyzed free radicals. Albumin and ceruloplasmin are the copper ion sequesters, hepatoglobin is hemoglobin sequester, and transferrin acts as free iron sequester.

Nonprotein Antioxidants Bilirubin, uric acids, and ubiquinol are nonprotein antioxidants which inhibit the oxidation processes by scavenging free radicals (Papas 1998).

Bilirubin Bilirubin is an end product of heme catabolism. It is a lipid-soluble cytotoxic product that needs to be excreted. However, bilirubin efficiently scavenges peroxyl radical at micromolar concentrations in in vitro model (Stocker et al. 1987) and is regarded as the best antioxidant against lipid peroxidation.

Uric Acid Uric acid is a powerful antioxidant and is a scavenger of singlet oxygen and radicals. Urate reduces the oxo-heme oxidant formed by peroxide reaction with hemoglobin and protects erythrocytes from peroxidative damage. The plasma-urate levels in humans are about 300 μM, making it one of the major antioxidants in humans (Ames et al. 1981).

Coenzyme Q Coenzyme Q is also known as ubiquinol (Co Q) and is an oil-soluble antioxidant. This is produced in the body through monovalent pathway, in heart, liver, kidney, pancreas, etc. The mechanism of the action may occur in two ways:

In the first mechanism, reduced form of ubiquinol (CoQH) acts as chain-breaking antioxidant and reduces peroxyl (ROO·) and alcoxyl radicals (LO·) (Papas 1998) (6.5 and 6.6).

$$\mathrm{CoQH + ROO^{\cdot} \rightarrow Q^{\cdot} + ROOH} \tag{6.5}$$

In the second mechanism, it reacts with vitamin E radical (TO·) and regenerating vitamin E.

$$\mathrm{CoQH + TO^{\cdot} \rightarrow Q^{\cdot} + ROOH} \tag{6.6}$$

6.2.2 Synthetic Antioxidants

Synthetic antioxidants are artificially produced or synthesized using various techniques. Generally, they are polyphenolic compounds mainly that capture the free

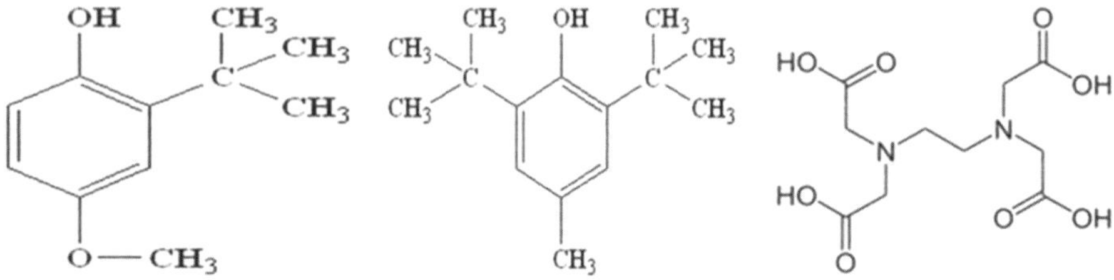

Fig. 6.3 (1) BHA, (2) BHT, (3) EDTA, (4) Ethoxyquin, (5) PG, (6) TBHQ

radicals and stop the chain reactions. Polyphenolic derivatives usually contain more than one hydroxyl or methoxy group. Ethoxy quinine is the only heterocyclic, N-containing compound reported to be used as antioxidant in the food, especially animal feed. Mostly reported synthetic phenolic antioxidants are *p*-substituted, whereas the natural phenolic compounds are mostly *o*-substituted. The *p-substituted* substances are preferred because of their lower toxicity. Synthetic phenolic antioxidants are always substituted with alkyl groups to improve their solubility in fats and oils and to reduce their toxicity. These synthetic compounds possessing antioxidant activity are commonly used in pharmaceuticals, as preservatives for cosmetics and to stabilize the fat, oil, and lipid in food (Gupta and Sharma 2006).

These new findings about the synthetic antioxidants have led the researches to develop new synthetic antioxidants in terms of their water solubility, stability, and non-toxicity. Characteristics of some of the known synthetic antioxidants, such as butylated hydroxyanisole (BHA), butylated hydroxytoluene (BHT), ethylenediaminetetraacetic acid (EDTA), 6-ethoxy-1,2-dihydro-2,2,4-trimethylquinoline (ethoxyquin), propyl gallate (PG), and tertiary butylhydroquinone (TBHQ), are given below (Hamid et al. 2010) (structures of these antioxidants are shown in Fig. 6.3).

6.2.2.1 BHA

It is a monophenolic, lipid-soluble antioxidant, better used for the lipid oxidation in animal fat compared to vegetable oil.

6.2.2.2 BHT

It is also a monophenolic fat-soluble antioxidant but is more stable than BHA at high temperature, and both act synergistically. Many commercially available antioxidant formulations contain both of these antioxidants. BHA interacts with peroxy radicals to produce a BHA phenoxy radical which in turn may remove a hydrogen atom from the hydroxyl group of BHT. BHA is regenerated by the hydrogen radical provided by BHT. The BHT radicals so formed can react with a peroxy radical and act as a chain terminator.

6.2.2.3 EDTA

EDTA is a common sequestrant, water-soluble antioxidant added to foods, body care, and household products. It binds with trace minerals, such as copper, iron, and nickel, that may be present in the food product. If not inactivated, these minerals may lead to discoloration, rancidity, and textural breakdown. When added as an antioxidant, EDTA prevents oxygen from causing color changes and rancidity.

6.2.2.4 Ethoxyquin

It is as an antioxidant primarily used to protect carotenoid oxidation in animal feeds, vegetables, and fruits during storage.

6.2.2.5 PG

It is an ester formed by the condensation of gallic acid and propanol. It acts as an antioxidant which is used as a food additive to protect mainly oils and fat in the food products.

6.2.2.6 TBHQ

TBHQ is a highly effective diphenolic antioxidant. In foods, it is used as a preservative for unsaturated vegetable oils and many edible animal fats. It does not cause discoloration even in the presence of iron and does not even change flavor or odor of the material to which it is added. It is used industrially as a stabilizer to inhibit auto-polymerization of organic peroxides. It is also used as a corrosion inhibitor in biodiesel. In perfumery, it is used as a fixative to lower the evaporation rate and improve stability. It is also added to varnishes, lacquers, resins, and oil field additives. It can be used alone or in combination with BHA or BHT (Said et al. 2002).

6.3 Sources of Antioxidants

Antioxidants can be derived from two main sources: natural source such as fruits, vegetables, cereals, legumes, beverages, spices, and animals (Table 6.1) and from agro-industry, e.g., waste processing industry (Table 6.2).

Table 6.1 List of natural sources of antioxidants

Source	Name	Type of antioxidants	References
Fruits	Blackcurrant	Vitamin C, carotene (lutein and β-carotene), phenolic compound (anthocyanins), and phenolic acids (hydroxycinnamic acid)	Hägg et al. (1995a), Benvenuti et al. (2004a), Zadernowski et al. (2005a)
	Strawberry	Vitamin C and phenolic compound (anthocyanins and ellagic acid with its derivatives)	Hägg et al. (1995b), Anttonen and Karjalainen (2005)
	Grapes	*Seed*: gallic acid, catechins, and epicatechins *Peel*: ellagic acid, myricetin, quercetin, kaempferol, *trans-resveratrol*, anthocyanins, flavonols, etc.	Pastrana-Bonilla et al. (2003)
	Bilberries	Anthocyanin, vitamin C, carotenoids, etc.	Kähkönen et al. (1999)
	Cranberries	Anthocyanins (peonidin and cyanidin), flavanones, procyanidin, flavonols (quercetin and myricetin), derivatives of hydroxycinnamic acid, etc.	Kähkönen et al. (1999, 2001), Taruscio et al. (2004), and Määttä-Riihinen et al. (2004)
	Crowberry fruits	Vitamin C, carotenoids (lutein, β-carotene), and phenolic compounds	Kähkönen et al. (1999, 2001), Halvorsen et al. (2002), and Olsson et al. (2004)
	Blackberry	*Pulp*: anthocyanins, flavonols, and ellagic acid *Seed*: procyanidins and epicatechins	Benvenuti et al. (2004b), Siriwoharn et al. (2004), Reyes-Carmona et al. (2005), and Zadernowski et al. (2005b)
	Citrus food (lemons, oranges, grapefruits, etc.)	Vitamin C, polyphenolic compounds such as ferulic acid, *p*-coumaric acid, and caffeic acid	Gorinstein et al. (2001)
	Apple (peel contains seven times higher polyphenols than in pulp)	*Polyphenols*: epicatechin and its dimmer procyanidin B2 *Phenolic acid*: chlorogenic acid *Dihydrochalcones*: phlorizin, phloretin-2-xyloglucoside, etc.	Lu and Foo (1997, 2000)
	Cherries	Hydroxycinnamic acid and anthocyanins	Chaovanalikit and Wrolstad (2004)
	Plums, prunes, pears, kiwi	Hydroxycinnamic acid, catechins	Belitz and Grosch (1999), Yanishlieva-Maslarova and Heinonen (2001), and Mannach et al. (2004)
Vegetables	Tomatoes	Lycopene, quercetin, etc.	Stewart et al. (2000) and Knoblich et al. (2005)
	Onion	Flavonoids	Lachman et al. (2003)
	Parsley roots, carrot, and pumpkin	Vitamin C and β-carotene	Sikora et al. (2008)
	Parsley	Flavones	Beecher (2003)
	Brassica vegetables (white cabbage, kale, broccoli sprouts, or cauliflower)	Vitamin C, carotenoids, derivatives of hydrocinnamic acids such as chlorogenic acid, ferulic acid, and flavonols	Kurilich et al. (1999), Kopsell et al. (2004), and Vallejo et al. (2003)
	Spinach	Flavonoids, *p*-coumaric acid, etc.	Lomnitski et al. (2003)

Cereals and legumes	Wheat and rye	Phenolic acid: ferulic, vanillic, and *p*-coumaric acids	Yanishlieva-Maslarova and Heinonen (2001), and Mannach et al. (2004)
	Oats	Avertramidin (polyphenols of oats), catechins, etc.	Nie et al. (2006) and Peterson et al. (2001)
	Buckwheat	Polyphenols: rutin, catechins, etc.	Holasova et al. (2002)
	Rice	Anthocyanins, gallic acid, ferulic acids, etc.	Laokuldilok et al. (2011)
	Beans	Seed: flavanols	Mannach et al. (2004)
	Soybeans	Isoflavones and phenolic acid (*p*-hydroxybenzoic acid, salicylic acid, *p*-coumaric acid, and ferulic acid) along with tocopherol, sterols, phospholipids, etc.	Kim et al. (2006)
	Cocoa seeds	Procyanidins, quercetin and its glycosides, caffeic acid, etc.	Sikora et al. (2008) and Jolić et al. (2011)
Beverages: alcoholic drink	Red wine	Polyphenols: flavan-3-ols, flavanols, anthocyanins, etc.	Mannach et al. (2004) and Beecher (2003)
	Cider	Hydrocinnamic acid	Mannach et al. (2004)
	Beer	*Phenolic acid*: cinnamic, chlorogenic, vanillic, ferulic, and gallic acids *Flavan-3-ol*: catechin, epicatechin, procyanidin, prodelphinidin, etc.	Sikora et al. (2008) and Walters et al. (1997)
Other drinks	Orange juice	Flavanols etc.	Mannach et al. (2004)
	Tea	Catechins, theaflavins, etc.	Sikora et al. (2008) and Friedman et al. (2006)
	Coffee	Hydrocinnamic acid and chlorogenic acid	Sikora et al. (2008) and Mannach et al. (2004)
	Chocolate	Flavanols etc.	Mannach et al. (2004); Beecher (2003)
Spices	Ginger	Gingerol	Wang et al. (2003)
	Rosemary	Carnosol, carnosic acid, rosmanul, etc.	Senorans et al. (2000)
	Sage	Carnosol, carnosic acid, luteolin, rosmanul, rosmarinic acid, etc.	Yanishlieva-Maslarova and Heinonen (2001)
	Thyme	Thymol, flavonoids, luteolin, etc.	Zheng and Wang (2001) and Exarchou et al. (2002)
	Summer savory	Flavonoids, carnosol, carvacrol, etc.	Dimitrios (2006)
	Black pepper	Piperine	Srinivasan (2007)
	Red pepper	Vitamin C, luteolin, apigenin, etc.	Sikora et al. (2008) and Materska and Perucka (2005)
	Clove	Eugenol, gallic acid, etc.	Pathak et al. (2004)
Animal derived food	Red crabs	Carotenoid, astaxanthin, etc.	Pokorny et al. (2001)
	Blue crab	Canthaxanthin, 4-hydroxyechinenone, 3-hydroxycanthaxanthin, echinenone, isocryptoxanthin, β-carotene, and astaxanthin	Pokorny et al. (2001)
	Crustacea	Carotenoids etc.	Pokorny et al. (2001)

Table 6.2 List of source of antioxidants from agro industry (waste and processing industry)

Source	Antioxidant compounds	References
Liquid state culture of *Phanerochaete chrysosporium* ATCC 24275 with apple pomace sludge and synthetic medium	Polyphenolic compounds	Gassara et al. (2012)
Aspergillus niger NRRL 567 cultivated on apple pomace as a solid substrate	Citric acid	Dhillon et al. (2013)
Olive mill wastewater (OMW)	*Derivative of benzoic acid*: 4-hydroxybenzoic, protocatechuic, vanillic acids *Derivative hydroxycinnamic acid*: ferulic acid caffeic acids *Tyrosol*: 4-hydroxyphenethyl alcohol, homovanillyl alcohol: 4-hydroxy-3-methoxyphenethyl alcohol *Hydroxytyrosol*: 3,4-dihydroxyphenethyl alcohol	Federici et al. (2009)
Grape pomace	*Flavonol glycosides*: quercetin 3-*O*-glucoside and quercetin 3-*O*-glucuronide *Anthocyanin*: malvidin 3-*O*-glucoside *Methanolic extract*: flavonols, flavonols glucosides, flavanols and their gallate esters, anthocyanins, and low molecular weight proanthocyanins *Ethanolic extract*: triterpenes lupeol, oleanolic acid, flavonol quercetin, and daucosterol	Spatafora and Tringali (2012)
Tomato peel and seed by-products	*Peel byproduct as carotenoids*: lycopene, lutein, β-carotene, and *cis*-β-carotene *Seed byproduct as carotenoid*: lycopene and other carotenoids	Knoblich et al. (2005)

6.3.1 *Natural Sources*

6.3.2 *Agro-industry*

In recent past, agro industry is also found to be one of the major sources for the production of antioxidants. They can be derived either from the waste produced from the agro industries or as a by-product during the processing of the food material.

Food processing processes generally produce large amount of waste as well as by-products along with reasonable quantity of effluent. These food processing by-products, agro-industrial waste, and effluents typically consist of high amounts of proteins, sugars, and lipids along with specific organic compounds as well. Therefore, this could be used as a cheap and abundant source of fine chemicals and secondary metabolites. Valuable natural antioxidants, antimicrobial agents, vitamins, etc., along with macromolecules, can be produced by pretreating the waste by physical and biological agents followed by tailored recovery procedures. Efficient pretreatment is necessary for the optimal recovery of the main classes of products. Here we have discussed a case study related to the agro-industrial waste management for the recovery of the natural antioxidants. Some of the known sources of the agro-industrial waste and the recovery vis-à-vis by-production of antioxidants are summarized in Table 6.2

6.3.2.1 Some Examples

There are very few scientific groups involved in this kind of studies worldwide. Some of the examples for such studies are briefly described below.

The use of liquid state culture of *Phanerochaete chrysosporium* ATCC 24275 for the production of polyphenolic compounds by employing apple pomace sludge and synthetic medium has been studied. Increased polyphenol content was observed by acetone extraction (383–720 mg Gallic Acid Equivalent/l) (GAE/l) during the fermentation of apple pomace and it is further increased by ~1.5-fold until 67 h of fermentation by ethanolic extraction (Gassara et al. 2012).

The bioproduction of citric acid and optimization of extraction from *Aspergillus niger* NRRL 567 cultivated on apple pomace as a solid substrate with inducers, ethanol and methanol, by rotating drum-type solid-state bioreactor has been carried out. Results showed that optimum conditions achieved for higher citric acid bioproduction (220.6 ± 13.9 g/kg dry solids, DS) were 3 % (v/v) methanol, intermittent agitation of 1 h after every 12 h at 2 rpm and 1 vvm (volume per volume per minute) of aeration rate and 120 h incubation time. Highest production of citric acid was of 294.19 g/kg DS (dry substrate) (Dhillon et al. 2013). In this study, scientists have reviewed the results of various studies carried out for the recovery of value-added products from the Olive mill wastewater (OMW). OMW is supposed to be a rich source of polyphenolic compounds, such as benzoic acid derivatives (4-hydroxybenzoic, protocatechuic, vanillic acids), hydroxycinnamic acid derivatives (ferulic acid, caffeic acids), tyrosol (4-hydroxyphenethyl alcohol), homovanillyl alcohol (4-hydroxy-3-methoxyphenethyl alcohol), and hydroxytyrosol (3,4-dihydroxyphenethyl alcohol). Methods have been optimized for the maximum yield of hydroxytyrosol (a polyphenols). The process for the recovery of these antioxidants involved filtration to eliminate the suspended solids followed by physicochemical processes, such as ultrafiltration, nanofiltration, and reverse osmosis. However, hydroxytyrosol is one of the main polyphenol recovered from OMW (Federici et al. 2009).

The antioxidant phenolic compounds were isolated and identified by HPLC, LC-MS, and flash chromatography in the fractions of methanolic and ethanolic extracts of destemmed grape pomace. The analytical studies showed the presence of the main flavonols, flavonol glucosides and their gallate esters, anthocyanins, and low molecular weight proanthocyanins. Five pyranoanthocyanins were also identified for the first time in grape pomace. Quercetin 3-*O*-glucoside and quercetin

3-*O*-glucuronide resulted the most abundant flavonol glycosides, and malvidin 3-*O*-glucoside is the main anthocyanin. Triterpenes lupeol, oleanolic acid, flavonol quercetin, and daucosterol were the main constituents identified for the ethanolic extract (Spatafora and Tringali 2012).

In the study, carried out by Knoblich et al. (2005), tomato peel and seed by-products were used for the isolation of antioxidants (carotenoids). The lycopene content of peel byproduct was found to be 734 μg/g of dry material. Significant amount of lutein, β-carotene, and *cis*-β-carotene were also identified. Seed by-product mainly contained lycopene while the other carotenoids were approximately half of that present in the peels (Knoblich et al. 2005).

6.4 Mechanism of Antioxidant Activity

There are mainly three types of mechanism known for the antioxidant activity, viz., chain breaking, preventive, and synergetic. Schematic representation of these mechanisms is given in Fig. 6.4a–c.

6.5 Techniques for Measurement of Antioxidant Activity

There are three major techniques mostly used for the measurement of antioxidant activity in various samples.

6.5.1 Chemical Assays for Antioxidant Activity

There are many chemical assays used for the assessment of antioxidant activity in the products (herbal, nutraceuticals, and food items). Some of the well-documented and most practiced methods are described below.

6.5.1.1 Oxygen Radical Absorption Capacity

Oxygen radical absorption capacity (ORAC) method uses dichlorofluorescein as the fluorescent probe and an azo-compounds, such as 2,2′-azobis(2-amidinopropane) dihydrochloride (AAPH) as the radical generator. It measures the inhibition of the peroxyl radical induced oxidation initiated by thermal decomposition of AAPH. Over time, the free radical generated from the thermal decomposition of AAPH quenches the signal from the fluorescent probe fluorescein. The subsequent addition of an antioxidant produces a more stable fluorescence signal due to the inhibition of fluorescein decay by single antioxidant and/or complex mixture. Rate of decay of fluorescence measures the antioxidant's capacity (Číž et al. 2010).

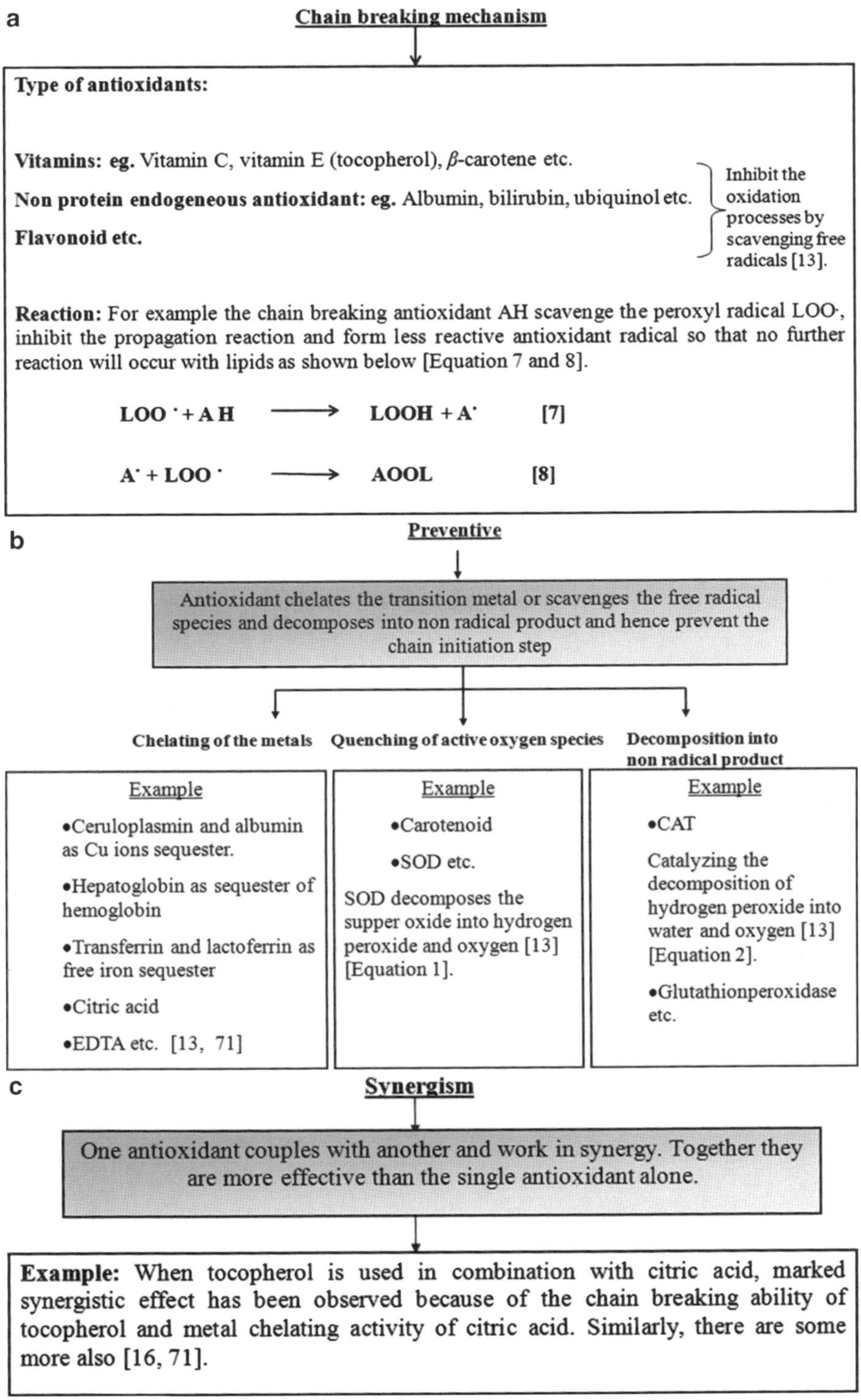

Fig. 6.4 (**a–c**) showing schematic representation of mechanism of chain breaking, preventive and synergetic action of antioxidants respectively

6.5.1.2 Determination of Total Phenolic Content (TPC)

Total phenolic content of the extracts are determined using Folin–Ciocalteu (FC) reagent using spectrophotometer, measured at 725 nm. This method is based on reduction ability of phenolic functional group. Oxidation and reduction reaction of phenolate ion takes place at base condition. The reduction of phosphotungstate–phosphomolybdenum complex (Folin–Ciocalteu reagent) by phenolat ion will change its color to blue. The reduction of complex will increase when the extract contains more phenolic compounds. Thus the color will be darker and the absorbance will be higher, showing higher antioxidant activity (Prior et al. 2005).

6.5.1.3 1,1′-Diphenyl-2-Picrylhydrazyl

DPPH (1,1′-diphenyl-2-picrylhydrazyl) assay is carried out as per the reported method of Brand-Williams et al. (1995). $DPPH^-$ free radical is obtained by dissolving DPPH in methanol and is stable when placed under the dark at −20 °C until used. As $DPPH^-$ reacts with antioxidants present in the sample, color changes from violet to yellow and absorbance of the solution so obtained is measured spectrophotometrically at 515 nm (Brand-Williams et al. 1995).

6.5.1.4 Trolox Equivalent Antioxidant Capacity

In this assay, ABTS {2,2′-azino-bis (3-ethylbenzothiazoline-6-sulphonic acid)} is used to measure the antioxidant capacity of the substance (food stuffs). Trolox equivalent antioxidant capacity (TEAC) is also known as ABTS assay and the procedure is based on the reported method of Arnao et al. (2001). When ABTS reacts with potassium persulfate, it becomes a free radical ($ABTS^+$) which gives blue color to the solution. The phenolics, thiols, or vitamin C present in the food stuffs scavenge this $ABTS^+$ free radical and convert it into its neutral colorless form which is measured spectrophotometrically. $ABTS^+$ absorbs light at 734 nm (Arnao et al. 2001).

6.5.1.5 Ferric Reducing Antioxidant Power

Ferric reducing antioxidant power (FRAP) assay is carried out using the earlier reported method as described by Benzie and Strain (1996). When ferric chloride reacts with 2,4,6-tripyridyl-*s*-triazine (TPTZ) at low pH, ferric is converted into ferrous causing formation of ferrous tripyridyl triazine complex. FRAP values are obtained by comparing the absorbance change at 593 nm in reaction mixture with those containing ferrous ions in known concentration (Benzie and Strain 1996).

6.5.1.6 Determination of Total Reducing Power (TRP)

TRP is determined following the method of Negi et al. (2005). It is measured spectrophotometrically in terms of their capacity to reduce the potassium ferricyanide (Fe^{3+}) to the potassium ferrocyanide (Fe^{2+}), depending upon the concentration of the antioxidant compounds present in the sample, which in turn reacts with ferric chloride to form ferric ferrous complex that has an absorption maximum at 700 nm (Negi et al. 2005).

6.5.2 Biochemical Assays for Antioxidant Activity Assessment

Antioxidant activity may also be measured in biological system, i.e., in vivo and in vitro models. These include measurement of oxidative stress marker of the adduct or end product of ROS with the molecules, such as lipid, protein, DNA, and other molecules. These methods include thiobarbituric acid reactive substances (TBARS), SOD, CAT, GPx, GSH, and ferrous oxidation-xylenol orange (FOX) assay. These assays may be carried out in blood, urine, breath and tissues. Some of the examples are described below:

6.5.2.1 TBARS

TBARS method determines the extent of lipid peroxydation in sample. TBARS is the reaction product of thiobarbituric acid (TBA) and malondialdehyde (MDA) which results from the decomposition of lipid hydroperoxide in the sample which is read spectrophotometrically at 532 nm (Ohkawa et al. 1979).

6.5.2.2 Protein Carbonyl

Protein carbonyl content results from the oxidative cleavage of protein. In this case, 2,4-dinitrophenylhydrazine (DNPH) reacts with protein carbonyl and forms a Schiff base to produce corresponding hydrazone. The amount of protein hydrazone produced is quantified spectrophotometrically at an absorbance between 360 and 380 nm (Levine et al. 1990).

6.5.2.3 FOX

Hydroperoxide content of the lipid can be determined from its ability to oxidize ferric (Fe^{2+}) to ferrous (Fe^{3+}). Ferrous (Fe^{3+}) formed a complex with xylenol orange reagent (bluish-purple color) which is measured at 560 nm (Nourooz-Zadeh et al. 1994).

6.5.2.4 CAT

Catalase activity can be measured by using H_2O_2 as a substrate according to the method of Aebi (1984).

6.5.2.5 SOD

SOD is measured using the method of Kakkar et al. (1984) where nicotinamide adenine dinucleotide (NADH) is used as a substrate. The color intensity of the chromogen (purple color) in butanol layer is measured against butanol (blank) on spectrophotometer at 560 nm (Kakkar et al. 1984).

6.5.2.6 ROS

In this assay, 2′,7′-dichlorofluorescein diacetate (DCFDA) is used to measure ROS level. It undergoes cellular oxidation by ROS and gets converted into fluorescent dichlorofluorescein (DCF) which is highly fluorescent at 530 nm (Moein and Moein 2012).

6.5.3 Instrumental Technique (Antioxidant Analyzer)

Recently, an instrument named PHOTOCHEM Antioxidant Analyzer developed by Analytik Jena UK is being used for the measurement of antioxidant property of different products. It is capable of measuring both water-soluble and lipid-soluble antioxidants in a single system. It is based on the principle of photochemiluminescence with luminometric detection. It can measure the antioxidant capacity of lyophilized vegetables, fruits juices, beer, and water; lipid-soluble antioxidative capacity in baker's yeast, cheese, tea, and coffee; and lipid-soluble antioxidative capacity in edible oil and salami extracts (http://www.selectscience.net/product-news/rapid-and-accurate-antioxidant-measurement-in-foods).

6.6 Conclusions

Antioxidant defense system is always maintained in the body to counter the adverse effect of oxidative stress developed in the biological system due to the formation of reactive oxygen species. Oxidative stress is a key factor which plays an important role in the progression of various pathological diseases. There are many reactive species produced in the body as a result of metabolic functions. These active oxygen species are crucial for life because they are also responsible for various physiological activities, such as production of energy, synthesis of essential compounds, and in signal transduction. They may also attack the macromolecules including protein, DNA, and lipid, causing cell damage. Each body type maintains a particular ratio of antioxidants and ROS and the imbalance causes oxidative stress. Therefore, antioxidants are widely supplemented in diets for the maintenance of proper health and prevention of various pathological diseases. Besides, they also have many industrial uses, such as for the preservation of food and cosmetics as well as in preventing the degradation of rubber and gasoline.

It is further concluded that production of antioxidant compounds from the agro-industrial waste should be highly encouraged and practiced more and more in order to minimize the waste production and reduce the adverse effect on the environment.

Acknowledgement The authors are thankful to the Director of DIPAS, Delhi, for her constant support and encouragement. One of the authors, Ms. Mamta, is thankful to the University Grants Commission, Delhi, India, for getting the senior research fellowship.

References

Aebi H (1984) Catalase invitro. Methods Enzymol 105:121–126

Ajila CM, Brar SK, Verma M, Tyagi RD, Godbout S, Valéro JR (2011) Extraction and analysis of polyphenols: recent trends. Crit Rev Biotechnol 31(3):227–249

Ames B, Cathcart R, Schwiers E, Hochstein P (1981) Uric acid provides an antioxidant defense in humans against oxidant- and radical-caused aging and cancer: a hypothesis. Proc Natl Acad Sci U S A 78(11):6858–6862

Amit K, Priyadarsini KI (2011) Free radicals, oxidative stress and importance of antioxidants in human health. J Med Allied Sci 1(2):53–60

Anttonen MJ, Karjalainen RO (2005) Environmental and genetic variation of phenolic compounds in red raspberry. J Food Compost Anal 18:759–769

Arnao MB, Cano A, Acosta M (2001) The hydrophilic and lipophilic contribution to total antioxidant activity. Food Chem 73:239–244

Beecher GR (2003) Overview of dietary flavonoids: nomenclature, occurrence and intake. J Nutr 133(10):3248S–3254S

Belitz HD, Grosch W (1999) Phenolic compounds. Food chemistry. Springer, Berlin, pp 764–775

Benvenuti S, Paellati F, Melegari M, Bertelli D (2004) Polyphenols, anthocyanins, ascorbic acid, and radical scavenging activity of *Rubus*, *Ribes*, and Aronia. J Food Sci 69:164–169

Benzie IFF, Strain JJ (1996) The ferric reducing ability of plasma (FRAP) as a measure of "antioxidant power": the FRAP assay. Anal Biochem 239:70–76

Biswas SK, McClure D, Jimenez LA, Megson IL, Rahman I (2005) Curcumin induces glutathione biosynthesis and inhibits NF-kappaB activation and interleukin-8 release in alveolar epithelial cells: mechanism of free radical scavenging activity. Antioxid Redox Signal 7(1–2):32–41

Brand-Williams W, Cuvelier ME, Berset C (1995) Use of free radical method to evaluate antioxidant activity. Lebensm Wiss Technol 28:25–30

Chakraborty P, Kumar S, Dutta D, Gupta V (2009) Role of antioxidants in common health diseases. Res J Pharm Technol 2(2):238–244

Chaovanalikit A, Wrolstad RE (2004) Anthocyanin and polyphenolic composition of fresh and processed cherries. J Food Sci 69(1):73–83

Číž M, Čížová H, Denev P, Kratchanova M, Slavov A, Lojek A (2010) Different methods for control and comparison of the antioxidant properties of vegetables. Food Control 21:518–523

Dabbagh AJ, Mannion T, Lynch SM, Frei B (1984) The effect of iron overload on rat plasma and liver oxidant status in vivo. Biochem J 300(Pt 3):799–803

Dhillon GS, Brar SK, Kaur S, Verma M (2013) Bioproduction and extraction optimization of citric acid from *Aspergillus niger* by rotating drum type solid-state bioreactor. Ind Crops Prod 41:78–84

Dimitrios B (2006) Sources of natural phenolic antioxidants. Trends Food Sci Technol 17:505–512

Exarchou V, Nenadis N, Tsimidou M, Gerothanassis IP, Troganis A, Boskou D (2002) Antioxidant activities and phenolic composition of extracts from Greek oregano, Greek sage and summer savory. J Agric Food Chem 50:5294–5299

Fang YZ, Yang S, Wu G (2002) Free radicals, antioxidants, and nutrition. Nutrition 18(10):872–879

Federici F, Fava F, Kalogerakis N, Mantzavinos D (2009) Valorisation of agro-industrial by-products, effluents and waste: concept, opportunities and the case of olive mill wastewaters. J Chem Technol Biotechnol 84:895–900. doi:10.1002/jctb.2165

Friedman M, Henika PR, Levin CE, Mandrell RE, Kozukue N (2006) Antimicrobial activities of tea catechins and theaflavins and tea extracts against Bacillus cereus. J Food Prot 69(2):354–361

Gassara F, Ajila CM, Brar SK, Verma M, Tyagi RD, Valero JR (2012) Liquid state fermentation of apple pomace sludge for the production of ligninolytic enzymes and liberation of polyphenolic compounds. Process Biochem 47(6):999–1004

Gorinstein S, Martin-Belloso O, Park Y, Haruenkit R, Lojek A, Ciž M, Capi A, Libman I, Trakhtenberg S (2001) Comparison of some biochemical characteristics of different citrus fruits. Food Chem 74:309–315

Gupta VK, Sharma SK (2006) Plant as natural antioxidant. Nat Product Radiance 5(4):326–334

Hägg M, Ylikoski S, Kumpulainen J (1995) Vitamin C content in fruits and berries consumed in Finland. J Food Compost Anal 8:12–20

Halvorsen BL, Holte K, Myhrstad MCW, Barikmo I, Hvattum E, Remberg SF, Wold A, Haffner K, Baugerod H, Andersen LF, Moskaug J, Jacobs DR, Blomhoff R (2002) A systematic screening of total antioxidants in dietary plants. J Nutr 132:461–471

Hamid AA, Aiyelaagbe OO, Usman LA, Ameen OM, Lawal A (2010) Antioxidants: its medicinal and pharmacological applications. Afr J Pure Appl Chem 4(8):142–151

Hissin PJ, Hilf R (1976) A fluorometric method for determination of oxidized and reduced glutathione in tissues. Anal Biochem 74(1):214–226

Holasova M, Fiedlerova V, Smrcinova H, Orsak M, Lachman J, Vavreinova S (2002) Buckwheat the source of antioxidant activity in functional foods. Food Res Int 35:207–211

http://www.selectscience.net/product-news/rapid-and-accurate-antioxidant-measurement-in-foods. Accessed 19 Oct 2012

Jolić SM, Redovniković IR, Marković K, Šipušić ĐI, Delonga K (2011) Changes of phenolic compounds and antioxidant capacity in cocoa beans processing. Int J Food Sci Technol 46(9):1793–1800

Kähkönen MP, Hopia AI, Vuorela HJ, Rauha J, Pihlaja K, Kujala TS, Heinonen M (1999) Antioxidant activity of plant extracts containing phenolic compounds. J Agric Food Chem 47:3954–3962

Kähkönen MP, Hopia AI, Heinonen M (2001) Berry phenolics and their antioxidant activity. J Agric Food Chem 49(8):4076–4082

Kakkar PS, Das BB, Viswanathan PN (1984) A modified spectrophotometric assay of superoxide dismutase. Indian J Biochem Biophys 21:130–132

Kim EH, Kim SH, Chung JI, Chi HY, Kim JA, Chung IM (2006) Analysis of phenolic compounds and isoflavones in soybean seeds (Glycine max (L.) merill) and sprouts grown under different conditions. Eur Food Res Technol 222:201–208

Knoblich M, Anderson B, Latshaw D (2005) Analyses of tomato peel and seed byproducts, and their use as a source of carotenoids. J Sci Food Agric 85:1166–1170

Kopsell DA, Kopsell DE, Lefsrud MG, Curran-Celentano J, Dukach LE (2004) Variation in lutein, β-carotene and chlorophyll concentrations among *Brassica oleracea* cultigens and seasons. HortScience 39(2):361–364
Kurilich AC, Tsaui GJ, Brown A, Howard L, Klein BP, Jeffrey EH, Kushad M, Wallig MA, Juvik J (1999) Carotene, tocopherol, and ascorbate contents in subspecies of *Brassica oleracea*. J Agric Food Chem 47:1576–1681
Lachman J, Proněk D, Hejtmánková A, Dudjak J, Pivec V, Faitová K (2003) Total polyphenol and main flavonoid antioxidants in different onion (Allium cepa L.) varieties. Hortic Sci (Prague) 30(4):142–147
Laokuldilok T, Shoemaker CF, Jongkaewwattana S, Tulyathan V (2011) Antioxidants and antioxidant activity of several pigmented rice brans. J Agric Food Chem 59(1):193–199
Levine RL, Garland D, Oliver CN, Amici A, Climent I, Lenz AG (1990) Determination of carbonyl content in oxidatively modified proteins. Methods Enzymol 186:464–478
Lomnitski L, Bergman M, Nyska A, Ben-Shaul V, Grossman S (2003) Composition, efficacy, and safety of spinach extracts. Nutr Cancer 46(2):222–231
Lu Y, Foo LY (1997) Identification and quantification of major polyphenols in apple pomace. Food Chem 59(2):187–194
Lu Y, Foo LY (2000) Antioxidant and radical scavenging activities of polyphenols from apple pomace. Food Chem 68:81–85
Määttä-Riihinen KR, Kamal-Eldin A, Mattila PH, Gonzalez-Paramas AM, Törrönen AR (2004) Distribution and contents of phenolic compounds in eighteen Scandinavian berry species. J Agric Food Chem 52:4477–4486
Mannach C, Scalbert A, Morand C, Jimenez L (2004) Polyphenols: food sources and bioavailability. Am J Clin Nutr 79:727–747
Materska M, Perucka I (2005) Antioxidant activity of the main phenolic compounds isolated from hot pepper fruit (*Capsicum annuum* L.). J Agric Food Chem 53:1750–1756
Moein S, Moein MR (2012) New usage of a fluorometric method to assay antioxidant activity in plant extracts. Iran J Pharm Sci 8(1):71–78
Negi PS, Chauhan AS, Sadia GA, Rohinishree YS, Ramteke RS (2005) Antioxidant and antibacterial activities of various Seabuckthorn (Hippophae rhamnoides L.) seed extracts. Food Chem 92:119–124
Nie L, Wise ML, Peterson DM, Meydani M (2006) Avenanthramide, a polyphenol from oats, inhibits vascular smooth muscle cell proliferation and enhances nitric oxide production. Atherosclerosis 186:260–266
Nourooz-Zadeh J, Tajaddini-Sarmadi J, Wolff SP (1994) Measurement of plasma hydroperoxide concentrations by the ferrous oxidation-xylenol orange assay in conjunction with triphenylphosphine. Anal Biochem 220(2):403–409
Ohkawa H, Ohisi N, Yagi K (1979) Assay for lipid peroxides in animal tissues by thiobarbituric acid reaction. Anal Biochem 95:351–358
Olsson ME, Gustavsson K, Andersson S, Nilsson A, Duan RD (2004) Inhibition of cancer cell proliferation in vitro by fruit and berry extracts and correlation with antioxidant levels. J Agric Food Chem 52:7264–7271
Papas AM (1998) Antioxidant status, diet, nutrition, and health. CRC Series, Boca Raton, FL
Pastrana-Bonilla E, Akoh CC, Sellappan S, Krewer G (2003) Phenolic content and antioxidant capacity of muscadine grapes. J Agric Food Chem 51:5497–5503
Pathak SB, Niranjan K, Padh H, Rajani M (2004) TLC densitometric method for the quantification of eugenol and gallic acid in clove. Chromatographia 60(3–4):241–244
Peterson DM, Emmons CL, Hibbs AH (2001) Phenolic antioxidants and antioxidant activity in pearling fractions of oat groats. J Cereal Sci 33:97–103
Pokorny J, Yanishlieva N, Gordon MH (2001) Antioxidants in food: practical applications. CRC, Boca Raton, FL
Prior RL, Wu X, Schaich K (2005) Standardized methods for the determination of antioxidant capacity and phenolics in foods and dietary supplements. J Agric Food Chem 53(10):4290–4302

Raygani V, Rahimi Z, Zahraie M, Noroozian M, Pourmotabbed A (2007) Enzymatic and non-enzymatic antioxidant defense in Alzheimer's disease. Acta Med Iran 45(4):271–276
Reyes-Carmona J, Youseg GG, Martinez-Peniche RA, Lila MA (2005) Antioxidant capacity of fruit extracts of blackberry (*Rubus* sp.) produced in different climatic regions. J Food Sci 70:497–503
Said S, Allam M, Moustafa H, Mohamedz A (2002) A thermal stability of some commercial natural and synthetic antioxidants and their mixtures. J Food Lipids 9:277–293
Senorans FJ, Ibanez E, Cavero S, Tabera J, Reglero G (2000) Liquid chromatographic–mass spectrometric analysis of supercritical-fluid extracts of rosemary plants. J Chromatogr A 870: 491–499
Sikora E, Cieslik E, Topolska K (2008) The sources of natural antioxidants. Acta Sci Pol Technol Aliment 7(1):5–17
Siriwoharn T, Wrolstad RE, Finn CE, Pereira CB (2004) Influence of cultivar, maturity, and sampling on blackberry (*Rubus* L. Hybrids) anthocyanins, polyphenolics, and antioxidant properties. J Agric Food Chem 52:8021–8030
Spatafora C, Tringali C (2012) Valorization of vegetable waste: identification of bioactive compounds and their chemo-enzymatic optimization. Open Agric J 6:9–16
Srinivasan K (2007) Black pepper and its pungent principle-piperine: a review of diverse physiological effects. Food Sci Nutr 47(8):735–774
Stewart AJ, Bozonnet S, Mullen W, Jenkins GI, Lean ME, Crozier A (2000) Occurrence of flavonols in tomatoes and tomato-based products. J Agric Food Chem 48:2663–2669
Stocker R, Yamamoto Y, McDonagh AF, Glazer AN, Ames BN (1987) Bilirubin is an antioxidant of possible physiological importance. Science 235(4792):1043–1046
Taruscio TG, Barney DL, Exon J (2004) Content and profile of flavanoid and phenolic acid compounds in conjunction with the antioxidant capacity for a variety of northwest *Vaccinium* berries. J Agric Food Chem 52:3169–3176
Teresa MM, Magdalena W, Andrzej KA (2011) Systematic review of the effect of vitamin C infusion and vitamin E-coated membrane on hemodialysis-induced oxidative stress. Intech doi:10.5772/22542
Touriño S, Lizárraga D, Carreras A, Matito C, Ugartondo V, Mitjans M, Centelles JJ, Vinardell MP, Juliá L, Cascante M, Torres JL (2008) Antioxidant/prooxidant effects of bioactive polyphenolics. Electron J Environ Agric Food Chem 7(8):3348–3352
Vallejo F, Tomas-Barberan FA, Garcia-Viguera C (2003) Phenolic compound contents in edible parts of broccoli inflorescences after domestic cooking. J Sci Food Agric 83:1511–1616
Walters MT, Heasman AP, Hughes PS (1997) Comparison of (+)-catechina and ferulic acid as natural antioxidants and their impact on beer flavor stability. Part 1: Forced-aging. J Am Soc Brew Chem 55(2):83–89
Wang CC, Chen LG, Lee LT, Yang LL (2003) Effects of 6-gingerol, an antioxidant from ginger, on inducing apoptosis in human leukemic HL-60 cells. In Vivo 17(6):641–645
Yanishlieva-Maslarova NN, Heinonen M (2001) Sources of natural antioxidants. In: Gordon M, Pokorny J, Yanishlieva N (eds) Antioxidants in food. CRC, Boca Raton, FL
Zadernowski R, Naczk M, Nesterowicz J (2005) Phenolic acid profiles in some small berries. J Agric Food Chem 53:2118–2124
Zheng W, Wang SY (2001) Antioxidant activity and phenolic compounds in selected herbs. J Agric Food Chem 49:5165–5170

Chapter 7
Solid-State Fermentation of Agricultural Residues for the Production of Antibiotics

Ganesh Kumar Arumugam, Venkatesh Selvaraj, Dharani Gopal, and Kirubagaran Ramalingam

7.1 Introduction

The agro-industrial residues are generated globally and a major portion is left unutilised, leading to loss of biomass and an environmental pollution problems. Wide agricultural practices and activities of agro-based industries produce tonnes of by-products, viz., sugarcane bagasse, sweet sorghum, citrus and agave, seeds and peels, rice, barley, wheat and oat straw, corn straw and corncobs. Based on the nutritional properties, the agro-industry residues are classified into two major groups: fibrous residues (higher and lower digestibility) and brans. High digestibility fibrous residues include citrus pulp, corn gluten bran, soy husk and brewing residues, and those with low digestibility include sugarcane bagasse, cereal, wheat, corn, cotton, soy and peanut husk. Brans include rice, peanut, soy and cotton (Graminha et al. 2008). Considering the properties and chemical constitution, these agricultural residues can be used as a natural bioresource for the production of bioactive compounds such as secondary metabolites from various selected microorganisms.

In microbial cultures, the end of primary growth phase initiates the synthesis of secondary metabolites. Various groups of secondary metabolites can play many different roles, such as antibiotics, toxins, ionophores and bioregulators, and involve in intra- and interspecific signalling. These metabolites represent some of the most important industrial products and possess tremendous economic importance. Most certainly, the antibiotics are the best-known subdivision of this group of metabolites. Several species of filamentous fungi and actinomycetes followed by other bacteria, such as *Bacillus*, *Pseudomonas*, myxobacteria and cyanobacteria, have been used in utilising the agricultural residues through fermentation techniques due to their ability to grow on particle surfaces as sources of carbon and energy and

G.K. Arumugam (✉) • V. Selvaraj • D. Gopal • K. Ramalingam
Marine Biotechnology, National Institute of Ocean Technology (Ministry of Earth Sciences, Govt. of India), Pallikaranai, Chennai 600 100, Tamil Nadu, India
e-mail: microganesh@yahoo.com; microganesh@niot.res.in

S.K. Brar et al. (eds.), *Biotransformation of Waste Biomass into High Value Biochemicals*, DOI 10.1007/978-1-4614-8005-1_7,

Table 7.1 Examples of bacterial and fungal biosynthesis of secondary metabolite, antibiotics

Class	Antibiotics	Producing microorganisms	Reference
Aminoglycoside	Streptomycin	*Streptomyces griseus*	Miyake et al. (1990)
Aminonucleoside	Puromycin	*Streptomyces alboniger*	Tercero et al. (1996)
β-Lactams	Cephalosporin	*Acremonium chrysogenum*	Teijeira et al. (2011)
	Cephamycin C	*Streptomyces clavuligerus*	Devi and Padma (2000)
	Cephabacin	*Lysobacter lactamgenus*	Sohn et al. (2001)
	Penicillin	*Penicillium chrysogenum*	Barrios-Gonzalez et al. (1993)
	Nocardicin A	*Nocardia uniformis*	Jeanne and Craig (2009)
Polyketide	Actinorhodin	*Streptomyces coelicolor*	Elibol (2004)
	Enterocin	*Enterococcus faecium*	Kumar and Srivastava (2011)
	Rifamycin B	*Amycolatopsis mediterranei*	Venkateswarlu et al. (2000)
	Tetracenomycin	*Streptomyces glaucescens*	Gramajo et al. (1991)
Polypeptide	Actinomycin	*Streptomyces chrysomallus*	Haese and Keller (1988)
Macrolide	Carbomycin	*Streptomyces thermotolerans*	Epp et al. (1987)
	Erythromycin	*Saccharopolyspora erythraea*	El-Enshasy et al. (2008)
	Oleandomycin	*Streptomyces antibioticus*	Quirs and Salas (1995)
	Oxytetracycline	*Streptomyces rimosus*	Yang and Swei (1996)
	Pimaricin	*Streptomyces natalensis*	Recio et al. (2006)
	Spiramycin	*Streptomyces ambofaciens*	Karray et al. (2010)
	Tetracycline	*Streptomyces aureofaciens*	Asanza et al. (1997)
	Tylosin	*Streptomyces fradiae*	Bate et al. (2000)
Lipopeptide	Iturin	*Bacillus subtilis*	Shih et al. (2008)
	Surfactin	*Bacillus subtilis*	Wei et al. (2007)
Anthracyclines	Daunorubicin	*Streptomyces peucetius*	Otten et al. (1995)
Streptogramin	Pristinamycin	*Streptomyces pristinaespiralis*	Mehmood et al. (2012)
Aminocyclitol	Spectinomycin	*Streptomyces spectabilis*	Hyun et al. (2000)
Cyclopentanoid	Methylenomycin	*Streptomyces coelicolor*	Obanye et al. (1996)

produce important secondary metabolites including antibiotics (Marinelli and Marcone 2011). Initial researches are in the view that secondary metabolites have no functional importance in the growth of the producing cultures. However, the current opinions have changed that every secondary metabolite is synthesised to confer a survival advantage for the producing organism in a particular habitat (Brakhage et al. 2009). One of the predominant secondary metabolites, antibiotic, have greater importance in human health and possess a high pharmaceutical and commercial importance (Table 7.1). The alarming rise of multidrug-resistant pathogens and its infections has prompted a desperate search for novel antibiotics. It is noteworthy that microorganisms have the ability to meet the challenges of change, e.g. a new streptogramin antibiotic etamycin produced by an actinomycetes species against methicillin-resistant *Staphylococcus aureus* (Haste et al. 2010).

A wide variety of agricultural residues are readily available as underutilised resources, which can be considered as inexpensive renewable carbon source for the commercial production of secondary metabolites (Poonam Singh and Pandey 2009). Bioconversion of agricultural residues for antibiotic production would hold a prominent position in future fermentation technologies, mainly because of its cost-effectiveness, eco-friendliness and feasibility in both developed and developing countries. Finally, the catabolism and utilisation of agricultural residues represent an important contribution to the implementation of biotechnology concepts and to the reduction of environmental problems associated with the disposal of solid wastes.

7.2 Antibiotics

One of the greatest achievements of medical science was the discovery of antibiotics which have profound importance on human health. Antibiotics are one of the best-known groups of the secondary metabolites synthesised by microorganisms, which are active against other microorganisms. Antibiotics affect a multitude of targets and essential cellular functions, which include DNA replication (actinomycin and griseofulvin), transcription (rifamycin), translation by 70S (S-Svedberg sedimentation mass value) ribosomes (chloramphenicol, tetracycline, erythromycin and streptomycin), transcription by 80S ribosomes (cyclohexamide), transcription by 70S and 80S ribosomes (puromycin and fusidic acid), cell wall synthesis (cycloserine, bacitracin, penicillin, cephalosporin and vancomycin) and cell membrane disruption (polymyxin and amphotericin, ionophores such as gramicidin, lonomycin and monensin) (Indu 2006).

Due to its enormous importance in human health care, demand for antibiotics is increasing worldwide. Moreover, continuous efforts are being made to decrease its production cost by process optimisation using raw materials like agricultural residues through different fermentation processes like SmF and SSF. Inexpensive substrates, such as agricultural residues and agro-industrial waste products have been found to be very valuable for economy and appropriate for biotechnological process. Their usage as substrate has widely opened the potential to reduce production costs up to 60 % by reducing the cost of raw material during fermentations (Lotfy 2007). The importance of agro-industrial residues in SSF system for the production of antibiotics and other secondary metabolites has gained much recognition in recent years (Mahalaxmi et al. 2010). Antibiotic production using SSF requires very minimum energy and less investment cost, and recently it has gained increased importance due to its higher productivity through fermentation, eco-friendliness and lesser disadvantages when compared to SmF (Poonam Singh and Pandey 2009).

7.2.1 *Secondary Metabolites and Growth*

Metabolites are organic compounds produced by organisms using multitude of enzyme-catalysed biochemical reactions called metabolic pathways. Metabolites can be the initial, the intermediary or the end products of these biochemical reactions. A variety of metabolites and reactions combine and work together allowing an organism to sustain life. Primary metabolites are found in almost all species within broad phylogenetic groups, produced by nearly the same pathway and are normally involved in growth, development and reproduction. The secondary metabolites are often restricted to a very narrow set of species within the same phylogenetic group. Secondary metabolites secreted by organisms are the chemical components that do not involve or interact in normal growth and development, but usually have a potent function. The growth and metabolism of many microorganisms in fermentation generally implies a series of phases. In microorganisms, secondary metabolites are not usually produced during the log or exponential phase of a culture (trophophase), but are synthesised in culture medium after a period of complete consumption of key nutrients, such as carbon, nitrogen, phosphate and mineral sources. Although primary and secondary metabolism share the common transcriptional and translational machinery, the nutrient depletion initiates the actions of precursors to accumulate products other than primary metabolites called the secondary metabolites during a subsequent production stage (idiophase).

7.2.2 *Biosynthesis of Microbial Secondary Metabolites*

The evolution of exciting and new secondary metabolic pathways is likely to have been driven by the ecological robustness, abiotic and biotic stresses and survival in unique ecosystem. Microbial secondary metabolites show enormous diversity of chemical structures. The biosynthetic pathways have emerged from network of primary metabolisms at a relatively small number of points (Barrios-Gonzalez et al. 2003) and evolve later independently.

Most secondary metabolites are synthesised from one or a combination of different biosynthetic pathways (Fig. 7.1):

1. Metabolites derived from sugars (streptomycin, neomycin and kanamycin)
2. Metabolites derived from shikimic acid pathway; shikimic acid is one of the major source for the biosynthesis of antibiotics (ansamycin and rifamycin)
3. Metabolites derived from aliphatic amino acid pathway for the biosynthesis of β-lactam antibiotics (penicillin, cephalosporins and cephamycins)
4. Metabolites derived from chorismic acid pathway (candicidin, nystatin and chloramphenicol)
5. Metabolites derived from aromatic amino acid pathway (actinomycin, indolmycin, novobiocin, lincomycin and polymyxin)
6. Metabolites derived from the acetyl-CoA and malonyl-CoA (erythromycin, vancomycin and tetracyclines)

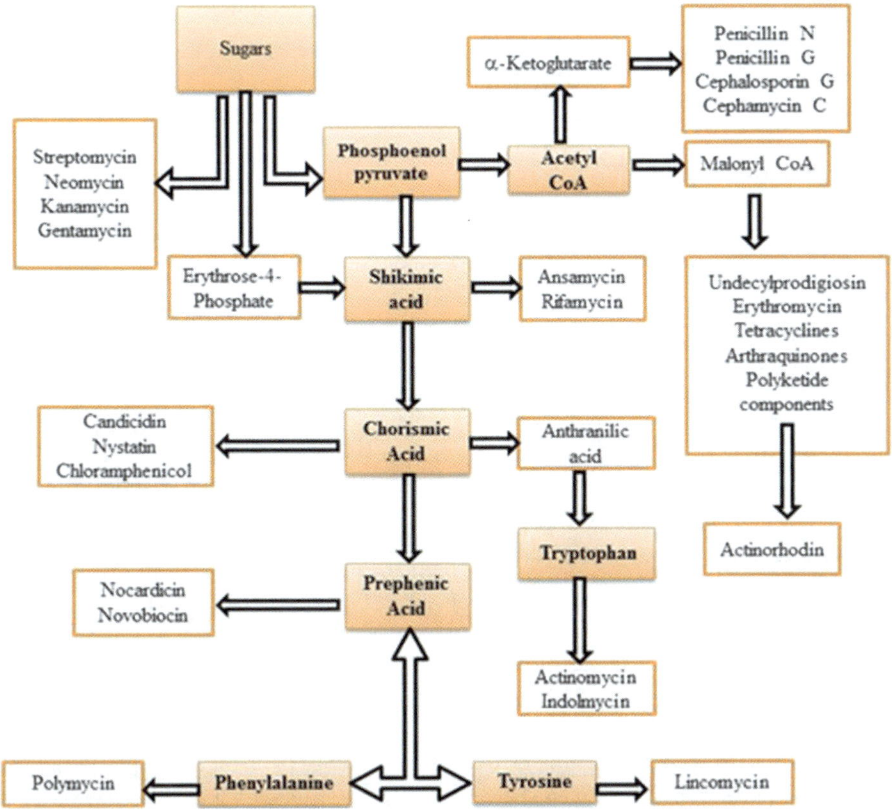

Fig. 7.1 Metabolic pathways leading to biosynthesis of secondary metabolite, antibiotics

7.3 Regulatory Mechanisms Involved in the Biosynthesis of Antibiotics

Antibiotic production is a multi-complex and highly regulated process, controlled by different physicochemical, biological and environmental factors. Although a wide range of primary and secondary metabolites from microorganisms has been identified, these products cannot be easily distinguished on the origin of precursors, chemical structures, functional analysis or its synthesis. In common, products of the primary metabolism serve as precursors of secondary metabolic pathways and increase its production.

7.3.1 Inducer

Microorganisms have evolved the ability to survive and proliferate to a constantly changing physical and chemical environment. The ability of microbial cells to live, function and replicate in a suitable environment depends on the presence of a specific compound. These compounds, the inducer, initiate the production of biochemical metabolites or intermediates based on their requirements. In certain pathways, amino acids stimulate the production of secondary metabolites either by increasing the quantity of a limiting precursor or by inducing a biosynthetic enzyme. Inducers include valine for valine dehydrogenase of the tylosin process in *Streptomyces fradiae* (Nguyen et al. 1995), lysine for lysine-ε-aminotransferase in the cephamycin pathway of *Streptomyces clavuligerus* (Rius et al. 1996) and methionine for δ-(L-α-aminoadipyl)-L-cysteinyl-D-valine (ACV) synthetase, isopenicillin-N and deacetylcephalosporin-C synthase in the cephalosporin pathway of *Acremonium chrysogenum* (Juan and Arnold 2002). In the production of cephamycin C by *Nocardia lactamdurans* using SSF, addition of 1,3-diaminopropane acted as an inducer having a beneficial effect on production and increased the yield to 27.64 mg/g dry substrate (Kagliwal et al. 2009).

7.3.2 Autoregulator

Autoinducers are one of the best-known regulators of secondary metabolism in bacteria (Recio et al. 2006) and fungi (Martin et al. 2011). There are different types of autoregulatory molecules having ability to trigger wide ranges of antibiotic production (Recio et al. 2006). The regulatory factors of antibiotic biosynthesis are of great interest. One of the most established family of autoregulators consists of γ-butyrolactones, which are active even at nano-level concentrations and elicit antibiotic production by modulating the DNA-binding activity of cognate receptor proteins. Thus, the γ-butyrolactones have been referred to as bacterial hormones (Takano 2006). In several *Streptomyces* species, γ-butyrolactone autoregulator-receptor systems are well known to regulate antibiotic production (Arakawa et al. 2007). In *Streptomyces coelicolor*, a furan-type autoregulator, methyl furan was seen to induce antibiotic production (Corre et al. 2008). In *Amycolatopsis mediterranei*, B-factor [3′-(1-butylphosphoryl)adenosine] induced rifamycin production (Kawaguchi et al. 1988). In certain cases, modified peptides (Kleerebezem et al. 1997) and other small molecules, such as 2,3-diamino-2, 3-bis(hydroxymethyl)-1,4-butanediol (Recio et al. 2006) and 1,3-diaminopropane (Martin et al. 2011), serve as autoregulators. In *Pseudomonas fluorescens*, pyoluteorin serves as an autoregulator, positively influencing its own production. In addition to its autoregulatory role, pyoluteorin influenced the production of another secondary metabolite, 2,4-diacetylphloroglucinol. This findings elucidate that pyoluteorin establishes its contribution to regulation of at least two metabolic pathways within the bacterial cell (Brodhagen et al. 2004).

7.3.3 Carbon Catabolite Repression

The concentrations of carbon, nitrogen and phosphate imply an important regulatory effect on primary and secondary metabolism in different microorganisms. Antibiotic production rate can be influenced by manipulating the type and concentration of nutrients formulating the production medium. Carbon sources, such as glucose and other carbohydrates, are excellent sources for growth and metabolism in microorganisms, but they interfere with antibiotic synthesis and this effect depends on the rapid utilisation of the preferred carbon source. Once the preferred carbon source is completely utilised, the next available carbon source is used for the production phase, known as the "idiophase". The readily available carbon source exerts some negative effect on the production of antibiotics. Different mechanisms have been described in bacteria and fungi to explain the negative effects of carbon catabolites on antibiotic production. This carbon catabolite regulation mechanism is widely distributed among microbial systems. In this regulation, the microorganisms catabolise the readily assimilable carbon source through biochemical pathways suppressing the secondary metabolite biosynthesis. In the filamentous fungus *Aspergillus nidulans* and *A. chrysogenum*, the biosynthesis of the β-lactam antibiotic penicillin and cephalosporin is repressed by glucose, respectively (Espeso et al. 1995; Jekosch and Kuck 2000).

7.3.4 Nitrogen Regulation

Many secondary metabolic pathways are influenced by nitrogen sources available for growth of microorganisms. Various nitrogen sources, inorganic (ammonium and nitrate) and organic (different amino acids), are used to enhance the production of secondary metabolites. The complex media used in fermentation often include a protein source (soybean meal) and the defined media usually contain a slowly assimilated amino acid (proline) as the nitrogen source for production of antibiotics (Gupte and Kulkarni 2002). Production of some aminoglycoside antibiotics is unfavourably affected by ammonium, e.g. neomycin and kanamycin (Shapiro 1989), whereas nitrate and certain amino acids possess stimulatory effect. Doull and Vining (1990) observed the nitrogen catabolite regulation during the actinorhodin production in *S. coelicolor*. Ammonium, either directly supplied as a nitrogen source or originating from the breakdown of amino acids, plays a vital role in nitrogen catabolic repression of pristinamycin production by *Streptomyces pristinaespiralis* (Voelker and Altaba 2001).

7.3.5 Phosphate Regulation

Phosphate, an essential component of the energy dynamics of cells, regulates the biosynthesis of many different types of antibiotics and other secondary metabolites. High concentration of phosphate had a negative effect on the biosynthesis of streptomycin, oxytetracycline, clavulanic acid, tylosin, echinomycin, cephalosporin, cephamycin and thienamycin (Juan 2004). In certain cases, negative phosphate control is exerted at the transcriptional level. Recently, it was shown that phosphate control of antibiotic biosynthesis in *Streptomyces lividans* and *S. coelicolor* is mediated by the two-component PhoR–PhoP system that also controls the (*phoA*) alkaline phosphatase gene (Juan 2004). Phosphate limiting nutritional condition regulates biosynthesis of two antibiotic secondary metabolites, prodigiosin and carbapenem, through a multiple biochemical pathways, incorporating transcriptional control mediated by three regulators, PhoB, SmaR and Rap, in *Serratia* 39006 (Gristwood et al. 2009). Several mechanisms have been proposed and illustrated to explain the effect of phosphate on the biosynthesis of antibiotic, such as phosphate stimulating primary metabolic pathway, phosphate transferring carbohydrate catabolic pathways and phosphate inhibiting the formation of precursors required for antibiotic synthesis (Juan 2004).

7.3.6 Feedback Regulation

Many secondary metabolites inhibit or repress their own biosynthetic and catabolic pathways. Secondary metabolites can regulate and/or operate the activity of preexisting enzymes (feedback inhibition) or stop their synthesis (feedback repression). The role of feedback regulation in controlling primary and secondary metabolism is well established. In tylosin synthesis, feedback control of polyketide metabolism is observed (Butler et al. 2001). Bacitracin represses enzyme bacitracin synthetase involved in its biosynthetic process (Froyshov et al. 1980). In tetracycline synthesis, enzyme anhydrotetracycline oxygenase is inhibited by tetracycline, chlortetracycline and oxytetracycline (Behal et al. 1983).

7.4 Genetic Regulation of Antibiotic Production

The genetic regulation of secondary metabolite biosynthesis includes multitude layers of cellular control. These genetic elements, though poorly understood, are competent of influencing rates of biosynthesis significantly in microorganisms.

Malik (1979) groups the genes controlling antibiotic production into five classes. These are as follows:

1. Structural genes coding for enzymes that specify the biosynthesis of the secondary metabolite
2. Regulatory genes that determine the onset and extent of repression of the structural genes for biosynthesis
3. Genes that determine the resistance of the producing organism to the product (product toxicity)
4. Genes controlling the permeability to the compound (transport/excrete complex metabolites)
5. Regulatory genes that control primary pathways (precursors and cofactors) needed for antibiotic synthesis

7.4.1 Gene Clusters

Details of the pathways involved in the biosynthesis of antibiotics have been widely studied in microorganisms. Furthermore, sufficient information is available on the regulation of genes, important for antibiotic production. Many researches showed that the genes encoding specialised and myriad functions for the biosynthesis of many antibiotics are often coded by clustered genes on chromosomal DNA and less frequently on plasmid DNA (Brakhage and Schroeckh 2011). In operonic organisation of genes, the groups of functional and closely related genes are expressed as a single polycistronic mRNA, regulating the process of transcription and subsequent translation (Koonin et al. 2001). This is the governing principle of genomic organisation and expression in most microorganisms.

Karray et al. (2007) analysed the biosynthetic gene cluster for spiramycin production in *Streptomyces ambofaciens* and demonstrated the origin of precursors used by polyketide synthase. The 12-membered macrolide antibiotics, namely, methymycin and neomethymycin, and 14-membered macrolide antibiotics, namely, narbomycin and pikromycin, produced by *Streptomyces venezuelae* have gene cluster which is a polyketide synthase that encodes a six-module enzyme system (Xue et al. 1998). Gutierrez et al. (1999) elucidated the biosynthetic gene cluster in *Penicillium chrysogenum* consisting of three genes pcbAB, pcbC and penDE responsible for penicillin production. Brautaset et al. (2000) observed the biosynthetic gene cluster consisting of six genes coding for polyene antibiotic nystatin in *Streptomyces noursei*. The cloning and functional analysis of nucleoside antibiotic, polyoxin, and the biosynthetic gene cluster showed 20 different genes involved in antibiotic biosynthesis in *Streptomyces cacaoi* (Chen et al. 2009). Keller et al. (2010) identified a gene cluster encompassing 50 kb of contiguous DNA containing 28 genes with the biosynthetic functions on the chromosome of *Streptomyces chrysomallus* involved in the biosynthesis of actinomycin. Gene cluster, gene order representation for the biosynthesis of antibiotic and manipulation of the genes identified may potentially lead to the generation of novel antibiotics as well as yield enhancements in the microbial strains. Expression of the secondary metabolite gene clusters in wild-type microbial strains in natural habitats is frequently very low and is modulated in response to different physical, chemical or environmental stimuli (Laich et al. 1999).

In some cases most of the genes required for the secondary metabolite biosynthesis remain dormant, although it is likely that they may be expressed under unknown conditions (Scherlach and Hertweck 2009). The expression levels of several genes involved directly or indirectly in cephalosporin C (CPC) biosynthesis in *A. chrysogenum* are studied under a variable pH in SmF and SSF. Differences in intermediate and certain biosynthetic gene expression levels are observed predominantly and they evidence the relationships between physiological features and gene expression that open important advancement perspectives for fermentation systems (Lopez-Calleja et al. 2012).

7.5 Solid-State Fermentation Production Systems

Antibiotics are usually produced, using SmF, which require high energy and capital investment. Continuous requirements for huge amounts of antibiotics with minimum production costs are the need of the hour for the constant fight against dreadful diseases and multidrug-resistant pathogens. Recently, SSF has gained much importance due to higher productivity, shorter production time, lower energy and less capital investment using agricultural wastes as low cost carbohydrate sources and with much lesser disadvantages when compared with SmF (Table 7.2).

In the SSF process, the solid substrate supports the growth of microorganisms by providing nutrients and essential cofactors. The substrate that provides all the essential nutrient sources to the microorganisms should be considered as the ideal substrate for process optimisation. However, in certain cases, single natural substrate can provide the nutrient and it is essential to provide them externally. In general, SSF can be distinguished into two types, depending on the nature of solid phase matrix (Barrios-Gonzalez and Mejia 1996).

1. Solid culture of single support-substrate phase: solid phase addressing the functions of support and nutrient source. Examples include agriculture residues or animal wastes as support substrate.
2. Solid culture of dual substrate-support phase: solid phase is constituted by an inert matrix impregnated with a liquid medium, which serves as a reservoir for nutrients, water and additional supplements. Examples include polyurethane as inert support.

In both cases, the success of the fermentation and antibiotic production process is related to the physical characteristics (particle size and shape, porosity and consistency) of the support, which favours gases, nutrients and metabolites diffusion.

7.5.1 *SSF on Natural Support Systems*

Cultivation on natural substrates, SSF system uses natural materials that could serve both as a support matrix and as a nutrient source. Agricultural residues are one of the ideal substrates for the cultivation of the culture in SSF. These materials are typically starch or lignocellulose-based agricultural products or agro-industrial sources such as grains and grain by-products, cassava, soybean, sugar beet, sweet potato, potato and sweet sorghum; crop residues such as bran and straw of wheat and rice, bran of maize, ragi, green gram, black gram and red gram, hull of soy, corn and rice, sago and bagasse of sugarcane and cassava; residues of the coffee processing industry such as coffee pulp, coffee husk, coffee spent ground; residues of fruit-processing industries such as pomace of apple and grapes, wastes of pineapple and carrot processing, banana waste, waste of oil-processing mills such as coconut cake, soybean cake, peanut cake, canola meal and palm oil mill waste; and others such as sawdust, corncobs, carob pods, tea waste, chicory roots and bread.

Table 7.2 Application of agro-industrial residues as a substrate for antibiotic production

Antibiotic	Species	SSF substrate	Reference
Bacitracin	*Bacillus licheniformis*	Soya bean meal, sunflower meal, wheat bran	Farzana et al. (2005)
Cephalosporin C	*Cephalosporium* sp.	Wheat bran, wheat grains, rice grains, barley and rice bran	Ellaiah et al. (2002)
	Acremonium chrysogenum	Wheat bran and rawa, bombay rawa, barley, rice bran	Adinarayana et al. (2003)
	Acremonium chrysogenum	Sugarcane bagasse, sugarcane molasses, corn steep waste	Cuadra et al. (2008)
Cephamycin C	*Nocardia lactamdurans*	Soybean flour	Kagliwal et al. (2009)
	Streptomyces clavuligerus	Soybean meal	Bussari et al. (2008)
	Streptomyces clavuligerus	Wheat rawa, cotton seed deoiled cake, sunflower cake	Kota and Sridhar (1999)
Compactin	*Penicillium brevicompactum*	Groundnut oil cake, wheat bran, soybean meal	Shaligram et al. (2009)
Cyclosporin A	*Tolypocladium* sp.	Wheat bran	Sekar et al. (1997)
	Tolypocladium inflatum	Wheat bran flour and coconut oil cake	Survase et al. (2009)
Iturin A	*Bacillus subtilis*	Soya bean curd residue, okara	Mizumoto et al. (2006)
Griseofulvin	*Penicillium griseofulvum*	Rice bran	Saykhedkar and Singhal (2004)
Meroparamycin	*Streptomyces* sp.	Rice, wheat bran quaker, bread, ground corn	El-Naggar et al. (2009)
Mevastatin	*Penicillium citrinum*	Wheat bran	Ahamad et al. (2006)
Neomycin	*Streptomyces fradiae*	Apple pomace, cotton seed meal, soy bean powder, wheat bran	Vastrad and Neelagund (2011)
	Streptomyces marinensis	Wheat, rice, maize and ragi bran, red, green and black gram bran, wheat and rice rawa, rice husk rice straw, corn and jowar flour, sago and sugar cane bagasse	Ellaiah et al. (2004)

(continued)

Table 7.2 (continued)

Antibiotic	Species	SSF substrate	Reference
Tetracycline	*Streptomyces rimosus* *Streptomyces alboflavus* *Streptomyces aureofaciens* *Streptomyces vendagensis*	Peanut shells, corncob, corn pomace, cassava peels	Asagbra et al. (2005a)
Oxytetracycline	*Streptomyces rimosus*	Corncob	Yang and Swei (1996)
	Streptomyces rimosus *Streptomyces alboflavus* *Streptomyces aureofaciens* *Streptomyces vendagensis*	Peanut shells, corncob, corn pomace, cassava peels	Asagbra et al. (2005b)
Penicillin	*Penicillium chrysogenum*	Sugarcane bagasse	Barrios-Gonzalez et al. (1993)
Rifamycin B	*Amycolatopsis* sp.	Cornhusk corncob and wheat bran	Mahalaxmi et al. (2010)
	Amycolatopsis mediterranean	Coconut oil cake, groundnut oil cake, groundnut shell and rice husk	Vastrad and Neelagund (2012)
	Amycolatopsis mediterranei	Wheat bran	Venkateswarlu et al. (2000)
Surfactin	*Bacillus subtilis*	Soybean curd residue (okara)	Ohno et al. (1995)

The classical method of optimisation of fermentation medium involves changing one independent variable (nutrient, pH, temperature, etc.) while maintaining all others at a constant level. This is a time-consuming and expensive process for analysing a large number of fermentation variables. In order to overcome this difficulty, the fermentation parameters were further optimised by experimental factor design (EFD) and response surface methodology (RSM). The nutritional parameters for neomycin production by *Streptomyces marinensis* under SSF is optimised using EFD and RSM. The maximum productivity of antibiotic was 17,150 mg/kg of wheat rawa under optimum conditions of dextrin 14.1 g/kg, raspberry seed powder 64.91 g/kg and mineral salt solution 172.6 ml/kg (Adinarayana et al. 2003).

RSM was employed to optimise the cultivation conditions of *Bacillus subtilis* S3 in SSF for the enhancement of iturin A, a lipopeptide antibiotic. Maximum production of iturin A reached 11.44 mg/g when *B. subtilis* S3 was cultivated at 25 °C for 5 days in SSF containing high gluten flour and rice bran (Shih et al. 2008). Soybean flour was used as substrate in SSF for the production of cephamycin C by using *N. lactamdurans* and fermentation parameters were optimised by RSM. Under optimal SSF conditions, maximum production of 15.75 ± 0.27 mg/g of cephamycin C was observed when compared to 8.37 ± 0.23 mg/g dry substrate before optimisation (Kagliwal et al. 2009).

7.5.2 *SSF on Inert Support Systems*

One of the negative factors of SSF processes, which utilise agricultural materials as substrate, is the impurity of the product. These impurities complicate the downstream processing when an end product of high purity is demanded. One of the potential alternatives to tackle these problems is to use an inert carrier as a supporting system. Recent results also show that the artificial inert support matrix enhanced the production of antibiotics. The use of inert supports for SSF resulted in the production of novel antibiotics, pyrrocidins A and B from *Cylindrocarpon* sp. and acremonidins A–E from *Acremonium* sp. In these experiments, the *Cylindrocarpon* sp. was cultured on a polyester-cellulose support on malt extract agar wherein pyrrocidin, which contains an unusual 13-membered macrocyclic ring, was produced. In contrast, a simple liquid version of the same medium failed to support the synthesis of the antibiotic. In the second case, an *Acremonium* strain produced the polyketides acremonidins A–E when cultured on a polyester-cellulosic support in malt extract medium, significantly in elevated levels over those produced in culture without the support (Bigelis et al. 2006).

7.6 Optimisation of Fermentation

7.6.1 *Selection of Supplements*

Research studies have proven that the optimisation of supplements has played a vital role in increasing the yields from different metabolites and solid substrates. The big advantage of selection of supplements is their uniqueness, since they often provide some variations in the secondary metabolite synthesis. Apart from usage of C and N sources and minerals in the substrate, the additional supplements can act as inducers or precursors in the process of synthesis.

In the process of cephamycin C production in SSF, wheat rawa enhanced the growth of *S. clavuligerus* and gave the highest titre value. Supplementation of sunflower deoiled cake, cotton seed deoiled cake and corn steep liquor enhanced production and at 0.5, 1.0 and 50 % weight of support, respectively, gave highest titres of (10 mg/g) cephamycin C (Kota and Sridhar 1999). External carbon sources like glycerol addition to the solid substrate resulted in the maximum Cyclosporin A (CyA) production of 4,659 ± 58 mg/kg followed by the sugars, dextrin and maltose. Addition of nitrogen sources like ammonium sulphate resulted in maximum production of 5,014 ± 65 mg/kg followed by 4,858 ± 45 mg/kg bacto-peptone and 4,827 ± 47 mg/kg casein peptone, respectively. The combination of glycerol (1 %)

and ammonium sulphate (1 %) gave a remarkable CyA production of 5,454 ± 44 mg/kg (Survase et al. 2009).

Yang and Swei (1996) utilised corncob, as SSF substrate in the production of oxytetracycline by *Streptomyces rimosus*, supplemented with 20 % (w/w) rice bran or 1.5–2.5 % ammonium sulphate and showed a yield of 10–11 mg/g substrate. Sircar et al. (1998) used supplementary sources including soya flour, KH_2PO_4 and sunflower oil cake and improved the production of clavulanic acid by *S. clavuligerus*. The maximum productivity of cephalosporin C (22,281 µg/g) employing *A. chrysogenum* ATCC 48272 was achieved by utilising wheat rawa with optimised process parameters including 1 % w/w soluble starch and 1 % w/w yeast extract as additives (Adinarayana et al. 2003).

The use of ammonium oxalate as a supplementary nitrogen source for cephamycin C production using *S. clavuligerus* NT4, under the optimised conditions, yielded 21.68 ± 0.76 mg/g of cephamycin C as compared to 10.50 ± 1.04 mg/g dry substrate before optimisation (Bussari et al. 2008). Supplementing the solid substrate with 0.1 % of choline chloride served as precursor and produced a 76 % increase in the yield of griseofulvin in SSF (Saykhedkar and Singhal 2004). Haloduracin, a bacteriocin, was produced by *Bacillus halodurans* when cultivated on wheat bran as a solid-state substrate, at 245 AU per wheat bran. Under the optimum conditions, supplementation of the bran with 10 % (w/w) sodium carbonate, the organism produced about 3,000 AU per gram dry bran (Danesh et al. 2011). Compactin production validation studies by *Penicillium brevicompactum* under SSF using statistical model-defined conditions resulted in an improved yield of 1,250 µg/g. Further improvement in yield was obtained using carbon supplementation in fed-batch mode. The feeding of glycerol (20 % v/v) on day 3 resulted in the much improved compactin yield of 1,406 µg/g dry substrate. This demonstrates usage of statistical experiment design as an easy tool to improve the process conditions for secondary metabolites production (Shaligram et al. 2009).

7.6.2 Substrate Pretreatment

Natural substrates and even inert supports generally require some kind of physical or chemical pretreatments. This modulates the support to attain more accessibility to microbial colonisation and penetration through adhesion to more susceptible physical structure along with their chemical constituents. Moreover, it also contributes to the improvement of its moisture-holding capacity. Wheat bran supplemented at an initial moisture content of 55 % was pretreated in autoclave for 1 h at 121 °C and used for cyA production by *Tolypocladium inflatum* (Murthy et al. 1999). The commercial solid substrates such as rice bran, wheat bran and ground corns were used for the production of meroparamycin. These solid substrates were washed and soaked in starch-nitrate medium at room temperature overnight. The soaked substrate was sterilised and used for production of meroparamycin by *Streptomyces* sp. strain MAR01 in SSF (El-Naggar et al. 2009).

7.6.3 Effect of Moisture Content

Moisture level is not considered as a vital factor in all submerged fermentation as water occupies a major percentage of the medium. Whereas in SSF, where the process is carried out on a solid medium with low moisture content, it is a critical parameter vital for the microorganism to grow on the surface of the solid substrate particles. Moisture content has a predominant role in enhancing the diffusion of extracellular enzymes, nutrients and metabolic products through the solid matrix. In SSF, the initial moisture content value depends primarily on the water retention capacity of the substrate.

The initial moisture content of the substrate played an eminent role on cephamycin C production by SSF. The range of moisture content from 60 to 80 % in wheat rawa showed substantial growth of *S. clavuligerus* and cephamycin C production (7–10 mg/g substrate). Below 60 % or above 80 % moisture concentration, there was no appreciable growth and further the decrease of cephamycin C concentration was observed (Kota and Sridhar 1999). At low moisture levels of 1:1 (2 gds:2 ml), 1:1.5 (2 gds:3 ml) and 1:2 (2 gds:4 ml), no rifamycin B production by isolated *Amycolatopsis* strain was observed. A gradual increase in antibiotic production from 0.89 to 3.47 g/kgds is observed at moisture level increase from 1:2.5 (2 gds:5 ml) to 1:4.5 (2 gds:9 ml) and further increase in moisture levels reduced the antibiotic yield. Interestingly, the specific antibiotic production (44.91 mg/g biomass) remained constant at all moisture levels. The observed differences in antibiotic production values may be attributed to moisture dependent mass transfer and related variations during SSF (Mahalaxmi et al. 2010). It is reported that higher substrate moisture in SSF resulted in suboptimal product formation due to reduced mass transfer process such as diffusion of solutes and gas to cell during fermentation. Interestingly, some researchers have observed optimised antibiotic production at higher substrate moisture in SSF. The highest neomycin production (5,227 μg/g) was achieved at 80 % initial moisture content of wheat rawa (Ellaiah et al. 2004), and the high cephalosporin C antibiotic titre (4,445 μg/g) was attained when the initial moisture level was 80 % in comparison with that at low or high moisture levels (Adinarayana et al. 2003).

The critical importance of moisture level in SSF media and its influence on the biosynthesis and secretion of antibiotics can be attributed to the interference of moisture in the physical properties of the solid particles. Increase in moisture level reduce the porosity of the solid substrates, thus limiting oxygen transfer, and low moisture content causes reduction in the solubility of nutrients of the substrate and causes low degree of swelling, resulting in decreased secondary metabolites production.

7.6.4 Effect of Particle Size

Among the several factors in SSF processes, Selection of proper particle size of substrates is one of the essential requirements for optimum production in SSF with different microbial strains. Generally, smaller substrate particles will provide a larger surface area for microbial attachment and thus it should be considered as a desirable factor favouring SSF. However, too small substrate particles may result in substrate agglomeration in most cases, which may interfere with aeration and may result in poor growth rate. At the same time, larger particles provide better aeration efficiency but provide limited surface for microbial adherence. Thus, it may be necessary to provide an optimised particle size (Pandey et al. 2000). In rifamycin B production, a 30 % (3.55–4.53 g/kg) improvement was observed on optimised corn-husk particle size (6×4 mm) and further variation of substrate particle size resulted in reduction of antibiotic production (Mahalaxmi et al. 2010). In the neomycin production, wheat rawa of coarse size 0.84 mm gave the best results (4,478 μg/g) compared to the intermediate and fine size substrates, which yielded 4,043 and 3,427 μg/g substrate, respectively (Ellaiah et al. 2004). Some researchers used larger particle size in SSF for antibiotic synthesis; e.g. the sugarcane bagasse of particle size 14 mm increased the penicillin production by 37 %. However, this effect was due to higher sugar concentration in bagasse fraction (Barrios-Gonzalez et al. 1993).

7.6.5 Effect of pH and Temperature

The metabolic activities of the microorganisms are very much sensitive to the pH change, and antibiotic production by microorganism is found to be affected if pH level of the substrate is higher or lower compared with optimum value. Many researchers demonstrated the strain-dependent variation of pH for optimum antibiotic production in SSF. Cuadra et al. (2008) revealed pH as a key parameter in cephalosporin C production in solid-state fermentation by *A. chrysogenum* C10 using sugarcane bagasse as support. The production of cephalosporin C reached 3,200 μg/g of dry matter at an optimised pH between 6.4 and 7.8. In neomycin production by *S. marinensis*, when the initial pH was 6.0, there was less production and as the pH increased, its production reached the maximum (5,780 μg/g) at pH 7.5 (Ellaiah et al. 2004).

Strain-dependent variation of pH and temperature was reported for optimum rifamycin B production by *Amycolatopsis* sp. (Venkateswarlu et al. 2000) suggesting screening and determination of the optimum levels of fermentation parameters, which are very important for overall economic feasibility of the production process. Rifamycin B production by isolated *Amycolatopsis* sp. RSP 3 under SSF showed pH conditions pH 7–9 are favourable for production and maximal production (2.33 g/kgds or 40 mg/g biomass) was observed at pH 8.0. Maximum antibiotic yield (3.0 g/kgds or 41.3 mg/g biomass) is observed at 28 °C (Mahalaxmi et al. 2010). The optimal pH for tetracycline production by *Streptomyces viridifaciens* exactly matched the pH of sweet potato residue solid substrate, between pH 5.8 and 6.0. Each gram of dry substrate produced 1,570 μg total tetracycline equivalent potency (Yang and Ling 1989). In the course of cephamycin C production, when the initial pH was 5.0, there was very little growth in *S. clavuligerus* and no production of antibiotic is observed. Substantial increase of the pH increased the cephamycin C production and reached the maximum (15 mg/g) at 6.5 pH. A further increase in pH resulted in a decrease in cephamycin C production and no production was observed at 8.0 pH. At 20 °C, cephamycin C production was 6 mg/g, at 30 °C production was 15 mg/g and it decreased to zero when the incubation temperature was 37 °C. The pH 6.5 and temperature 28 °C were found to be optimal both for SSF and SmF for cephamycin C production (Kota and Sridhar 1999). Ohno et al. (1995) elucidated temperature dependency for the production of lipopeptide antibiotics, namely, iturin and surfactin by *B. subtilis* RB14, in the solid-state fermentation of okara and observed the optimal temperature for iturin is 25 °C and for surfactin it is 37 °C. The highest neomycin production (5,760 μg/g) was attained at 30 °C, and a decrease in the yield of neomycin was observed when the incubation temperature was higher or lower than the optimum incubation temperature (Ellaiah et al. 2004). Higher temperatures were found to have adverse effects on the metabolic activities of the microorganism and it is also reported that the metabolic activities of the microorganisms become slower at lower temperature. Hence, incubation temperature and its control in SSF process are crucial as the heat evolved during SSF processes is accumulated due to poor heat dissipation in solid media.

7.6.6 Inoculum Level

Irrespective of the type of fermentation, whether it is SSF or SmF, inoculum level affects the yield of antibiotic. A progressive increase in rifamycin B yield is observed with increase in *Amycolatopsis* sp. RSP 3 starter inoculum from 2.4 to 7.2 %. Maximum antibiotic production of 4.84 g/kg dry substrate was observed with 7.2 % inoculum. However, higher inoculum level (12 %) resulted in more than 50 % reduced antibiotic production (Mahalaxmi et al. 2010). During the course of cephamycin C production, wheat rawa was found to be most suitable substrate in SSF for production by *S. clavuligerus*. When solid support was mixed with 1×10^2 cell per gram, cephamycin C production was very low (1 mg/g) but as the concentration reached 10^8 cells per gram, its concentration reached a maximum of 10 mg/g of substrate. It is important to note that further increase in cell concentration did not affect the concentration of cephamycin C (Kota and Sridhar 1999). Optimum neomycin production (6,880 mg/g) was observed at 0.5 % w/w dry cell mass of inoculum. At lower and higher inoculum levels, poor neomycin production was observed (Ellaiah et al. 2004). Inoculum level was also important factors for the production of cephalosporin C. High inoculum levels are inhibitory in nature. Higher antibiotic production (5,596 μg/g) was obtained at 10 % (v/w) inoculum level as compared to low or high inoculum levels (Adinarayana et al. 2003). It is important to provide an optimum inoculum level in fermentation process. A lower inoculum density may give insufficient biomass causing reduced product formation, whereas a higher inoculum may produce too much biomass and deplete the substrate of nutrients leading to poor product formation.

7.7 Industrial Strain Development

Industrial strain development has been the main focus of research in the commercial development of microbial fermentation processes. Discoveries in mutation, protoplast fusion, genetic manipulations, recombinant DNA technology and operation of commercial large-scale fermenters have revolutionised the concept of microbial strain development. Although greater improvements in overproduction of metabolites and antibiotics of specific microbes have resulted from essentially random empirical approaches to mutation and strain development, future strain development technology will be supplemented by more knowledge-based scientific methods. With the advances in understanding biosynthetic pathways, elucidation of regulatory mechanisms related to induction and repression of genes and bioengineering design, it will be possible to apply new strategies and limitless combinations for isolating improved strains. Furthermore, tailoring genes through the avenue of in vitro DNA recombination techniques in both bacteria and fungi has been shown to be feasible. Perhaps these areas will facilitate new strategies and have higher impact on industrial strain improvement. Khaliq et al. (2009) developed SSF system for hyperproduction of tylosin using a mutant γ-1 of *S. fradiae* NRRL-2702 and its parent strain. Various types of agro-industrial wastes were screened to study their effect on tylosin production in SSF. Wheat bran is the ideal solid substrate giving highest production of 2,500 μg of tylosin per gram substrate by mutant γ-1 against parent strain which gave 300 μg tylosin per gram substrate. Fermentation optimisation (70 % moisture, 10 % inoculum (v/w), pH 9.2, 30 °C, supplemental lactose and sodium glutamate on day 9) further improved the tylosin yield to 4,500 μg/g substrate. Wild parent strain displayed less production of tylosin (655 μg/g substrate) in SSF even after fermentation optimisation. This study evidenced tylosin yield enhancement by γ-1 *S. fradiae* strain under solid-state fermentation system.

Production of lipopeptide antibiotic, namely, surfactin in SSF on okara (soybean curd residue) as a solid substrate was carried out using *B. subtilis* MI113 with a recombinant plasmid pC112, which contains lpa-14, a gene related to surfactin production cloned from a wild-type surfactin producer, *B. subtilis* RB14. The amount of surfactin produced by MI113 (pC112) was 2.0 g/kg wet weight, which was eight times as high as that of the original *B. subtilis* RB14 at the optimal fermentation conditions. Further, the stability of the plasmid studied both under SSF and SmF system showed a similar pattern; however, the production of surfactin in SSF was four to five times more efficient than in SMF (Ohno et al. 1995).

Clones of four industrial strains of *P. chrysogenum* producing higher penicillin were subjected to assess its capacity to produce high quantity of antibiotic in SSF (Barrios-Gonzalez et al. 1993). *S. rimosus* TM-55 is treated with 3 % EMS, and 29 auxotrophic mutants (AM-1 to AM-29) are isolated from 5,457 colonies. Three sets of the auxotrophic mutants were chosen for protoplast fusion with 50 % PEG 1000 for 30 min at 25 °C, and 25 fusants were isolated. In solid substrate tested for oxytetracycline production, 20 % of fusants production is higher than that of the wild strain (Yang and Kao 1991).

7.8 Future Prospects

Penicillin, the first commercial antibiotic produced by fermentation in large-scale process, has stimulated the development of the fermentation biotechnology significantly. To develop an economically viable SSF that can be scaled up to industrial level, production process must address the path integrated with the following two critical stages: upstream and downstream processing. The commercial scale production process development is much difficult, owing to limitations in control of operations (heat, mass transfer and cooling) and variable factors (temperature, nutrient concentration, pH and moisture) influencing fermentations, which are essential to the system.

Fermentation biotechnology plays a major role in development of both broad-spectrum and narrow-spectrum antibiotics. The production of antibiotics using SSF has addressed two important aspects. Firstly, the amounts of antibiotics obtained by SSF are manyfold higher than SmF. Secondly the products obtained have enhanced vital properties when produced in SSF. In addition, the SSF approaches provide an alternative technology platform for the production of potent metabolites, with the growth environment presenting different physiological challenges to the microorganisms inducing corresponding differences in the organism's biochemistry. Recent researches showed the use of inert supports for fungal SSF resulting in the production of novel antibiotics (pyrrocidins and acremonidins). Recently, the exploitation of marine microorganisms for the discovery of secondary metabolites has led to numerous new antibiotics discoveries. Future pertinence of marine microorganisms in SSF could revolutionise the production of novel antibiotics. Therefore, if SSF variables are well controlled at optimised culture conditions and the purity of the product is defined, this may be a lucrative technology for commercial production of antibiotics than any process available currently.

Acknowledgements The authors gratefully acknowledge the financial support given by the Earth System Science Organization, Ministry of Earth Sciences, Government of India. The authors are thankful to the Director, National Institute of Ocean Technology (NIOT), Ministry of Earth Sciences, Govt. of India, for his constant support and encouragement for preparation of this chapter. The authors are also thankful to all the scientific and supporting staffs of Marine Biotechnology, NIOT, Chennai, for their support.

References

Adinarayana K, Prabhakar T, Srinivasulu V, Anitha Rao M, Jhansi Lakshmi P, Ellaiah P (2003) Optimization of process parameters for cephalosporin C production under solid state fermentation from *Acremonium chrysogenum*. Process Biochem 39:171–177

Ahamad MZ, Panda BP, Javed S, Ali M (2006) Production of mevastatin by solid state fermentation using wheat bran as substrate. Res J Microbiol 1(5):443–447

Arakawa K, Mochizuki S, Yamada K, Noma T, Kinashi H (2007) γ-Butyrolactone autoregulator-receptor system involved in lankacidin and lankamycin production and morphological differentiation in *Streptomyces rochei*. Microbiology 153(6):1817–1827

Asagbra AE, Sanni AI, Oyewole OB (2005a) Solid-state fermentation production of tetracycline by *Streptomyces* strains using some agricultural wastes as substrate. World J Microbiol Biotechnol 21(2):107–114

Asagbra AE, Oyewole OB, Odunfa SA (2005b) Production of oxytetracycline from agricultural wastes using *Streptomyces* species. Niger Food J 23:174–182

Asanza TML, Gontier E, Bienaime C, Nava Saucedo JE, Barbotin JN (1997) Response surface analysis of chlortetracycline and tetracycline production with K-carrageenan immobilized *Streptomyces aureofaciens*. Enzyme Microb Technol 21(5):314–320

Barrios-Gonzalez J, Mejia A (1996) Production of secondary metabolites by solid-state fermentation. Biotechnol Annu Rev 2:85–121

Barrios-Gonzalez J, Castillo TE, Mejia A (1993) Development of high penicillin producing strains for solid state fermentation. Biotechnol Adv 11(3):525–537

Barrios-Gonzalez J, Fernandez FJ, Tomasini A (2003) Microbial secondary metabolites production and strain improvement. Ind J Biotechnol 2:322–333

Bate N, Butler AR, Smith IP, Cundliffe E (2000) The mycarose-biosynthetic genes of *Streptomyces fradiae*, producer of tylosin. Microbiology 146:139–146

Behal V, Neuzil J, Hostalek Z (1983) Effect of tetracycline derivatives and some cations on the activity of anhydrotetracycline oxygenase. Biotechnol Lett 5:537–542

Bigelis R, He H, Yang HY, Chang LP, Greenstein M (2006) Production of fungal antibiotics using polymeric solid supports in solid-state and liquid fermentation. J Ind Microbiol Biotechnol 33(10):815–826

Brakhage AA, Schroeckh V (2011) Fungal secondary metabolites—strategies to activate silent gene clusters. Fungal Genet Biol 48:15–22

Brakhage AA, Thon M, Sprote P, Scharf DH, Al-Abdallah Q, Wolke SM, Hortschansky P (2009) Aspects on evolution of fungal β-lactam biosynthesis gene clusters and recruitment of trans-acting factors. Phytochemistry 70:1801–1811

Brautaset T, Sekurova ON, Sletta H, Ellingsen TE, Strom AR, Valla S, Zotchev SB (2000) Biosynthesis of the polyene antifungal antibiotic nystatin in *Streptomyces noursei* ATCC 11455: analysis of the gene cluster and deduction of the biosynthetic pathway. Chem Biol 7:395–403

Brodhagen M, Henkels MD, Loper JE (2004) Positive autoregulation and signaling properties of pyoluteorin, an antibiotic produced by the biological control organism *Pseudomonas fluorescens* Pf-5. Appl Environ Microbiol 70(3):1758–1766

Bussari B, Parag SS, Nikhil SS, Survase AS, Rekha SS (2008) Production of cephamycin C by *Streptomyces clavuligerus* NT4 using solid-state fermentation. J Ind Microbiol Biotechnol 35(1):49–58

Butler AR, Flint SA, Cundliffe E (2001) Feedback control of polyketide metabolism during tylosin production. Microbiology 147:795–801

Chen W, Huang T, He X, Meng Q, You D, Bai L, Li J, Wu M, Li R, Xie Z, Zhou H, Zhou X, Tan H, Deng Z (2009) Characterization of the polyoxin biosynthetic gene cluster from *Streptomyces cacaoi* and engineered production of polyoxin H. J Biol Chem 284(16):10627–10638

Corre C, Song L, O'Rourke S, Chater KF, Challis GL (2008) 2-Alkyl-4-hydroxymethylfuran-3-carboxylic acids, antibiotic production inducers discovered by *Streptomyces coelicolor* genome mining. Proc Natl Acad Sci U S A 105:17510–17515

Cuadra T, Fernandez FJ, Tomasini A, Barrios-Gonzalez J (2008) Influence of pH regulation and nutrient content on cephalosporin C production in solid-state fermentation by *Acremonium chrysogenum* C10. Lett Appl Microbiol 46(2):216–220

Danesh A, Mamo G, Mattiasson B (2011) Production of haloduracin by *Bacillus halodurans* using solid-state fermentation. Biotechnol Lett 33(7):1339–1344

Devi S, Padma S (2000) Production of cephamycin C in repeated batch operations from immobilized *Streptomyces clavuligerus*. Process Biochem 36(3):225–231

Doull JL, Vining LC (1990) Nutritional control of actinorhodin production by *Streptomyces coelicolor* A3(2): suppressive effects of nitrogen and phosphate. Appl Microbiol Biotechnol 32(4):449–454

El-Enshasy HA, Mohamed NA, Farid MA, El-Diwany AI (2008) Improvement of erythromycin production by *Saccharopolyspora erythraea* in molasses based medium through cultivation medium optimization. Bioresour Technol 99(10):4263–4268

Elibol M (2004) Optimization of medium composition for actinorhodin production by *Streptomyces coelicolor* A3(2) with response surface methodology. Process Biochem 39(9):1057–1062

Ellaiah P, Premkumar J, Kanthachari PV, Adinarayana K (2002) Production and optimization studies of cephalosporin C by solid state fermentation. Hindustan Antibiot Bull 44(1–4):1–7

Ellaiah P, Shrinivasulu B, Adinarayana K (2004) Optimization studies on neomycin production by a mutant strain of *Streptomyces marinensis* in solid-state fermentation. Process Biochem 39:529–534

El-Naggar MY, El-Assar SA, Abdul-Gawad SM (2009) Solid-state fermentation for the production of meroparamycin by *Streptomyces* sp. strain MAR01. J Microbiol Biotechnol 19(5):468–473

Epp JK, Burgett SG, Schoner BE (1987) Cloning and nucleotide sequence of a carbomycin-resistance gene from *Streptomyces thermotolerans*. Gene 53(1):73–83

Espeso EA, Fernandez-Canon JM, Penalva MA (1995) Carbon regulation of penicillin biosynthesis in *Aspergillus nidulans*: a minor effect of mutations in creB and creC. FEMS Microbiol Lett 126:63–68

Farzana K, Shah SN, Butt FB, Awan SB (2005) Biosynthesis of bacitracin in solid-state fermentation by *Bacillus licheniformis* using defatted oil seed cakes as substrate. Pak J Pharm Sci 18(1):55–57

Froyshov O, Mathiesen A, Haavik HI (1980) Regulation of bacitracin synthetase by divalent metal ions in *Bacillus licheniformis*. J Gen Microbiol 117:163–167

Gramajo HC, White J, Hutchinson CR, Bibb MJ (1991) Overproduction and localization of components of the polyketide synthase of *Streptomyces glaucescens* involved in the production of the antibiotic tetracenomycin C. J Bacteriol 173:6475–6483

Graminha EBN, Goncalves AZL, Pirota RDPB, Balsalobre MAA, Da Silva R, Gomes E (2008) Enzyme production by solid-state fermentation: application to animal nutrition. Anim Feed Sci Technol 144(1–2):1–22

Gristwood T, Fineran PC, Everson L, Williamson NR, Salmond GP (2009) The PhoBR two-component system regulates antibiotic biosynthesis in *Serratia* in response to phosphate. BMC Microbiol 9:112–126

Gupte MD, Kulkarni PR (2002) A study of antifungal antibiotic production by *Streptomyces chattanoogensis* MTCC 3423 using full factorial design. Lett Appl Microbiol 35(1):22–26

Gutierrez S, Fierro F, Casqueiro J, Martin JF (1999) Gene organization and plasticity of the beta-lactam genes in different filamentous fungi. Antonie Van Leeuwenhoek 75:81–94

Haese A, Keller U (1988) Genetics of actinomycin C production in *Streptomyces chrysomallus*. J Bacteriol 170(3):1360–1368

Haste NM, Perera VR, Maloney KN, Tran DN, Jensen P, Fenical W, Nizet V, Hensler ME (2010) Activity of the streptogramin antibiotic etamycin against methicillin-resistant *Staphylococcus aureus*. J Antibiot 63:219–224

Hyun C, Kim SS, Sohng JK, Hahn J, Kim J, Su J (2000) An efficient approach for cloning the dNDP-glucose synthase gene from *actinomycetes* and its application in *Streptomyces spectabilis* a spectinomycin producer. FEMS Microbiol Lett 183(1):183–189

Indu ST (2006) Environmental biotechnology: Basic concepts and applications. In: Antibiotic industry, 2nd edn. IK International Pvt Ltd, India, pp 435–443

Jeanne MD, Craig AT (2009) Identification and characterization of NocR as a positive transcriptional regulator of the β-Lactam nocardicin A in *Nocardia uniformis*. J Bacteriol 191:1066–1077

Jekosch K, Kuck U (2000) Loss of glucose repression in an *Acremonium chrysogenum* β-lactam producer strain and its restoration by multiple copies of the cre1 gene. Appl Microbiol Biotechnol 54:556–563

Juan FM (2004) Phosphate control of the biosynthesis of antibiotics and other secondary metabolites is mediated by the PhoR–PhoP system: an unfinished story. J Bacteriol 186:5197–5201

Juan FM, Arnold LD (2002) Unraveling the methionine–cephalosporin puzzle in *Acremonium chrysogenum*. Trends Biotechnol 20(12):502–507

Kagliwal LD, Survase SA, Singhal RS (2009) A novel medium for the production of cephamycin C by *Nocardia lactamdurans* using solid-state fermentation. Bioresour Technol 100(9):2600–2606

Karray F, Darbon E, Oestreicher N, Dominguez H, Tuphile K, Gagnat J, Blondelet-Rouault MH, Gerbaud C, Pernodet JL (2007) Organization of the biosynthetic gene cluster for the macrolide antibiotic spiramycin in *Streptomyces ambofaciens*. Microbiology 153(12):4111–4122

Karray F, Darbon E, Nguyen HC, Gagnat J, Pernodet JL (2010) Regulation of the biosynthesis of the macrolide antibiotic spiramycin in *Streptomyces ambofaciens*. J Bacteriol 192:5813–5821

Kawaguchi T, Azuma M, Horinouchi S, Beppu T (1988) Effect of B-factor and its analogues on rifamycin biosynthesis in *Nocardia sp.*. J Antibiot 41:360–365

Keller U, Lang M, Crnovcic I, Pfennig F, Schauwecker F (2010) The actinomycin biosynthetic gene cluster of *Streptomyces chrysomallus*: a genetic hall of mirrors for synthesis of a molecule with mirror symmetry. J Bacteriol 192(10):2583–2595

Khaliq S, Rashid N, Akhtar K, Ghauri MA (2009) Production of tylosin in solid-state fermentation by *Streptomyces fradiae* NRRL-2702 and its gamma-irradiated mutant (γ-1). Lett Appl Microbiol 49(5):635–640

Kleerebezem M, Quadri LEN, Kupers OP, De Vos WM (1997) Quorum sensing by peptide pheromones and two-component signal-transduction systems in Gram-positive bacteria. Mol Microbiol 24:895–904

Koonin EV, Wolf YI, Aravind L (2001) Prediction of the archaeal exosome and its connections with the proteasome and the translation and transcription machineries by a comparative-genomic approach. Genome Res 11:240–252

Kota KP, Sridhar P (1999) Solid state cultivation of *Streptomyces clavuligerus* for cephamycin C production. Process Biochem 34:325–328

Kumar M, Srivastava S (2011) Effect of calcium and magnesium on the antimicrobial action of enterocin LR/6 produced by *Enterococcus faecium* LR/6. Int J Antimicrob Agents 37(6):572–575

Laich F, Fierro F, Cardoza RE, Martín JF (1999) Organization of the gene cluster for biosynthesis of penicillin in *Penicillium nalgiovense* and antibiotic production in cured dry sausages. Appl Environ Microbiol 65:1236–1240

Lopez-Calleja AC, Cuadra T, Barrios-Gonzalez J, Fierro F, Fernandez FJ (2012) Solid-state and submerged fermentations show different gene expression profiles in cephalosporin C production by *Acremonium chrysogenum*. J Mol Microbiol Biotechnol 22(2):126–134

Lotfy WA (2007) Production of cephalosporin C by *Acremonium chrysogenum* grown on beet molasses: optimization of process parameters through statistical experimental designs. Res J Microbiol 2:1–12

Mahalaxmi Y, Sathish T, Subba Rao C, Prakasham RS (2010) Corn husk as a novel substrate for the production of rifamycin B by isolated *Amycolatopsis sp.* RSP3 under SSF. Process Biochem 45(1):47–53

Malik VS (1979) Genetics of applied microbiology. Adv Genet 20:37–126

Marinelli F, Marcone GL (2011) Microbial secondary metabolites. In: Comprehensive biotechnology, 2nd edn. Elsevier, Netherlands, pp 285–297

Martin J, Garcia-Estrada C, Rumbero A, Recio E, Albillos SM, Ullan RV, Martin JF (2011) Characterization of an autoinducer of penicillin biosynthesis in *Penicillium chrysogenum*. Appl Environ Microbiol 77(16):5688

Mehmood N, Olmos E, Goergen JL, Blanchard F, Marchal P, Klockner W, Buchs J, Delaunay S (2012) Decoupling of oxygen transfer and power dissipation for the study of the production of pristinamycins by *Streptomyces pristinaespiralis* in shaking flasks. Biochem Eng J 68:25–33

Miyake K, Kuzuyama T, Horinouchi S, Beppu T (1990) The A-factor-binding protein of *Streptomyces griseus* negatively controls streptomycin production and sporulation. J Bacteriol 172:3003–3008

Mizumoto S, Hirai M, Shoda M (2006) Production of lipopeptide antibiotic iturin A using soybean curd residue cultivated with *Bacillus subtilis* in solid-state fermentation. Appl Microbiol Biotechnol 72:869–875

Murthy MVR, Mohan EVS, Sadhukhan AK (1999) Cyclosporin A production by *Tolypocladium inflatum* using solid state fermentation. Process Biochem 34:269–280

Nguyen KT, Nguyen LT, Behal V (1995) The induction of valine dehydrogenase activity from *Streptomyces* by l-valine is not repressed by ammonium. Biotechnol Lett 17:31–34

Obanye AIC, Hobbs G, Gardner DCJ, Oliver SG (1996) Correlation between carbon flux through the pentose phosphate pathway and production of the antibiotic methylenomycin in *Streptomyces coelicolor* A3(2). Microbiology 142:133–137

Ohno A, Ano T, Shoda M (1995) Production of a lipopeptide antibiotic, surfactin, by recombinant *Bacillus subtilis* in solid state fermentation. Biotechnol Bioeng 47(2):209–214

Otten SL, Ferguson J, Hutchinson CR (1995) Regulation of daunorubicin production in *Streptomyces peucetius* by the dnrR2 locus. J Bacteriol 177(5):1216–1224

Pandey A, Soccol CR, Mitchell D (2000) New developments in solid state fermentation: 1-bioprocess and products. Process Biochem 35(10):1153–1169

Poonam Singh N, Pandey A (2009) Biotechnology for agro-industrial residues utilisation. Utilisation of agro-residues, vol XVIII. Springer, New York, p 466

Quirs LM, Salas JA (1995) Biosynthesis of the macrolide oleandomycin by *Streptomyces antibioticus*. Purification and kinetic characterization of an oleandomycin glucosyltransferase. J Biol Chem 270:18234–18239

Recio E, Aparicio JF, Rumbero A, Martin JF (2006) Glycerol, ethylene glycol and propanediol elicit pimaricin biosynthesis in the PI-factor-defective strain *Streptomyces natalensis* npi287 and increase polyene production in several wild-type actinomycetes. Microbiology 152:3147–3156

Rius N, Maeda K, Demain AL (1996) Induction of l-lysine ε-aminotransferase by l-lysine in *Streptomyces clavuligerus*, producer of cephalosporins. FEMS Microbiol Lett 144:207–211

Saykhedkar SS, Singhal RS (2004) Solid-state fermentation for production of griseofulvin on rice bran using *Penicillium griseofulvum*. Biotechnol Prog 20(4):1280–1284

Scherlach K, Hertweck C (2009) Triggering cryptic natural product biosynthesis in microorganisms. Org Biomol Chem 7(9):1753–1760

Sekar C, Rajasekar VW, Balaraman K (1997) Production of Cyclosporin A by solid state fermentation. Bioprocess Biosyst Eng 17:257–259

Shaligram NS, Singh SK, Singhal RS, Szakacs G, Pandey A (2009) Effect of precultural and nutritional parameters on compactin production by solid-state fermentation. J Microbiol Biotechnol 19(7):690–697

Shapiro S (1989) Nitrogen assimilation in Actinomycetes and the influence of nitrogen nutrition on Actinomycete secondary metabolism. In: Shapiro S (ed) Regulation of secondary metabolism in actinomycetes. CRC Press, Boca Raton, FL, pp 135–212

Shih IL, Kuo CY, Hsieh FC, Kao SS, Hsieh C (2008) Use of surface response methodology to optimize culture conditions for iturin A production by *Bacillus subtilis* in solid-state fermentation. J Chin Ins Chem Eng 39(6):635–643

Sircar A, Sridhar P, Das PK (1998) Optimization of solid state medium for the production of clavulanic acid by *Streptomyces clavuligerus*. Process Biochem 33(3):283–289

Sohn YS, Nam DH, Ryu DD (2001) Biosynthetic pathway of cephabacins in *Lysobacter lactamgenus*: molecular and biochemical characterization of the upstream region of the gene clusters for engineering of novel antibiotics. Metab Eng 3(4):380–392

Survase SA, Shaligram NS, Pansuriya RC, Annapure US, Singhal RS (2009) A novel medium for the enhanced production of cyclosporin A by *Tolypocladium inflatum* MTCC 557 using solid state fermentation. J Microbiol Biotechnol 19(5):462–467

Takano E (2006) γ-butyrolactones: Streptomyces signalling molecules regulating antibiotic production and differentiation. Curr Opin Microbiol 9(3):287–294

Teijeira F, Ullan RV, Fernandez-Aguado M, Martin JF (2011) CefR modulates transporters of beta-lactam intermediates preventing the loss of penicillins to the broth and increases cephalosporin production in *Acremonium chrysogenum*. Metab Eng 13(5):532–543

Tercero JA, Espinosa JC, Lacalle RA, Jimenez A (1996) The biosynthetic pathway of the aminonucleoside antibiotic puromycin as deduced from the molecular analysis of the pur cluster of *Streptomyces alboniger*. J Biol Chem 271:1579–1590

Vastrad BM, Neelagund SE (2011) Optimization and production of neomycin from different agro industrial wastes in solid state fermentation. Int J Pharm Sci Drug Res 3(2):104–111

Vastrad BM, Neelagund SE (2012) Optimization of process parameters for rifamycin b production under solid state fermentation from *Amycolatopsis mediterranean* MTCC14. Int J Curr Pharm Res 4(2):101–108

Venkateswarlu G, Murali Krishna PS, Pandey A, Venkateshwar Rao L (2000) Evaluation of *Amycolatopsis mediterranei* VA18 for production of rifamycin-B. Process Biochem 36(4):305–309

Voelker F, Altaba S (2001) Nitrogen source governs the patterns of growth and pristinamycin production in '*Streptomyces pristinaespiralis*'. Microbiology 147(9):2447–2459

Wei YH, Lai CC, Chang JS (2007) Using Taguchi experimental design methods to optimize trace element composition for enhanced surfactin production by *Bacillus subtilis* ATCC 21332. Process Biochem 42(1):40–45

Xue Y, Zhao L, Liu HW, Sherman DH (1998) A gene cluster for macrolide antibiotic biosynthesis in *Streptomyces venezuelae*: architecture of metabolic diversity. Proc Natl Acad Sci U S A 95(21):12111–12116

Yang SS, Kao CY (1991) Oxytetracycline production in solid state and submerged fermentation by protoplast fusants of *Streptomyces rimosus*. Proc Natl Sci Counc Repub China B 15(1):20–27

Yang SS, Ling MY (1989) Tetracycline production with sweet potato residue by solid state fermentation. Biotechnol Bioeng 33:1021–1028

Yang SS, Swei WJ (1996) Cultural condition and oxytetracycline production by *Streptomyces rimosus* in solid state fermentation of corncob. World J Microbiol Biotechnol 12:43–46

Chapter 8
Plant Growth Hormones and Other Phytochemicals

Luciana Porto de Souza Vandenberghe, Cristine Rodrigues, Juliana de Oliveira, and Carlos Ricardo Soccol

8.1 Introduction

Plants need light, carbon dioxide, water, and minerals, including nitrogen in soil, for its growth. With these conditions, the plant has the ability to transform some simple materials into complex organic substances that compose all living organisms. In this way, plant hormones and phytohormones have a very important function or activity in the growth regulation. Hormones are organic substances that are produced in a tissue and transported to another, where they provoke a physiological response. They are active in very low concentrations. The term hormone comes from the Greek and means "impetus."

It is known that in many instances, plant development can be dramatically influenced by a set of five structurally simple phytohormones, auxins, ethylene, cytokinin, abscisic acid (ABA), and gibberellins, each of which can elicit different responses. Experimental and theoretical approaches to this problem have prompted a long-standing debate concerning the relative importance of variations in phytohormone concentration versus differential sensitivity of different plant cells to particular phytohormones (Palme et al. 1991).

This chapter intends to show the most important information, such as definition, structure, action, and lastly the advances, about the main groups of plant hormones auxins, ethylene, cytokinin, ABA, and gibberellins.

L.P. de S. Vandenberghe (✉) • C. Rodrigues • J. de Oliveira • C.R. Soccol
Department of Biotechnology and Bioprocess Engineering, Federal University of Paraná—UFPR, Av. Cel. Francisco H. dos Santos, 100., Box 19011 Zip code 81531-990, Curitiba, Paraná, Brazil
e-mail: lvandenberghe@ufpr.br

S.K. Brar et al. (eds.), *Biotransformation of Waste Biomass into High Value Biochemicals*, DOI 10.1007/978-1-4614-8005-1_8, © Springer Science+Business Media New York 2014

8.2 Auxins

The term auxin (from Greek "auxein," meaning "to increase" or "to grow") includes a spectrum of compounds that differ structurally and bring about a variety of auxin-type responses, albeit to varying degrees.

Since the original discovery of auxin as an indole compound that gave the grass coleoptile curvature (or growth) is tested, the definition of auxins has been broadened to include not only indole-3-acetic acid (IAA) but several other indole as well as non-indole compounds. Simon and Petrasek (2011) presented that many heterogeneous synthetic substances have auxin activity, complicating studies of structure–activity and the search for a common mode of action (Ferro et al. 2010). Even the most frequently used synthetic auxins, 2,4-dichlorophenoxyacetic acid (2,4-D) and naphthalene-1-acetic acid (NAA), do not completely share their mechanism of action with native IAA. Only indole-3-butyric acid (IBA), phenylacetic acid (PAA), and 4-chloroindole-3-acetic acid (4-Cl-IAA) (Fig. 8.1) are synthesized by plants and therefore qualify as "endogenous auxins," but their roles and mechanisms of action have not been satisfactorily described (Simon and Petrasek 2011).

8.2.1 Indole-3-Acetic Acid

IAA is the most widely distributed, naturally occurring auxin in vascular plants, dicots, monocots, gymnosperms, and ferns. There are also reports of IAA being present in mosses and liverworts, as well as in some green algae (e.g., *Caulerpa*). IAA is a weak acid with a pH of 4.85. It occurs in dissociated state at neutral pH solutions. IAA is involved in nearly every aspect of plant growth and development, from embryo to adult reproductive plant. The processes regulated include pattern formation in embryo development, induction of cell division, stem and coleoptile elongation, apical dominance, induction of rooting, vascular tissue differentiation, fruit development, and tropic movements such as bending of shoots toward light or of roots toward gravity.

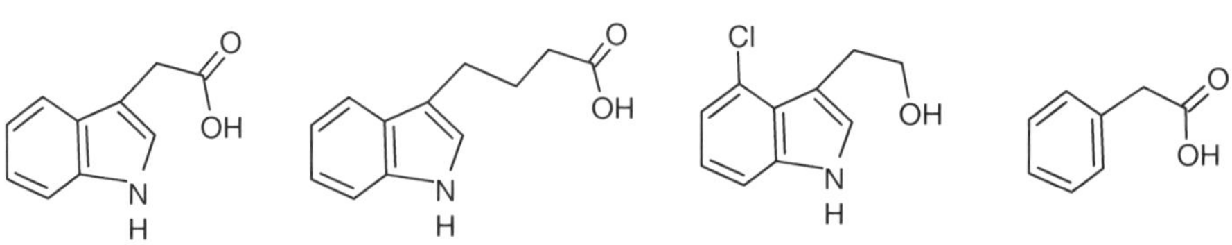

Indole-3-acetic acid (IAA) | Indole-3-butyric acid (IBA) | 4-chloroindole-3-acetic acid (4-Cl-IAA) | Phenylacetic acid (PAA)

Fig. 8.1 Chemical structure of four endogenous auxins. Indole-3-acetic acid (IAA), indole-3-butyric acid (IBA), 4-chloroindole-3-acetic acid (4-Cl-IAA), and phenylacetic acid (PAA)

It is difficult to unambiguously define typical "auxin activity." Auxin displays morphogenic properties that are modulated by the environment and defined by dynamic changes in its perception and signal transduction. This machinery has been intensively studied during the past decade and includes effects that are either dependent or independent of gene expression (Bhalerao and Bennett 2003). Thus, "auxin action" may be understood as the sum of all these processes (Simon and Petrasek 2011). Later research convincingly demonstrated that auxin is required together with other plant hormones for both cell division and oriented cell expansion (Perrot-Rechenmann 2010), influencing all aspects of plant development (Vanneste and Friml 2009).

The isolation of plant mutants related to auxin showed that the modification of the regulation of auxin biosynthesis, transport, or signaling generates severe alterations in many aspects of plant development. For example, the auxin overproducer mutant *Yucca* leads to defects in vascular tissue formation (Cheng et al. 2007). Disruption in auxin transport, in the mutant *pin1*, leads to defects in floral development (Okada et al. 1991). Finally, mutation in auxin signaling can trigger a global dwarfism as for the auxin-resistant *axr112* mutant (Lincoln et al. 1990), the absence of root formation as for the monopteros (*mp*) mutant (Hamann et al. 2002), or even embryo lethality as for the *abp1* null mutant (Chen et al. 2001). This demonstrates that in plants, the phytohormone auxin plays a central role in plant growth and development.

Auxin is considered as a morphogen since it regulates the development in a dose-dependent manner (Bhalerao and Bennett 2003). It highlights the importance of auxin gradients and the necessity of a subtle regulation of auxin concentration at the scale of organ, tissues, or even cells. To achieve such regulation, plants have developed various mechanisms aimed at controlling auxin homeostasis and the dynamics of auxin redistribution. In addition, various tissues exhibit distinct sensitivity to auxin, thus reflecting that the responsiveness (perception and signaling) is also tightly modulated (Tromas and Perrot-Rechenmann 2010).

8.2.2 Auxin-Binding Soluble Proteins

Shishova and Lindberg (2010) reported that for more than 100 years, the most intriguing question in plant physiology has been how IAA might trigger such enormous variety in physiological responses. According to recent knowledge, a broad spectral activity is observed, which might correlate with changes in the number and properties of auxin receptors. These proteins are responsible for recognition of the hormone and the initiation of further signal transduction chains, resulting in a specific physiological response. Thus, one of the main properties of the auxin receptor is its capability to bind auxin. An investigation of auxin-binding sites in plant cells started almost 30 years ago (Hertel et al. 1972). It has showed a heterogeneity of these sites both in affinity and localization. So, the pool of plant cell auxin-binding proteins (ABPs) consists of two groups: soluble and membrane-bound proteins.

Early biochemical investigations identified a number of auxin-binding soluble proteins such as 1,3-glucanase (MacDonald et al. 1991), β-glucosidase (Campos et al. 1992), glutathione S-transferase (Bilang et al. 1993), and superoxide dismutase (Feldwisch et al. 1994). Two soluble ABPs with a relatively low affinity for IAA were purified and reported to stimulate RNA synthesis in isolated nuclei (Kikuchi et al. 1989). Later, it was shown that one of these protein-bound RNA polymerase II had DNA-binding activity (Sakai 1992). Another polypeptide, a 65-kDa protein, was also found to have a nuclear localization (Prasad and Jones 1991). A soluble ABP 44-kDa protein showed a close link to auxin effects on elongation growth and high affinity labeling of chlorinated auxins (Reinard and Jacobsen 1995; Reinard et al. 1998).

8.3 Gibberellins

The first reports about gibberellins (GAs) came from a group of Japanese scientists who focused some of their studies on a disease called bakanae, which particularly affected rice produced by local farmers (Hori 1903). Bakanae disease is caused by one or more *Fusarium* species. This disease produces a myriad of symptoms, including seedling blight, root and crown rot, stunting, and the classic symptoms of etiolation and abnormal elongation induced by the fungal production of gibberellins (Sun and Snyder 1981; Webster and Gunnell 1992; Nicholson et al. 1998).

Although bakanae disease was described and identified more than 100 years ago in Japan, it is still not clear which *Fusarium* species are associated with the various symptoms of the disease (Ou 1985). Earlier studies in Japan contributed to the identification of a pathogen known as "*Fusarium moniliforme*" in a broad sense (Ou 1985). However, this taxon comprises a number of distinct species, now collectively named the *Gibberella fujikuroi* species complex. The formation of *Gibberella's* sexual stage can distinguish mating populations, or biological species, within this group (Hsieh et al. 1977; Kuhlman 1982; Leslie 1995).

In the 1950s, the British firm Imperial Chemical Industry (ICI) began a program to select for strains of *F. moniliforme* that had a greater capacity to produce gibberellins, and attempts were made to optimize liquid and surface fermentation studies. After some purification steps, a gibberellin was isolated and called "gibberellic acid." These gibberellins had the molecular formula $C_{19}H_{22}O_6$ and had various physiological properties of the gibberellin A that was previously discovered in Japan. This fact stimulated Japanese researchers to produce gibberellins and separate them into three products: gibberellins A_1, A_2, and A_3. Gibberellin A_3 was identified as the gibberellic acid produced by ICI (Takahashi 1986).

In the 1960s, a series of studies were carried out on the application of GAs on various plants, poultry, animals, and microorganisms (Mees and Elson 1978). Since then, several production techniques have been developed in order to make the process more reproducible and economically viable.

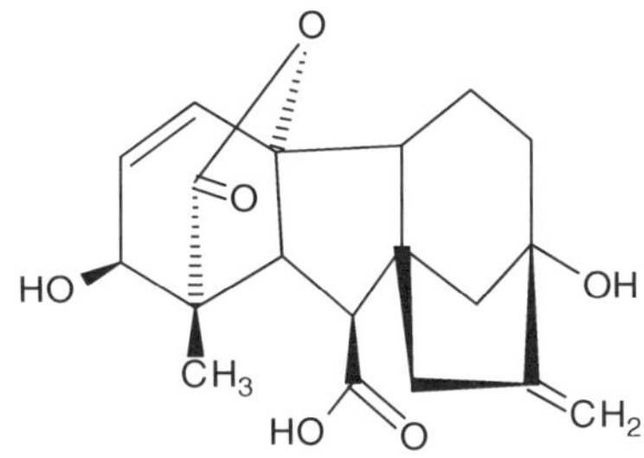

Fig. 8.2 Basic structure of the GAs

Currently, there are 136 gibberellins (GAs) isolated from plants, produced by microorganisms such as fungi and bacteria or obtained synthetically (Blake et al. 2000; Bömke and Tudzynsk 2009). Gibberellins are designated by GAn where "n" corresponds, approximately, to the order of its discovery.

All gibberellins have an *ent*-gibberellane (Fig. 8.2) ring system and are divided in two main types based on the number of carbon atoms, the C_{20}GAs which have a full complement of 20 carbon atoms and C_{19}GAs in which the twentieth carbon atom has been lost by metabolism. Besides the carbon number, the gibberellins differ in the number and position of hydroxyl groups, on the oxidation state of C_{20}, and the presence or absence of lactone bridge between C_{10} and C_{19}.

The effects of GAs on plant growth and development are mediated through gene expression modulation, as RNA and protein synthesis inhibitors interfere with these processes. To further understand the molecular mechanism by which GA regulates the growth and development of plants, it is necessary to identify and analyze more genes that are controlled by GA. Microarrays provide high-throughput, simultaneous analysis of mRNA for hundreds and thousands of genes (Aharoni and Vorst 2002); however, there are only a few reports on the microarray analysis of GA-regulated gene expression in *Arabidopsis* and rice (Ogawa et al. 2003; Yamauchi et al. 2004; Yang et al. 2004; Yazaki et al. 2004). A throughput analysis of transcript profiles in GA-regulated gene expression using different plant tissues and organs remains pertinent, and further characterization of the individual genes will help in understanding how GA regulates the growth and development of plants.

8.3.1 Gibberellic Acid

Gibberellic acid (GA_3) is an important member of the gibberellin family and acts as a natural plant growth hormone, controlling many developmental processes, and is gaining great attention all over the world due to its effective use in agriculture, nurseries, tissue culture, tea gardens, etc. (Davies 2004; Shukla et al. 2005). GA_3 is used in ppm levels, and its use results in various physiologic effects. Some of its applications are presented in Table 8.1.

Table 8.1 Applications of GA_3

Application	Action/benefit	Reference
Application on flowering of *Helleborus niger* and *Helleborus* x *ericsmithii*	Progressive decrease of the time to flower	Christiaens et al. (2012)
Effects of GA_3 and calcium chloride in restoring the metabolic alterations resulting from salt stress in linseed	Increased plant height, number of branches, number of leaves, leaf area, fresh and dry weights	Khan et al. (2010)
Influenced all the vegetative parameters of "Chandler" strawberry	Crown height, crown spread, petiole length, leaf number, and leaf area	Sharma and Singh (2009)
Pea seeds	Stimulate shoot growing	Baumgartner et al. (2008)
Potato cultivation	Promote cell multiplication and elongation, breaking of dormancy	Alexopoulos et al. (2007)
Fruitful tree "Yu Her Pau" litchi	Raise fruit weight	Chang and Lin (2006)
Passiflora nitida Khunt	Effect on germination—breaking of dormancy	Passos et al. (2004)
Grapes of cultivars Vênus	Raise berries mass and number, decrease seeds number	Botelho et al. (2003)

The cost of GA_3 has precluded its use in promoting plant growth, except for certain high-value plants. A reduction of its production costs could lead to wider applications for a variety of crops (Linnemannstons et al. 2002).

Recently, studies have been carried out to decrease GA_3 production costs using several approaches, such as screening fungi, optimizing the nutrients and culture conditions, using agro-industrial residues as a substrate, developing new processes (immobilized cells, fed-batch culture, airlift bioreactor) and minimizing the extraction procedure costs (Rodrigues et al. 2012).

8.4 Cytokinins

Cytokinins (CKs) are a class of phytohormones that play an important role at all phases of plant development from seed germination to senescence. At the organism level, CKs take part in the control of many biological processes throughout the life of plants. They act on induction expression of some genes, promotion of mitosis, and chloroplast development and by releasing buds from apical dominance or by inhibiting root growth (Riefler et al. 2006; Kakimoto 2003; Sakakibara 2006; Werner et al. 2001). Moreover, CKs were found to negatively regulate stress signaling (Nishiyama et al. 2011) and iron accumulation in the leaf (Werner et al. 2010). In tomatoes, salt and drought stresses were linked to reduced CK content (Albacete et al. 2008; Ghanem et al. 2008; Kudoyarova et al. 2007).

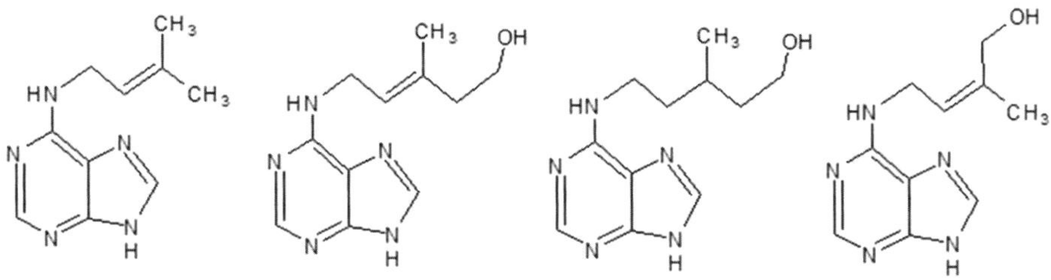

iP tZ DZ cZ

Fig. 8.3 Chemical structures of isoprenoids CKs: N^6-(Δ^2-isopentenyl) adenine (iP), *trans*-zeatin (tZ), *cis*-zeatin (cZ) and dihydrozeatin (DZ)

The CKs was discovered during the 1950s because of the powerful ability of these purine derivatives to trigger plant cell division in vitro. Rapidly thereafter, a variety of additional activities of the hormone were described, including the capability to induce the formation of shoots from unorganized callus tissue, to retard leaf senescence, to stimulate pigment accumulation, and to support plastid development (Heyl et al. 2012).

Currently, there are numerous studies about the possible uses and functions of CKs, such as delay of senescence (Gan and Amasino 1995; Kim et al. 2006), root proliferation (Werner et al. 2001, 2003), apical dominance (Shimizu-Sato et al. 2009; Tanaka et al. 2006), nutritional signaling (Samuelson and Larsson 1993; Takei et al. 2001), and shoot meristem function (Higuchi et al. 2004; Kurakawa et al. 2007; Nishimura et al. 2004; Miyawaki et al. 2006). The CKs also mediates the responses to variable extrinsic factors, such as light conditions in the shoot and availability of nutrients and water in the root, and has a role in the response to biotic and abiotic stress. Together, these activities contribute to the fine-tuning of quantitative growth regulation in plants.

The CKs has been targeted in many plant species to improve their tolerance to different environmental stresses (Barna et al. 1996; Huynh et al. 2005; Havlova et al. 2008; Zhang and Ervin 2008). Exogenous application of CK has been shown to have potential in alleviating heat injury in various higher plants (Skogqvist 1974; Liu et al. 2002; Schrader 2005). For example, retarded leaf senescence and reduced cell membrane lipid peroxidation in creeping bentgrass were observed via exogenous zeatin riboside application and enhanced antioxidant response. This fact was suggested as a possible mechanism for the observed reductions in heat injury (Liu et al. 2002; Zhang and Ervin 2008), also indicating that CK is thought to protect plants under stress via its antioxidant properties (Wang et al. 2012).

Naturally occurring CKs are adenine derivatives that carry either an isoprene-derived side chain or an aromatic side chain at the N^6 terminus, called isoprenoid CKs or aromatic CKs, respectively (Mok and Mok 2001; Sakakibara 2006). The isoprenoid CKs are classified into one of the four basic molecules: N^6-(Δ^2-isopentenyl) adenine (iP), *trans*-zeatin (tZ), *cis*-zeatin (cZ), and dihydrozeatin (DZ) (Fig. 8.3). Each CK molecule is distinguished by characteristics of the side chain, namely, the presence or the absence of a hydroxyl group at the end of the prenyl chain and the stereoisomeric position.

Among isoprenoid CKs, *trans*-zeatin is considered central due to its general occurrence and high activity in the most bioassays. Its stereoisomer, *cis*-zeatin, is characterized by weak activity in bioassays. Dihydrozeatin and N^6-(Δ^2-isopentenyl)-adenine are also commonly present in lower and vascular plants (Emery et al. 1998; Sakakibara 2006; Stirk et al. 2008).

N^6-Benzyladenine and its derivatives, representing aromatic CKs, have been detected in a number of plant species as minor components of the total CKs. Hydroxylated derivatives of N^6-benzyladenine in meta or ortho position of benzyl group are commonly named as meta- and ortho-topolin, respectively (Strnad et al. 1997). Kinetin, the most known CK, has furfuryl ring at the N^6-position of adenine and was identified in both animal cellular DNA and plant tissue extracts (Barciszewski et al. 2000).

All forms of CKs may be reversible or irreversible conjugated with sugars and amino acids. In most bioassays, CK bases are the most active, and therefore, CK conjugation contributes to the regulation of their activity. CK conjugates seem to serve as storage, transport, and deactivate forms because they are resistant to degradation by cytokinin oxidase/dehydrogenase (Auer 2002; Blagoeva et al. 2004).

8.5 Abscisic Acid

Absicisic acid (ABA) belongs to a class of metabolites known as isoprenoids, also called terpenoids. ABA, which contains 15 carbon atoms, belongs to a class of metabolites known as isoprenoids also called terpenoids (Fig. 8.4). ABA was discovered in the early 1960s. Originally it was believed to be involved in the abscission of fruit and dormancy of woody plants. However, in later studies, it became evident that ABA is necessary for seed development, adaptation to several abiotic stress, and sugar sensing (Zeevaart and Creelman 1988; Nambara and Marion-Poll 2005; Schwartz and Zeevaart 2010).

ABA is not only synthesized by higher plants, it is also produced by certain algae (Zeevaart and Creelman 1988), by several phytopathogenic fungi (Zeevaart and Creelman 1988; Tudzynski and Sharon 2002) and by bacteria (Karadeniz et al. 2006).

The metabolism pathway of ABA has been studied in plants and fungi. Since 1987, there were many in vivo and in vitro biochemical studies of ABA metabolism, including studies of ABA-deficiency mutants, that have established an outline of the probable biosynthesis route. It was thought that all isoprenoids were synthesized

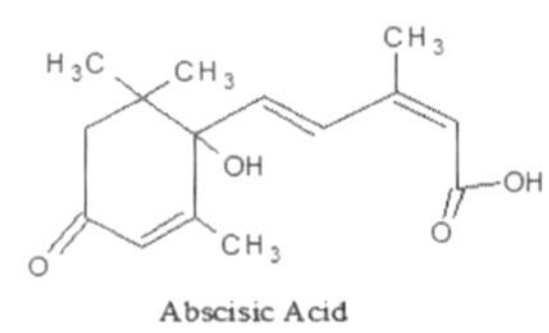

Fig. 8.4 Abscisic acid

Table 8.2 Application of ABA in plants

Application	Action/benefit	Reference
Blueberries (*Vaccinium darrowii*)	Increase the firmness	Buran et al. (2012)
Arachis hypogaea L.	Inhibits lateral root primordial initiation	Guo et al. (2012)
Wheat	Effect of cold acclimation and ABA on amino acid content and composition	Kovacs et al. (2011)
Plant defense against pathogens	Processes of plant defense against pathogens such as virus, fungi, and bacteria	Song et al. (2011); Iriti and Faoro (2008); Vysotskaya et al. (2008); De Torres-Zabala et al. (2007)
Arabidopsis roots	Have a negative effect to gravitropic response as regard to root growth in Arabidopsis roots	Han et al. (2009)
Tobacco plants	Increase the hydraulic conductance of whole tobacco roots and stimulated aquaporin expression	Mahdieh and Mostajeran (2009)
Maize seedlings	Induction of cytosolic Ca^{2+} concentration of mesophyll cells	Guo et al. (2008)
Rice	Involvement of H_2O_2 (hydrogen peroxide) in ABA-induced anthocyanin accumulation in rice leaves	Hung et al. (2008)
Canarian laurel trees	Response of gas exchange and osmotic adjustment capacity in drought-treated trees	Sánchez-Díaz et al. (2008)
Carrot (*Daucus carota*)	Somatic embryo development	Shiota et al. (2008)

from mevalonic acid (MVA), but recently, it was shown that carotenoids and ABA are formed by the "non-mevalonate" triose-pyruvate pathway (from pyruvate to isopentenyl diphosphate) in chloroplasts (Milborrow and Lee 1998; Tudzynski and Sharon 2002; Nambara and Marion-Poll 2005).

8.5.1 ABA Effects in Plants

ABA seems to act as a general inhibitor of growth and metabolism, but, like other plant hormones, ABA has multiple roles during the life cycle of a plant (Zeevaart and Creelman 1988; Srivastava 2002). It plays an important role in modifying transpiration of drought-treated plants. It also acts in stomatal conductance affecting the water supply of the plant (Larcher 2003; Guo et al. 2008; Aasamaaa and Sõberb 2011). Other applications of ABA in plants are presented in Table 8.2.

Table 8.3 ABA production by bacteria

Microorganism	Culture medium	ABA production	Reference
Azospirillum brasilense strain Sp 245	NFb medium with NH_4Cl and NaCl	235 ± 17 ng/ml	Cohen et al. (2008)
Azospirillum brasilense strain Sp 245	NFb medium with NH_4Cl	73 ± 8 ng/ml	Cohen et al. (2008)
Isolated endophytic bacteria from sunflower	LB supplemented with polyethylene glycol	20–45 pmol/ml	Forchetti et al. (2007)
P. mirabilis	Brain heart broth	4.20 ± 1.75 μg/100 ml	Karadeniz et al. (2006)
P. vulgaris	Brain heart broth	0.44 ± 0.02 μg/100 ml	Karadeniz et al. (2006)
B. megaterium	Brain heart broth	0.07 ± 0.00 μg/100 ml	Karadeniz et al. (2006)
B. cereus	Brain heart broth	0.03 ± 0.01 μg/100 ml	Karadeniz et al. (2006)
K. pneumoniae	Brain heart broth	0.91 ± 0.10 μg/100 ml	Karadeniz et al. (2006)
E. coli	Brain heart broth	1.45 ± 1.00 μg/100 ml	Karadeniz et al. (2006)

8.5.2 ABA Production by Microorganisms

The presence of ABA was demonstrated in several fungi, such as *Cercospora rosicola*, *C. cruenta*, *Botrytis cinerea*, and other phytopathogenic fungi, as a secondary metabolism product (Zeevaart and Creelman 1988; Tudzynski and Sharon 2002). Budková et al. (2000) demonstrated that micromycetes from soil, *Aspergillus niger* and *Cladosporium cladosporioides*, produced ABA into the culture medium Czapek-Dox and a liquid medium.

Yurekli et al. (1999) utilized white-rot fungi *Funalia trogii* ATCC 200800 and *Trametes versicolor* ATCC 200801 to produce ABA, and as substrate, they utilized olive oil mill wastewater (OOMW) and vinasse, with a dilution ratio of 20:80, v/v (wastewater/distilled water). The higher ABA concentration was obtained utilizing *F. trogii* and vinasse (170.41 μg/ml) and utilizing OOMW as a substrate concentration of 16.28 μg/ml was obtained. *T. versicolor* fungus produced 34.95 μg/ml using vinasse and 5.32 μg/ml utilizing OOMW.

Plant growth-promoting bacteria (PGPB) such as *Azospirillum* produce phytohormones including ABA (Forchetti et al. 2007; Cohen et al. 2008). Bacteria commonly found in the human body which also live in the soil and in water (*Proteus mirabilis*, *P. vulgaris*, *Bacillus megaterium*, *B. cereus*, *Klebsiella pneumoniae*, *Escherichia coli*) also produce ABA (Karadeniz et al. 2006). Some studies that report the production of ABA by bacteria can be seen in Table 8.3.

8.5.3 ABA Analyses

Several methods have been established for measurement of ABA, such as thin layer chromatography (TLC). It has been used with ultraviolet (UV) lamp and the mobile phase of benzene, ethyl acetate, and acetic acid (Budková et al. 2000) and isopropanol, ammonia, and distilled water (Karadeniz et al. 2006). High-performance liquid chromatography (HPLC) coupled to UV detector also has been utilized (Budková et al. 2000; Karadeniz et al. 2006).

Gas chromatography coupled to mass spectrometry (GC–MS) was reported for analysis of ABA (Forchetti et al. 2007; Cohen et al. 2008).

Recently, liquid chromatography–electrospray ionization tandem mass spectrometry (LC–ESI-MS/MS) has been applied to the determination of phytohormones including ABA (Hou et al. 2008; López-Carbonell et al. 2009).

8.6 Ethylene

Ethylene is an unsaturated hydrocarbon which plays various important functions in plant growth and development, such as seed germination, flower and fruit development, dormancy, abscission, senescence, certain plant defense mechanisms, and a number of interactions with other plant hormones (Abeles et al. 1992; Arteca 1996; Binder 2008).

Ethylene is produced by higher plants and also by bacteria and fungi (Kanellis et al. 1999; Al-Masri et al. 2006). In higher plants, ethylene is biosynthesized from methionine by a well-defined pathway in which methionine is first converted to *S*-adenosyl methionine (SAM) which is then used for the production of 1-aminocyclopropanecarboxylic acid (ACC) by the enzyme ACC synthase. ACC oxidase converts ACC to ethylene (Yang and Hoffman 1984).

Ethylene is a gaseous hormone, and because of this, gas chromatography has been utilized for ethylene analysis. Orberá Ratón et al. (2012) and Thuler et al. (2003) utilized GC equipped with a PORAPAK-N 80/100—INOX column that operated isothermally at 70 °C with nitrogen as gas carrier and flame ionization detector. Visible/short-wave near-infrared (Vis/SW NIR) spectroscopy technique was proposed for the determination of ethylene content in tomatoes by Xie et al. (2009).

8.6.1 Effects of Ethylene in Plants

One of the most studied ethylene effects were reported in fruit ripening. Fruit ripening can also be diverse; however, it involves many closely related changes such as color change, softening of walls, and conversion of starch to sugar (Barry and Giovannoni 2007). Fruits known as climacteric, such as banana, apple, avocado, and tomato, produce ethylene which influences the fruit ripening process. Other fruits that are known as nonclimacteric, such as grape, citrus, and strawberry, show no climacteric and no significant production of ethylene. Even though, they show ripening-related changes (Srivastava 2002; Barry and Giovannoni 2007).

Applications of ethylene in plants are shown in Table 8.4.

Table 8.4 Applications of ethylene in plants

Application	Action/benefit	Reference
Arabidopsis thaliana	Ethylene in the response of root hair cells	Galland et al. (2012)
Gentiana scabra flowers	Ethylene production of unpollinated flowers was very low, but pollination increases ethylene production	Shimizu-Yumoto and Ichimura (2012)
Mexican "Ataulfo" mango	Improving the uniformity in ripening	Tovar et al. (2011)
Spinach (*Spinacia oleracea* L. cv Bison)	Effect of ethylene on ascorbic acid (antioxidant) metabolism during dark-induced leaf senescence	Gergoff et al. (2010)
Tomato and pepper fruits	Existence of extensive common regulons suggests the conservation of ripening mechanisms in climacteric and nonclimacteric fruits	Lee et al. (2010)
Carrot	Exposure to methyl jasmonate and ethylene treatments enhanced the accumulation of bioactive phenolic compounds and phenylalanine ammonia lyase enzyme activity in carrot tissue	Heredia and Cisneros-Zevallos (2009)
Sand verbenas (*Abronia* spp.)	Effects on the germination	Drennan (2008)
Guzmania lingulata Mez. "Anita"	Endogenous ethylene production contributes substantially to floral induction. Ethylene treatment on a single young leaf induced flowering as well	Dukovski et al. (2006)
Hydroponically grown strawberry plants	Ethylene levels from leaves are useful as an early indicator of stress conditions within the system	Hogan et al. (2006)
Potato tuber (*Solanum tuberosum* L.)	Determination of hormonal requirements for wound-induced suberization	Lulai and Suttle (2004)
Carnation cultivars (*Dianthus caryophyllus* L.)	Continuous and short exposures of ethylene reduce the vase life of flowers	Wu et al. (1991)

8.6.2 Ethylene Production by Microorganisms

Beside plants, ethylene can be produced by microorganisms and can be obtained by cracking the petroleum, a process that requires crude oil and has severe effects on the environment (Tudzynski and Sharon 2002).

Ethylene production by fungus *Sclerotinia sclerotiorum* was observed by Al-Masri et al. (2006). Zhu et al. (2012) analyzed ethylene production by six strains of *B. cinerea* (a mass-destructive, necrotrophic plant pathogen that causes grey mold) and all were confirmed to produce ethylene. On grape juice agar (GJA) medium, *B. cinerea* produced ethylene without methionine (Met) addition. When Met was added, the fungus produced more ethylene than that on Czapek and potato dextrose agar (PDA) media.

Orberá Ratón et al. (2012) isolated bacteria and fungi from sugarcane rhizosphere and the isolated microorganisms. Four isolates produced ethylene, which correspond to *Bacillus* sp. B63 (134.12 ng/mL), *Brevibacillus* sp. (B65 279.44 ng/mL), *Bacillus* sp. B90 (870.8 ng/mL), and *Paenibacillus* sp. B100 (166.88 ng/mL).

A comparison of plant growth-promoting potential of rhizospheric bacteria from endophytic bacteria, both isolated from sugar cane, was studied by De Santi Ferrara et al. (2012). In the study, the ability of these bacteria to produce amino acids IAA and ethylene was assessed. The putative endophytes released significantly higher amounts of amino acids than the rhizospheric bacteria, while the latter produced higher quantities of ethylene and were more actively antagonistic to fungi. Both types of bacteria released similar amounts of IAA.

8.7 Phytochemicals

Phytochemicals are non-nutritive plant chemicals that have protective or disease-preventive properties. They are nonessential nutrients, meaning that they are not required by the human body for sustaining life (Karthishwaran et al. 2010; Srivastava et al. 2010; Badugu 2012; Neerati et al. 2012). Plant produces these chemicals as a natural defense against disease and infection, and they have been used throughout human history for various purposes. It has been recognized that natural compounds play an important role in modern pharmaceutical care. Numerous studies of folk medical practices have been undertaken to verify the real properties of some plants used in ancestral treatments (Kamanzi Atindehou et al. 2002; Neerati et al. 2012; Brusotti et al. 2013).

There are many phytochemicals known and each works differently. These phytochemicals have various health benefits such as antioxidant, antimicrobial, anti-inflammatory, cancer-preventive, antidiabetic, and anti-hypertensive effects (Kumar et al. 2009; Nisha Raj and Radhamany 2010; Srivastava et al. 2010). Antioxidant properties of plants have been studied to reduce oxidative stress which may cause various degenerative diseases (Bektas et al. 2005; Pereira et al. 2009). Antioxidant activity can be attributed to phenolic compounds (Girones-Vilaplana et al. 2012; Lv et al. 2012; Silva et al. 2012; Sousa et al. 2013; Jaberian et al. 2013), flavonoids (Girones-Vilaplana et al. 2012; Sousa et al. 2013), and vitamins (Girones-Vilaplana et al. 2012; Jaberian et al. 2013).

Extracts of plants were tested against fungi and bacteria and demonstrated antimicrobial activity. Brusotti et al. (2013) showed that tannin resulted more active against bacteria, while the saponin showed a pronounced activity against fungus *Pyricularia grisea*. Others studies have demonstrated a correlation between antibacterial activity and phytochemicals (alkaloid, saponin, phenol) (Akgul and Gulshen 2005; Doughari and Manzara 2008; Jaberian et al. 2013).

Phytochemicals have been utilized in other studies such as antiproliferative activities (Lv et al. 2012; Kontogianni et al. 2013), antipyretic activity (saponin, tannins, and flavonoids) (Sasmal et al. 2012), and anti-inflammatory activity (phenolic compounds and betacyanins) (Silva et al. 2012).

8.8 Conclusions

Auxins, gibberellins, ABA, cytokinins, and ethylene are the main groups of plant hormones. There are numerous reports that present the main applications and the way of action of these hormones, each of them having their specificities and potentialities. The transport and mode of action of each hormone are being elucidated. However, there are some technological barriers to surpass concerning the economical production and purification of these important biomolecules. In this way, many researchers are seriously involved to find best conditions for these bioprocesses.

References

Aasamaaa K, Sõberb A (2011) Stomatal sensitivities to changes in leaf water potential, air humidity, CO_2 concentration and light intensity, and the effect of abscisic acid on the sensitivities in six temperate deciduous tree species. Environ Exp Bot 71:72–78

Abeles FB, Morgan PW, Saltveit ME (1992) Ethylene in plant biology. Academic, New York

Aharoni A, Vorst O (2002) DNA microarrays for functional plant genomics. Plant Mol Biol 48:99–118

Akgul C, Gulshen S (2005) Antibacterial activity of crude methanolic extract and its fractions of aerial parts of *Anthemis tinctoria*. Indian J Biochem Biophys 42:395–397

Albacete A, Ghanem ME, Martínez-Andújar C, Acosta M, Sanchez-Bravo J, Martinez V (2008) Hormonal changes in relation to biomass partitioning and shoot growth impairment in salinized tomato (*Solanum lycopersicum* L.) plants. J Exp Bot 59:4119–4131

Alexopoulos AA, Akoumianakis KA, Vemmos SN, Passam HC (2007) The effect of postharvest application of gibberellic acid and benzyl adenine on the duration of dormancy of potatoes produced by plants grown from TPS. Postharvest Biol Technol 46:54–62

Al-Masri MI, Elad Y, Sharon A, Barakat R (2006) Ethylene production by *Sclerotinia sclerotiorum* and influence of exogenously applied hormone and its inhibitor aminoethoxyvinylglycine on white mold. Crop Prot 25:356–361

Arteca R (1996) Plant growth substances: principles and applications. Chapman & Hall, New York

Auer CA (2002) Discoveries and dilemmas concerning cytokinin metabolism. J Plant Growth Regul 21:24–31

Badugu LR (2012) Phytochemical screening, quantitative estimation total phenolics and total flavonoids, anti microbial evaluation of *Cyamopsis tetragonoloba*. Int J Res Pharm Biomed Sci 3(3):1139–1142

Barciszewski J, Siboska G, Clark BFC, Rattan SIS (2000) Cytokinin formation by oxidative metabolism. J Plant Physiol 158:587–588

Barna B, Adam AL, Kiraly Z (1996) Increased levels of cytokinin induce tolerance to necrotic diseases and various oxidative stress-causing agents in plants. In: 2nd International conference on oxygen, free radicals and environmental stress in plants, pp 25–29

Barry CS, Giovannoni JJ (2007) Ethylene and fruit ripening. J Plant Growth Regul 26:143–159

Baumgartner S, Shah D, Schaller J, Kämpfer U, Thurneysen A, Heusser P (2008) Reproducibility of dwarf pea shoot growth stimulation by homeopathic potencies of gibberellic acid. Complement Ther Med 16:183–191
Bektas T, Sokmen M, Akpulat HA, Sokmen A (2005) In-vitro antioxidant activities of the methanol extracts of five *Allium* species from Turkey. Food Chem 92(1):89–92
Bhalerao RP, Bennett MJ (2003) The case for morphogens in plants. Nat Cell Biol 5:939–943
Bilang J, MacDonald H, King PJ, Sturm A (1993) A soluble auxin-binding protein from *Hyoscyamus muticus* is a glutathione S-transferase. Plant Physiol 102:29–34
Binder BM (2008) The ethylene receptors: complex perception for a simple gas. Plant Sci 175:8–17
Blagoeva E, Dobrev P, Malbeck J, Motyka V, Gaudinova A, Vankova R (2004) Effect of exogenous cytokinins, auxins and adenine on cytokinin Nglucosylation and cytokinin oxidase/dehydrogenase activity in de-rooted radish seedlings. Plant Growth Regul 44:15–23
Blake PS, Taylor DR, Crisp CM (2000) Identification of endogenous gibberellins in strawberry, including the novel gibberellins GA123, GA124 and GA125. Phytochemistry 55:887–890
Bömke C, Tudzynsk B (2009) Diversity, regulation, and evolution of the gibberellin biosynthetic pathway in fungi compared to plants and bacteria. Phytochemistry 70:1876–1893
Botelho RV, Pires EJP, Terra MM (2003) Efeitos do thidiazuron e do ácido giberélico nas características dos cachos e bagas de uvas 'niagara rosada' na região de Jundiaí-SP. Rev Bras Frutic 25:96–99
Brusotti G, Tosi S, Tava A, Picco AM, Grisoli P, Cesari I, Caccialanza G (2013) Antimicrobial and phytochemical properties of stem bark extracts from *Piptadeniastrum africanum* (Hook f.) Brenan. Ind Crops Prod 43:612–616
Budková M, Vizárová G, Simonovicová A, Chalanyová M (2000) The possibility of soil micromycetes produced the abscisic acid. Acta Physiol Plant 22(2):179–184
Buran TJ, Sandhu AK, Azeredo AM, Bent AH, Williamson JG, Gu L (2012) Effects of exogenous abscisic acid on fruit quality, antioxidant capacities, and phytochemical contents of southern high bush blueberries. Food Chem 132:1375–1381
Campos N, Bako L, Feldwisch J, Schell J, Palme K (1992) A protein from maize labeled with azido-IAA has novel b-glucosidase activity. Plant J 2:675–684
Chang JC, Lin TS (2006) GA_3 increases fruit weight in 'Yu Her Pau' litchi. Sci Hortic 108: 442–443
Chen JG et al (2001) ABP1 is required for organized cell elongation and division in Arabidopsis embryogenesis. Genes Dev 15:902–911
Cheng Y, Dai X, Zhao Y (2007) Auxin synthesized by the YUCCA flavin monooxygenases is essential for embryogenesis and leaf formation in Arabidopsis. Plant Cell 19:2430–2439
Christiaens A, Dhooghe E, Pinxteren D, Van Labeke MC (2012) Flower development and effects of a cold treatment and a supplemental gibberellic acid application on flowering of *Helleborus niger* and *Helleborus* x *ericsmithii*. Sci Hortic 136:145–151
Cohen AC, Bottini R, Piccoli PN (2008) *Azospirillum brasilense* Sp 245 produces ABA in chemically-defined culture medium and increases ABA content in Arabidopsis plants. Plant Growth Regul 54:97–103
Davies PJ (ed) (2004) Plant hormones: biosynthesis, signal transduction, action! Kluwer, The Netherlands
De Santi Ferrara FI, Oliveira ZM, Gonzales HHS, Floh EIS, Barbosa HR (2012) Endophytic and rhizospheric enterobacteria isolated from sugar cane have different potentials for producing plant growth-promoting substances. Plant Soil 353:409–417
De Torres-Zabala M, Truman W, Bennett MH, Lafforgue G, Mansfield JW, Rodriguez Egea P, Bögre L, Grant M (2007) *Pseudomonas syringae* pv. tomato hijacks the Arabidopsis abscisic acid signaling pathway to cause disease. EMBO J 26:1434–1443
Doughari J, Manzara S (2008) In vitro antibacterial activity of crude leaf extracts of *Mangifera indica* Linn. Afr J Microbiol Res 2:67–72
Drennan PM (2008) Sand verbenas (*Abronia spp.*, Nyctaginaceae) germinate in response to ethylene. J Arid Environ 72:847–852

Dukovski D, Bernatzky R, Han S (2006) Flowering induction of Guzmania by ethylene. Sci Hortic 110:104–108
Emery RJN, Leport L, Barton JE, Turner NC, Atkins CA (1998) cis-Isomers of cytokinins predominate in chickpea seeds throughout their development. Plant Physiol 117:1515–1523
Feldwisch J, Zettl R, Campos N, Palme K (1994) Identification of a 23 kDa protein from maize photo affinity labeled with azido-IAA. Biochem J 305:853–857
Ferro N, Bredow T, Jacobsen HJ, Reinard T (2010) Route to novel auxin: auxin chemical space toward biological correlation carriers. Chem Rev 110:4690–4708
Forchetti G, Masciarelli O, Alemano S, Alvarez D, Abdala G (2007) Endophytic bacteria in sunflower (*Helianthus annuus L.*): isolation, characterization, and production of jasmonates and abscisic acid in culture medium. Appl Microbiol Biotechnol 76:1145–1152
Galland M, Gamet L, Varoquaux F, Touraine B, Touraine B, Desbrosses G (2012) The ethylene pathway contributes to root hair elongation induced by the beneficial bacteria *Phyllobacterium brassicacearum* STM196. Plant Sci 190:74–81
Gan S, Amasino RM (1995) Inhibition of leaf senescence by autoregulated production of cytokinin. Science 270:1986–1988
Gergoff G, Chaves A, Bartoli CG (2010) Ethylene regulates ascorbic acid content during dark-induced leaf senescence. Plant Sci 178:207–212
Ghanem ME, Albacete A, Martínez-Andújar C, Acosta M, Romero-Aranda R, Dodd IC, Lutts S, Pérez-Alfocea F (2008) Hormonal changes during salinity-induced leaf senescence in tomato (*Solanum lycopersicum* L.). J Exp Bot 59:3039–3050
Girones-Vilaplana A, Valentao P, Andrade PB, Ferreres F, Moreno DA, Garcia-Viguera C (2012) Phytochemical profile of a blend of black chokeberry and lemon juice with cholinesterase inhibitory effect and antioxidant potential. Food Chem 134:2090–2096
Guo XL, Ma YY, Liu ZH, Liu BH (2008) Effects of exterior abscisic acid on calcium distribution of mesophyll cells and calcium concentration of guard cells in maize seedlings. Agric Sci China 7(4):438–446
Guo D, Liang J, Qiao Y, Yan Y, Li L, Dai Y (2012) Involvement of G1-to-S transition and AhAUX-dependent auxin transport in abscisic acid-induced inhibition of lateral root primodia initiation in *Arachis hypogaea* L. J Plant Physiol 169:1102–1111
Hamann T et al (2002) The Arabidopsis BODENLOS gene encodes an auxin response protein inhibiting MONOPTEROS-mediated embryo patterning. Genes Dev 16:1610–1615
Han W, Rong H, Zhang H, Wang MH (2009) Abscisic acid is a negative regulator of root gravitropism in *Arabidopsis thaliana*. Biochem Biophys Res Commun 378:695–700
Havlova M, Dobrev PI, Motyka V et al (2008) The role of cytokinins in responses to water-deficit in tobacco plants over-expressing transzeatin O-glucosyltransferase gene under 35S or SAG12 promoters. Plant Cell Environ 31:341–353
Heredia JB, Cisneros-Zevallos L (2009) The effect of exogenous ethylene and methyl jasmonate on pal activity, phenolic profiles and antioxidant capacity of carrots (*Daucus carota*) under different wounding intensities. Postharvest Biol Technol 51:242–249
Hertel R, K-St T, Russo VEA (1972) In vivo auxin binding to particulate cell fractions from corn coleoptiles. Planos 107:325–340
Heyl A, Riefler M, Romanov GA, Schmulling T (2012) Properties, functions and evolution of cytokinin receptors. Eur J Cell Biol 91:246–256
Higuchi M, Pischke MS, Mahonen AP et al (2004) In planta functions of the Arabidopsis cytokinin receptor family. Proc Natl Acad Sci USA 101:8821–8826
Hogan JD, Murray EE, Harrison MA (2006) Ethylene production as an indicator of stress conditions in hydroponically-grown strawberries. Sci Hortic 110:311–318
Hori S (1903) Bakanae disease of rice: lectures on plant disease, 1st edn. Seibido, Tokyo, p 114
Hou S, Zhu J, Ding M, Lv G (2008) Simultaneous determination of gibberellic acid, indole-3-acetic acid and abscisic acid in wheat extracts by solid-phase extraction and liquid chromatography–electrospray tandem mass spectrometry. Talanta 76:798–802
Hsieh WH, Smith SN, Zinder WC (1977) Mating groups in *Fusarium moniliforme*. Phytopathology 67:1838–1842

Hung KT, Cheng DG, Hsu YT, Kao CH (2008) Abscisic acid-induced hydrogen peroxide is required for anthocyanin accumulation in leaves of rice seedlings. J Plant Physiol 165:1280–1287

Huynh LN, VanToai T, Streeter J, Banowetz G (2005) Regulation of flooding tolerance of SAG12:ipt Arabidopsis plants by cytokinin. J Exp Bot 56:1397–1407

Iriti M, Faoro F (2008) Abscisic acid is involved in chitosan-induced resistance to tobacco necrosis virus (TNV). Plant Physiol Biochem 46:1106–1111

Jaberian H, Piri K, Nazari J (2013) Phytochemical composition and in vitro antimicrobial and antioxidant activities of some medicinal plants. Food Chem 136(1):237–244

Kakimoto T (2003) Perception and signal transduction of cytokinins. Annu Rev Plant Biol 54: 605–627

Kamanzi Atindehou K, Koné M, Terreaux C, Traore D, Hostettmann K, Dosso M (2002) Evaluation of the antimicrobial potential of medicinal plants from the Ivory Coast. Phytother Res 16: 497–502

Kanellis AK, Chang C, Klee H, Bleecker AB, Pech JC, Grierson D (eds) (1999) Biology and biotechnology of the plant hormone ethylene II. Kluwer Academic Publishers, Dordrecht

Karadeniz A, Topcuoglu SF, Inan S (2006) Auxin, gibberellin, cytokinin and abscisic acid production in some bacteria. World J Microbiol Biotechnol 22:1061–1064

Karthishwaran K, Mirunalini S, Dhamodharan G, Krishnaveni M, Arulmozhi V (2010) Phytochemical investigation of methanolic extract of the leaves of *Pergularia daemia*. J Biol Sci 10(3):242–246

Khan MN, Siddiqui MH, Mohammad F, Naeem M, Khan MMA (2010) Calcium chloride and gibberellic acid protect linseed (*Linum usitatissimum* L.) from NaCl stress by inducing antioxidative defence system and osmoprotectant accumulation. Acta Physiol Plant 32:121–132

Kikuchi M, Imaseki H, Sakai S (1989) Modulation of gene expression in isolated nuclei by auxin-binding proteins. Plant Cell Physiol 30:765–773

Kim HJ, Ryu H, Hong SH, Woo HR, Lim PO, Lee IC, Sheen J, Nam HG, Hwang I (2006) Cytokinin-mediated control of leaf longevity by AHK3 through phosphorylation of ARR2 in Arabidopsis. Proc Natl Acad Sci USA 103:814–819

Kontogianni VG, Tomic G, Nikolic I, Nerantzaki AA, Sayyad N, Stosic-Grujicic S, Stojanovic I, Gerothanassis IP, Tzakos AG (2013) Phytochemical profile of *Rosmarinus officinalis* and *Salvia officinalis* extracts and correlation to their antioxidant and anti-proliferative activity. Food Chem 136:120–129

Kovacs Z, Simon-Sarkadi L, Sovany C, Kirsch K, Galiba G, Kocsy G (2011) Differential effects of cold acclimation and abscisic acid on free amino acid composition in wheat. Plant Sci 180: 61–68

Kudoyarova GR, Vysotskaya LB, Cherkozyanova A, Dodd IC (2007) Effect of partial rootzone drying on the concentration of zeatin-type cytokinins in tomato (*Solanum lycopersicum* L.) xylem sap and leaves. J Exp Bot 58:161–168

Kuhlman EG (1982) Varieties of *Gibberella fujikuroi* with anamorphs in *Fusarium* section *Liseola*. Mycologia 74:759–768

Kumar A, Ilavarasan R, Jayachandran T, Decaraman M, Aravindhan P, Padmanabhan N, Krishnan MRV (2009) Phytochemicals investigation on a tropical plant, *Syzygium cumini* from Kattuppalayam, Erode District, Tamil Nadu, South India. Pakistan J Nutr 8(1):83–85

Kurakawa T, Ueda N, Maekawa M, Kobayashi K, Kojima M, Nagato Y, Sakakibara H, Kyozuka J (2007) Direct control of shoot meristem activity by a cytokinin-activating enzyme. Nature 445:652–655

Larcher W (2003) Physiological plant ecology, 4th edn. Springer, Berlin

Lee S, Chung EJ, Joung YE, Choi D (2010) Non-climacteric fruit ripening in pepper: increased transcription of EIL-like genes normally regulated by ethylene. Funct Integr Genomics 10: 135–146

Leslie JF (1995) *Gibberella fujikuroi*: available populations and variable traits. Can J Bot 73(S1): 282–291

Lincoln C, Britton JH, Estelle M (1990) Growth and development of the axr1 mutants of Arabidopsis. Plant Cell 2:1071–1080

Linnemannstons P, Schulte J, Prado MM, Proctor RH, Avalos J, Tudzynski B (2002) The polyketide synthase gene *pks4* from *Gibberella fujikuroi* encodes a key enzyme in the biosynthesis of the red pigment bikaverin. Fungal Genet Biol 37:134–148

Liu XZ, Huang BR, Banowetz G (2002) Cytokinin effects on creeping bentgrass responses to heat stress. I: shoot and root growth. Crop Sci 42:457–465

López-Carbonell M, Gabasa M, Jáuregui O (2009) Enhanced determination of abscisic acid (ABA) and abscisic acid glucose ester (ABA-GE) in *Cistus albidus* plants by liquid chromatography–mass spectrometry in tandem mode. Plant Physiol Biochem 47:256–261

Lulai EC, Suttle JC (2004) The involvement of ethylene in wound-induced suberization of potato tuber (*Solanum tuberosum* L.): a critical assessment. Postharvest Biol Technol 34:105–112

Lv J, Yu L, Lu Y, Niu Y, Liu L, Costa J, Yu LL (2012) Phytochemical compositions, and antioxidant properties, and antiproliferative activities of wheat flour. Food Chem 135:325–331

MacDonald H, Jones AM, King PJ (1991) Photoaffinity labeling of soluble auxin-binding proteins. J Biol Chem 266:7393–7399

Mahdieh M, Mostajeran A (2009) Abscisic acid regulates root hydraulic conductance via aquaporin expression modulation in *Nicotiana tabacum*. J Plant Physiol 166:1993–2003

Mees GC, Elson GW (1978) In: Peacock FC (ed) Jealott's Hill-fifty year of agricultural research, 1928–1978. Kynoch Press, Birmingham, pp 55–60

Milborrow BV, Lee HS (1998) Endogenous biosynthetic precursors of (+)-abscisic acid. VI. Carotenoids and ABA are formed by the 'non-mevalonate' triose-pyruvate pathway in chloroplasts. Aust J Plant Physiol 25(5):507–512

Miyawaki K, Tarkowski P, Matsumoto-Kitano M, Kato T, Sato S, Tarkowska D, Tabata S, Sandberg G, Kakimoto T (2006) Roles of Arabidopsis ATP/ADP isopentenyltransferases and tRNA isopentenyltransferases in cytokinin biosynthesis. Proc Natl Acad Sci USA 103:16598–16603

Mok DW, Mok MC (2001) Cytokinin metabolism and action. Annu Rev Plant Physiol Plant Mol Biol 52:89–118

Nambara E, Marion-Poll A (2005) Abscisic acid biosynthesis and catabolism. Annu Rev Plant Biol 56:165–185

Neerati P, Mohammad R, Bangaru R, Devde R, Kanwar JR (2012) The effects of verapamil, curcumin, and capsaicin pretreatments on the BBB uptake clearance of digoxin in rats. J Pharm Res 5(4):2126–2133

Nicholson P, Simpson DR, Weston G, Rezanoor HN, Lees AK, Parry DW, Joyce D (1998) Detection and quantification of *Fusarium culmorum* and *Fusarium graminearum* in cereals using PCR assay. Plant Pathol 53:17–37

Nisha Raj RS, Radhamany PM (2010) Preliminary phytochemical and in vitro antioxidant properties of *Brunfelsia americana* L. J Pharm Res 3(11):2712–2713

Nishimura C, Ohashi Y, Sato S, Kato T, Tabata S, Ueguchi C (2004) Histidine kinase homologs that act as cytokinin receptors possess overlapping functions in the regulation of shoot and root growth in Arabidopsis. Plant Cell 16:1365–1377

Nishiyama R, Watanabe Y, Fujita Y, Le DT, Kojima M, Werner T, Vankova R, Yamaguchi-Shinozaki K, Shinozaki K, Kakimoto T et al (2011) Analysis of cytokinin mutants and regulation of cytokinin metabolic genes reveals important regulatory roles of cytokinins in drought, salt and abscisic acid responses, and abscisic acid biosynthesis. Plant Cell 23:2169–2183

Ogawa M, Hanada A, Yamauchi Y, Kuwahara A, Kamiya Y, Yamaguchi S (2003) Gibberellin biosynthesis and response during *Arabidopsis* seed germination. Plant Cell 15:1591–1604

Okada K et al (1991) Requirement of the auxin polar transport system in early stages of arabidopsis floral bud formation. Plant Cell 3:677–684

Orberá Ratón TM, Yano R, Rodríguez Gámez O, Floh EIS, Jesús Serrat Díaz M, Barbosa HR (2012) Isolation and characterisation of aerobic endospore forming Bacilli from sugarcane rhizosphere for the selection of strains with agriculture potentialities. World J Microbiol Biotechnol 28:1593–1603

Ou SH (1985) Rice diseases. Commonwealth Mycological Institute, Kew

Palme K, Hesse T, Moore I, Campos N, Fedwisch J, Garbers C, Hesse F, Schell J (1991) Hormonal modulation of plant growth: the role of auxin perception. Mech Dev 33:97–106

Passos IRS, Matos GVC, Meletti LMM, Scott MDS, Bernacci LC, Vieira MAR (2004) Utilização do ácido giberélico para a quebra de dormência de sementes de Passiflora nitida Kunth germinadas in vitro. Rev Bras Frutic 26:380–381

Pereira RP, Fachinetto R, Prestes AS, Puntel RL, Silva GNS, Heinzmann BM, Boschetti TK, Athayde ML, Bürger ME, Morel AF, Morsch VM, Rocha JBT (2009) Antioxidant effects of different extracts from *Melissa officinalis*, *Matricaria recutita* and *Cymbopogon citratus*. Neurochem Res 34:973–983

Perrot-Rechenmann C (2010) Cellular responses to auxin: division versus expansion. Cold Spring Harb Perspect Biol 2:a001446

Prasad PV, Jones AM (1991) Putative receptor for the plant growth hormone auxin identified and characterized by anti-idiotypic antibodies. Proc Natl Acad Sci USA 88:5479–5483

Reinard T, Jacobsen H-J (1995) A soluble high affinity auxin binding protein from Pea Apex. J Plant Physiol 147:132–138

Reinard T, Achmus H, Waltner A, Rescher U, Klambt D, Jacobsen H-J (1998) Assignment of the auxin binding abilities of ABP44 ingel. Plant Cell Physiol 39:874–878

Riefler M, Novak O, Strnad M, Schmülling T (2006) Arabidopsis cytokinin receptor mutants reveal functions in shoot growth, leaf senescence, seed size, germination, root development, and cytokinin metabolism. Plant Cell 18:40–54

Rodrigues C, Vandenberghe LPS, Oliveira J, Soccol CR (2012) New perspectives of gibberellic acid production: a review. Crit Rev Biotechnol 32(3):1–11

Sakai S (1992) Regulatory functions of soluble auxin-binding proteins. Int Rev Cytol 135: 239–267

Sakakibara H (2006) Cytokinins: activity, biosynthesis and translocation. Ann Rev Plant Biol 57:431–449

Samuelson ME, Larsson C-M (1993) Nitrate regulation of zeatin riboside levels in barley roots: effects of inhibitors of N assimilation and comparison with ammonium. Plant Sci 93:77–84

Sánchez-Díaz M, Tapia C, Antolín MC (2008) Abscisic acid and drought response of Canarian laurel forest tree species growing under controlled conditions. Environ Exp Bot 64:155–161

Sasmal S, Majumdar S, Gupta M, Mukherjee A, Mukherjee PK (2011) Pharmacognostical, phytochemical and pharmacological evaluation for the antipyretic effect of the seeds of *Saraca asoca* Roxb. Asian Pac J Trop Biomed 2(10):782–786

Schrader WL (2005) Effects of plant growth regulators on heat stress in annual artichoke production. In: Bianco VV, Rubatzky V, (ed) IV International congress on artichoke ishs acta horticulturae

Schwartz SH, Zeevaart JAD (2010) Abscisic acid biosynthesis and metabolism. In: Davies PJ (ed) Plant hormones. Springer Publications, pp 137–155

Sharma RR, Singh R (2009) Gibberellic acid influences the production of malformed and button berries, and fruit yield and quality in strawberry (Fragaria x ananassa Duch.). Sci Hortic 119:430–433

Shimizu-Sato S, Tanaka M, Mori H (2009) Auxin-cytokinin interactions in the control of shoot branching. Plant Mol Biol 69:429–435

Shimizu-Yumoto H, Ichimura K (2012) Effects of ethylene, pollination, and ethylene inhibitor treatments on flower senescence of gentians. Postharvest Biol Technol 63:111–115

Shiota H, Ko S, Wada S, Otsu CT, Tanaka I, Kamada H (2008) A carrot G-box binding factor-type basic region/leucine zipper factor DcBZ1 is involved in abscisic acid signal transduction in somatic embryogenesis. Plant Physiol Biochem 46:550–558

Shishova M, Lindberg S (2010) A new perspective on auxin perception. J Plant Physiol 167: 417–422

Shukla R, Chand S, Srivastava AK (2005) Batch kinetics and modeling of gibberellic acid production by *Gibberella fujikuroi*. Enzyme Microb Technol 36:492–497

Silva LR, Valentão P, Faria J, Ferreres F, Sousa C, Gil-Izquierdo A, Pinho BR, Andrade PB (2012) Phytochemical investigations and biological potential screening with cellular and non-cellular models of globe amaranth (*Gomphrena globosa* L.) inflorescences. Food Chem 135:756–763

Simon S, Petrasek J (2011) Why plants need more than one type of auxin. Plant Sci 180:454–460

Skogqvist NGR (1974) Induced heat sensitivity of wheat roots and protecting effect of ethanol and kinetin. Physiol Plant 32:166–169

Song W, Ma X, Tan H, Zhou J (2011) Abscisic acid enhances resistance to *Alternaria solani* in tomato seedlings. Plant Physiol Biochem 49:693–700

Sousa EO, Rocha JBT, Barros LM, Barros ARC, Costa JGM (2013) Phytochemical characterization and in vitro antioxidant properties of *Lantana camara* L. and *Lantana montevidensis* Briq. Ind Crops Prod 43:517–522

Srivastava LM (2002) Chapter 10: Abscisic acid. In: Srivastava LM (ed) Plant growth and development: hormones and environment. Academic, San Diego

Srivastava M, Kumar A, Pal M (2010) Phytochemical investigation on *Jatropha curcas* seed cake. Int J Pharm Life Sci 1(6):357–362

Stirk WA, Novak O, Vaclavikova K, Tarkowski P, Strnad M, van Staden J (2008) Spatial and temporal changes in endogenous cytokinins in developing pea roots. Planta 227:1279–1289

Strnad M, Hanuš J, Vanek T, Kaminek M, Ballantine J, Fussell B, Hanke DE (1997) meta-Topolin, a highly active aromatic cytokinin from poplar leaves (Populus _ canadensis Moench., cv. Robusta). Phytochemistry 45:213–218

Sun SK, Snyder WC (1981) The bakanae disease of the rice plant. In: Nelson PE, Tourssoun TA, Cook RJ (eds) *Fusarium*: diseases, biology and taxonomy. The Pennsylvania State University Press, University Park, pp 104–113

Takahashi Y (1986) Chemistry of plant hormones. CRC, Boca Raton, 277

Takei K, Sakakibara H, Taniguchi M, Sugiyama T (2001) Nitrogen-dependent accumulation of cytokinins in root and the translocation to leaf: implication of cytokinin species that induces gene expression of maize response regulator. Plant Cell Physiol 42:85–93

Tanaka M, Takei K, Kojima M, Sakakibara H, Mori H (2006) Auxin controls local cytokinin biosynthesis in the nodal stem in apical dominance. Plant J 45:1028–1036

Thuler DS, Floh EIS, Handro W, Barbosa HR (2003) Plant growth regulators and amino acids released by *Azospirillum sp.* in chemically defined media. Lett Appl Microbiol 37:174–178

Tovar B, Montalvo E, Damián BM, García HS, Mata M (2012) Application of vacuum and exogenous ethylene on Ataulfo mango ripening. Food Sci Technol 10:2040–2046

Tromas A, Perrot-Rechenmann C (2010) Recent progress in auxin biology. C R Biol 333: 297–306

Tudzynski B, Sharon A (2002) Biosynthesis, biological role and application of fungal phytohormones. In: Osiewacz HD (ed) The mycota X: industrial application. Springer, Berlin

Vanneste S, Friml J (2009) Auxin a trigger for change in plant development. Cell 136:1005–1016

Vysotskaya LB, Korobova AV, Kudoyarova GR (2008) Abscisic acid accumulation in the roots of nutrient-limited plants: its impact on the differential growth of roots and shoots. J Plant Physiol 165:1274–1279

Wang K, Zhang X, Ervin E (2012) Antioxidative responses in roots and shoots of creeping bentgrass under high temperature: effects of nitrogen and cytokinin. J Plant Physiol 169:492–500

Webster RK, Gunnell PS (1992) Compendium of rice diseases. The American Phytopathological Society Press, St. Paul, MN

Werner T, Motyka V, Strnad M, Schmülling T (2001) Regulation of plant growth by cytokinin. Proc Natl Acad Sci USA 98:10487–10492

Werner T, Motyka V, Laucou V, Smets R, Van Onckelen H, Schmülling T (2003) Cytokinin-deficient transgenic Arabidopsis plants show multiple developmental alterations indicating opposite functions of cytokinins in the regulation of shoot and root meristem activity. Plant Cell 15:2532–2550

Werner T, Nehnevajova E, Köllmer I, Novák O, Strnad M, Krämer U, Schmülling T (2010) Root-specific reduction of cytokinin causes enhanced root growth, drought tolerance, and leaf mineral enrichment in Arabidopsis and tobacco. Plant Cell 22:3905–3920

Wu MJ, Zacarias L, Reid MS (1991) Variation in the senescence of carnation (*Dianthus caryophyllus* L.) cultivars. II. Comparison of sensitivity to exogenous ethylene and of ethylene binding. Sci Hortic 48:109–116

Xie L, Ying Y, Ying T (2009) Rapid determination of ethylene content in tomatoes using visible and short-wave near-infrared spectroscopy and wavelength selection. Chemom Intell Lab Syst 97:141–145

Yamauchi Y, Ogawa M, Kuwahara A, Hanada A, Kamiya Y, Yamaguchi S (2004) Activation of gibberellin biosynthesis and response pathways by low temperature during imbibition of *Arabidopsis thaliana* seeds. Plant Cell 16:367–378

Yang S, Hoffman N (1984) Ethylene biosynthesis and its regulation in higher plants. Annu Rev Plant Physiol 35:155–189

Yang GX, Jan A, Shen S-H, Yazaki J, Ishikawa M, Shimatani Z (2004) Microarray analysis of brassinosteroids and gibberellin-regulated gene expression in rice seedings. Mol Genet Genomics 271:468–478

Yazaki J, Shimatani Z, Hashimoto A, Nagata Y, Fujii F, Kojima K (2004) Transcriptional profiling of genes responsive to abscisic acid and gibberellin in rice: phenotyping and comparative analysis between rice and Arabidopsis. Physiol Genomics 17:87–100

Yurekli F, Yesilada O, Yurekli M, Topcuoglu SF (1999) Plant growth hormone production from olive oil mill and alcohol factory wastewaters by white rot fungi. World J Microbiol Biotechnol 15:503–505

Zeevaart JAD, Creelman RA (1988) Metabolism and physiology of abscisic acid. Annu Rev Plant Physiol Plant Mol Biol 39:439–473

Zhang XZ, Ervin EH (2008) Impact of seaweed extract-based cytokinins and zeatin riboside on creeping bentgrass heat tolerance. Crop Sci 48:364–370

Zhu P, Xu L, Zhang C, Toyoda H, Gan SS (2012) Ethylene produced by Botrytis cinerea can affect early fungal development and can be used as a marker for infection during storage of grapes. Postharvest Biol Technol 66:23–29

Part III
Natural Functional Food Products

Chapter 9
Probiotics

Galina Novik, Anastasiya Sidarenka, Elena Kiseleva, Emily Kolomiets, and Estera Szwajcer Dey

9.1 Introduction

The human gastrointestinal tract (GIT) is a complex ecosystem and its bacteria inhabitants can achieve very high densities. A delicate balance exists between human intestinal microflora and its host. Upset in this community structure may lead toward the symptoms of acute gastroenteritis, inflammatory bowel disease and colon cancer. It is therefore important to sustain gut microflora in an optimal manner (Gibson and Fuller 2000; Vaughan et al. 2005).

In recent years, the advances in understanding the relationship between human gut microbiota and health have resulted in the development of the concept of probiotics. Probiotics are defined as "live microorganisms, as they are consumed in adequate numbers confer a health benefit on the host" (FAO/WHO 2001). Bacterial strains selected as probiotics are predominantly from the genera *Bifidobacterium* and *Lactobacillus* (Saarela et al. 2000), which form a part of normal human intestinal ecosystem (Backhead et al. 2005) and play a pivotal role in maintenance of healthy human gut (Gomes and Malcata 1999).

G. Novik (✉) • A. Sidarenka • E. Kiseleva • E. Kolomiets
Collection of Microorganisms, Institute of Microbiology, Belarus National Academy of Sciences, Acad. V.F. Kuprevich Street, 2, 220141, Minsk, Belarus
e-mail: galina_novik@mbio.bas-net.by

E.S. Dey
Pure and Applied Biochemistry, Chemical Centre, Lund University, Paradisgatan 2, Box 117, SE-221 00, Lund, Sweden

S.K. Brar et al. (eds.), *Biotransformation of Waste Biomass into High Value Biochemicals*,
DOI 10.1007/978-1-4614-8005-1_9, © Springer Science+Business Media New York 2014

9.2 Morphology and Physiology of Probiotic Bacteria

Bifidobacteria are gram-positive, strictly anaerobic, non-motile, non-spore-forming pleomorphic rods with a particular cell morphology ranging from regular rods to various branched and club-shaped forms (Leahy et al. 2005). Morphologic heterogeneity of cells from developing populations of bifidobacteria correlates with ultrastructure peculiarities (Fig. 9.1).

Active proliferating cells in exponential phase are characterised by formation of intracytoplasmatic membrane complex represented by lamellar, myelinoform, vesicular structures. Nucleoid is localised as the central polybranched or disperse osmophobic zone. Nucleoid distribution is determined by morphogenesis processes—exobudding, branching or multiseptation. Electronograms reveal multiple polyphosphate and polysaccharide inclusions. Ageing of bifidobacterial populations is accompanied with ultrastructural changes: cell wall hypersynthesis, reorganisation and increased size of intracytoplasmatic membrane complex, altered morphology and compactness of nuclei and formation and dissimilation of inclusions (Novik et al. 1994). Populations of *Bifidobacterium*, growing on liquid and agar media, are represented by highly ordered mycelial structures. Their topography depends on mutual arrangement of polymorphic cells and the way of their daughter cells' separation after division. Evidence obtained by scanning electron microscopy (SEM) of total preparations and by transmission electron microscopy (TEM) of ultrathin sections correlate well. The data showed the existence of morphologically varied intercellular contacts that ensure the stability of such microbial consortia during adaptation to ambient conditions (Fig. 9.2).

Intercellular contacts with the aid of different extracellular structures—microfibrillae, knob-like juts, cell wall evaginations and capsule form stuff (glycocalyx)—are the result of genetically determined self-regulating development of microbial populations as multicellular systems (Novik and Vysotskii 1995). Figure 9.3 shows a scheme of morphological transformations of bifidobacterial cells in the developmental cycle of populations. Multiplication occurs via reversion of transitory rod-shaped and coccoid forms into repeatedly budding and dichotomously branching multiseptate filaments, which, under certain conditions, fragment with the formation of differentiated reproductive forms (Novik 1998).

Bifidobacteria are generally described as strictly anaerobic, although some strains can tolerate oxygen. The sensitivity to oxygen, however, can differ between species and between different strains within a species (de Vries and Stouthamer 1969; Talwalkar et al. 2001). Most human isolates of bifidobacteria grow at an optimum temperature of 36–38 °C, whereas animal strains appear to have slightly higher optimum growth temperature of 41–43 °C. The notable exceptions are *Bifidobacterium thermacidophilum* which exhibits a maximal growth temperature of 49.5 °C and *Bifidobacterium psychraerophilum* which has been shown to grow at temperatures as low as 4 °C (Dong et al. 2000; Leahy et al. 2005; Simpson et al. 2004).

Bifidobacteria are acid-tolerant microbes and their optimum pH for growth is between 6.5 and 7.0. Strains of *Bifidobacterium animalis* ssp. *animalis* and

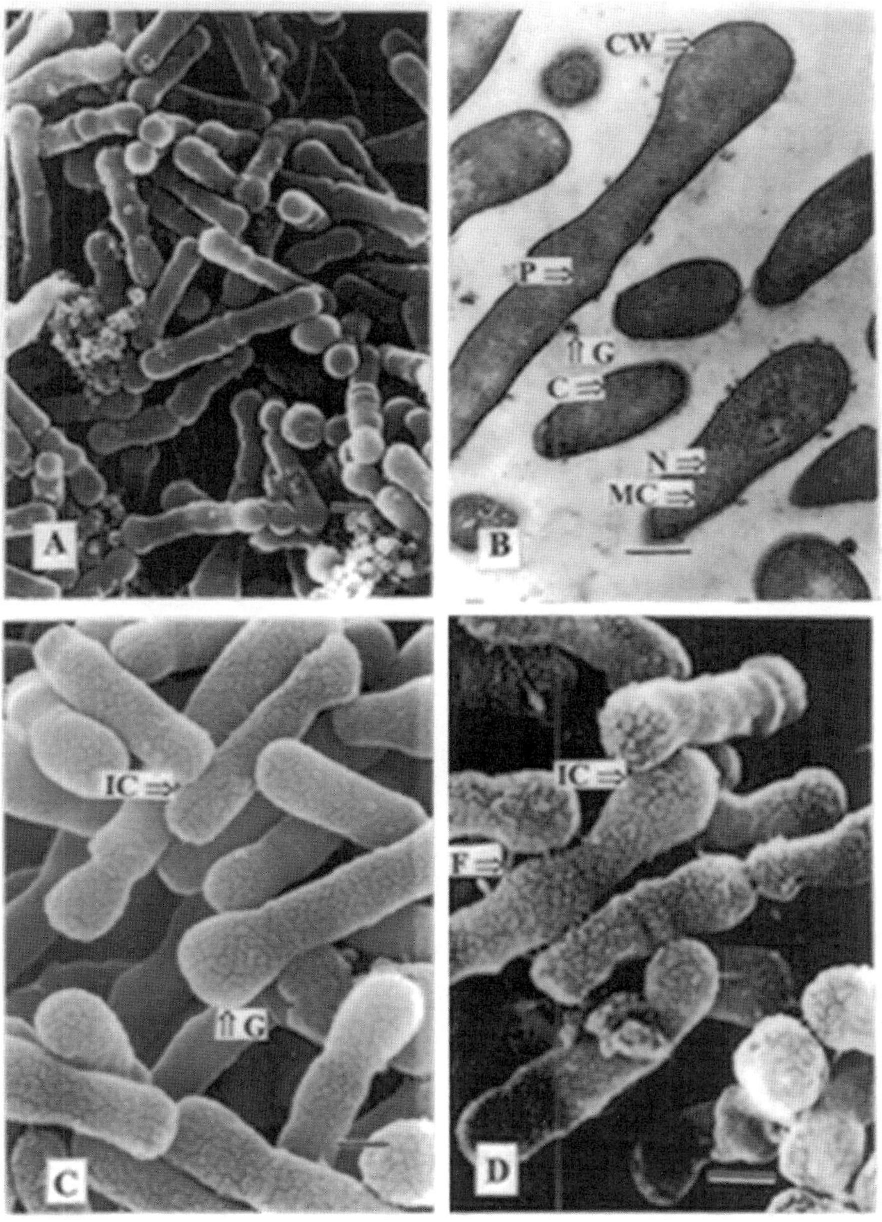

Fig. 9.1 Morphological analysis of bifidobacterial cells by means of electron microscopy; (**a**) rod-shaped forms, having wide ends; (**b**) fine structure of cells; (**c**, **d**) visualisation of intercellular contacts in some cells. *CW* cell wall, *C* cytoplasm, *MC* membrane complex, *N* nucleoid, *P* polyphosphate granules, *IC* intercellular contacts, *F* microfibrillae, *G* capsule (Photo by Galina Novik)

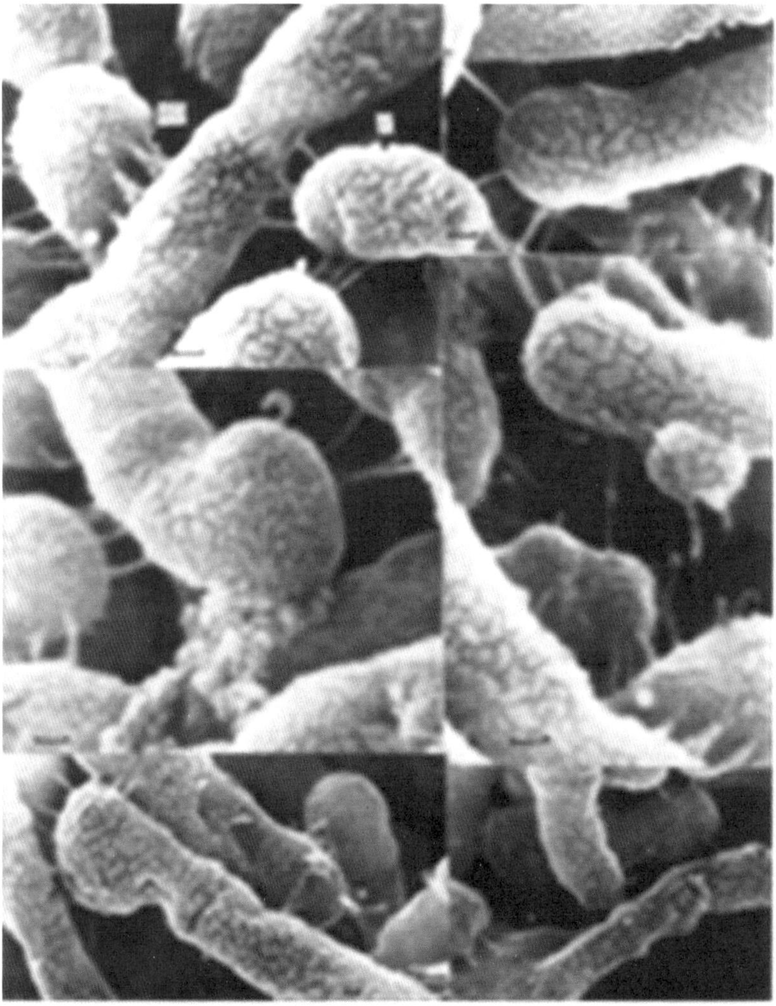

Fig. 9.2 Morphology of bifidobacterial cells by means of electron microscopy with visualisation of intercellular links in some cells (Photo by Galina Novik)

B. animalis ssp. *lactis* can survive exposure to pH 3.5, whereas most of *Bifidobacterium* strains do not survive at pH 8.5 (Leahy et al. 2005).

Bifidobacteria are strictly fermentative bacteria, in which hexose metabolism occurs through a unique fructose-phosphate pathway also called "bifidus shunt" (Biavati and Mattarelli 2006; Leahy et al. 2005). They ferment glucose to lactic acid and acetic acid in molar ratio 2:3 without carbon dioxide. Variations in growth conditions, such as quality and quantity of carbon source, may result in the production of varying amounts of fermentation products. Bifidobacteria possess an array of enzymes that allow them to utilise a great variety of monosaccharides, disaccharides

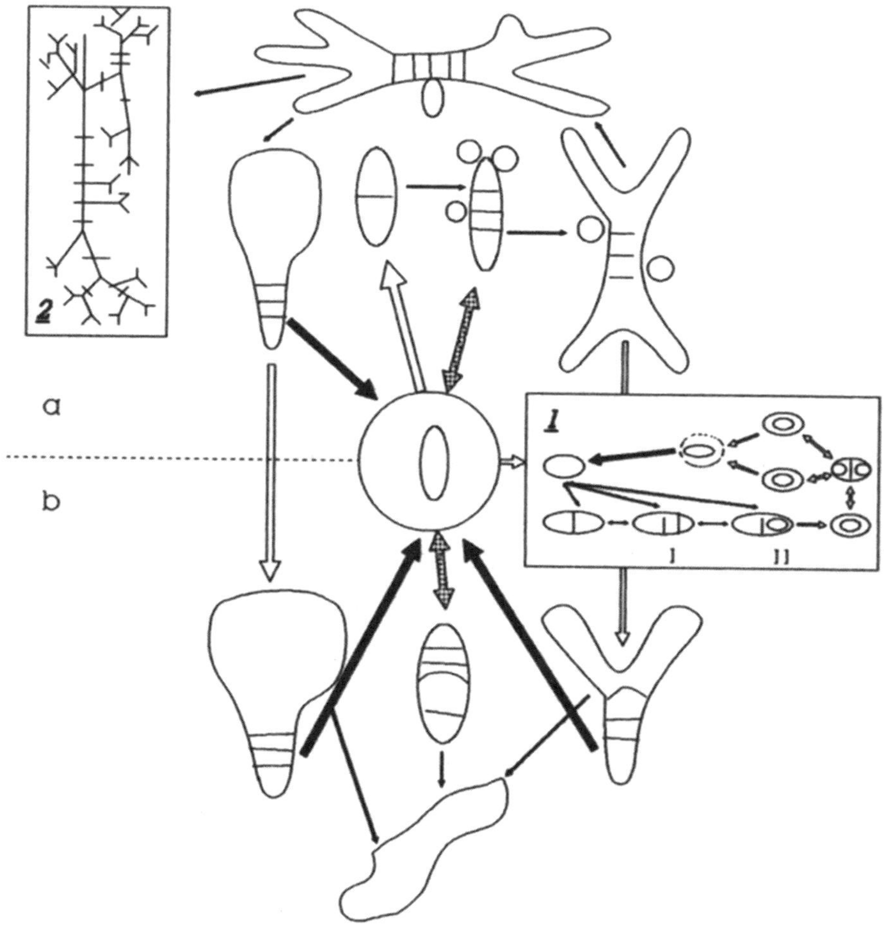

Fig. 9.3 Morphological and structural differentiation of cells in a cycle of development of bifidobacteria populations: (**a**) reproductive forms; (**b**) ageing of bifidobacterial cells; *1*—stage of transitory rod-shaped and coccoid forms; *2*—stage of branching filaments (Scheme by Galina Novik)

and complex carbohydrates as carbon sources. This feature should give bifidobacteria an ecological advantage to colonise the intestinal environment where complex carbohydrates, such as mucin, are present in large quantities either because of production by the host epithelium or introduction through the diet. The selective stimulation of the growth of bifidobacteria by simple or complex carbohydrates is the basis of prebiotic concept (Biavati and Mattarelli 2006).

Lactobacilli are a broad, morphologically defined group of gram-positive, catalase-negative, non-spore-forming bacteria. They usually occur as rods that may differ in length between the various species. Some species grow as coccobacilli or appear curved or coryneform. Some heterofermentative lactobacilli may appear coccoid and can be confused with leuconostocs. Some homofermentative anaerobic lactobacilli from intestinal sources may resemble morphologically certain bifidobacteria. Morphological variations may occur within some *Lactobacillus* species depending on growth conditions (Hammes and Hertel 2003).

Lactobacilli are strictly fermentative, aerotolerant or anaerobic, aciduric or acidophilic bacteria (Kandler and Weiss 1986). Based on the type of fermentation, *Lactobacillus* species are divided into three groups: homofermentative, facultatively heterofermentative and obligately heterofermentative. Homofermentative lactobacilli are able to ferment glucose almost exclusively to lactic acid via the Embden-Meyerhof-Parnas (EMP) pathway, while pentoses and gluconate are not fermented as they lack phosphoketolase. Facultatively heterofermentative lactobacilli degrade hexose to lactic acid by the EMP pathway and are also able to degrade pentoses and often gluconate as they possess both aldolase and phosphoketolase. Obligately heterofermentative lactobacilli degrade hexoses by the phosphogluconate pathway producing lactate, ethanol, acetic acid and carbon dioxide; moreover, pentoses are also fermented through this pathway (Hammes and Vogel 1995; Pot et al. 1994).

Generally, lactobacilli grow well at temperatures above 20 °C and below 42 °C, though some strains of these microorganisms can grow at up to 44 °C and down to 15 °C (Hammes and Hertel 2003; Savoie et al. 2007). Lactobacilli are mostly microaerophilic, but many strains of these microorganisms grow better either anaerobically or in the presence of increased CO_2 tension, particularly on first isolation (Hammes and Hertel 2003).

Lactobacilli have complex nutrient requirements—they grow in the presence of carbohydrates, amino acids, peptides, nucleic acid derivatives and vitamins. Lactobacilli grow in a variety of habitats, wherever high levels of soluble carbohydrate, protein breakdown products, vitamins and a low oxygen tension occur. Large amounts of lactic acid and small amounts of other compounds are the products of their carbohydrate metabolism, which lowers the pH of the substrate and suppresses the growth of many other bacteria (Kandler and Weiss 1986).

9.3 Taxonomy of Probiotic Bacteria

According to current taxonomy, bifidobacteria belong to the phylum *Actinobacteria*, class *Actinobacteria*, order *Bifidobacteriales*, family *Bifidobacteriaceae*, genus *Bifidobacterium* and are represented by over 32 species (Biavati and Mattarelli 2006; Holzapfel et al. 2001; Leahy et al. 2005) with type species *Bifidobacterium bifidum*. In the phylogenetic tree of bacteria, *Bifidobacterium* cluster is in the subdivision of high G+C gram-positive bacteria together with other genera such as *Propionibacterium*, *Actinomyces* and *Streptomyces*, so they form a part of the so-called *Actinomycetes* branch (Leahy et al. 2005). The species belonging to the genus *Bifidobacterium* form a coherent phylogenetic unit and show generally over 93 % similarity of 16S rRNA sequences with other members of the genus. A number of phylogenetic studies carried out during the past decade, mainly based on sequencing of 16S rRNA gene and housekeeping genes, have grouped the bifidobacterial species in six groups, namely, *B. boum* group, *B. asteroids* group, *B. adolescentis* group, *B. pullorum* group, *B. longum* group and *B. pseudolongum* group (Matsuki et al. 2003; Sakata et al. 2006; Ventura et al. 2004; 2006).

All the currently known *Bifidobacterium* species were isolated from limited number of habitats, including human and animal gut, insect intestine, food, sewage and breast milk (Felis and Dellaglio 2007; Ventura et al. 2004). Strains which are the most typical for human GIT belong to species *B. catenulatum*, *B. pseudocatenulatum*, *B. adolescentis*, *B. longum*, *B. breve*, *B. angulatum* and *B. bifidum*, and the most frequent species isolated from dairy products is *B. animalis* ssp. *lactis* (Masco et al. 2005). Therefore these species are the most widely used probiotics (Biavati et al. 2001).

The genus *Lactobacillus* belongs to the phylum *Firmicutes* (gram-positive bacteria with low G+C content), class *Bacilli*, order *Lactobacillales*, family *Lactobacillaceae*, and its closest relatives are the genera *Paralactobacillus* and *Pediococcus*, being grouped within the same family (Felis and Dellaglio 2007). Lactobacilli form the largest group among the lactic acid bacteria (LAB), containing at present more than 120 species; the type species is *Lactobacillus delbrueckii* (Felis and Dellaglio 2007; Vaughan et al. 2005).

Phylogenetic structure of the genus *Lactobacillus* is quite complicated. According to the results of the first phylogenetic analysis of lactobacilli, they were divided into three groups, *L. delbrueckii* group, *L. casei–Pediococcus* group and *Leuconostoc* group, which also contained some lactobacilli (Felis and Dellaglio 2007). In 1995 *L. delbrueckii* group was given the name of *L. acidophilus* group and the *L. casei–Pediococcus* group was split into further four subclusters (Schleifer and Ludwig 1995). Recently, due to the description of a large number of species and the following re-examination of the genus, this strategy of grouping was updated. Nowadays, genus *Lactobacillus* includes *L. delbrueckii* group, *L. salivarius* group, *L. reuteri* group, *L. buchneri* group, *L. alimentarius–L. farciminis* group, *L. casei* group, *L. sakei* group, *L. fructivorans* group, *L. coryniformis* group, *L. plantarum* group, *L. perolens* group, *L. brevis* group, *Pediococcus dextrinicus* group, *Pediococcus* group, couples (e.g. *L. rossiae–L. siliginis*) and single species (*L. kunkeei*, *L. malefermentans*, *L. pantheris*, etc.) (Felis and Dellaglio 2007; Hammes and Hertel 2003).

The dominant species isolated from human gut are those belonging to the *L. casei* group (*L. casei, L. paracasei, L. rhamnosus, L. zeae*); obligatory homofermentative species *L. gasseri, L. crispatus and L. johnsonii*; and heterofermentative species *L. reuteri* (Dunne et al. 1999; Morelli et al. 1998; Song et al. 2000; Tannock et al. 2000).

9.4 Identification of Probiotic Bacteria

The correct identification of probiotic bacteria is the first prerequisite of their microbiological safety. Thus, the use of adequate tools to provide proper strain identification is strictly necessary (FAO/WHO 2002; Saarela et al. 2000).

Traditionally, bifidobacteria have been identified on the basis of phenotype characteristics. Cell morphology, determination of metabolites, enzyme activities and ability to ferment sugars are used for routine bifidobacteria identification. Genus *Bifidobacterium* can be distinguished from other bacterial groups such as lactobacilli, actinomycetes and anaerobic corynebacteria by the peculiar metabolic pathway of glucose fermentation, the bifidus shunt, the key enzyme of which is fructose-6-phosphate phosphoketolase (F6PPK). This enzyme was considered a taxonomic marker for identification on the genus level (Tannock 1999; Vlkova et al. 2002), but, due to the reclassification of *Bifidobacterium* species into new genera, it can be used as taxonomic character of the family *Bifidobacteriacea* (Felis and Dellaglio 2007). Since fermentation of glucose by the bifidus pathway produces acetic and lactic acids in a theoretical ratio of 3:2, gas liquid chromatography of fermentation products provides another means of differentiating bifidobacteria from other bacterial types (Tannock 1999).

Currently, biochemical tests for the identification of members of the genus *Bifidobacterium* are largely superseded by the use of the genus-specific PCR primers, which amplify 523 bp or 1.35 kbp regions of the 16S rRNA gene. Genus-specific probes have proved useful in the detection and identification of bifidobacteria in faecal and food samples (Kaufmann et al. 1997; Kok et al. 1996).

With regard to *Bifidobacterium* species identification, they can sometimes be differentiated using the results of fermentation tests together with the electrophoretic mobilities of enzymes such as transaldolase (14 types) or 6-phosphogluconate dehydrogenase (19 types) (Tannock 1999). The most important species may be distinguished to some degree by the fermentation of L-arabinose, D-xylose, D-mannose, salicin, D-mannitol, D-sorbitol and D-mellesitose (Holzapfel et al. 2001). In many cases, phenotypic characterisation is not enough to identify *Bifidobacterium* strains at the species level. So, genotypic approaches hold the most promise for the rapid and accurate identification of bifidobacteria (Gomes and Malcata 1999; O'Sullivan 2000; Satokari et al. 2003).

DNA–DNA reassociation studies have been widely used in the taxonomy of bifidobacteria and currently it is the most reliable method for *Bifidobacterium* species identification (O'Sullivan 2000; Satokari et al. 2003). Species identification by 16S rRNA gene sequence analysis is hampered by the high level of sequence relatedness between closely related bifidobacterial species. Comparison of 16S rRNA gene sequences from 18 species of *Bifidobacterium* showed that they ranged in similarity from 92 to 99 %. This high level of relatedness makes it impossible to differentiate between some species on the basis of 16S rDNA sequence analysis (Leblond-Bourget et al. 1996; McCartney et al. 1996). However, in many cases subtle differences in the 16S rRNA gene sequences have been successfully utilised to design species-specific probes or PCR primers that can be applied in species identification (Langendijk et al. 1995; Matsuki et al. 1999; 2003; Welling et al. 1997; Yamamoto et al. 1992).

The sequence analysis of conserved genes other than 16S rRNA such as *recA*, major enzyme involved in recombination, and *ldh*, coding for L-lactate dehydrogenase, has been proposed as a method for identification of closely related bifidobacteria

(Kullen et al. 1997; Roy and Sirois 2000). Currently, the gene sequence of *hsp60*, heat-shock protein of 60 kDa, is preferentially used to distinguish between different species of *Bifidobacterium* (Jian et al. 1991). Method based on PCR targeting the transaldolase gene and subsequent separation of the amplicons by denaturing gradient gel electrophoresis (DGGE) was developed for the identification of *Bifidobacterium* species (Requena et al. 2002). Multilocus sequencing—sequencing of 16S rRNA and housekeeping genes, such as *tuf, recA, xfp, atpD, groEL, groES, dnaK, hsp60, dnaB* and *dnaJ*—is the highly discriminatory method for bifidobacteria identification, providing unambiguous results (Ventura et al. 2006).

Methods based on the PCR are widely used to differentiate species and even strains of bifidobacteria. Randomly amplified polymorphic DNA (RAPD) profiling is successfully applied to distinguish between strains of *Bifidobacterium* (Määttö et al. 2004; Vincent et al. 1998). Different modifications of repetitive extragenic palindromic PCR (REP-PCR), such as enterobacterial repetitive intergenic consensus sequence PCR (ERIC-PCR), BOX-PCR $(GTG)_5$-PCR, can be considered as promising methods for the identification of bifidobacteria at species, subspecies and even strain level (Krizova et al. 2008; Masco et al. 2003, 2004; Šrůtkova et al. 2011; Ventura et al. 2003). Pulsed field gel electrophoresis (PFGE) protocols have been established for bifidobacteria and the techniques have shown superior discriminatory power in comparison to other typing methods in species and strain differentiation (O'Riordan and Fitzgerald 1997; Roy et al. 1996). A complete survey of methods for bifidobacteria identification has been compiled in reviews (Sidarenka et al. 2008; Ward and Roy 2005).

Members of the genus *Lactobacillus* for a long time were identified on the base of their phenotypic features, including cell morphology, fermentation of carbohydrates, growth at different temperatures and salt concentrations. However, it has been widely recognised that *Lactobacillus* species and strains display a high level of phenotypic variability, making classical microbiological methods of identification unreliable (Hammes and Hertel 2003). Recently, it was demonstrated that the API 40 identification system failed to identify 7 reference strains and 86 freshly isolated *Lactobacillus* strains (Boyd et al. 2005).

Comparative analysis of complete or at least sufficiently informative part (approximately the first 900 bases) of the 16S rRNA gene can be used for the reliable identification of *Lactobacillus* species (Mori et al. 1997; Tannock 1999). It should be noted, however, that in some cases the 16S rRNA gene may be too well conserved to reliably identify closely related species, such as *L. plantarum*, *L. pentosus* and *L. paraplantarum* (99.7–99.9 %); *L. kimchii* and *L. paralimentarius* (99.9 %); and *L. mindensis* and *L. farciminis* (99.9 %) (Fox et al. 1992). Analysis of 16S–23S rRNA spacer region sequences reveals that this region is less conserved compared to 16S rRNA gene and can be considered as powerful tool for genus and species differentiation of lactobacilli (Tannock et al. 1999). Based on the nucleotide sequences of 16S rRNA gene and 16S–23S spacer region, species-specific primers for lactobacilli identification have been derived. Currently, specific primers are available for most *Lactobacillus* species (Berthier and Ehrlich 1998; Kwon et al. 2004; Settanni et al. 2005). Nucleotide differences in 16S rRNA gene can also be used for the separation by denaturing gradient gel electrophoresis (DGGE) or temporal temperature gradient electrophoresis (TTGE), which are promising tools for the identification of lactobacilli at strain level (Fasoli et al. 2003; Vasquez et al. 2001).

Genes *recA, groES* and *groEL*, coding for highly conserved proteins, are also utilised to identify lactobacilli species (Felis et al. 2001; Torriani et al. 2001; Walker et al. 1999), providing phylogenetic resolution comparable to that of 16S rRNA gene at all taxonomic levels. Comparative analysis of fructose-1,6-biphosphatase (*fbp*) gene has been successfully used for identifying food, newborn and clinical strains of *L. rhamnosus* (Roy and Ward 2004). The powerful multilocus sequencing technique based on the analysis of six genes (*ddl, gyrB, purK1, gdh, mutS, pgm*) has been applied for analysis of *L. plantarum* strains (De las Rivas et al. 2006). Recently, multilocus sequencing variant called multilocus variable-number tandem repeat analysis has been developed for subtyping of *L. casei/L. paracasei* strains (Diancourt et al. 2007).

Many PCR-based typing methods are used for identification of lactobacilli at strain level, including RAPD-PCR (Khaled et al. 1997; Du Plessis and Dicks 1995; Nigatu et al. 2001; Schillinger et al. 2003), REP-PCR (Gevers et al. 2001; Ventura and Zink 2002), PFGE (Tynkkynen et al. 1999; Ventura and Zink 2002; Weiss et al. 2005). Molecular approaches available for *Lactobacillus* identification are described in reviews (Mohania et al. 2008; Singha et al. 2009).

9.5 Criteria for Selection of Probiotic Bacteria

Different in vitro and in vivo approaches have been used to select potentially probiotic strains of bifidobacteria and lactobacilli, as well as to measure their efficacy (Gibson and Fuller 2000). Criteria for the selection of probiotic bacteria have been defined in several reviews (Adams 1999; Bhadoria and Mahapatra 2011; Gibson and Fuller 2000; Saarela et al. 2000; Salminen et al. 1998). They indicate that many aspects, including safety and functional and technological characteristics, have to be taken into consideration in the selection process of probiotic microorganisms.

9.5.1 Safety of Probiotic Bacteria

The safety of probiotic strains is of prime importance. Although vigorous debates continue on what constitutes appropriate safety testing for novel probiotic strains proposed for human use, it generally includes such characteristics as origin, non-pathogenicity and antibiotic-resistance characteristics.

Strains for human use are preferably of human origin, isolated from healthy GIT (Saarela et al. 2000). Probiotic bacteria must be non-pathogenic, with no history of association with diseases such as infective endocarditis or gastrointestinal disorders. Knowledge on survival of the probiotics within the GIT, their translocation and colonisation properties, is also important for the evaluation of possible positive or negative effect of probiotic consumption (Marteau et al. 1995). From this point of view, lactic acid bacteria and bifidobacteria are widely used in fermented food and dairy products with no case of local or systemic infections occurred, which confirms their GRAS ("generally regarded as safe") status (Sleator 2010). Many findings indicate that the general human population is not at risk from exposure to probiotic bacteria of *Bifidobacterium* and *Lactobacillus* genera. Although the rare cases of infection associated with probiotics have occurred in groups of people whose conditions predispose them to opportunistic infections, in many cases people with serious underlying diseases have benefited from probiotics (Benchimol and Mack 2005; Reid 2006; von Wright 2005).

Another aspect of safety consideration is antibiotic resistance of probiotic bacteria strains. The resistance of bacteria to antibiotics is an increasingly important public health problem worldwide. There is a pressing need to limit the spread of resistance genes, since these could be transferred to opportunistic and pathogenic bacteria (Ammor et al. 2008; Blazquez et al. 2002). Antibiotic resistance could be "intrinsic" and "acquired." Intrinsic resistance is inherent to bacteria species and involves the absence of the target, presence of low-affinity target, low cell permeability or presence of efflux mechanisms. The acquisition of antibiotic resistance occurs through the mutation of pre-existing genes or by horizontal transmission, i.e. acquisition of foreign DNA from other bacteria. Therefore attention is currently being paid to probiotic LAB and bifidobacteria with respect to their potential role in the spread and transmission of antibiotic-resistance determinants (Ammor et al. 2008; Saarela et al. 2000).

Most bifidobacteria are intrinsically resistant to nalidixic acid, neomycin, polymyxin B, kanamycin, gentamicin, streptomycin and metronidazole (Charteris et al. 1998). Their resistance to other antibiotics differs depending on strain and in some cases may be due to the presence of genetic determinants. Indeed, microarray analysis revealed presence of *tet(W)* genes in *B. longum* and *B. bifidum* strains, as well as *aph(E)* and/or *sat(3)* genes in *B. bifidum*, *B. longum*, *B. catenulatum* and *B. pseudocatenulatum* strains (Ammor et al. 2008). Screening of 26 *B. animalis* subsp. *lactis* strains isolated from different sources revealed the presence of *tet*(W) in all isolates. Moreover, in all strains a transposase gene upstream of *tet*(W) gene was detected, which is cotranscribed in tandem. Transposases have been found to be involved in the horizontal gene transfer of genetic elements among bacteria, but to date there is no evidence that *tet*(W) in *B. animalis* subsp. *lactis* is transmissible (Gueimonde et al. 2010). Presence of the resistance determinant *erm(X)* was demonstrated in six erythromycin- and clindamycin-resistant *B. thermophilum* strains during investigation of a large collection of bifidobacteria that could be potential probiotics (Mayrhofer et al. 2007). Analysis of additional bifidobacteria revealed that this antibiotic-resistance gene was also present in *B. animalis* subsp. *lactis* strains (Määttö et al. 2007). It was demonstrated that the *erm(X)* gene from erythromycin-resistant *Bifidobacterium* strains was part of transposon Tn5432 and was nearly identical to *erm(X)* determinants present in several opportunistic pathogenic corynebacteria and propionibacteria (van Hoek et al. 2008). Although most of the antibiotic-resistance genes were located on bacterial chromosome, studies on the genetics of antibiotic resistance of bifidobacteria are guarantee their safe application.

Lactobacilli display a wide range of antibiotic resistance, and antibiotic susceptibility patterns vary greatly between different species of these microorganisms (Charteris et al. 1998). Thus, *L. delbrueckii* strains as components of yogurt cultures showed intrinsic resistance toward mycostatin, nalidixic acid, neomycin, polymyxin B, trimethoprim, colimycin and sulphonamides. Susceptibility to cloxacillin, dihydrostreptomycin, doxycycline, novobiocin, oleandomycin, oxacillin and streptomycin was prominent while resistance to kanamycin and streptomycin varied. Many lactobacilli carry intrinsic resistance toward vancomycin (Marthur and Singh 2005). In most cases antibiotic resistance of lactobacilli is not of the transmissible type (Saarela et al. 2000), and such strains do not usually form a safety concern. Although plasmid-linked antibiotic resistance is not very common among lactobacilli, they do occur (Rinckel and Savage 1990). R-plasmids encoding tetracycline, erythromycin, chloramphenicol or macrolide–lincomycin–streptomycin resistance have been reported in *L. reuteri*, *L. fermentum*, *L. acidophilus* and *L. plantarum*, isolated from raw meat, silage and faeces. Most of these R-plasmids had a size smaller than 10 kb (Marthur and Singh 2005). The presence of 5.7 kb plasmid carrying *erm* gene conferring high-level erythromycin resistance was demonstrated in *L. fermentum* isolated from pig faeces (Fons et al. 1997). Plasmid-encoding tetracycline-resistance gene *tet(M)* was detected in *Lactobacillus* isolates from fermented dry sausages (Gevers et al. 2002). The 10,877 bp tetracycline-resistance plasmid pMD5057 from *L. plantarum* 5057 was completely sequenced and the sequence revealed that tetracycline-resistant region contains a *tet(M)* gene with high homology to sequences of this gene from *Clostridium perfringens* and *Staphylococcus aureus* (Danielsen 2002). Since transfer of antibiotic-resistance genes may occur between phylogenetically distant bacteria, *Lactobacillus* strains that harbour mobile elements carrying resistance genes should not be used either as human or animal probiotics (Saarela et al. 2000).

9.5.2 Technological Properties of Probiotic Bacteria

Potential probiotic strains of *Lactobacillus* and *Bifidobacterium* should fulfil many technological criteria, such as simple large-scale production of a viable culture concentrate, survival during preparation and storage of the career of the food and survival in the intestinal ecosystem of the host (Bhadoria and Mahapatra 2011).

Bifidobacteria are fastidious and noncompetitive organisms. They are very sensitive to environmental parameters and require expensive media for propagation, as well as the addition of growth-promoting factors due to their stringent growth requirements. Growth of bifidobacteria in milk is often slow or limited compared with lactic acid bacteria, and this appears to be partially due to low proteolytic activities. Bifidobacteria generally have a low survival rate in common processes used to prepare microbial food adjuncts such as freeze-drying or spray-drying. Survival of most bifidobacteria is also low in many dairy products due to acidic pH and exposure to oxygen (Roy 2005). Furthermore, bifidobacteria require strict anaerobiosis in the early phase of growth and long fermentation times due to their weak growth and acid production (de Vuyst 2000). Finally, strains of *Bifidobacterium* differ greatly in their survival in the gastrointestinal tract and in their ability to adhere to epithelial cells (Doleyres and Lacroix 2005). The adhesion of bifidobacteria might be strain specific and depends on the surface properties of bacterial cells (Canzi et al. 2005). The probiotic *Bifidobacterium* species most commonly used in food is *B. animalis* ssp. *lactis*. This species is significantly more robust than human intestinal species *B. longum, B. bifidum, B. breve* and *B. adolescentis* also utilised in probiotics and food (Biavati et al. 2001; Crittenden 2004; Roy 2005).

Lactobacilli are more technologically suitable than bifidobacteria. Lactobacilli can utilise a wide range of carbon substrates, with differences in the carbon substrate profiles occurring between species and strains. They are able to grow and survive in fermented milk and yogurts with pH values between 3.7 and 4.3. Lactobacilli are mostly microaerophilic; thus oxygen levels are rarely an important consideration in maintaining the survival of lactobacilli during manufacturing and storage of probiotics and food products. Lactobacilli are less sensitive than bifidobacteria to acidic conditions of stomach and high concentrations of bile in gut, although this property seems to be strain specific (Hammes and Hertel 2003). There is a wide range of *Lactobacillus* species technologically suitable for application in probiotics and foods. Common examples include *L. acidophilus, L. johnsonii, L. rhamnosus, L. casei, L. paracasei, L. fermentum, L. reuterii* and *L. plantarum* (Bergamini et al. 2005; Gokavi et al. 2005; Phillips et al. 2006; Sameshima et al. 1998).

It is generally believed that probiotics must endure a harsh transit through the intestinal tract with different conditions depending on the location, which affect their viability. Different in vitro and in vivo studies have been performed to determine survival of probiotic lactobacilli and bifidobacteria during GIT transit. In one such study, two strains *Bifidobacterium* sp. were exposed to model stomach conditions for 90 min. A notable 4 log unit decrease of viability was observed for one strain, whereas viability of the another one decreased by only 0.5 log units (Berrada et al. 1991). In another study, 6 *Lactobacillus* and 9 *Bifidobacterium* strains were maintained at pH 1.5–3.0 for 3 h and demonstrated different survival ability depending on the pH, the duration of exposure to acid, and the species and strains used (Pochart et al. 1992).

It is important for probiotic strains to show antagonism against pathogenic and opportunistic microorganisms via antimicrobial substance production and competitive exclusion. Therefore, enormous research efforts have been focused on bacteriocin production. Although probiotic strains of *Lactobacillus* and *Bifidobacterium* may produce bacteriocins, their role in pathogen inhibition in vivo could be very limited, since traditionally bacteriocins have an inhibitory effect only against closely related species. Low molecular weight metabolites, such as hydrogen peroxide, lactic and acetic acid and secondary metabolites, may be more important since they show wide inhibitory spectrum against many harmful organisms like *Salmonella, Escherichia, Clostridium* and *Helicobacter* (Saarela et al. 2000). Generally, probiotic strains of bifidobacteria demonstrate inhibition of a wide range of pathogenic and opportunistic bacteria, including *Escherichia coli, Klebsiella ozaenae, Listeria monocytogenes, Staphylococcus aureus, Salmonella enteritidis, Enterococcus faecalis, Pseudomonas aeruginosa* and *Gardnerella vaginalis* (Korshunov et al. 1999; Biavati et al. 2001; Bevilacqua et al. 2003; Zinedine and Faid 2007; Vanegas et al. 2010). Strains of *Lactobacillus*, primarily due to the production of organic acids, ethanol, H_2O_2 and bacteriocin-like substances, inhibit growth of certain enteropathogens such as *Salmonella, Listeria, Escherichia, Campylobacter,* as well as *Clostridium difficile* and *Helicobacter pylori*, without interfering with the normal microbiota of the gastrointestinal tract (Chen et al. 2010; Coconnier et al. 2000; Fernandez et al. 2003; Naaber et al. 2004).

9.5.3 Enhancing Stress Resistance of Probiotic Bacteria

Most of currently used *Bifidobacterium* and *Lactobacillus* probiotic strains are fastidious organisms, nutritionally demanding and sensitive to environmental conditions. Therefore, product manufacturing and storage reduce viability of probiotic bacteria, causing an economic burden for manufacturers and compromising the efficacy of probiotic products. The intrinsic stress tolerance of the *Bifidobacterium* and *Lactobacillus* strains seems to be a crucial factor in the overall resistance to manufacture and storage of probiotic products, and enhancing of this property is of great importance. Different strategies are applied to enhance resistance of probiotic *Bifidobacterium* and *Lactobacillus* strains to environmental stresses, including selection of naturally resistant strains, stress adaptation and genetic modification of the strains. The two former approaches used already existing diversity and genetic potential, and the latter one implies genetic manipulations leading to genetically modified organism (Gueimonde and Sanchez 2012).

Different strains of bifidobacteria and lactobacilli demonstrate large differences in their ability to survive different manufacturing and storage conditions. Therefore, the initial screening and selection of the most stress-resistant *Bifidobacterium* and *Lactobacillus* strains is considered the primary target for enhancing their stability in probiotics. In this regard, exopolysaccharide-producing strains may show better stress tolerance and, therefore, could be initially selected (Gueimonde and Sanchez 2012; Stack et al. 2010).

Probiotic strains can be adapted to better tolerate stressful conditions. Three main approaches have been used for this aim: stress pretreatment, mutagenesis and selective pressure. While the first one is limited to physiological changes, the last two usually involved changes in genetic content of the strain.

Stress pretreatment includes the subjecting of strains to the sublethal stress before exposing them to the harsh conditions affecting the viability of microorganisms during product manufacture and storage. Random mutagenesis induced by UV light and chemicals has been successfully used for increasing the stability of *B. animalis* ssp. *animalis* in low pH products (Saarela et al. 2011) and to obtain strains of bifidobacteria producing low amounts of acetic acid (Sánchez and Margolles 2012). Stress-resistant strains of probiotic bacteria can be obtained by exposing sensitive strains to a selective pressure of stress factor. Usually, such derivatives present stable phenotype and cross-resistance to other stresses, which is advantageous in terms of stability in industrial process. This approach has been applied to obtain both *Bifidobacterium* and *Lactobacillus* strains with improved acid, bile, heat and oxygen tolerance (Collado and Sanz 2006; Li et al. 2010; Noriega et al. 2005; Park et al. 1995; Ruas-Madiedo et al. 2005).

An alternative for increasing stability of probiotic bacteria is genetic engineering. However, genetically modified microorganisms are not well accepted by consumers and this strategy has not found wide application (Sánchez and Margolles 2012).

Different approaches used to enhance stability of probiotic bacteria are the subject of several recent reviews (Betoret et al. 2011; Sánchez and Margolles 2012; Sánchez et al. 2012).

9.6 Mechanisms of Probiotic Bacteria Positive Action

The beneficial effects of probiotics may be classified in three modes of action. Modulation of host's defences including integrity of the epithelial border and immunomodulation is most important for the prevention and therapy of infectious diseases, treatment of chronic inflammation of the digestive tract, eradication of neoplastic host cells and treatment of non-intestinal autoimmune disorders. Direct effect of probiotics on other microorganisms, commensal and/or pathogenic ones, is important for the prevention and therapy of infections and restoration of the microbial equilibrium in the gut. Finally, probiotic effects may be based on detoxification of microbial products' host metabolites (e.g. bile salts) and food components in the gut.

In general, the list of health claims made for probiotics is much longer than the list of probiotic effects, for which clinical evidence is available. According to the known review (De Vrese and Schrezenmeir 2008), well-established probiotic effects are as follows: (a) prevention and/or reduction of rotavirus-induced or antibiotic-associated diarrhea as well as alleviation of complaints due to lactose intolerance, (b) beneficial effects on inflammatory diseases of the gastrointestinal tract, (c) normalisation of passing stool and stool consistency in subjects suffering from obstipation or an irritable colon, (d) prevention and alleviation of unspecific and irregular complaints of the gastrointestinal tracts in healthy people, (e) reduction of the concentration of cancer-promoting enzymes and/or putrefactive (bacterial) metabolites in the gut, (f) prevention or alleviation of allergies and atopic diseases in infants and (g) prevention of respiratory tract infections (common cold, influenza) and other infectious diseases as well as treatment of urogenital infections. The preliminary evidence exists with respect to cancer prevention, a so-called hypocholesterolemic effect, improvement of the mouth flora and caries prevention, prevention or therapy of ischemic heart diseases and amelioration of autoimmune diseases (De Vrese and Schrezenmeir 2008). The molecular processes underlying host–microbe interactions in general and probiotic effects in particular are far from clarifying. Gaining insight into the mechanisms of probiotic action could not only help to improve the credibility of the probiotic concept but also to develop tailor-made strategies for the prevention or treatment of various diseases.

The main constituents of "human part" of a complex ecosystem that include intestinal mucosa, gut-associated immune tissue and resident microbiota are listed briefly as targets of probiotic action. The mucosal surface of the intestinal tract is the largest body surface in contact with the external environment (200–300 m^2). The host is protected from attack by potentially harmful enteric microorganisms by the physical and chemical barriers created by the intestinal epithelium. The intestinal epithelium is composed of four epithelial cell lineages, including the enterocytes, enteroendocrine, goblet and Paneth cells. In addition, M cells that sample bacteria and present them to gut-associated immune tissue are in lymphoid follicle-associated epithelium.

The intestinal epithelium is covered by mucus layer; thickness of the layer is relatively small in the small intestine and gradually increases from the colon to the rectum (Atuma et al. 2001; Matsuo et al. 1997). Intestinal mucus layer secreted by goblet cells consists mainly of compact mesh-like network of viscous, permeable, gel-forming secreted MUC2 mucin, which associates with the secreted mucins (MUC 1, MUC 3A, MUC 3B, etc.) by both covalent and noncovalent bonds. Mucin-type molecules consist of a core protein moiety (apomucin) attached to carbohydrate chains by glycoside bonds. O-linked and N-linked oligosaccharides form up to 80 % of the molecule, and the lengths of the carbohydrate side chains range from 1 to more than 20 residues (Seregni et al. 1997). Main functions of mucins (and especially their oligosaccharide chains) are effect of stoichiometric power that excludes larger molecules and microorganisms, hygroscopic effect that influences the degree of hydration at the epithelial cell surface, ion exchange effect and effect of an area that contains bioactive molecules that are listed below. Additionally, mucin type oligosaccharides provide binding sites for lectins, selectins and adhesion molecules.

Besides mucins, intestinal mucus layer contains other goblet-cell products including trefoil peptides, resistin-like molecule β (Th2 cytokine immune effector molecule, an inhibitor of hemotaxis of parasites and regulator of Muc2 transcription and secretion) and Fc-γ binding protein (substance that binds IgG antibodies and stabilises the mucin network through covalent attachment to MUC2). Other components of mucus layer are Paneth cells products including antimicrobial peptides β-defensins (Ayabe et al. 2000; Bevins 2004), two from six known β-defensins, actually HD-5 and HD-6 (Cunliffe 2003), cathelicidins (Zanetti 2004) and antimicrobial molecules, such as lysozyme. Additionally, mucus layer contains immune molecules synthesised by gut-associated immune cells and enterocytes (secretory IgA, growth factors, cytokines and chemokines). Thus, intestinal epithelium covered by mucus layer together with resident microbiota provides the front line of defence against pathogenic microorganisms.

Note that the role of mucus layer is controversial because it plays a generally accepted role in cytoprotection (Van Klinken et al. 1995) and simultaneously offers ecological advantages for bacterial growth of both the indigenous enteric microbiota (Lupp and Finlay 2005) and the pathogens that adhere to the mucus (Helander et al. 1997; Lillehoj et al. 2001; Rajkumar et al. 1998; Vimal et al. 2000) through providing of energy source and numerous attachment sites. Consistent with this, bacteria associate with the outer layer of mucus and interact with the diverse oligosaccharides of mucin glycoproteins, whereas an "inner" adherent mucus layer is largely devoid of bacteria.

It is considered that resident intestinal bacteria are able to inhibit the adherence of pathogenic bacteria to intestinal epithelial cells as a result of (a) their ability to increase the production of intestinal mucins and antimicrobial substances and (b) competition for the sites of adhesion; both mechanisms are implicated in probiotic effect known as gut epithelium defence.

9.6.1 Increasing the Production of Intestinal Mucins and Antimicrobial Substances

Studies have shown that germ-free mice can exhibit changes in the number of rectal goblet cells and mucin composition in response to oral administration of microorganisms prepared from the faeces of genetically identical mice (Fukushima et al. 1999). However, the data on the role of the probiotics in the induction of mucin synthesis are very limited. For example, *L. plantarum* strain 299v increases the levels of expression of the mRNA of the secretory mucins MUC2 and MUC3, thus in turn inhibiting the cell attachment of enteropathogenic *Escherichia coli* (EPEC), an effect that can be mimicked by adding purified exogenous MUC2 and MUC3 mucins (Mack et al. 1999, 2003). Note that a spontaneous mutant of *L. plantarum* 299v with reduced adhesion capabilities to such a cell line was unable to induce mucin secretion (Mack et al. 2003). The data suggest that adhesion of probiotic bacteria to host cells could be a mechanism for the induction of mucin secretion through the action of certain bacterial surface proteins. However, the bacteria contained in the VSL#3 probiotic formula which consists of four *Lactobacillus* spp., three *Bifidobacterium* spp. and *Streptococcus thermophilus* and is manufactured by Seaford Pharmaceuticals secrete soluble compounds that are able to induce mucin secretion and muc2 gene expression in murine colonic epithelial cells (Caballero-Franco et al. 2007). How the microbiota can influence antimicrobial peptides production also remains controversial. Some reports suggest that in fact the microbiota has no influence. In contrast, it has been shown that *E. coli* strain Nissle 1917, a human faecal isolate and widely used probiotic, induces the human β-defensin 2 (hBD-2) in Caco-2 cells (Schlee et al. 2007; Wehkamp et al. 2004), and flagellin is the main hBD-2-inducing factor (Schlee et al. 2007). Note that the ability to increase the production of intestinal mucins and antimicrobial substances is common for probiotics, symbionts and intestinal pathogens. For example, the expression of angiogenin-4 (Ang4), a molecule produced by mouse Paneth cells, which is active against a number of gram-negative and gram-positive bacteria can be triggered by lipopolysaccharides from *Salmonella* and unidentified substances of *Bacteroides thetaiotaomicron*, a dominant member of the gut microbiota that currently is not used as probiotic (Hooper et al. 2001, 2003).

9.6.2 Competitive Exclusion of Pathogenic Bacteria Through Probiotics Adhesion to Mucus, Epithelial Cells, Extracellular Matrix Proteins and Plasma Components

It is believed that to be effective, the probiotic bacteria must possess a number of functional characteristics, including the ability to adhere to the epithelium. For example, *L. gasseri* and *L. reuteri* are autochthonous lactobacilli which are able to colonise the mucosal surface of the gastrointestinal tract, but *L. plantarum, L. casei* and *L. rhamnosus* are transient organisms. Despite this, the last three bacteria are used as probiotics mainly due to good technological properties (Reuter 2001).

Adhesion is believed not only to play a role in the persistence of a particular strain in the digestive tract but also to participate in pathogen exclusion by competition and blocking of their binding sites at the mucosa (Collado et al. 2007). Additionally, probiotic adhesion may contribute to immunomodulation (Galdeano et al. 2007). However, some authors have hypothesised that attachment factors in lactic acid bacteria are risk factors that might be indicative of their pathogenic potential (Vesterlund et al. 2007).

Assessment of bacterial adhesion is conventionally performed by using in vitro models, based on tissue-cultured cells and intestinal mucus preparations. Mainly used Caco-2 or HT29 cell lines only mimic enterocytes, thereby underestimating the role of the mucus layer, and mucus-producing cell lines such as HT29-MTX are a more appropriate way of studying the mechanism of adhesion and estimation of binding potential of probiotic bacteria (Turpin et al. 2012). Unfortunately, the known in vitro models do not account for all factors involved in probiotic adhesion to the human intestine and show the data that significantly differ from the data obtained in vivo (Larsen et al. 2009).

It was shown in vitro that probiotics prevent gut colonisation by *B. vulgatus*, *Clostridium difficile*, *Clostridium histolyticum*, *Listeria monocytogenes*, *Salmonella choleraesuis*, *St. aureus* and certain *E. coli* strains (Collado et al. 2007; Lin et al. 2008; Sherman et al. 2005). EcN 1917 protects epithelial cells from the invasion by *Salmonella enterica*, *Yersinia enterocolitica*, *Shigella flexneri*, *Legionella pneumophila*, *L. monocytogenes* and *E. coli* (Boudeau et al. 2003; Altenhoefer et al. 2004).

Competitive exclusion based on binding to the same receptor sites on the intestinal surface (including mucus, epithelial cells and extracellular matrix) by probiotics and pathogenic bacteria appears to be one of the underlying mechanisms explaining these observations (Mukai et al. 2004; Sun et al. 2007). In the last decade, the increasing amount of data dealing with the molecular origin of adhesion has improved our understanding of binding mechanisms.

Though specific mechanisms are not yet well understood, evidence suggests that carbohydrate–protein interactions play a key role in the adhesion of bacterial proteins to mucin-bound oligosaccharides, especially taking into account that numerous mucus-binding proteins contain regions homologous with binding domains of lectins. Besides carbohydrate moiety of mucins, adhesins of commensals are able to interact with the host's extracellular matrix proteins (EMPs), such as fibronectin, collagen and laminin. The latter three substances are shed into the mucus or can be exposed to the intestinal lumen in case of trauma, infection or inflammation. Moreover, some probiotic bacteria can bind to plasminogen just as it was found for pathogens that exploit the proteolytic activity of the plasminogen system to overcome barriers formed by the host's extracellular matrix proteins. Bacterial proteins involved in the adhesion mechanism can be separated into five classes: LPXTG-motif proteins, transporter proteins, surface layer proteins, anchorless housekeeping proteins, and "other" proteins (Ljungh and Wadström 2009). Additionally, adhesion of probiotics to mucus/epithelial surfaces is facilitated by exopolysaccharides and lipoteichoic acid as it was found for lactobacilli (Lebeer et al. 2008; Sánchez et al. 2008). The substances generally play a role in nonspecific interactions of lactobacilli with abiotic surfaces and biotic surfaces by contributing to the bacterial cell surface physicochemical properties. Besides, exopolysaccharides could also act as ligands for host lectins mediating adhesion (Ruas-Madiedo et al. 2006).

The most studied example of mucus-targeting bacterial adhesins is the mucus-binding protein (MUB) produced by *L. reuteri* 1063 (Roos and Jonsson 2002). MUB contains C-terminal sortase recognition motif (LPXTG) for anchoring the protein to peptidoglycan, repeated functional domains and an N-terminal region signalling the protein for secretion. Actually repeated functional domains (referred to as MUB domains) are responsible for the protein adhesive properties and allow including the protein to mucin-binding protein (MucBP) domain family. Numerous MUB homologues and MucBP domain-containing proteins have been found, but almost exclusively in lactobacilli (Van Tassell and Miller 2011). Some of them are listed below: Mub of *L. acidophilus* NCFM (Buck et al. 2005), the mannose lectin (Msa) of *L. plantarum* WCFS1 (Pretzer et al. 2005), the *Lactobacillus* surface protein A (LspA) of *L. salivarius* UCC118 (van Pijkeren et al. 2006) and the mucin adhesion-promoting protein (MapA) of *Lactobacillus fermentum* 104R, recently reclassified as *L. reuteri* 104R (Miyoshi et al. 2006; Rojas et al. 2002). Certain other surface proteins contributed to adhesion of lactobacilli to mucus but are otherwise not well characterised. For instance, a 32 kDa protein associated with adhesion to porcine mucus in *L. fermentum*, named 32-Mmubp, was identified as a homologue of the substrate-binding domains of the OpuAC ABC-transport protein family (Macías-Rodríguez et al. 2009).

A majority of the known EMPs-targeting bacterial adhesins have been identified as surface layer proteins (SLPs) of lactobacilli. Briefly, SLPs of *Lactobacillus* species are highly basic proteins (with computed isoelectric point values ranging from 9.4 to 10.4 and a molecular weights ranging from 25 to 71 kDa), and its only known functional role is adhesion to host tissues (Ävall-Jääskeläinen and Palva 2005). SLPs are non-covalently attached to the cell surface through N-terminal domains, which are responsible for their binding to accessory molecules (such as teichoic acids, lipoteichoic acids and neutral polysaccharides) embedded in the peptidoglycan matrix (Mesnage et al. 2000). Direct experimental evidence of SLPs binding to EMPs and to epithelial cell lines has already been obtained for the protein SlpA of *L. acidophilus* NCFM (protein that additionally carries two mucin-binding domains) (Buck et al. 2005), SlpA of *L. brevis* ATCC 8287 (de Leeuw et al. 2006), CbsA of *L. crispatus* JCM 5810 (Antikainen et al. 2002), SlpB of *L. crispatus* ZJ001 (Chen et al. 2007) and the SlpA of *L. helveticus* R0052 (Johnson-Henry et al. 2007). To date, a number of studies associating SLPs of probiotic bacteria with competitive exclusion of pathogens and pathogen adhesion to mucus have been carried out (Chen et al. 2007; Sánchez et al. 2009; Zhang et al. 2010).

In addition, at least two small surface-associated proteins (with a molecular mass 3 kDa) have been shown to be responsible for the adhesion of *L. fermentum* to Caco-2 cells (Baccigalupi et al. 2005). Other examples of probiotic adhesins are non-covalently surface attached proteins including the fibronectin-binding protein (FbpA) of *L. acidophilus* NCFM (Buck et al. 2005), the collagen-binding protein from *L. reuteri* NCIB 11951 (Roos et al. 1996) and its homologous p29 of *L. fermentum* RC-14 (Heinemann et al. 2000).

It was shown that species of the genus *Lactobacillus* have moonlighting proteins (Jeffery 2009) that carry out the function of adhesion. The term "moonlighting protein" means that the protein performs multiple functions and the additional activity may occur only when the protein is in a different location from that which it normally occupies. Moonlighting proteins are anchorless proteins and they do not possess any export motifs or surface-attachment domains. In particular, glycolytic enzymes of *L. crispatus* strain ST1, namely enolase, glyceraldehyde-3-phosphate dehydrogenase, glutamine synthetase and glucose-6-phosphate isomerase that are known as cytosolic proteins were found in cell wall where they moonlight either as adhesins with affinity for basement membrane and EMP or as plasminogen receptors. The proteins were bound onto the bacterial surface at acidic pH, whereas suspension of the cells to pH 8 caused their release into the buffer (Antikainen et al. 2007a, b; Kainulainen et al. 2012). This could be one of the mechanisms by which probiotic bacteria respond to the physicochemical changes of the gastrointestinal environment. The glyceraldehyde-3-phosphate dehydrogenase of *L. plantarum* LA 318 also acts as an adhesin and is able to bind to human colonic mucin (Kinoshita et al. 2008). Note that the host plasminogen activation by enolase and glyceraldehyde-3-phosphate dehydrogenase secreted by probiotics might interfere in the interaction between plasminogen and gastrointestinal pathogens such as *Helicobacter pylori* and *Salmonella sp.* (Hurmalainen et al. 2007; Jönsson et al. 2004; Lähteenmaki et al. 2005).

Another moonlighting protein known as elongation factor Tu was found on the cell surfaces of *L. johnsonii* NCC 533 and identified as the substance-mediating attachment to intestinal epithelial cells and mucin (Granato et al. 2004). Expression of elongation factor Tu was upregulated in the presence of mucus (Ramiah et al. 2007), and its adhesion to epithelial cells or mucus was pH dependent (Granato et al. 2004). Chaperonin GroEL was detected at the surface of *L. johnsonii* NCC 533 and its moonlighting as an adhesin was proved by detection of attachment of recombinant GroEL expressed in *E. coli* to mucus as well as to the HT29 cell line (Bergonzelli et al. 2006). Both elongation factor Tu and chaperonin GroEL belong to group of anchorless housekeeping proteins implicated in adhesion (Ljungh and Wadström 2009).

It is known that fimbriae, also referred to as pili, are thin proteinaceous extensions from bacterial cells, predominantly in gram-negative bacteria, that promote adhesion (Nakamura et al. 1997). The direct visualisation of pili on cells of *L. rhamnosus* GG (Kankainen et al. 2009) proved for the first time that fimbrial interaction with mucus can mediate adhesion of lactobacilli to host epithelium.

At least 20 genes are reported to be functionally important in the binding of *Lactobacillaceae* to the digestive tract. The genetic screening could have been an ideal tool to assess potential bacterial adhesion, but proved to be inadequate, since there was a gap between the potential identified by screening and the results obtained by functional analysis using tissue-cultured cells (Turpin et al. 2012).

By contrast with lactobacilli, very little is known on the mechanisms of bifidobacterial adhesion. Adhesion of *B. breve* strain 4 to intestinal epithelial cells is mediated by a proteinaceous component present on the cell surface and in spent culture supernatant (Bernet et al. 1993). Binding of human plasminogen in vitro was shown for *B. longum, B. bifidum, B. breve* and *B. lactis*. Chaperone protein DnaK also as the key glycolytic enzyme enolase expressed in cell wall as moonlight proteins was identified as plasminogen receptors implicated in the interaction of the bacteria with host tissues (Candela et al. 2009, 2010). A cell surface lipoprotein named BopA was shown to be involved in adhesion of *B. bifidum* MIMBb75 to Caco-2 and HT-29 cells, and its adhesion strongly depended on the environmental conditions, including the presence of sugars and bile salts and the pH (Guglielmetti et al. 2009). More recently BopA was identified as a *B. bifidum*-specific lipoprotein involved in adhesion to intestinal epithelial cells (Gleinser et al. 2012). Expression of BopA in *B. longum/infantis* E18 allows to enhance adhesion to epithelial cells suggesting possibility to create recombinant bifidobacteria with improved adhesive properties (Gleinser et al. 2012). Another example demonstrating improved probiotic properties of recombinant lactic acid bacteria is a recombinant *L. paracasei* strain expressing the gene coding for the *Listeria* adhesion protein (Lap), which was shown to protect Caco-2 cells from infection with *Listeria monocytogenes* by interaction with host cell receptor Hsp60 (Koo et al. 2012).

While adhesion might play an important role in establishing administered probiotic bacteria in the intestinal tract, the data on correlation between the health-promoting properties of probiotics and their adhesion to intestinal epithelial cells/mucus are limited. For example, in different murine models of intestinal inflammation, it was shown that animals treated with *B. bifidum* S17, a highly adherent strain, were protected from weight loss, had a normalised colonic weight to length ratio and showed improved histological scores. By contrast, the weakly adherent *B. longum/infantis* E18 had no protective effect (Preising et al. 2010).

9.6.3 Protection and Restoring of Epithelial Integrity

The intestinal epithelial cells are tightly bound together by intercellular junctional complexes that regulate the paracellular permeability and are crucial for the maintenance of barrier integrity. The junctional complexes consist of the tight junctions (TJ), gap junctions, adherens junctions and desmosomes (Farquhar and Palade 1963). Actually TJ are most important because TJ forms a seal between adjacent epithelial cells near the apical surface (Schneeberger and Lynch 2004). TJ are complex structures comprising over 50 proteins. Briefly, transmembrane proteins such as occludin and claudin (tetra-span proteins) and junctional adhesion molecules (JAM) (single-span proteins) form fibrils that cross the plasma membrane and interact with proteins in the adjoining cells (Chiba et al. 2008). Plaque proteins, such as the zonula occludens (ZO) proteins, act as cytoplasmic adaptors that connect transmembrane proteins to several cytoplasmic regulatory proteins and the actin cytoskeleton within the cell (Fanning et al. 1998). TJ are highly dynamic structures that are constantly being remodelled due to interactions with external stimuli, such as food residues and pathogenic and commensal bacteria.

Currently, it is known that probiotics promote intestinal barrier integrity in mouse models of colitis (Madsen et al. 2001) and reduce intestinal permeability in Crohn's disease patients (Gupta et al. 2000) and in rats subjected to psychological stress (Zareie et al. 2006). The known mechanisms of promoting intestinal barrier integrity by probiotics include regulation of TJ structure through (a) changes in TJ protein expression and distribution and (b) changes in activity of the kinases that regulate contraction of the perijunctional actomyosin ring. It is known that treatment of epithelial cells with EcN 1917 leads to increased expression of ZO-2 protein and redistribution of ZO-2 from the cytosol to cell boundaries in vitro (Zyrek et al. 2007). *L. plantarum* regulates human epithelial TJ proteins in vivo and to confer protective effects against chemically induced dislocation of ZO-1 and occludin from Caco-2 monolayers (Karczewski et al. 2010). Furthermore, treatment of Caco-2 cells with the probiotic *L. plantarum* MB452 from the probiotic product VSL#3 results in increased transcription of occludin and cingulin genes, suggesting that bacteria-induced improvements in intestinal barrier integrity may also be regulated at the gene expression level (Anderson et al. 2010).

Interestingly that some probiotics and commensals prevent and even reverse the adverse effects of pathogens on intestinal barrier function. For example, enteroinvasive *E. coli* (EIEC) strain O124:NM induces loss of expression and distribution of TJ-associated proteins in Caco-2 monolayer, but these effects are absent when EIEC and *L. plantarum* strain CGMCC 1258 are incubated with Caco-2 cells simultaneously.

Moreover, the disruption and disorganisation of the actin cytoskeleton induced by EIEC can then be reversed by incubating the epithelial cells with *L. plantarum* (Qin et al. 2009).

Pretreatment with metabolites from probiotic bacteria may also be protective against pathogen-induced changes in intestinal barrier function. The treatment of Caco-2 cells with the cell-free supernatant of *B. lactis* 420 before adding the supernatant of enterohemorrhagic *E. coli* (EHEC) strain O124:H7 increased transepithelial electrical resistance (TEER) which is used as a measure of paracellular ion permeability, whereas adding the supernatant of EHEC alone decreased TEER (Putaala et al. 2008). The increase in TEER was not seen, however, if the supernatant was added with or after pathogen treatment. The data suggests that supernatant of *B. lactis* 420 protects Caco-2 cells against changes induced by EHEC but does not repair TJ integrity after damage. Regulation of TJ structure is achieved via myosin light chain II (MLC) phosphorylation and contraction of the perijunctional actomyosin ring. The enzymes of enterocytes namely protein kinase C (PKC), MLC kinase (MLCK), mitogen-activated protein kinases (MAPK) ERK1/2 and p38 and Rho kinase (ROCK; activated by Rho GTPases) phosphorylate MLC and induce contraction of the actomyosin ring causing increased permeability (Kimura et al. 1996). It was shown that pathogens can alter barrier function by activation of MLCK (*Helicobacter pylori*), inactivation or activation of small Rho GTPases (*Clostridium difficile* and *Salmonella typhimurium*) and enhancement of actin polymerisation by PKC (*Vibrio cholerae*) (Ohland and MacNaughton 2010). The same proteins are possible targets for probiotics that enhance epithelial barrier integrity (Ulluwishewa et al. 2011). For example, the ability of the probiotics *S. thermophilus* and *L. acidophilus* to preserve phosphorylation of occludin in cells infected with EIEC can be reduced by treating the cells with ROCK inhibitors (Trivedi et al. 2003), suggesting that these bacteria employ Rho family GTPases to protect against EIEC-induced TJ disruption. EcN 1917 uses a PKCζ-dependent signalling pathway to reduce epithelial barrier disruption caused by EPEC (Zyrek et al. 2007); activation of PKCζ by the probiotic leads to phosphorylation of ZO-2, thus reducing ZO-2-PKCζ colocalisation and allowing association of ZO-2 with the cytoskeleton. *B. infantis* Y1 secretes metabolites which increase TEER in cultured epithelial monolayer through MAPK-dependent pathways including a transient phosphorylation of ERK1/2 and a decrease in phosphorylation of MARK p38 (Ewaschuk et al. 2008). *B. infantis-induced* TEER increase can be prevented by inhibition of extracellular signal regulated kinases (ERK), a group of MAPK (Ewaschuk et al. 2008). However, it has also been shown that the ability of *S. thermophilus* and *L. acidophilus* to protect against EIEC infection, which is reduced by ROCK inhibitors, does not seem to be affected by inhibition of ERK1/2 or p38 (Trivedi et al. 2003). In general, consumption of live probiotics promotes TJ integrity and prevents pathogenic bacteria and their effectors from entering via the paracellular pathway to cause further damage; different species of probiotics may use multiple pathways to modulate TJ integrity.

Commensals and probiotics are also known to preserve epithelial barrier function by interfering with pro-inflammatory cytokine signalling. Treatment of cell monolayers with the cytokines TNFα and IFNγ leads to a decrease in TEER and an increase in epithelial permeability (Resta-Lenert and Barrett 2006), and TEER decrease can be prevented by preincubation of the cells with the probiotics *S. thermophilus* ATCC19258 and *L. acidophilus* ATCC4356 or the commensal *Bacteroides thetaiotaomicron* ATCC29184 (Resta-Lenert and Barrett 2006). The reversal of cytokine-induced decrease in TEER was shown to be dependent on activation of kinases ERK and p38, a group of MAPK, and phosphatidylinositol 3-kinase (PI3K) (Resta-Lenert and Barrett 2006). In was shown that following *L. rhamnosus* GG inoculation, IFNγ priming and TNFα stimulation, Caco-2bbe cells maintained TEER and ZO-1 distribution. The signalling interaction between the probiotic and Caco-2bbe cells included suppression of cytokine-induced nuclear factor kappa-light-chain enhancer of activated B cells (NF-κB) inhibition and ERK1/2 response (Donato et al. 2010). DNA from the commensal bacteria *L. rhamnosus* GG and *B. longum* SP 07/3 have also been shown to induce a signal transduction cascade via an epithelial cell surface receptor, which reduces TNFα-induced p38 phosphorylation (Ghadimi et al. 2010).

Note that increased permeability of the epithelial barrier can also be caused by apoptosis via caspase-3 activation (Chin et al. 2002) and probiotics modulate apoptosis initiation by harmful stimuli. Two proteins (p40 and p75) secreted from *L. rhamnosus* inhibited cytokine-induced apoptosis in epithelial cell lines by activating the epidermal growth factor (EGF) receptor and its downstream target serine/threonine kinase Akt (also known as protein kinase B), as well as inhibiting p38 MAPK activation, in vitro and ex vivo (Yan et al. 2007). Akt promotes cell survival by inactivating proapoptotic proteins, including caspases 3 and 9 (Hanada et al. 2004). Expression of p40 and p75 is strain specific because *L. casei*, but not *L. acidophilus*, also produces these proteins (Yan et al. 2007). Additionally, apical or basolateral pretreatment with either p40 or p75 protected several cell lines from hydrogen peroxide-induced disruption of barrier function, as measured by TEER and paracellular permeability. This effect was via inhibition of hydrogen peroxide-induced cytosolic relocalisation of the TJ proteins occludin and ZO-1 and the AJ proteins E-cadherin and β-catenin. These effects were all dependent on activation of PKCε, PKCβI, and the MAP kinases ERK1/2 (Seth et al. 2008). Therefore, bacterial proteins isolated from *L. rhamnosus* cultures effectively block the induction of apoptosis, helping to enhance epithelial barrier function.

Interestingly, that conditioned media from the probiotic *L. rhamnosus* GG induce expression of cytoprotective heat-shock proteins (Hsps) Hsp25 and Hsp72 in intestinal epithelial cells and the effect is mediated by a low-molecular-weight peptide that is acid and heat stable. Inhibitors of MAP kinases block the expression of Hsp72 normally induced by the probiotic (Tao et al. 2006). Similarly, VSL#3 produces soluble factors that induce the expression of cytoprotective Hsps in young adult mouse colonic epithelial cells (Petrof et al. 2004). It is known that Hsps are involved in protein folding, assembly, degradation and intracellular localisation, acting as molecular chaperones, and their overexpression represents a ubiquitous molecular mechanism to cope with stress. Thus, induction of Hsp is another mechanism of probiotic action that provides cellular protection and improves epithelial integrity.

Future investigations into the bacterial factors (such as *Lactobacillus* p40 and p75) involved in improvement of intestinal epithelium integrity would be useful in developing probiotic-derived products for therapy in immunocompromised individuals who cannot consume live probiotics.

9.6.4 Modulation of Host Immune Functions

The gut-associated immune system recognises intestinal microorganisms by pattern-recognition receptors such as the Toll-like receptors (TLRs). The TLRs recognise molecular signatures of different bacteria such as cell wall components or specific DNA motifs (CpG-DNA). Activation of the TLRs results in the induction of complex intracellular signal transduction cascades and finally in the modulation of pro- and anti-inflammatory cytokine expression (Cario 2005). Probiotic bacteria may act through the stimulation of TLRs and it appears that certain effects exerted by some probiotic strains or preparations are mediated through interactions with distinct TLRs. In dextran sodium sulphate ((DSS)-treated mice, γ-irradiated VSL#3 is capable of decreasing the severity of inflammation through TLR9 that is activated by non-methylated bacterial DNA (Rachmilewitz et al. 2004). In contrast, EcN 1917 clearly exerts its effects on DSS-induced colitis in mice through TLR2 and TLR4 (Grabig et al. 2006).

An important aspect of the probiotic immune modulation is the regulation of pro- and anti-inflammatory cytokine production by direct interactions with immune cells. In healthy subjects, *L. rhamnosus* GG triggers the synthesis of the anti-inflammatory interleukin IL-10 and decreases the release of pro-inflammatory IFN-γ, IL-6 and TNF-α from CD4+ T-cells pre-stimulated with intestinal bacteria (Schultz et al. 2003). Co-cultivation of inflamed mucosa explants from celiac disease patients with *L. bulgaricus* LB 10 and *L. casei* DN-114001 reduces the number of TNF-a-secreting CD4+ T-cells and the TNF-a expression by intraepithelial lymphocytes (Borruel et al. 2002). It can be assumed that probiotic bacteria stimulate dendritic cells which in turn produce anti-inflammatory cytokines. This has been demonstrated for *L. reuteri*, *L. casei* and VSL#3, all of which are capable of stimulating IL-10 production by human dendritic cells (Hart et al. 2004; Smits et al. 2005). In addition to the examples given here, probiotics display many other immune modulatory functions, which have extensively been reviewed elsewhere (Shida and Nanno 2008; Vanderpool et al. 2008).

9.6.5 Influence on Host Microbiota and Pathogenic Bacteria

It has been proposed that probiotics exert their effect by modulating gut microbiota composition (Fuller 1989). Indeed, a considerable number of studies support this assumption by demonstrating changes in a number of bacterial groups in response to the consumption of probiotics. For example, numbers of bifidobacteria and lactobacilli increase in healthy subjects after ingestion of *L. casei* Shirota or *L. johnsonii* La1 while those of enterobacteria or clostridia decrease (Spanhaak et al. 1998; Yamano et al. 2006). In patients suffering from intestinal bowel disease, in particular ulcerative colitis, the intestinal microbiota composition can differ substantially from that of healthy subjects (Sartor 2006, 2008) as reflected by high titers of *Bacteroides vulgatus* and *E. coli* (Fujita et al. 2002; Kotlowski et al. 2007). High proportions of *B. vulgatus* are reduced in the gut of patients suffering from ulcerative colitis by the consumption of fermented milk containing bifidobacteria (Ishikawa et al. 2003). The probiotic preparation VSL#3 is effective in elevating the number of total gut bacteria and in restoring the intestinal microbiota diversity in the patients that have inflammation of the lining of the internal pouch (Kuhbacher et al. 2006). In addition, VSL#3 increases caecal bifidobacteria numbers and modifies the metabolic activity of caecal bacteria in mice with chronic colitis induced by DSS (Gaudier et al. 2005). Thus, one possible mechanism by which probiotics can alleviate the severity of ulcerative colitis is the reduction of bacterial species involved in the pathogenesis.

Proposed mechanisms involved in the modification of the intestinal microbiota composition by probiotics include competition and cooperation for nutrients (Lebeer et al. 2008) and production of antibacterial substances including lactic and acetic acid, hydrogen peroxide and antibacterial peptides. Lactic acid as the end product of LAB metabolism lowers the local pH and thereby inhibits the growth of bacteria sensitive to acidic conditions (Alakomi et al. 2000; De Keersmaecker et al. 2006; Makras et al. 2006). The same effects are typical for acetic acid that is one of the bifidobacterial end products. Hydrogen peroxide production by lactobacilli is an important antimicrobial mechanism, especially in the vagina of healthy women (Servin 2004). Recently it was shown that *L. johnsonii* NCC533 produces up to millimolar quantities of hydrogen peroxide when resting cells are incubated in the presence of oxygen, and the role for hydrogen peroxide in the anti-*Salmonella* activity of the probiotic strain was proved in vitro. The genetic base for this hydrogen peroxide production is not clear, but at least four enzymes are implicated in the effect (Pridmore et al. 2008).

Many LAB produce antibacterial peptides (bacteriocins) that vary in spectrum of activity, mode of action, molecular weight, genetic origin and biochemical properties. According to Klaenhammer (1993), the major classes of bacteriocins produced by LAB include (a) lantibiotics, (b) small heat stable peptides, (c) large heat labile proteins and (d) complex proteins whose activity requires the association of carbohydrate or lipid moieties. The existence of the fourth class was supported mainly by the observation that some bacteriocin activities obtained in cell-free supernatant, exemplified by the activity of *L. plantarum* LPCO 10, were abolished not only by protease treatments but also by glycolytic and lipolytic enzymes (Jimenez-Diaz et al. 1993).

Most recently, bacteriocins were classified mainly into two classes: the lanthionine-containing bacteriocins (lantibiotics) (class I) and the non-lanthionine-containing bacteriocins (class II) (Cotter et al. 2005). The lantibiotics are small peptides (19–38 amino acids in length) and contain posttranslationally modified amino acids such as lanthionine, beta-methyllanthionine, dehydroalanine and dehydrobutyrine. Covalent bridge formation, as a result of these unusual residues, leads to the formation of internal "rings" which give the lantibiotics their characteristic structural features. The class II bacteriocins are also relatively small (<10 kDa) but unlike the class I bacteriocins are not subject to extensive posttranslational modification. Being a rather heterogeneous group, they have been further classified into the class IIa pediocin-like or *Listeria*-active bacteriocins, the class IIb two-peptide bacteriocins (e.g. lactococcin G (Oppegård et al. 2007)), the class IIc cyclic bacteriocins and the class IId linear non-pediocin-like one-peptide bacteriocins (Nissen-Meyer et al. 2009). Pediocin-like bacteriocins are the most important and well-studied group of class II bacteriocins that includes now more than 20 items, including the "classical" member pediocin AcH from *Pediococcus acidilactici* AcH identified in 1991; plantaricin 423, curvacin A, few sakacins and curvacin A named in accordance with the species of lactobacilli in which they were found; lactococcin MMFII from *Lactococcus lactis*; few enterocines from *Enterococcus faecium*; and bifidocin B from *B. bifidum* (Drider et al. 2006; Nissen-Meyer et al. 2009). They are all cationic and partly amphiphilic and/or hydrophobic and have between 37 and 48 residues, and in the N-terminal region (up to about residue 17), they all contain the conserved Y-G-N-G-V/L "pediocin box" motif and two cysteine residues joined by a disulfide bridge. They also contain more hydrophobic C-terminal region (from about residue 18). Pediocin-like bacteriocins are unstructured in aqueous solution but become structured upon contact with membrane of target bacterial cell. The cationic N-terminal β-sheetlike domain mediates binding to the target cell surface through electrostatic interactions and the hydrophobic C-terminal hairpin-like domain penetrates into the hydrophobic core of target membranes, which induces leakage of ions and leads to cell death.

The two-peptide (class IIb) bacteriocins consist of two very different peptides and optimal activity requires both peptides in about equal amounts. Since the first class IIb bacteriocin (lactococcin G) was identified in 1992, at least 15 two-peptide bacteriocins have been isolated and characterised including thermophilin 13 from *S. thermophilus*, lactococcin G from *Lactococcus lactis*, plantaricin E/F from *L. plantarum C11* (Fimland et al. 2008) and lactacin B from *L. acidophilus* (Tabasco et al. 2009). The individual peptides of two-peptide bacteriocins share characteristics with one-peptide bacteriocins in that they are usually cationic, 30–50 residues long, hydrophobic and/or amphiphilic and are all synthesised with a 15–30 residue N-terminal leader sequence that is cleaved before export of the peptides from cells. Interestingly, the two peptides of class IIb bacteriocins function together as one antimicrobial entity. As a rule, both peptides contain GxxxG motifs which allow forming of membrane-penetrating helix–helix structures interacting with integrated membrane proteins, which induces leakage of ions and leads to cell death.

The cyclic bacteriocins (Maqueda et al. 2008) whose N- and C-termini are covalently linked are placed in class IIc (Nissen-Meyer et al. 2009). At least seven cyclic bacteriocins produced by gram-positive bacteria have been characterised, including gassericin A, reutericin 6, acidocins B and D20079 from *L. gasseri* LA39, *L. reuteri*

LA6 and two strains of *L. acidophilus*, respectively. They are all cationic and relatively hydrophobic, and they range in size from 3,400 to 7,200 Da. All cyclic bacteriocins render the target cell membrane permeable to small molecules, which eventually results in cell death.

The linear non-pediocin-like one-peptide bacteriocins are placed in class IId according to the classification proposed by Cotter et al. (Cotter et al. 2005). The extensive group includes at least 30 items (Nissen-Meyer et al. 2009), in particular, enterocins EJ97, B, L50A, L50B, etc. from the strains of *Enterococcus faecalis* and *E. faecium*; acidocins A, 1B, CH5 from the strains of *L. acidophilus*; aureocins A70 and A53 from the strains of *St. aureus*; and so on. The properties of the bacteriocins belonging to the class IId are heterogeneous and it is difficult to draw their general properties. Interestingly, aureocin A53 functions at micromolar concentrations and acts through the membrane disruption rather than formation of target-mediated pores upon binding with high affinity to specific receptors or docking molecules as is typical for class IIa and class IIb bacteriocins that function at nanomolar concentrations.

Note that the bacteria that synthesize bacteriocins have the so-called immunity proteins that associate with membranes within cells, recognise and bind bacteriocin–permease complex and thus prevent cell self-killing.

Generally, bacteriocins of bifidobacteria are under the initial study and a list of known bacteriocins includes bifidin from *B. bifidum*, bifidocin B from *B. bifidum* NCFB 1454, biflong Bb-46 from *B. longum* Bb-46 and biflact Bb-12 from *B. lactis* Bb-12 (Cheikhyoussef et al. 2008). The most investigated is bifidocin B (Cheikhyoussef et al. 2008) that inhibits the growth of selected species of the genera *Listeria*, *Bacillus*, *Enterococcus*, *Lactobacillus*, *Leuconostoc* and *Pediococcus* but was not active against gram-negative bacteria due to its interaction with teichoic acids that are absent in cell wall of the group of bacteria. In the sensitive gram-positive cells, bifidocin B molecules bind to specific or lethal receptor(s) and form pores leading to cell death with or without lysis. Bifidocin B consists of one polypeptide chain of 36 amino acid residues with a molecular mass of 4432.9 Da and shares significant homology with other class IIa LAB bacteriocins. Production of bifidocin B by *B. bifidum* NCFB 1454 was associated with an 8 kb size plasmid which may be used in the construction of food-grade vectors for improvement of bifidobacteria.

Currently bacteriocin-like substances from bifidobacteria are under investigation. For example, six selected *Bifidobacterium* strains produce bacteriocin-like substances that are active against gram-positive and gram-negative bacteria and yeasts (Collado et al. 2005). The substances are active at pH values between 3 and 10, stable at 100 °C for 10 min, resistant to alpha-amylase and lipase A, sensitive to proteinases (trypsin, proteinase K, protease A, pepsin and cathepsin B) and have molecular weighs less than 30 kDa (Collado et al. 2005).

The potential of bacteriocin-producing gut isolates as bioprotective agents against pathogenic bacteria both in vitro and in vivo has been well documented in the literature (Gillor et al. 2008). For example, the ability of a five-strain *Lactobacillus/Pediococcus* combination from porcine intestines has been shown to protect against *Salmonella* infection in a porcine model (Casey et al. 2007). In a mouse model, production of the bacteriocin Abp118 was shown to be responsible for inhibition of *L. monocytogenes* infection (Corr et al. 2007). In a similar type of study, human isolates of *Pediococcus acidilactici* producing pediocin PA-1 and *L. lactis* producing nisin Z were shown to reduce vancomycin-resistant enterococci intestinal colonisation in a mouse model (Millette et al. 2008). In view of the widespread resistance to currently available antibiotic treatments, bacteriocins which inhibit pathogenic bacteria offer a highly favourable alternative. Moreover, bacteriocin-producing microorganisms in the gut may provide a viable mechanism for therapeutic delivery to the site of infection, an approach likely to be more effective than the bacteriocins themselves, which would undoubtedly be broken down during passage through the gastrointestinal tract (Gardiner et al. 2007).

In recent years, interest in the bacteriocins has grown substantially due to their potential usefulness as friendly food biopreservatives either in the form of protective cultures and as additives (Ross et al. 2010; Settanni and Corsetti 2008). Note that fermentation of various foods by LAB is one of the oldest forms of biopreservation practised by mankind. Currently, nisin is the only bacteriocin that is applied as a food additive in European countries and the USA (Delves-Broughton et al. 1996; FAO/WHO 2007; Vandenberg 1993). Nisin prevents clostridial spoilage of processed and natural cheeses, inhibits the growth of some psychrotrophic bacteria in cottage cheese, extends the shelf life of milk, prevents the growth of spoilage lactobacilli in beer and wine fermentations and provides additional protection against spores of *Bacillus* and *Clostridia* in canned foods.

Besides bacteriocins probiotics produce also certain antibiotics. The production of the antibiotic reuterin (3-hydroxypropionaldehyde) by *L. reuteri* strain ATCC55730 has been reported. Reuterin is a broad-spectrum antibiotic active not only against gram-positive and gram-negative bacteria but also against yeast, fungi, protozoa and viruses (Cleusix et al. 2008). Additionally, probiotic bacteria are able to produce so-called deconjugated bile acids that are derivatives of bile salts synthesised by the host, show a stronger antimicrobial activity compared to the bile salts (Kurdi et al. 2006) and are interesting in the context of cholesterol-lowering effects of the bacteria.

9.7 Supercritical Carbon Dioxide (scCO_2) for the Extraction of Polar Lipids from Probiotic Bacteria

Extensive investigations of biologically active compounds from probiotic bacteria are primarily directed at the development of immunostimulants (Sekine et al. 1985). Among these substances glycolipids and phospholipids are the most abundant components in bacterial cell (Poupard et al. 1973). We described a novel isolation procedure for polar lipids from probiotic bacteria with the aid of supercritical fluid extraction (Novik et al. 2006). The extraction of polar lipids from biomass was performed by supercritical carbon dioxide using SFE-2X100F system (Fig. 9.4).

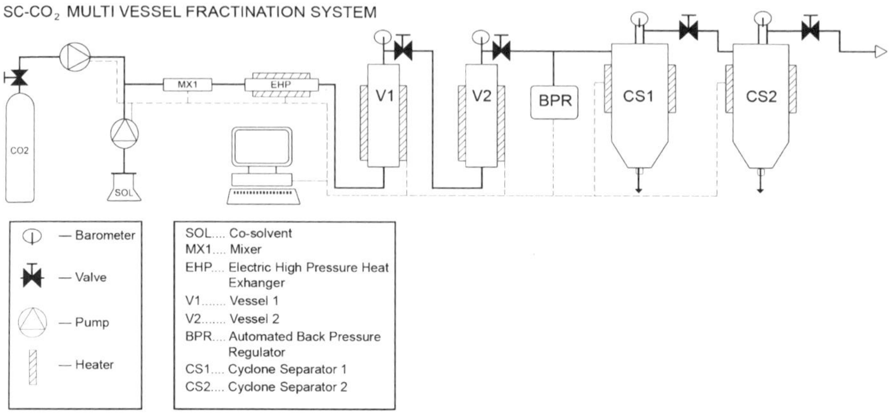

Fig. 9.4 Schematic diagram of the $scCO_2$ extraction system from Thar supercritical fluid extraction system Manual, April 2003, Thar Technologies, INC. 100 Beta Dr. Pittsburgh, PA 15238, USA, section "System Schematic" (Adapted by Estera Szwajcer Dey)

Temperature and flow rate of carbon dioxide are the major factors effecting the polar lipids extraction from probiotic bacteria (Novik et al. 2006). Conditions such as pressure 250 bar, temperature 45 °C, flow rate of CO_2 5 g/min and concentration of co-solvent (methanol/water 9:1) 10 % make possible isolation of major and minor glycolipids, as well as significant amounts of phospholipids (Fig. 9.5). Double or triple amounts of glycolipids and phospholipids in comparison with classical methods were found in the lipid extracts from bifidobacteria and lactobacteria (Izhyk et al. 2012; Novik et al. 2006; Rakhuba et al. 2009). ELISA of SFE lipid fractions from probiotic bacteria showed that the glycolipids are more immunoreactive compared with phospholipids. By employing the modification of $scCO_2$ settings, the high purity polar lipids of probiotic bacteria can be effectively extracted. The $scCO_2$ isolation technology can be combined with metabolic engineering and immunological studies in biotechnologies.

9.8 Brewing Waste as Media for Growth of Probiotic Bacteria

The brewing process involves specially prepared raw materials and yeast. The classical starting raw material is barley which has to be malted. The malting process consists of steeping, germination and drying (kilning). The outcome is malt, which further undergoes mashing to produce wort used for brewing of beer. The leftover solid material after mashing is referred to as "brewers' spent grains (BSG)." The BSG is a waste enriched in proteins and fibres (Ishiwaki et al. 2000; Lasztity 1984).

Modern technologies of probiotic production require not only active strains but low-cost media for their cultivation. Media with protein and carbohydrate

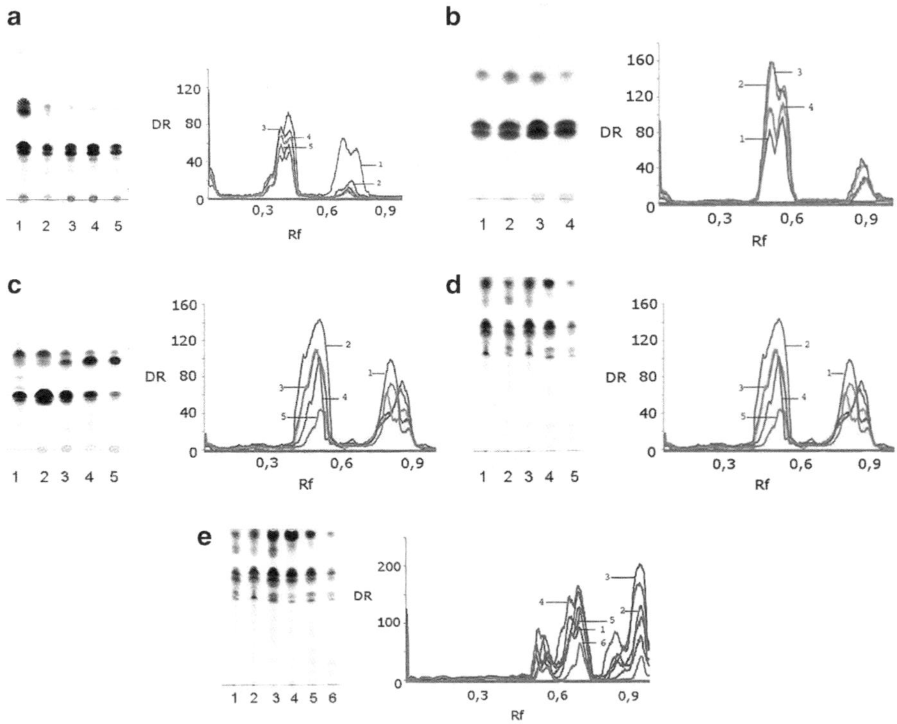

Fig. 9.5 Examples of HPTLC chromatograms of glycolipids and densitometric spectra thereof (*right*) showing lanes from chromatograms. *Bifidobacterium longum* (**a**) *lane 1*—classical method fraction, *2–5 lanes*—SFE fractions, (**b**) *1–4 lanes*—SFE fractions, (**c**) *1–5 lanes*—SFE fractions; *Bifidobacterium angulatum* (**d**) *lane 1*—classical method fraction, *2–5 lanes*—SFE fractions and (**e**) *1–6 lanes*—SFE fractions. *Rf* relative flow, *DR* density response (Izhyk et al. 2012)

components readily utilised by bifidobacteria and LAB allow recycling of a waste product and cultivation of probiotic biomass. Components of growth media act as sources of carbon, nitrogen and phosphorus and can also have beneficial effects on human health. The products of grain processing are known to be efficient substrates for cultivation of LAB and bifidobacteria; they provide for high growth potential, metabolite production, and cell viability during a long-term storage (Bamba et al. 2002; Charalampopoulos et al. 2002b; Kanauchi et al. 2003). These products can act as prebiotics which selectively stimulate growth of LAB and bifidobacteria in the intestines. Cereal grains contain water-soluble carbohydrate polymers (beta-glucan and arabinoxylan), oligosaccharides (galacto- and fructooligosaccharides) and water-insoluble polysaccharides (xylan, cellulose and starch), which presumably act as prebiotics. Moreover, these alimentary fibres can be included in the diet as sources of carbohydrates with multiple beneficial physiological effects (Charalampopoulos et al. 2002a). Recently, food additives containing the species of Lactobacillus and Bifidobacterium and the products of cereal grain processing, which serve as prebiotic components, have appeared in the market.

Germinated barley products and their polysaccharide fractions are known to prevent diarrhoea and enteritis and can be used for prophylaxis of colitis (Bamba et al. 2002). The components of BSG have traditionally been used as additives to ruminant fodder and ingredients of bakery products. The BSG protein and polysaccharide fractions are of growing interest as dietary supplements for treatment of dyspepsia and as alternative sources of protein and carbohydrates (Dongowski et al. 2002; Schrezenmeir and de Vrese 2001). Protein content of BSG varies from 8 to 55 % depending on the initial protein level in the barley and subsequent processes of malting and wort preparation (Crittenden et al. 2001; Das and Singh 2004; Hosono et al. 1997; Ishiwaki et al. 2000; Janer et al. 2005; Kanauchi et al. 1999; Kleyn and Hough 1971; Lasztity 1984; Lauer and Kandler 1976; Shukla 1998; Szponar et al. 2003; Szwajcer Dey et al. 1992). Earlier, we studied application of protein and polysaccharide fractions of BSG as a basis of growth media for probiotic bacteria (Novik et al. 2007). The protein fraction is mainly composed of hydrophobic peptides and proteins showing poor water solubility. It is known that some species of Bifidobacterium and Lactobacillus are able to produce extracellular proteinases, allowing the cells to utilise casein, albumin and some immunoglobulins (Novik et al. 2001). Probiotic bacteria were shown to decompose hardly hydrolysable protein compounds (Janer et al. 2005). These bacteria were characterised by synthesis of proteolytic enzymes which break peptide bonds between an amino acid and proline or glutamic acid in the P1 position. Our results demonstrated production of similar enzymes by the studied LAB and bifidobacteria (Szwajcer Dey et al. 1992). The findings indicated that protein and polysaccharide fractions of BSG can be used as components of media for cultivation of probiotic bacteria (Novik et al. 2007). High values of biomass yield, cell viability and organic acid production were observed in the variants of media containing BSG supplemented with lactose, ascorbic acid, yeast extract and mineral salts. Cells of LAB and bifidobacteria showed the typical rod-shaped morphology. We also recommended protein fraction from BSG as the main component of the media for isolation and cultivation of actinobacteria, production of biologically active substances and intense sporulation (Szponar et al. 2003). These results agree well with the data on the use of media containing protein fraction of BSG for isolation of Xanthomonas sp., which produced proline-specific endopeptidase (Szwajcer Dey et al. 1992). Since both fractions are suitable for bacterial growth, upgraded BSG can also be applied for probiotic production. The fraction containing poorly soluble polysaccharides (alimentary fibres) supplemented with protein from yeast extract may be used as a food additive for prophylaxis of diarrhoea and colitis (Bamba et al. 2002). The BSG media supported active bacterial growth, high biomass accumulation and formation of organic acids, which opens new prospects for the brewing waste in optimising technologies of low-cost probiotics production. Biomass of probiotic bacteria can serve as a promising source of immunostimulating compounds, such as polysaccharides and glycolipids used in vaccine production. A method for isolation of immunostimulating substances using fluidised carbon dioxide has been described (Novik et al. 2002, 2006). The obtained results can be used for development of low-cost manufacturing of an array of biopreparations, including probiotics, prebiotics and vaccines. Moreover, brewing wastes may be an integral part of industrial bioresource recycling scheme (Crittenden et al. 2001; Das and Singh 2004; Hosono et al. 1997; Kanauchi et al. 1999; Kleyn and Hough 1971; Shukla 1998).

9.9 Conclusion

Currently, probiotics are becoming an increasingly important part in the diet of industrialised countries, as their general and gastrointestinal beneficial effects are being gradually proven. Many people over the world have started to take probiotics, and various probiotics have been used in a wide variety of pharmaceutical forms similar to medicines, which are known as nutraceuticals, and food products such as dietary supplements, yogurt and infant formulas. Therefore, now consumers can choose what kind and form of probiotics they prefer. However, different probiotics have distinct properties and effects found in one species or strain of probiotics do not necessarily hold true for others. Thus, it is very important to select the appropriate probiotic strain. It has become necessary to harmonise marketing criteria, evaluate the efficacy of probiotics and correctly define the effective doses.

The safe use of probiotics is an absolutely crucial point. *Lactobacilli* and *Bifidobacteria* are considered as GRAS, although certain doubts have been raised regarding their use at massive doses in immunodepressed patients or in those who undergo intestinal resection due to benign or malignant disease. Therefore, there is a great need for controlled studies in humans to further document the health benefits of probiotics as part of the human diet. Important target groups for such studies include healthy people, elevated disease risk and people for developing a disease and people searching for dietary-management techniques to control symptoms. All these groups would benefit from publicly funded research of probiotics as foods or supplements. Strains of the same probiotic species can be different, which has been demonstrated both in vitro and in animals, although similar data in humans are rare. Thus, clinical results from one study are applicable only to the strain or strains being evaluated.

Taking into account that effects of probiotics are strain specific, strain identity is important to link a strain to a specific health effect as well as to enable accurate surveillance and epidemiological studies. Both phenotypic and genotypic tests using validated standard methodology should be conducted for accurate identification of probiotic bacteria at species and strain level. Nomenclature of the bacteria must conform to the current, scientifically recognised names.

Technological efficiency of probiotics must also be determined, such as the strains ability to be grown to high yields and concentrations, to be stable, both physiologically and genetically, through the end of the shelf life of the product and at the active site in the host.

While there have been numerous health benefits attributed to probiotic lactobacilli and bifidobacteria, some of which have been discussed above, the precise mechanisms by which these bacteria function as a probiotic are yet to be understood.

Additional research is therefore required to confirm a number of these health benefits credited to probiotic bacteria. The recent technological advances in the area of genomics and proteomics are now beginning to provide one important avenue of research along which the role of probiotic bacteria and the molecular mechanisms of probiotic action can be investigated.

References

Adam MR (1999) Safety of industrial lactic acid bacteria. J Biotechnol 68(2–3):171–178. doi:10.1016/S0168-1656(98)00198-9

Alakomi HL, Skytta E, Saarela M et al (2000) Lactic acid permeabilizes gram-negative bacteria by disrupting the outer membrane. Appl Environ Microbiol 66(5):2001–2005. doi:10.1128/AEM.66.5.2001-2005.2000

Altenhoefer A, Oswald S, Sonnenborn U et al (2004) The probiotic *Escherichia coli* strain Nissle 1917 interferes with invasion of human intestinal epithelial cells by different enteroinvasive bacterial pathogens. FEMS Immunol Med Microbiol 40(3):223–229. doi:10.1016/S0928-8244(03)00368-7

Ammor MS, Flórez AB, van Hoek AH et al (2008) Molecular characterization of intrinsic and acquired antibiotic resistance in lactic acid bacteria and bifidobacteria. J Mol Microbiol Biotechnol 14(1–3):6–15. doi:10.1159/000106077

Anderson RC, Cookson AL, McNabb WC et al (2010) *Lactobacillus plantarum* MB452 enhances the function of the intestinal barrier by increasing the expression levels of genes involved in tight junction formation. BMC Microbiol 10:316–326. doi:10.1186/1471-2180-10-316

Antikainen J, Anton L, Sillanpää J, Korhonen TK et al (2002) Domains in the S-layer protein CbsA of *Lactobacillus crispatus* involved in adherence to collagens, laminin and lipoteichoic acidsandinself-assembly.MolMicrobiol46(2):381–394.doi:10.1046/j.1365-2958.2002.03180.x

Antikainen J, Kaparinen V, Korhonen TK (2007a) Enolases from Gram-positive bacterial pathogens and commensal lactobacilli share functional similarity in virulence-associated traits. FEMS Immunol Med Microbiol 51(3):526–534. doi:10.1111/j.1574-695X.2007.00330.x

Antikainen J, Kuparinen V, Lähteenmäki K et al (2007b) pH-dependent association of enolase and glyceraldehydes-3-phosphate dehydrogenase of *Lactobacillus crispatus* with the cell wall and lipoteichoic acids. J Bacteriol 189(12):4539–4543. doi:10.1128/JB.00378-07

Atuma C, Strugala V, Allen A, Holm L (2001) The adherent gastrointestinal mucus gel layer: thickness and physical state in vivo. Am J Physiol Gastrointest Liver Physiol 280(5):G922–G929

Ävall-Jääskeläinen S, Palva A (2005) *Lactobacillus* surface layers and their applications. FEMS Microbiol Rev 29(3):511–529. doi:10.1016/j.femsre.2005.04.003

Ayabe T, Satchell DP, Wilson CL et al (2000) Secretion of microbicidal alpha-defensins by intestinal Paneth cells in response to bacteria. Nat Immunol 1(2):113–118. doi:10.1038/77783

Baccigalupi L, Di Donato A, Parlato M et al (2005) Small surface-associated factors mediate adhesion of a food-isolated strain of *Lactobacillus fermentum* to Caco-2 cells. Res Microbiol 156(7):830–836. doi:10.1016/j.resmic.2005.05.001

Backhead F, Ley RE, Sonnenburg JL, Peterson DA, Gordon JI (2005) Host-bacterial mutualism in the human intestine. Science 307(5717):1915–1919. doi:10.1126/science.1104816

Bamba T, Kanauchi O, Andoh A, Fujiyama Y (2002) A new prebiotic from germinated barley for nutraceutical treatment of ulcerative colitis. J Gastroenterol Hepatol 17(8):818–824. doi:10.1046/j.1440-1746.2002.02709.x

Benchimol EI, Mack DR (2005) Safety issues of probiotic ingestion. Pract Gastroenterol 29(11):23–34

Bergamini CV, Hynes ER, Quiberoni A et al (2005) Probiotic bacteria as adjunct starters: influence of the addition methodology on their survival in a semi-hard Argentinean cheese. Food Res Int 38(5):597–604. doi:10.1016/j.foodres.2004.11.013

Bergonzelli G, Granato D, Pridmore R et al (2006) GroEL of *Lactobacillus johnsonii* La1 (NCC533) is cell surface associated: potential role in interactions with the host and the gastric pathogen *Helicobacter pylori*. Infect Immun 74(1):425–434. doi:10.1128/IAI.74.1.425-434.2006

Bernet MF, Brassart D, Neeser JR et al (1993) Adhesion of human bifidobacterial strains to cultured human intestinal epithelial cells and inhibition of enteropathogen-cell interactions. Appl Environ Microbiol 59(12):4121–4128

Berrada N, Lemeland JE, Laroche G et al (1991) *Bifidobacterium* from fermented milks: survival during gastric transit. J Dairy Sci 74(2):409–413. doi:10.3168/jds.S0022-0302(91)78183-6

Berthier F, Ehrlich SD (1998) Rapid species identification within two groups of closely related lactobacilli using PCR primers that target the 16S/23S rRNA spacer region. FEMS Microbiol Lett 161(1):97–106. doi:10.1111/j.1574-6968.1998.tb12934.x

Betoret E, Betoret N, Vidal D, Fito P (2011) Functional food development: trends and technologies. Trends Food Sci Technol 22(9):498–508. doi:10.1016/j.tifs.2011.05.004

Bevilacqua L, Ovidi M, Di Mattia E et al (2003) Screening of *Bifidobacterium* strains isolated from human faeces for antagonistic activities against potentially bacterial pathogens. Microbiol Res 158(2):179–185. doi:10.1078/0944-5013-00192

Bevins CL (2004) The Paneth cell and the innate immune response. Curr Opin Gastroenterol 20(6):572–580. doi:10.1097/00001574-200411000-00012

Bhadoria PBS, Mahapatra SC (2011) Prospects, technological aspects and limitations of probiotics—a worldwide review. Eur J Food Res Rev 1(2):23–42

Biavati B, Mattarelli P (2006) The family *Bifidobacteriaceae*. In: Dworkin M (ed) The Procariotes: a handbook on the biology of bacteria, vol 3, 3rd edn. Springer, New-York, pp 322–382

Biavati B, Vescovo M, Torriani S, Bottazzi V (2001) Bifidobacteria: history, ecology, physiology and applications. Ann Microbiol 50(2):117–131

Blazquez J, Oliver A, Gomea-Gomez JM (2002) Mutation and evolution of antibiotic resistance: antibiotics as promoters of antibiotic resistance? Curr Drug Targets 3(4):345–349. doi:10.2174/1389450023347579

Borruel N, Carol M, Casellas F et al (2002) Increased mucosal tumour necrosis factor alpha production in Crohn's disease can be downregulated ex vivo by probiotic bacteria. Gut 51(5):659–664. doi:10.1136/gut.51.5.659

Boudeau J, Glasser AL, Julien S et al (2003) Inhibitory effect of probiotic *Escherichia coli* strain Nissle 1917 on adhesion to and invasion of intestinal epithelial cells by adherent-invasive *E. coli* strains isolated from patients with Crohn's disease. Aliment Pharmacol Ther 18(1):45–56. doi:10.1046/j.1365-2036.2003.01638.x

Boyd MA, Antonio MAD, Hillier SL (2005) Comparison of API 50 CH strips to whole chromosomal DNA probes for the identification of *Lactobacillus* species. J Clin Microbiol 43(10):5309–5311. doi:10.1128/JCM.43.10.5309-5311.2005

Buck BL, Altermann E, Svingerud T et al (2005) Functional analysis of putative adhesion factors in *Lactobacillus acidophilus* NCFM. Appl Environ Microbiol 71(12):8344–8351. doi:10.1128/AEM.71.12.8344-8351.2005

Caballero-Franco C, Keller K, De Simone C et al (2007) The VSL#3 probiotic formula induces mucin gene expression and secretion in colonic epithelial cells. Am J Physiol Gastrointest Liver Physiol 292(1):G315–G322

Candela M, Biagi E, Centanni M et al (2009) Bifidobacterial enolase, a cell surface receptor for human plasminogen involved in the interaction with the host. Microbiology 155(10):3294–3303. doi:10.1099/mic.0.028795-0

Candela M, Centanni M, Fiori J et al (2010) DnaK from *Bifidobacterium animalis* subsp. *lactis* is a surface-exposed human plasminogen receptor upregulated in response to bile salts. Microbiology 156(6):1609–1618. doi:10.1099/mic.0.038307-0

Canzi E, Guglielmetti S, Mora D et al (2005) Conditions affecting cell surface properties of human intestinal bifidobacteria. Antonie Van Leeuwenhoek 88(3–4):207–219. doi:10.1007/s10482-005-6501-3

Cario E (2005) Bacterial interactions with cells of the intestinal mucosa: toll-like receptors and NOD2. Gut 54(8):1182–1193. doi:10.1136/gut.2004.062794

Casey PG, Gardiner GE, Casey G et al (2007) A five-strain probiotic combination reduces pathogen shedding and alleviates disease signs in pigs challenged with *Salmonella enterica* Serovar *typhimurium*. Appl Environ Microbiol 73(6):1858–1863. doi:10.1136/gut.2004.062794
Charalampopoulos D, Pandiella SS, Webb C (2002a) Growth studies of potentially probiotic lactic acid bacteria in cereal-based substrates. J Appl Microbiol 92(5):851–859. doi:10.1046/j.1365-2672.2002.01592.x
Charalampopoulos D, Wang R, Pandiella SS, Webb C (2002b) Application of cereals and cereal components in functional foods: a review. Int J Food Microbiol 79(1–2):131–141
Charteris WP, Kelly PM, Morelli L, Collins JK (1998) Antibiotic susceptibility of potentially probiotic *Bifidobacterium* isolates from the human gastrointestinal tract. Lett Appl Microbiol 26(5):333–337. doi:10.1046/j.1472-765X.1998.00342.x
Cheikhyoussef A, Pogori N, Chen W et al (2008) Antimicrobial proteinaceous compounds obtained from bifidobacteria: from production to their application. Int J Food Microbiol 125(3):215–222. doi:10.1016/j.ijfoodmicro.2008.03.012
Chen X, Xu J, Shuai J et al (2007) The S-layer proteins of *Lactobacillus crispatus* strain ZJ001 is responsible for competitive exclusion against *Escherichia coli* O157:H7 and *Salmonella typhimurium*. Int J Food Microbiol 115(3):307–312. doi:10.1016/j.ijfoodmicro.2006.11.007
Chen X, Tian F, Liu X et al (2010) In vitro screening of lactobacilli with antagonistic activity against *Helicobacter pylori* from traditionally fermented foods. J Dairy Sci 93(12):5627–5634. doi:10.3168/jds.2010-3449
Chiba H, Osanai M, Murata M et al (2008) Transmembrane proteins of tight junctions. Biochim Biophys Acta 1778(3):588–600
Chin AC, Teoh DA, Scott KGE et al (2002) Strain-dependent induction of enterocyte apoptosis by *Giardia lamblia* disrupts epithelial barrier function in a caspase-3-dependent manner. Infect Immun 70(7):3673–3680. doi:10.1128/IAI.70.7.3673-3680.2002
Cleusix V, Lacroix C, Vollenweider S et al (2008) Glycerol induces reuterin production and decreases *Escherichia coli* population in an in vitro model of colonic fermentation with immobilized human feces. FEMS Microbiol Ecol 63(1):56–64. doi:10.1111/j.1574-6941.2007.00412.x
Coconnier MH, Lievin V, Lorrot M, Servin AL (2000) Antagonistic activity of *Lactobacillus acidophilus* LB against intracellular *Salmonella enterica* serovar *typhimurium* infecting human enterocyte-like Caco-2/TC-7 cells. Appl Environ Microbiol 66(3):1152–1157. doi:10.1128/AEM.66.3.1152-1157.2000
Collado MC, Sanz Y (2006) Method for direct selection of potentially probiotic *Bifidobacterium* strains from human feces based on their acid-adaptation ability. J Microbiol Methods 66(3):560–563. doi:10.1016/j.mimet.2006.01.007
Collado MC, Hernández M, Sanz Y (2005) Production of bacteriocin-like inhibitory compounds by human fecal *Bifidobacterium* strains. J Food Prot 68(5):1034–1040
Collado MC, Meriluoto J, Sa S (2007) Role of commercial probiotic strains against human pathogen adhesion to intestinal mucus. Lett Appl Microbiol 45(4):454–460. doi:10.1111/j.1472-765X.2007.02212.x
Corr SC, Li Y, Riedel CU et al (2007) Bacteriocin production as a mechanism for the antiinfective activity of *Lactobacillus salivarius* UCC118. Proc Natl Acad Sci U S A 104(18):7617–7621. doi:10.1073/pnas.0700440104
Cotter PD, Hill C, Ross RP (2005) Bacteriocins: developing innate immunity for food. Nat Rev Microbiol 3(10):777–788. doi:10.1038/nrmicro1273
Crittenden R (2004) An update on probiotic bifidobacteria. In: Salminen S, Von Wright A, Ouwehand A (eds) Lactic acid bacteria: microbiological and functional aspects, 3rd edn. Marcel Dekker, New York, pp 125–157
Crittenden R, Laitila A, Forssell P et al (2001) Adhesion of bifidobacteria to granular starch and its implications in probiotic technologies. Appl Environ Microbiol 67:3469–3475. doi:10.1128/AEM.67.8.3469-3475.2001
Cunliffe RN (2003) Alpha-defensins in the gastrointestinal tract. Mol Immunol 40(7):463–467
Danielsen M (2002) Characterization of the tetracycline resistance plasmid pMD5057 from *Lactobacillus plantarum* 5057 reveals a composite structure. Plasmid 48(2):98–103. doi:10.1016/S0147-619X(02)00118-X

Das H, Singh SK (2004) Useful byproducts from cellulosic wastes of agriculture and food industry—a critical appraisal. Crit Rev Food Sci Nutr 44(2):77–89. doi:10.1080/10408690490424630

De Keersmaecker SCJ, Verhoeven TLA, Desair J et al (2006) Strong antimicrobial activity of *Lactobacillus rhamnosus* GG against *Salmonella typhimurium* is due to accumulation of lactic acid. FEMS Microbiol Lett 259(1):89–96. doi:10.1111/j.1574-6968.2006.00250.x

De Las Rivas B, Marcobal A, Munoz R (2006) Development of a multilocus sequence typing meted for analysis of Lactobacillus plantarum strains. Microbiology 52:85–93. doi:10.1099/mic.0.28482-0

De Leeuw E, Li X, Lu W (2006) Binding characteristics of the *Lactobacillus brevis* ATCC 8287 surface layer to extracellular matrix proteins. FEMS Microbiol Lett 260(2):210–215. doi:10.1111/j.1574-6968.2006.00313.x

De Vrese M, Schrezenmeir J (2008) Probiotics, prebiotics, and synbiotics. Adv Biochem Eng Biotechnol 111:1–66. doi:10.1007/10_2008_097

De Vries W, Stouthamer AH (1969) Factors determining the degree of anaerobiosis of *Bifidobacterium* strains. Arch Microbiol 65(3):275–287. doi:10.1007/BF00407109

De Vuyst L (2000) Application of functional starter cultures. Food Technol Biotech 38(2):105–112

Delves-Broughton J, Blackburn P, Evans RJ et al (1996) Applications of the bacteriocin, nisin. Antonie Van Leeuwenhoek 69(2):193–202. doi:10.1007/BF00399424

Diancourt L, Passet V, Chervaux C et al (2007) Multilocus sequence typing of *Lactobacillus casei* reveals a clonal population structure with low levels of homologous recombination. Appl Environ Microbiol 73(20):6601–6611. doi:10.1128/AEM.01095-07

Doleyres Y, Lacroix C (2005) Technologies with free and immobilized cells for probiotic bifidobacteria production and protection. Int Dairy J 15(10):973–988. doi:10.1016/j.idairyj.2004.11.014

Donato KA, Gareau MG, Wang YJ et al (2010) *Lactobacillus rhamnosus* GG attenuates interferon-γ and tumour necrosis factor-α-induced barrier dysfunction and pro-inflammatory signaling. Microbiology 156(11):3288–3297. doi:10.1099/mic.0.040139-0

Dong X, Xin Y, Jian W et al (2000) *Bifidobacterium thermacidophilum* sp. nov., isolated from an anaerobic digester. Int J Syst Evol Microbiol 50(1):119–125. doi:10.1099/00207713-50-1-119

Dongowski G, Huth M, Gebhardt E, Flamme W (2002) Dietary fiber-rich barley products beneficially affect the intestinal tract of rats. J Nutr 132(12):3704–3714

Drider D, Fimland G, Hechard Y et al (2006) The continuing story of class IIa bacteriocins. Microbiol Mol Biol Rev 70(2):564–582. doi:10.1128/MMBR.00016-05

Du Plessis EM, Dicks LM (1995) Evaluation of random amplified polymorphic DNA (RAPD)-PCR as a method to differentiate *Lactobacillus acidophilus, Lactobacillus crispatus, Lactobacillus amylovorus, Lactobacillus gallinarum, Lactobacillus gasseri*, and *Lactobacillus johnsonii*. Curr Microbiol 31(2):114–118. doi:10.1007/BF00294286

Dunne C, Murphy L, Flynn S et al (1999) Probiotics: from myth to reality. Demonstration of functionality on animal models of disease and in human clinical trials. Antonie Van Leeuwenhoek 76(1–4):279–292

Ewaschuk JB, Diaz H, Meddings L et al (2008) Secreted bioactive factors from *Bifidobacterium infantis* enhance epithelial cell barrier function. Am J Physiol Gastrointest Liver Physiol 295(5):G1025–G1034. doi:10.1152/ajpgi.90227.2008

Fanning AS, Jameson BJ, Jesaitis LA et al (1998) The tight junction protein ZO-1 establishes a link between the transmembrane protein occludin and the actin cytoskeleton. J Biol Chem 273(45):29745–29753. doi:10.1074/jbc.273.45.29745

FAO/WHO Food and Agriculture Organization of the United Nations/World Health Organization (2001) Health and nutritional properties of probiotics in food including powder milk with live lactic acid bacteria. In: Report of a joint FAO/WHO expert consultation on evaluation of health and nutritional properties of probiotics in food including powder milk with live lactic acid bacteria, cyrdoba, 1–4 October 2001. http://www.who.int/foodsafety/publications/fs_management/en/probiotics.pdf

FAO/WHO Food and Agriculture Organization of the United Nations/World Health Organization (2002) Guidelines for the evaluation of probiotics in food. In: Report of a joint FAO/WHO working group on drafting guidelines for the evaluation of probiotics in food, London, Ontario, Canada, 30 April–1 May, 2002. http://www.fda.gov/ohrms/dockets/dockets/95s0316/95s-0316-rpt0282-tab-03-ref-19-joint-faowho-vol219.pdf

FAO/WHO Food and Agriculture Organization of the United Nations/World Health Organization (2007) Compendium of food additive specifications. In: FAO JECFA monographs. 68th meeting of joint FAO/WHO Expert Committee on Food Additives, Rome, 2007. http://www.fao.org/docrep/014/i2358e/i2358e00.pdf

Farquhar MG, Palade GE (1963) Junctional complexes in various epithelia. J Cell Biol 17(2):375–412

Fasoli S, Marzotto M, Rizzotti L et al (2003) Bacterial composition of commercial probiotics products as evaluated by PCR-DGGE analysis. Int J Food Microbiol 82(1):59–70. doi:10.1016/S0168-1605(02)00259-3

Felis E, Dellaglio F (2007) Taxonomy of lactobacilli and bifidobacteria. Curr Issues Intest Microb 8(2):44–61

Felis GE, Dellaglio L, Mizzi L, Torriani S (2001) Comparative sequence analysis of *recA* gene fragment brings new evidence for a change in the taxonomy of the *Lactobacillus casei* group. Int J Syst Evol Microbiol 51(6):2113–2117

Fernandez MF, Boris S, Barbes C (2003) Probiotic properties of human lactobacilli strains to be used in the gastrointestinal tract. J Appl Microbiol 94(3):449–455. doi:10.1046/j.1365-2672.2003.01850.x

Fimland N, Rogne P, Fimland G et al (2008) Three-dimensional structure of the two peptides that constitute the two-peptide bacteriocin plantaricin EF. Biochim Biophys Acta 1784(11): 1711–1719. doi:10.1016/j.bbapap.2008.05.003

Fons M, Hege T, Ladire M et al (1997) Isolation and characterization of a plasmid from *Lactobacillus fermentum* conferring erythromycin resistance. Plasmid 37(3):199–203. doi:10.1006/plas.1997.1290, DOI:10.1006%2fplas.1997.1290

Fox GE, Wisotzkey JD, Jurtshuk PJ (1992) How close is close: 16S rRNA sequence identity may not be sufficient to guarantee species identity. Int J Syst Bacteriol 42(1):166–170. doi:10.1099/00207713-42-1-166

Fujita H, Eish Y, Ishige I et al (2002) Quantitative analysis of bacterial DNA from *Mycobacteria* spp., *Bacteroides vulgatus*, and *Escherichia coli* in tissue samples from patients with inflammatory bowel diseases. J Gastroenterol 37(7):509–516. doi:10.1007/s005350200079

Fukushima K, Sasaki I, Ogawa H et al (1999) Colonization of microflora in mice: mucosal defense against luminal bacteria. J Gastroenterol 34(1):54–60. doi:10.1007/s005350050216

Fuller R (1989) Probiotics in man and animals. J Appl Bacteriol 66(5):365–378. doi:10.1111/j.1365-2672.1989.tb05105.x

Galdeano CM, de Moreno de LeBlanc A, Vinderola G et al (2007) Proposed model: mechanisms of immunomodulation induced by probiotic bacteria. Clin Vaccine Immunol 14(5):485–492. doi:10.1128/CVI.00406-06

Gardiner GE, Rea MC, O'Riordan B et al (2007) Fate of the two-component lantibiotic lacticin 3147 in the gastrointestinal tract. Appl Environ Microbiol 73(21):7103–7109. doi:10.1128/AEM.01117-07

Gaudier E, Michel C, Segain JP et al (2005) The VSL#3 probiotic mixture modifies microflora but does not heal chronic dextran sodium sulfate-induced colitis or reinforce the mucus barrier in mice. J Nutr 135(12):2753–2761

Gevers D, Danielson M, Huys G, Swings J (2002) Molecular characterization of *tet (M)* genes in *Lactobacillus* isolates from different types of fermented dry sausage. Appl Environ Microbiol 69(2):1270–1275. doi:10.1128/AEM.69.2.1270-1275.2003

Gevers D, Huys G, Swings J (2001) Applicability of rep-PCR fingerprinting for identification of *Lactobacillus* species. FEMS Microbiol Lett 205(1):31–36. doi:10.1111/j.1574-6968.2001.tb10921.x

Ghadimi D, Vrese MD, Heller KJ et al (2010) Effect of natural commensal-origin DNA on toll-like receptor 9 (TLR9) signaling cascade, chemokine IL-8 expression, and barrier integrity of polarized intestinal epithelial cells. Inflamm Bowel Dis 16(3):410–427. doi:10.1002/ibd.21057

Gibson GR, Fuller R (2000) Aspects of in vitro and in vivo research approaches directed toward identifying probiotics and prebiotics for human use. J Nutr 130:391–395

Gillor O, Etzion A, Riley MA (2008) The dual role of bacteriocins as anti- and probiotics. Appl Microbiol Biotechnol 81(4):591–606. doi:10.1007/s00253-008-1726-5

Gleinser M, Grimm V, Zhurina D et al (2012) Improved adhesive properties of recombinant bifidobacteria expressing the *Bifidobacterium bifidum*-specific lipoprotein BopA. Microb Cell Fact 11(1):80–94. doi:10.1186/1475-2859-11-80

Gokavi S, Zhang LW, Huang MK et al (2005) Oat-based symbiotic beverage fermented by *Lactobacillus plantarum*, *Lactobacilus paracasei* ssp. *casei*, and *Lactobacillus acidophilus*. J Food Sci 70(4):M216–M223. doi:10.1111/j.1365-2621.2005.tb07191.x

Gomes AM, Malcata FX (1999) *Bifidobacterium* spp. and *Lactobacillus acidophilus*: biological, biochemical, technological and therapeutical properties relevant for use as probiotics. Trends Food Sci Technol 10(4):139–157. doi:10.1016/S0924-2244(99)00033-3l

Grabig A, Paclik D, Guzy C et al (2006) *Escherichia coli* strain Nissle 1917 ameliorates experimental colitis via Toll-like receptor 2- and Toll-like receptor 4- dependent pathways. Infect Immun 74(7):4075–4082. doi:10.1128/IAI.01449-05

Granato D, Bergonzelli GE, Pridmore RD et al (2004) Cell surface-associated elongation factor Tu mediates the attachment of *Lactobacillus johnsonii* NCC533 (La1) to human intestinal cells and mucins. Infect Immun 72(4):2160–2169. doi:10.1128/IAI.72.4.2160-2169.2004

Gueimonde M, Flórez AB, van los Hoek AHAM et al (2010) Genetic basis of tetracycline resistance in Bifidobacterium animalis subsp. lactis. Appl Environ Microbiol 76(10):3364–3369. doi:10.1128/AEM.03096-09

Gueimonde M, Sanchez B (2012) Enhancing probiotic stability in industrial processes. Microb Ecol Health Dis 23:18562–18565

Guglielmetti S, Tamagnini I, Minuzzo M et al (2009) Study of the adhesion of *Bifidobacterium bifidum* MIMBb75 to human intestinal cell lines. Curr Microbiol 59(2):167–172. doi:10.1007/s00284-009-9415-x

Gupta P, Andrew H, Kirschner BS et al (2000) Is Lactobacillus GG helpful in children with Crohn's disease? Results of a preliminary, open-label study. J Pediatr Gastroenterol Nutr 31(4):453–457. doi:10.1097/00005176-200010000-00024

Hammes WP, Hertel C (2003) The genera Lactobacillus and Carnobacterium. In: Dworkin M (ed) Procariotes, vol 4, 3rd edn. Springer, New-York, pp 320–403

Hammes WP, Vogel RF (1995) The genus *Lactobacillus*. In: Wood BJB, Holzapfel WH (eds) The genera of lactic acid bacteria, vol 2. Blackie Academic and Professional, Glasgow, pp 19–54

Hanada M, Feng J, Hemmings BA (2004) Structure, regulation and function of PKB/AKT—a major therapeutic target. Biochim Biophys Acta 1697(1):3–16. doi:10.1016/j.bbapap.2003.11.009

Hart AL, Lammers K, Brigidi P et al (2004) Modulation of human dendritic cell phenotype and function by probiotic bacteria. Gut 53(11):1602–1609. doi:10.1136/gut.2003.037325

Heinemann C, Vlieg JETV, Janssen DB et al (2000) Purification and characterization of a surface-binding protein from *Lactobacillus fermentum* RC-14 that inhibits adhesion of *Enterococcus faecalis* 1131. FEMS Microbiol Lett 190(1):177–180. doi:10.1016/S0378-1097(00)00331-1

Helander A, Hansson GC, Svennerholm AM (1997) Binding of enterotoxigenic *Escherichia coli* to isolated enterocytes and intestinal mucus. Microb Pathog 23(6):335–346. doi:10.1006/mpat.1997.0163

Holzapfel WH, Haberer P, Geisen R et al (2001) Taxonomy and important features of probiotic microorganisms in food and nutrition. Am J Clin Nutr 73(2):365S–373S

Hooper LV, Stappenbeck TS, Hong CV et al (2003) Angiogenins: a new class of microbicidal proteins involved in innate immunity. Nat Immunol 4(3):269–273. doi:10.1038/ni888

Hooper LV, Wong MH, Thelin A et al (2001) Molecular analysis of commensal host–microbial relationships in the intestine. Science 291(5505):881–884. doi:10.1126/science.291.5505.881l

Hosono A, Lee J, Ametani A et al (1997) Characterization of a water-soluble polysaccharide fraction with immunopotentiating activity from *Bifidobacterium adolescentis* M101-4. Biosci Biotechnol Biochem 61(2):312–331

Hurmalainen V, Edelman S, Antikainen J et al (2007) Extracellular proteins of *Lactobacillus crispatus* enhance activation of human plasminogen. Microbiology 153(4):1112–1122. doi:10.1099/mic.0.2006/000901-0

Ishikawa H, Akedo I, Umesaki Y et al (2003) Randomized controlled trial of the effect of bifidobacteria-fermented milk on ulcerative colitis. J Am Coll Nutr 22(1):56–63. doi:10.1016/S0016-5085(00)85259-2

Ishiwaki N, Muruyama H, Awayama H et al (2000) Development of high value uses of spent grain by fractionation technology. MBAA Tech Quar 37(2):261–265

Izhyk A, Novik G, Szwajcer Dey E (2012) Extraction of polar lipids from bifidobacteria by supercritical carbon dioxide ($scCO_2$). J Supercrit Fluids 62:149–154. doi:10.1016/j.supflu.2011.10.013

Janer C, Arigoni F, Lee BH et al (2005) Enzymatic ability of *Bifidobacterium animalis* subsp. *lactis* to hydrolyze milk proteins: identification and characterization of endopeptidase O. Appl Environ Microbiol 71(12):8460–8465. doi:10.1128/AEM.71.12.8460-8465.2005

Jeffery CJ (2009) Moonlighting proteins—an update. Mol Biosyst 5(4):345–350. doi:10.1039/b900658n

Jian W, Zhu L, Dong X (1991) New approach to phylogenetic analysis of the genus *Bifidobacterium* based on partial HSP60 gene sequences. Int J Syst Bacteriol 41:548–557

Jimenez-Diaz R, Rios-Sanchez RM, Desmazeaud M et al (1993) Plantaricin S and T, two new bacteriocins produced by Lactobacillus plantarum LPCO 10 isolated from a green olive fermentation. Appl Environ Microbiol 59(5):1416–1424

Johnson-Henry KC, Hagen KE, Gordonpour M et al (2007) Surface-layer protein extracts from *Lactobacillus helveticus* inhibit enterohaemorrhagic *Escherichia coli* O157: H7 adhesion to epithelial cells. Cell Microbiol 9(2):356–367. doi:10.1111/j.1462-5822.2006.00791.x

Jönsson K, Guo B, Monstein H-J et al (2004) Molecular cloning and characterization of two *Helicobacter pylori* genes coding for plasminogen-binding proteins. Proc Natl Acad Sci USA 101(7):1852–1857. doi:10.1073/pnas.0307329101

Kainulainen V, Loimaranta V, Pekkala A et al (2012) Glutamine synthetase and glucose-6-phosphate isomerase are adhesive moonlighting proteins of *Lactobacillus crispatus* released by cathelicidin LL-37. J Bacteriol 194(10):2509–2519. doi:10.1128/JB.06704-11

Kanauchi O, Fujiyama Y, Mitsuyama K et al (1999) Increased growth of *Bifidobacterium* and *Eubacterium* by germinated barley foodstuff, accompanied by enhanced butyrate production in healthy volunteers. Int J Mol Med 3(2):175–179

Kanauchi O, Serizawa I, Araki Y et al (2003) Germinated barley foodstuff, a prebiotic product, ameliorates inflammation of colitis through modulation of the enteric environment. J Gastroenterol 38(2):134–141

Kandler O, Weiss N (1986) Genus *Lactobacillus* Beijerinck 1901, 212AL. In: Holt JG (ed) Bergey's manual of systematic bacteriology, vol 2. Williams and Wilkins, Baltimore, pp 1209–1234

Kankainen M, Paulin L, Tynkkynen S et al (2009) Comparative genomic analysis of *Lactobacillus rhamnosus* GG reveals pili containing a human-mucus binding protein. Proc Natl Acad Sci USA 106(40):17193–17198. doi:10.1073/pnas.0908876106

Karczewski J, Troost FJ, Konings I et al (2010) Regulation of human epithelial tight junction proteins by *Lactobacillus plantarum* in vivo and protective effects on the epithelial barrier. Am J Physiol Gastrointest Liver Physiol 298(6):G851–G859. doi:10.1152/ajpgi.00327.2009

Kaufmann P, Pfefferkorn A, Teuber M, Meile L (1997) Identification and quantification of *Bifidobacterium* species isolated from food with genus-specific 16S rRNA-targeted probes by colony hybridization and PCR. Appl Environ Microbiol 63(4):1268–1273

Khaled DAK, Neilan BA, Henriksson A, Conway PL (1997) Identification and phylogenetic analysis of *Lactobacillus* using multiplex RAPD-PCR. FEMS Microbiol Lett 153(1):191–197. doi:10.1016/S0378-1097(97)00260-7

Kimura K, Ito M, Amano M et al (1996) Regulation of myosin phosphatase by Rho and Rho-associated kinase (Rho-kinase). Science 273(5272):245–248. doi:10.1126/science.273.5272.245

Kinoshita H, Wakahara N, Watanabe M et al (2008) Cell surface glyceraldehyde-3-phosphate dehydrogenase (GAPDH) of *Lactobacillus plantarum* LA318 recognizes human A and B blood group antigens. Res Microbiol 159(9):685–691. doi:10.1016/j.resmic.2008.07.005

Klaenhammer TR (1993) Genetics of bacteriocins produced by lactic acid bacteria. FEMS Microbiol Rev 12(1–3):39–85. doi:10.1016/0168-6445(93)90057-G

Kleyn J, Hough J (1971) The microbiology of brewing. Annu Rev Microbiol 25:583–608. doi:10.1146/annurev.mi.25.100171.003055

Kok RG, de Waal A, Schut F et al (1996) Specific detection and analysis of a probiotic *Bifidobacterium* strain in infant faeces. Appl Environ Microbiol 62:3668–3672

Koo OK, Amalaradjou MAR, Bhunia AK (2012) Recombinant probiotic expressing *Listeria* adhesion protein attenuates *Listeria monocytogenes* virulence in vitro. PLoS One 7(1):e29277. doi:10.1371/journal.pone.0029277

Korshunov VM, Urtaeva ZA, Smeianov VV et al (1999) The antagonistic activity of bifidobacteria in vitro and in vivo studied by using gnotobiological technology. Zh Mikrobiol Epidemiol Immunobiol 5:72–77

Kotlowski R, Bernstein CN, Sepehri S et al (2007) High prevalence of *Escherichia coli* belonging to the B2+D phylogenetic group in inflammatory bowel disease. Gut 56(5):669–675. doi:10.1136/gut.2006.099796

Krizova J, Spanova A, Rittich B (2008) RAPD and rep-PCR fingerprinting for characterization of *Bifidobacterium* species. Folia Microbiol 53(2):99–104. doi:10.1007/s12223-008-0014-1

Kuhbacher T, Ott SJ, Helwig U et al (2006) Bacterial and fungal microbiota in relation to probiotic therapy (VSL#3) in pouchitis. Gut 55(6):833–841. doi:10.1136/gut.2005.078303

Kullen MJ, Linda JB, O'Sullivan DJ (1997) Evaluation of using a short region of the *recA* gene for the rapid and sensitive speciation of dominant bifidobacteria in the human large intestine. FEMS Microbiol Lett 154(2):377–383. doi:10.1016/S0378-1097(97)00356-X, DOI:10.1016%2fS0378-1097(97)00356-X

Kurdi P, Kawanishi K, Mizutani K (2006) Mechanism of growth inhibition by free bile acids in lactobacilli and bifidobacteria. J Bacteriol 188(5):1979–1986. doi:10.1128/JB.188.5.1979-1986.2006

Kwon HS, Yang EH, Yeon SW et al (2004) Rapid identification of probiotic *Lactobacillus* species by multiplex PCR using species-specific primers based on the region extending from16S rRNA through 23S rRNA. FEMS Microbiol Lett 239(2):267–275. doi:10.1016/j.femsle.2004.08.049

Lähteenmaki K, Edelman S, Korhonen TK (2005) Bacterial metastasis: the host plasminogen system in bacterial invasion. Trends Microbiol 13(2):79–85. doi:10.1016/j.tim.2004.12.003

Langendijk PS, Schuts F, Jansen GJ et al (1995) Quantitative fluorescence in situ hybridization of *Bifidobacterium* spp. with genus-specific 16S rRNA-targeted probes and its application in fecal samples. Appl Environ Microbiol 61(8):3069–3075

Larsen N, Michaelsen KF, Pærregaard A et al (2009) A comparative study on adhesion and recovery of potential probiotic strains of *Lactobacillus* spp. by in vitro assay and analysis of human colon biopsies. Microbiol Health Dis 21(2):95–99. doi:10.1080/08910600902907632

Lasztity R (1984) The chemistry of cereal proteins. CRC Press, Boca Raton

Lauer E, Kandler O (1976) Mechanismus der variation des verhalthisses acetat/lactat bei der vergarung von glucose durch bifidobacterien. Arch Microbiol 110(2–3):271–277. doi:10.1007/BF00690238

Leahy SC, Higgins DG, Fitzgerald GF, van Sinderen D (2005) Getting better with bifidobacteria. J Appl Microbiol 98(6):1303–1315. doi:10.1111/j.1365-2672.2005.02600.x

Lebeer S, Vanderleyden J, De Keersmaecker SC (2008) Genes and molecules of lactobacilli supporting probiotic action. Microbiol Mol Biol Rev 72(4):728–764. doi:10.1128/MMBR.00017-08

Leblond-Bourget N, Philippe H, Mangin I, Decaris B (1996) 16S rRNA and 16S to 23S internal transcribed spacer sequence analysis revealed inter- and intraspecific *Bifidobacterium* phylogeny. Int J Syst Bacteriol 46(1):102–111

Li Q, Chen Q, Hui R et al (2010) Isolation and characterisation of an oxygen, acid and bile resistant *Bifidobacterium animalis* subsp. *lactis* Qq08. J Sci Food Agric 90(8):1340–1346. doi:10.1002/jsfa.3942

Lillehoj EP, Hyun SW, Kim BT et al (2001) Muc1 mucins on the cell surface are adhesion sites for *Pseudomonas aeruginosa*. Am J Physiol Lung Cell Mol Physiol 280(1):L181–L187

Lin CK, Tsai HC, Lin PP et al (2008) *Lactobacillus acidophilus* LAP5 able to inhibit the *Salmonella choleraesuis* invasion to the human Caco-2 epithelial cell. Anaerobe 14(5):251–255. doi:10.1016/j.anaerobe.2008.07.003

Ljungh Å, Wadström T (2009) *Lactobacillus* molecular biology: from genomics to probiotics. Caister Academic Press, Norfolk

Lupp C, Finlay BB (2005) Intestinal microbiota. Curr Biol 15(7):R235–R236. doi:10.1016/j.cub.2005.03.032

Macías-Rodríguez ME, Zagorec M, Ascencio F et al (2009) *Lactobacillus fermentum* BCS87 expresses mucus- and mucin-binding proteins on the cell surface. J Appl Microbiol 107(6):1866–1874. doi:10.1111/j.1365-2672.2009.04368.x

Mack DR, Michail S, Wei L et al (1999) Probiotics inhibit enteropathogenic *E. coli* adherence in vitro by inducing intestinal mucin gene expression. Am J Physiol 276(4):G941–G950

Mack DR, Ahrne S, Hyde L et al (2003) Extracellular MUC3 mucin secretion follows adherence of *Lactobacillus* strains to intestinal epithelial cells in vitro. Gut 52(6):827–833. doi:10.1136/gut.52.6.827

Madsen K, Cornish A, Soper P et al (2001) Probiotic bacteria enhance murine and human intestinal epithelial barrier function. Gastroenterology 121(3):580–591. doi:10.1053/gast.2001.27224

Makras L, Triantafyllou V, Fayol-Messaoudi D et al (2006) Kinetic analysis of the antibacterial activity of probiotic lactobacilli towards *Salmonella enterica* serovar *typhimurium* reveals a role for lactic acid and other inhibitory compounds. Res Microbiol 157(3):241–247. doi:10.1016/j.resmic.2005.09.002

Maqueda M, Sánchez-Hidalgo M, Fernández M et al (2008) Genetic features of circular bacteriocins produced by Gram-positive bacteria. FEMS Microbiol Rev 32(1):2–22. doi:10.1111/j.1574-6976.2007.00087.x

Marteau P, Gerhardt MF, Myara A et al (1995) Metabolism of bile salts by alimentary bacteria during transit in the human small intestine. Microb Ecol Health Dis 8(4):151–157. doi:10.3109/08910609509140093

Marthur S, Singh R (2005) Antibiotic resistance in food lactic acid bacteria—a review. Int J Food Microbiol 105(3):281–295. doi:10.1016/j.ijfoodmicro.2005.03.008

Masco L, Huys G, Gevers D et al (2003) Identification of *Bifidobacterium* species using rep-PCR fingerprinting. Syst Appl Microbiol 26(4):557–563. doi:10.1078/072320203770865864

Masco L, Ventura M, Zink R et al (2004) Polyphasic taxonomic analysis of *Bifidobacterium animalis* and *Bifidobacterium lactis* reveals relatedness at the subspecies level: reclassification of *Bifidobacterium animalis* as *Bifidobacterium animalis* subsp. *animalis* subsp. nov. and *Bifidobacterium lactis* as *Bifidobacterium animalis* subsp. *lactis* subsp. nov. Int J Syst Evol Microbiol 54(4):1137–1143

Masco L, Huys G, De Brandt E et al (2005) Culture-dependent and culture-independent qualitative analysis of probiotic products claimed to contain bifidobacteria. Int J Food Microbiol 102(2):221–230. doi:10.1016/j.ijfoodmicro.2004.11.018

Matsuki T, Watanabe K, Tanaka R et al (1999) Distribution of bifidobacterial species in human intestinal microflora examined with 16S rRNA-gene-targeted species-specific primers. Appl Environ Microbiol 65:4506–4512

Matsuki T, Watanabe K, Tanaka R (2003) Genus- and species-specific PCR primers for the detection and identification of bifidobacteria. Curr Issues Intest Microb 4(2):61–69

Matsuo K, Ota H, Akamatsu T et al (1997) Histochemistry of the surface mucous gel layer of the human colon. Gut 40(6):782–789. doi:10.1136/gut.40.6.782

Määttö J, Malinen E, Suihko ML et al (2004) Genetic heterogeneity and functional properties of intestinal bifidobacteria. J Appl Microbiol 97(3):459–470. doi:10.1111/j.1365-2672.2004.02340.x

Määttö J, van Hoek AHAM, Domig KJ et al (2007) Susceptibility of human and probiotic *Bifidobacterium* spp. to selected antibiotics as determined by the E-test method. Int Dairy J 17(9):1123–1131. doi:10.1016/j.idairyj.2007.01.008

Mayrhofer S, Domig KJ, Amtmann E et al (2007) Antibiotic susceptibility of *Bifidobacterium thermophilum* and *Bifidobacterium pseudolongum* isolates from animal sources. J Food Prot 70(7):119–124

McCartney AL, Wang W, Tannock GW (1996) Molecular analysis of the composition of bifidobacterial and lactobacillus microflora of humans. Appl Environ Microbiol 62(12):4608–4613

Mesnage S, Fontaine T, Mignot T et al (2000) Bacterial SLH domain proteins are non-covalently anchored to the cell surface via a conserved mechanism involving wall polysaccharide pyruvylation. EMBO J 19(17):4473–4484. doi:10.1093/emboj/19.17.4473

Millette M, Cornut G, Dupont C et al (2008) Capacity of human nisin- and pediocin-producing lactic acid bacteria to reduce intestinal colonization by vancomycin-resistant enterococci. Appl Environ Microbiol 74(7):1997–2003. doi:10.1128/AEM.02150-07

Miyoshi Y, Okada S, Uchimura T, Satoh E (2006) A mucus adhesion promoting protein, MapA, mediates the adhesion of *Lactobacillus reuteri* to Caco-2 human intestinal epithelial cells. Biosci Biotechnol Biochem 70(7):1622–1628. doi:10.1271/bbb.50688

Mohania D, Nagpal R, Kumar M et al (2008) Molecular approaches for identification and characterization of lactic acid bacteria. J Dig Dis 9(4):190–198. doi:10.1111/j.1751-2980.2008.00345.x

Morelli L, Cesena C, de Haën C, Gozzini L (1998) Taxonomic *Lactobacillus* composition of feces from human newborns during the first few days. Microb Ecol 35(2):205–212. doi:10.1007/s002489900076

Mori K, Yamazaki K, Ishiyama T (1997) Comparative sequence analysis of the genes coding for 16S rRNA of *Lactobacillus casei*-related taxa. Int J Syst Bacteriol 47(1):54–57

Mukai T, Kaneko S, Matsumoto M, Ohori H (2004) Binding of *Bifidobacterium bifidum* and *Lactobacillus reuteri* to the carbohydrate moieties of intestinal glycolipids recognized by peanut agglutinin. Int J Food Microbiol 90(3):357–362. doi:10.1016/S0168-1605(03)00317-9

Naaber P, Smidt I, Stsepetova J et al (2004) Inhibition of *Clostridium difficile* strains by intestinal *Lactobacillus* species. J Med Microbiol 53(6):551–554. doi:10.1099/jmm.0.45595-0

Nakamura J, Ito D, Nagai K et al (1997) Rapid and sensitive detection of hiochi bacteria by amplification of hiochi bacterial common antigen gene by PCR method and characterization of the antigen. J Ferment Bioeng 83(2):161–167. doi:10.1016/S0922-338X(97)83576-3

Nigatu A, Ahrne S, Molin G (2001) Randomly amplified polymorphic DNA (RAPD) profiles for the distinction of *Lactobacillus* species. Antonie Van Leeuwenhoek 79(1):1–6. doi:10.1023/A:1010290403124

Nissen-Meyer J, Rogne P, Oppegård C et al (2009) Structure-function relationships of the non-lanthionine-containing peptide (class II) bacteriocins produced by gram-positive bacteria. Curr Pharm Biotechnol 10(1):19–37. doi:10.2174/138920109787048661

Noriega L, de los Reyes-Gavilán CG, Margolles A (2005) Acquisition of bile salt resistance promotes antibiotic susceptibility changes in Bifidobacterium. J Food Prot 68(9):1916–1919

Novik GI (1998) Structure-functional organization of bifidobacteria. Mikrobiologiia 67(3):376–383

Novik GI, Astapovich NI, Kubler J, Gamian A (2002) Characterization of the cell-bound polysaccharides of *Bifidobacterium adolescentis* 94 BIM. Mikrobiologiia 71(2):173–177

Novik GI, Astapovich NI, Samartsev AA (2001) Investigation of the physiological and biochemical characteristics of bifidobacteria at the late stages of their development. Mikrobiologiia 70(4):429–435

Novik G, Gamian A, Cruz Francisco J, Szwajcer Dey E (2006) A novel procedure for the isolation of glycolipids from *Bifidobacterium adolescentis* 94 BIM using supercritical carbon dioxide. J Biotechnol 121(4):555–562. doi:10.1016/j.jbiotec.2005.08.018

Novik GI, Vysotskii VV (1995) Architectonics of bifidobacteria populations: submicroscopic aspect of cell cohesion in *Bifidobacterium adolescentis* and *Bifidobacterium bifidum*. Mikrobiologiia 64(2):222–227

Novik GI, Vysotskii VV, Bogdanovskaia ZN (1994) Cellular ultrastructure of various species of the genus *Bifidobacterium*. Mikrobiologiia 63(3):515–522

Novik GI, Wawrzynczyk J, Norrlow O, Szwajcer Dey E (2007) Fractions of barley spent grain as media for growth of probiotic bacteria. Mikrobiologiia 76(6):804–808

O'Sullivan DJ (2000) Methods for analysis of the intestinal microflora. Curr Issues Intest Microb 1(2):39–50

Ohland CL, MacNaughton WK (2010) Probiotic bacteria and intestinal epithelial barrier function. Am J Physiol Gastrointest Liver Physiol 298(6):G807–G819. doi:10.1152/ajpgi.00243.2009

Oppegård C, Rogne P, Emanuelsen L et al (2007) The two-peptide class II bacteriocins: structure, production, and mode of action. J Mol Microbiol Biotechnol 13(4):210–219. doi:10.1159/000104750

O'Riordan K, Fitzgerald GF (1997) Determination of genetic diversity within the genus *Bifidobacterium* and estimation of chromosomal size. FEMS Microbiol Lett 156(2):259–264. doi:10.1016/S0378-1097(97)00435-7

Park HK, So JS, Heo TR (1995) Acid adaptation promotes survival of *Bifidobacterium breve* against environmental stresses. Food Biotechnol 4(4):226–230

Petrof EO, Kojima K, Ropeleski MJ et al (2004) Probiotics inhibit nuclear factor-kappaB and induce heat shock proteins in colonic epithelial cells through proteasome inhibition. Gastroenterology 127(59):1474–1487. doi:10.1053/j.gastro.2004.09.001

Phillips M, Kailasapathy K, Tran L (2006) Viability of commercial probiotic cultures (*L. acidophilus, Bifidobacterium* sp., *L. casei, L. paracasei* and *L. rhamnosus*) in cheddar cheese. Int J Food Microbiol 108(2):276–280. doi:10.1016/j.ijfoodmicro.2005.12.00

Pochart P, Marteau P, Bouhnik Y et al (1992) Survival of bifidobacteria ingested via fermented milk during their passage through the human small intestine: an in vivo study using intestinal perfusion. Am J Clin Nutr 55(1):78–80

Pot B, Ludwig W, Kersters K, Schleifer KH (1994) Taxonomy of lactic acid bacteria. In: de Vuyst L, Vandamme EJ (eds) Bacteriocins of lactic acid bacteria: microbiology, genetics and applications. Blackie Academic and Professional, Glasgow, pp 13–90

Poupard J, Husain I, Norris RF (1973) Biology of the bifidobacteria. Bacteriol Rev 37(2):136–165

Preising J, Philippe D, Gleinser M et al (2010) Selection of bifidobacteria based on adhesion and anti-inflammatory capacity in vitro for amelioration of murine colitis. Appl Environ Microbiol 76(9):3048–3051. doi:10.1128/AEM.03127-09

Pretzer G, Snel J, Molenaar D et al (2005) Biodiversity-based identification and functional characterization of the mannose-specific adhesin of *Lactobacillus plantarum*. J Bacteriol 187(17):6128–6136. doi:10.1128/JB.187.17.6128-6136.2005

Pridmore RD, Pittet AC, Praplan F et al (2008) Hydrogen peroxide production by *Lactobacillus johnsonii* NCC 533 and its role in anti-*Salmonella* activity. FEMS Microbiol Lett 283(2): 210–215. doi:10.1111/j.1574-6968.2008.01176.x

Putaala H, Salusjarvi T, Nordstrom M et al (2008) Effect of four probiotic strains and *Escherichia coli* O157:H7 on tight junction integrity and cyclooxygenase expression. Res Microbiol 159(9):692–698. doi:10.1016/j.resmic.2008.08.002

Qin H, Zhang Z, Hang X, Jiang Y (2009) *L. plantarum* prevents enteroinvasive *Escherichia coli-induced* tight junction proteins changes in intestinal epithelial cells. BMC Microbiol 9(1): 63–69. doi:10.1186/1471-2180-9-63

Rachmilewitz D, Katakura K, Karmeli F et al (2004) Toll-like receptor 9 signaling mediates the anti-inflammatory effects of probiotics in murine experimental colitis. Gastroenterology 126(2):520–528. doi:10.1053/j.gastro.2003.11.019

Rajkumar R, Devaraj H, Niranjali S (1998) Binding of *Shigella* to rat and human intestinal mucin. Mol Cell Biochem 178(1):261–268. doi:10.1023/A:1006844125976

Rakhuba D, Novik G, Szwajcer Dey E (2009) Application of supercritical carbon dioxide ($scCO_2$) for the extraction of glycolipids from *Lactobacillus plantarum* B-01. J Supercrit Fluids 49(1):45–51. doi:10.1016/j.supflu.2008.11.016

Ramiah K, van Reenen CA, Dicks LMT (2007) Expression of the mucus adhesion genes mub and MapA, adhesion-like factor EF-Tu and bacteriocin gene plaA of *Lactobacillus plantarum* 423, monitored with real-time PCR. Int J Food Microbiol 116(3):405–409. doi:10.1016/j.ijfoodmicro.2007.02.011

Reid G (2006) Safe and efficacious probiotics: what are they? Trends Microbiol 14(8):348–352. doi:10.1016/j.tim.2006.06.006

Requena T, Burton J, Matsuki T et al (2002) Identification, detection, and enumeration of human *Bifidobacterium* species by PCR targeting the transaldolase gene. Appl Environ Microbiol 68(5):2420–2427. doi:10.1128/AEM.68.5.2420-2427.2002

Resta-Lenert S, Barrett KE (2006) Probiotics and commensals reverse TNFalpha-and IFN-gamma-induced dysfunction in human intestinal epithelial cells. Gastroenterology 130(3):731–746. doi:10.1053/j.gastro.2005.12.015

Reuter G (2001) The *Lactobacillus and Bifidobacterium* microflora of the human intestine: composition and succession. Curr Issues Intest Microbiol 2(2):43–53

Rinckel LA, Savage DC (1990) Characterization of plasmids and plasmid-borne macrolide resistance from *Lactobacillus* sp. strain 100–33. Plasmid 23(2):119–125. doi:10.1016/0147-619X(90)90030-G

Rojas M, Ascencio F, Conway PL (2002) Purification and characterization of a surface protein from *Lactobacillus fermentum* 104R that binds to porcine small intestinal mucus and gastric mucin. Appl Environ Microbiol 68(5):2330–2336. doi:10.1128/AEM.68.5.2330-2336.2002

Roos S, Jonsson H (2002) A high-molecular-mass cell-surface protein from *Lactobacillus reuteri* 1063 adheres to mucus components. Microbiology 148(2):433–442

Roos S, Aleljung P, Robert N et al (1996) A collagen binding protein from *Lactobacillus reuteri* is part of an ABC transporter system? FEMS Microbiol Lett 144(1):33–38. doi:10.1016/0378-1097(96)00334-5

Ross RP, Mills S, Hill C et al (2010) Specific metabolite production by gut microbiota as a basis for probiotic function. Int Dairy J 20(4):269–276. doi:10.1016/j.idairyj.2009.12.003

Roy D, Sirois S (2000) Molecular differentiation of *Bifidobacterium* species with amplified ribosomal DNA restriction analysis and alignment of short regions of the *ldh* gene. FEMS Microbiol Lett 191(1):17–24. doi:10.1016/S0378-1097(00)00364-5

Roy D, Ward P (2004) Comparison of fructose-1,6-biphosphatase gene (fbp) sequences for the identification of *Lactobacillus rhamnosus*. Curr Microbiol 49(5):313–320. doi:10.1007/s00284-004-4355-y

Roy D, Ward P, Champagne G (1996) Differentiation of bifidobacteria by use of pulsed-field gel electrophoresis and polymerase chain reaction. Int J Food Microbiol 29(1):11–29. doi:10.1016/0168-1605(95)00013-5

Ruas-Madiedo P, Hernández-Barranco A, Margolles A, de los Reyes-Gavilán CG (2005) A bile-salt resistant derivative of *Bifidobacterium animalis* has an altered fermentation pattern when grown on glucose and maltose. Appl Environ Microbiol 71(11):6564–6570. doi:10.1128/AEM.71.11.6564-6570.2005

Ruas-Madiedo P, Gueimonde M, Margolles A et al (2006) Exopolysaccharides produced by probiotic strains modify the adhesion of probiotics and enteropathogens to human intestinal mucus. J Food Prot 69(8):2011–2015

Saarela M, Mogensen G, Fondén R et al (2000) Probiotic bacteria: safety, functional and technological properties. J Biotechnol 84(3):197–215. doi:10.1016/S0168-1656(00)00375-8

Saarela M, Alakomi HL, Mättö J et al (2011) Acid tolerant mutants of *Bifidobacterium animalis* subsp. *lactis* with improved stability in fruit juice. Food Sci Technol 44(4):1012–1018. doi:10.1016/j.lwt.2010.11.004

Sakata S, Ryu CS, Kitahara M et al (2006) Characterization of the genus *Bifidobacterium* by automated ribotyping and 16S rRNA gene sequences. Microbiol Immunol 50(1):1–10

Salminen S, von Wright A, Morelli L et al (1998) Demonstration of safety of probiotics—a review. Int J Food Microbiol 44(1):93–106

Sameshima T, Magome C, Takeshita K et al (1998) Effect of intestinal *Lactobacillus* starter cultures on the behaviour of *Staphylococcus aureus* in fermented sausage. Int J Food Microbiol 41(1):1–7. doi:10.1016/S0168-1605(98)00038-5

Sánchez B, Arias S, Chaignepain S et al (2009) Identification of surface proteins involved in the adhesion of a probiotic *Bacillus cereus* strain to mucin and fibronectin. Microbiology 155(5):1708–1716. doi:10.1099/mic.0.025288-0

Sánchez B, Bressollier P, Urdaci MC et al (2008) Exported proteins in probiotic bacteria: adhesion to intestinal surfaces, host immunomodulation and molecular cross-talking with the host. FEMS Immunol Med Microbiol 54(1):1–17. doi:10.1111/j.1574-695X.2008.00454.x

Sánchez B, Margolles A (2012) Selection of low-acetate producing *Bifidobacterium lactis* subsp. *lactis* strain. Appl Environ Microbiol 78(9):3338–3342. doi:10.1128/AEM.00129-12

Sánchez B, Ruiz L, Gueimonde M et al (2012) Toward improving technological and functional properties of probiotics in food. Trends Food Sci Technol 26(1):56–63. doi:10.1016/j.tifs.2012.02.002

Sartor RB (2006) Mechanisms of disease: pathogenesis of Crohn's disease and ulcerative colitis. Nat Clin Pract Gastroenterol Hepatol 3(7):390–407. doi:10.1038/ncpgasthep0528

Sartor RB (2008) Microbial influences in inflammatory bowel diseases. Gastroenterology 134:577–594. doi:10.1053/j.gastro.2007.11.059

Satokari RM, Vaughan EE, Smidt H et al (2003) Molecular approaches for the detection and identification of bifidobacteria and lactobacilli in the human gastrointestinal tract. Syst Appl Microbiol 26(4):572–584. doi:10.1078/072320203770865882

Savoie S, Champagne CP, Chiasson S, Audet P (2007) Media and process parameters affecting the growth, strain ratios and specific acidifying activities of a mixed lactic starter containing aroma-producing and probiotic strains. J Appl Microbiol 103(1):163–174. doi:10.1111/j.1365-2672.2006.03219.x

Schillinger U, Yousif NMK, Sesar L, Franz CMAP (2003) Use of group-specific and RAPD-PCR analyses for rapid differentiation of *Lactobacillus* strains from probiotic yogurts. Curr Microbiol 47(6):453–456. doi:10.1007/s00284-003-4067-8

Schlee M, Wehkamp J, Altenhoefer A et al (2007) Induction of human beta-defensin 2 by the probiotic *Escherichia coli* Nissle 1917 is mediated through flagellin. Infect Immun 75(5):2399–2407. doi:10.1128/IAI.01563-06

Schleifer KH, Ludwig W (1995) Phylogeny of the genus *Lactobacillus* and related genera. Syst Appl Microbiol 18(4):641–646. doi:10.1016/S0723-2020(11)80404-2

Schneeberger EE, Lynch RD (2004) The tight junction: a multifunctional complex. Am J Physiol Cell Physiol 286(6):C1213–C1228. doi:10.1152/ajpcell.00558.2003

Schrezenmeir J, de Vrese M (2001) Probiotics, prebiotics, and synbiotics—approaching a definition. Am J Clin Nutr 73(2):361S–364S

Schultz M, Linde HJ, Lehn N et al (2003) Immunomodulatory consequences of oral administration of *Lactobacillus rhamnosus* strain GG in healthy volunteers. J Dairy Res 70(2):165–173. doi:10.1017/S0022029903006034

Sekine K, Toida T, Saito M et al (1985) A new morphologically characterized cell wall preparation (whole peptidoglycan) from *Bifidobacterium infantis* with a higher efficacy on the regression of an established tumor in mice. Cancer Res 45(3):1300–1307

Seregni E, Botti C, Massaron S et al (1997) Structure, function and gene expression of epithelial mucins. Tumori 83(3):625–632

Servin AL (2004) Antagonistic activities of lactobacilli and bifidobacteria against microbial pathogens. FEMS Microbiol Rev 28(4):405–440. doi:10.1016/j.femsre.2004.01.003

Seth A, Yan F, Polk DB, Rao RK (2008) Probiotics ameliorate the hydrogen peroxide-induced epithelial barrier disruption by a PKC- and MAP kinase-dependent mechanism. Am J Physiol Gastrointest Liver Physiol 294(4):G1060–G1069. doi:10.1152/ajpgi.00202.2007

Settanni L, Corsetti A (2008) Application of bacteriocins in vegetable food biopreservation. Int J Food Microbiol 121(2):123–138. doi:10.1016/j.ijfoodmicro.2007.09.001

Settanni L, van Sinderen D, Rossi J, Corsetti A (2005) Rapid differentiation and in situ detection of 16 sourdough *Lactobacillus* species by multiplex PCR. Appl Environ Microbiol 71(6):3049–3059. doi:10.1128/AEM.71.6.3049-3059.2005

Sherman PM, Johnson-Henry KC, Yeung HP et al (2005) Probiotics reduce enterohemorrhagic *Escherichia coli* O157:H7- and enteropathogenic *E. coli* O127:H6-induced changes in polarized T84 epithelial cell monolayers by reducing bacterial adhesion and cytoskeletal rearrangements. Infect Immun 73(8):5183–5188. doi:10.1128/IAI.73.8.5183-5188.2005

Shida K, Nanno M (2008) Probiotics and immunology: separating the wheat from the chaff. Trends Immunol 29(11):565–573. doi:10.1016/j.it.2008.07.011

Shukla TP (1998) Nutraceutical novelties: trends in proprietary technology. Cereal Foods World 43(5):388–389

Sidarenka AV, Novik GI, Akimov VN (2008) Application of molecular methods to classification and identification of bacteria of *Bifidobacterium* genus. Mikrobiologiia 77(3):251–260

Simpson PJ, Ross RP, Fitzgerald GF, Santon C (2004) *Bifidobacterium psychraerophilum* sp. nov., and *Aeriscardovia aeriphila* gen. nov., sp. nov., isolated from a porcine caecum. Int J Syst Evol Microbiol 54(2):401–406. doi:10.1099/ijs.0.02667-0

Singha S, Goswami P, Singh R, Heller KJ (2009) Application of molecular identification tools for *Lactobacillus*, with a focus on discrimination between closely related species: a review. Food Sci Technol 42(2):448–457. doi:10.1016/j.lwt.2008.05.019

Sleator RD (2010) Probiotics — a viable therapeutic alternative for enteric infections especially in the developing world. Discov Med 10(51):119–124

Smits HH, Engering A, van der Kleij D et al (2005) Selective probiotic bacteria induce IL-10-producing regulatory T cells in vitro by modulating dendritic cell function through dendritic cell-specific intercellular adhesion molecule 3-grabbing nonintegrin. J Allergy Clin Immunol 115(6):1260–1267. doi:10.1016/j.jaci.2005.03.036

Song YL, Kato N, Liu CX et al (2000) Rapid identification of 11 human intestinal *Lactobacillus* species by multiplex PCR assays using group- and species-specific primers derived from the 16S-23S rRNA intergenic spacer region and its flanking 23S rRNA. FEMS Microbiol Lett 187(2):167–173. doi:10.1016/S0378-1097(00)00196-8

Spanhaak S, Havenaar R, Schaafsma G (1998) The effect of consumption of milk fermented by *Lactobacillus casei* strain Shirota on the intestinal microflora and immune parameters in humans. Eur J Clin Nutr 52(12):899–907. doi:10.1038/sj.ejcn.1600663

Šrůtkova D, Španova A, Špano M et al (2011) Efficiency of PCR-based methods in discriminating *Bifidobacterium longum* ssp. *longum* and *Bifidobacterium longum* ssp. *infantis* strains of human origin. J Microbiol Meth 87(1):10–16. doi:10.1016/j.mimet.2011.06.014

Stack HM, Kearney N, Stanton C et al (2010) Association of beta-glucan endogenous production with increased stress-tolerance of intestinal lactobacilli. Appl Environ Microbiol 76(2): 500–507. doi:10.1128/AEM.01524-09

Sun J, Le GW, Shi YH, Su GW (2007) Factors involved in binding of *Lactobacillus plantarum* Lp6 to rat small intestinal mucus. Lett Appl Microbiol 44(1):79–85. doi:10.1111/j.1472-765X.2006.02031.x

Szponar B, Pawlik KJ, Gamian A, Szwajcer Dey E (2003) Protein fraction of barley spent grain as a new simple medium for growth and sporulation of soil actinobacteria. Biotechnol Lett 25(20):1717–1721. doi:10.1023/A:1026046403010

Szwajcer Dey E, Rasmussen J, Meldal M, Breddam K (1992) Proline specific endopeptidases from microbial sources: isolation of an enzyme from *Xanthomonas* sp. J Bacteriol 174(8): 2454–2459

Tabasco R, García-Cayuela T, Peláez C, Requena T (2009) *Lactobacillus acidophilus* La-5 increases lactacin B production when it senses live target bacteria. Int J Food Microbiol 132 (2–3):109–116. doi:10.1016/j.ijfoodmicro.2009.04.004

Talwalkar A, Kailasapathy K, Peiris P, Arumugaswamy R (2001) Application of RBGR—a simple way for screening of oxygen tolerance in probiotic bacteria. Int J Food Microbiol 71(2): 245–248. doi:10.1016/S0168-1605(01)00563-3

Tannock GW (1999) Identification of lactobacilli and bifidobacteria. Curr Issues Mol Biol 1(1–2):53–64

Tannock GW, Munro K, Harmsen HJM et al (2000) Analysis of the fecal microflora of human subjects consuming a probiotic product containing *Lactobacillus rhamnosus* DR20. Appl Environ Microbiol 66(6):2578–2588. doi:10.1128/AEM.66.6.2578-2588.2000

Tannock GW, Tilsala-Timisjarvi A, Rodtong S et al (1999) Identification of *Lactobacillus* isolates from the gastrointestinal tract, silage, and yogurt by 16S-23S rRNA gene intergenic spacer region. Appl Environ Microbiol 65(9):4264–4276

Tao Y, Drabik KA, Waypa TS et al (2006) Soluble factors from *Lactobacillus* GG activate MAPKs and induce cytoprotective heat shock proteins in intestinal epithelial cells. Am J Physiol Cell Physiol 290(4):C1018–C1030. doi:10.1152/ajpcell.00131.2005

Torriani S, Felix GE, Dellaglio F (2001) Differentiation of *Lactobacillus plantarum*, *L. pentosus*, and *L. paraplantarum* by *recA* gene sequence analysis and multiplex PCR assay with *recA-derived* primers. Appl Environ Microbiol 67(8):3450–3454. doi:10.1128/AEM.67.8.3450-3454.2001

Trivedi K, Barrett KE, Resta-Lenert SC (2003) Probiotic inhibition of the entry of enteroinvasive *E. coli* into human intestinal epithelial cells involves both Rho-dependent and independent pathways. Gastroenterology 124(4):A106

Turpin W, Humblot C, Noordine ML et al (2012) *Lactobacillaceae* and cell adhesion: genomic and functional screening. PLoS One 7(5):e38034. doi:10.1371/journal.pone.0038034

Tynkkynen S, Satokari R, Saarela M et al (1999) Comparison of ribotyping, randomly amplified polymorphic DNA analysis, and pulsed-field gel electrophoresis in typing of Lactobacillus rhamnosus and L. casei strains. Appl Environ Microbiol 65(9):3908–3914

Ulluwishewa D, Anderson RC, McNabb WC et al (2011) Regulation of tight junction permeability by intestinal bacteria and dietary components. J Nutr 141(5):769–776. doi:10.3945/jn.110.135657

Van Hoek AH, Mayrhofer S, Domig KJ, Aarts HJ (2008) Resistance determinant *erm(X)* is borne by transposon Tn5432 in *Bifidobacterium thermophilum* and *Bifidobacterium animalis* subsp. *lactis*. Int J Antimicrob Agents 31(6):544–548. doi:10.1016/j.ijantimicag.2008.01.025

Van Klinken BJ, Dekker J, Büller HA, Einerhand AW (1995) Mucin gene structure and expression: protection vs. adhesion. Am J Physiol 269(5):G613–G627

Van Pijkeren J-P, Canchaya C, Ryan KA et al (2006) Comparative and functional analysis of sortase-dependent proteins in the predicted secretome of Lactobacillus salivarius UCC118. Appl Environ Microbiol 72(6):4143–4153. doi: 10.1128/AEM.03023-05

Van Tassell ML, Miller MJ (2011) *Lactobacillus* adhesion to mucus. Nutrients 3(5):613–636. doi:10.3390/nu3050613

Vandenberg PA (1993) Lactic acid bacteria, their metabolic products and interference with microbial growth. FEMS Microbiol Rev 12(1–3):221–237. doi:10.1111/j.1574-6976.1993.tb00020.x

Vanderpool C, Yan F, Polk DB (2008) Mechanisms of probiotic action: implications for therapeutic applications in inflammatory bowel diseases. Inflamm Bowel Dis 14(11):1585–1596. doi:10.1002/ibd.20525

Vanegas MC, Gonzalez LM, Arevalo SA (2010) Antibiotic activity of *Bifidobacterium* sp. isolated from breast milk and newborn faeces, against the main causes for foodborne illnesses. Infect 14(4):241–247

Vasquez A, Ahrne S, Pettersson B, Molin G (2001) Temporal temperature gradient gel electrophoresis (TTGE) as a tool for identification of *Lactobacillus casei*, *Lactobacillus paracasei*, *Lactobacillus zeae* and *Lactobacillus rhamnosus*. Lett Appl Microbiol 32:215–219

Vaughan EE, Heilig GHJ, Ben-Amor K, de Vos WM (2005) Diversity, vitality and activities of intestinal lactic acid bacteria and bifidobacteria assessed by molecular approaches. FEMS Microbiol Rev 29(3):477–490. doi:10.1016/j.femsre.2005.04.009

Ventura M, Zink R (2002) Specific identification and molecular typing analysis of *Lactobacillus johnsonii* by using PRC-based methods and pulsed-field gel electrophoresis. FEMS Microbiol Lett 217(2):141–154. doi:10.1111/j.1574-6968.2002.tb11468.x

Ventura M, Meylan V, Zink R (2003) Identification and tracing of *Bifidobacterium* species by use of enterobacterial repetitive intergenic consensus sequences. Appl Environ Microbiol 69(7):4296–4301. doi:10.1128/AEM.69.7.4296-4301.2003

Ventura M, van Sinderen D, Fitzgerald GF, Zink R (2004) Insights into the taxonomy, genetics and physiology of bifidobacteria. Antonie Van Leeuwenhoek 86(3):205–223. doi:10.1023/B:ANTO.0000047930.11029.ec

Ventura M, Canchaya C, Del Casale A et al (2006) Analysis of bifidobacterial evolution using a multilocus approach. Int J Syst Evol Microbiol 56(12):2783–2792. doi:10.1099/ijs.0.64233-0
Vesterlund S, Vankerckhoven V, Saxelin M et al (2007) Safety assessment of *Lactobacillus* strains: presence of putative risk factors in faecal, blood and probiotic isolates. Int J Food Microbiol 116(3):325–331
Vimal DB, Khullar M, Gupta S, Ganguly NK (2000) Intestinal mucins: the binding sites for *Salmonella typhimurium*. Mol Cell Biochem 204(1–2):107–117. doi:10.1023/A:1007015312036
Vincent D, Roy D, Mondou F, Dery C (1998) Characterization of bifidobacteria by random DNA amplification. Int J Food Microbiol 43(3):185–193. doi:10.1016/S0168-1605(98)00109-3
Vlkova E, Medkova J, Rada V (2002) Comparison of four methods for identification of bifidobacteria to the genus level. Czech J Food Sci 20(5):171–174
Von Wright A (2005) Regulating the safety of probiotics—the European approach. Curr Pharm Des 11(1):17–23. doi:10.2174/1381612053382322
Walker DC, Girgis HS, Klaenhammer TR (1999) The *groESL* chaperone operon of *Lactobacillus johnsonii*. Appl Environ Microbiol 65(7):3033–3041
Ward P, Roy D (2005) Review of molecular methods for identification, characterization and detection of bifidobacteria. Lait 85(1–2):23–32. doi:10.1051/lait:2004024
Wehkamp J, Harder J, Wehkamp K et al (2004) NF-kappa B- and AP-1-mediated induction of human beta defensin-2 in intestinal epithelial cells by *Escherichia coli* Nissle 1917: a novel effect of a probiotic bacterium. Infect Immun 72(10):5750–5758. doi:10.1128/IAI.72.10.5750-5758.2004
Weiss A, Lettner HP, Kramer W et al (2005) Molecular methods used for the identification of potentially probiotic *Lactobacillus reuteri* strains. Food Technol Biotech 43(3):295–300
Welling GW, Elfferich P, Raangs GC et al (1997) 16S ribosomal RNA-targeted oligonucleotide probes for monitoring of intestinal tract bacteria. Scand J Gastroenterol Suppl 222:17–19
Yamamoto T, Morotomi M, Tanaka R (1992) Species specific oligonucleotide probes for five *Bifidobacterium* species detected in human intestinal microflora. Appl Environ Microbiol 58(12):4076–4079
Yamano T, Iino H, Takada M et al (2006) Improvement of the human intestinal flora by ingestion of the probiotic strain *Lactobacillus johnsonii* La1. Br J Nutr 95(2):303–312. doi:10.1079/BJN20051507
Yan F, Cao H, Cover TL et al (2007) Soluble proteins produced by probiotic bacteria regulate intestinal epithelial cell survival and growth. Gastroenterology 132(2):562–575. doi:10.1053/j.gastro.2006.11.022
Zanetti M (2004) Cathelicidins, multifunctional peptides of the innate immunity. J Leukoc Biol 75(1):39–48. doi:10.1189/jlb.0403147
Zareie M, Johnson-Henry K, Jury J et al (2006) Probiotics prevent bacterial translocation and improve intestinal barrier function in rats following chronic psychological stress. Gut 55(11):1553–1560. doi:10.1136/gut.2005.080739
Zhang YC, Zhang LW, Tuo YF et al (2010) Inhibition of *Shigella sonnei* adherence to HT-29 cells by lactobacilli from chinese fermented food and preliminary characterization of s-layer protein involvement. Res Microbiol 161(8):667–672. doi:10.1016/j.resmic.2010.06.005
Zinedine A, Faid M (2007) Isolation and characterization of strains of bifidobacteria with probiotic properties in vitro. World J Dairy Food Sci 2(1):28–34
Zyrek AA, Cichon C, Helms S et al (2007) Molecular mechanisms underlying the probiotic effects of *Escherichia coli* Nissle 1917 involve ZO-2 and PKCzeta redistribution resulting in tight junction and epithelial barrier repair. Cell Microbiol 9(3):804–816. doi:10.1111/j.1462-5822.2006.00836.x

Chapter 10
Prebiotics

P.S. Panesar, Vandana Bali, Shweta Kumari, Neha Babbar, and Harinder Singh Oberoi

10.1 Introduction

Prebiotics are nondigestible oligosaccharides which are not broken down by the salivary and intestinal enzymes due to the presence of "osidic" bonds. These compounds beneficially affect the host by selectively stimulating the growth and/or activity of one or a limited number of bacteria in the colon (*Bifidobacteria* and *Lactobacilli*) by suppressing the activity of entero-putrefactive and pathogenic organism and also facilitate the absorption of nutrients (Roberfroid 1997). Some of the available prebiotic compounds have range of two to ten sugar moieties, and common examples are lactulose, galacto-oligosaccharides, lactosucrose, fructo-oligosaccharides, xylo-oligosaccharide, malto-oligosaccharides, inulin, and its hydrolysates (Fig. 10.1). The end products of these prebiotics, i.e., acetate, butyrate, and propionate, act as energy sources for host organisms.

Prebiotics have bifidus-stimulating ability, immunomodulatory effect, and antioxidant properties besides their role in reducing risks of cancer, acute gastroenteritis, osteoporosis, cholesterol, and hyperlipidemia (Conway 2001). These compounds can be used for the fortification of different food products for the development of functional foods with high nutritional and therapeutic properties. These can be supplemented as ingredients in the functional foods, cosmetics, pharmaceuticals, or agricultural products. This chapter provides a comprehensive overview on fundamentals of different prebiotics and their production using biotechnological strategies.

P.S. Panesar (✉) • V. Bali • S. Kumari
Biotechnology Research Laboratory, Department of Food Engineering & Technology,
Sant Longowal Institute of Engineering & Technology, Longowal 148 106, Punjab, India
e-mail: pspbt@yahoo.com

N. Babbar • H.S. Oberoi
Central Institute of Post Harvest Engineering and Technology, P.O. PAU, Ludhiana 141 004,
Punjab, India

S.K. Brar et al. (eds.), *Biotransformation of Waste Biomass into High Value Biochemicals*,
DOI 10.1007/978-1-4614-8005-1_10, © Springer Science+Business Media New York 2014

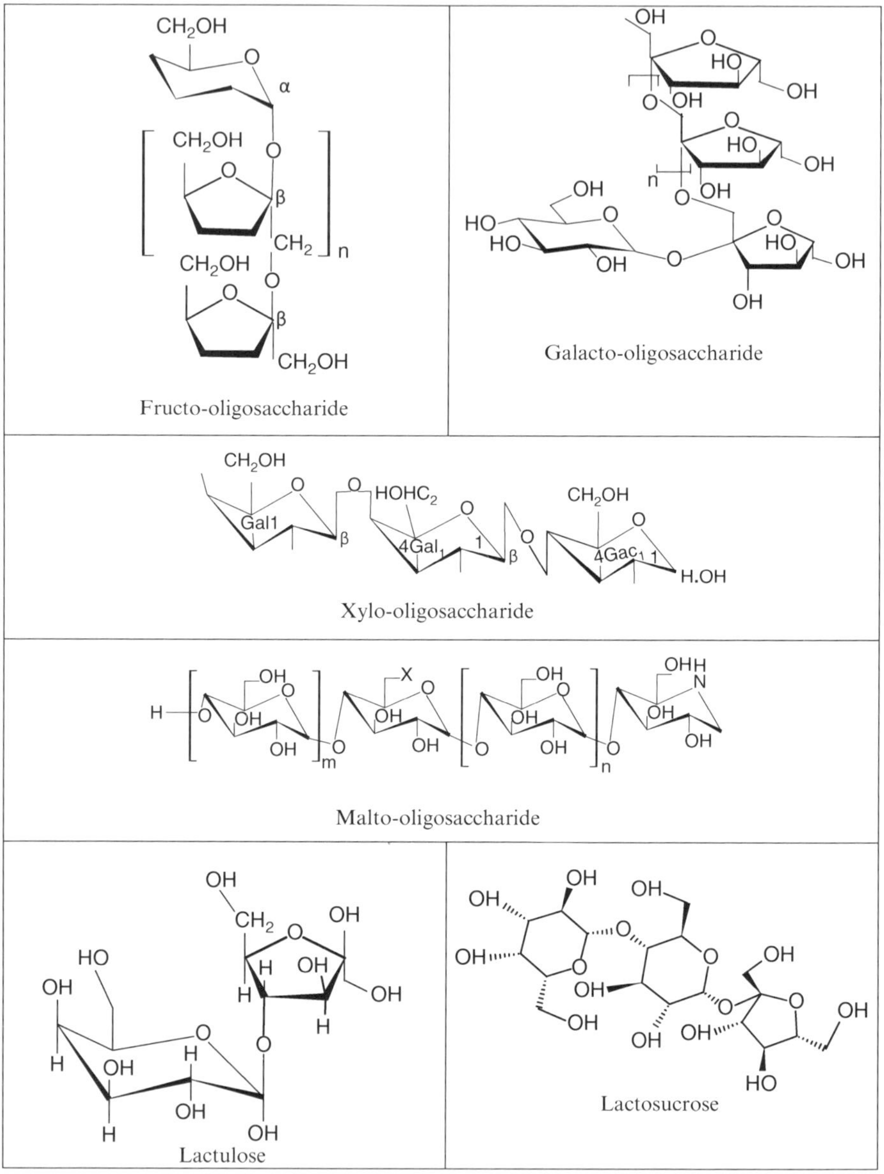

Fig. 10.1 Some of the commonly used prebiotics

10.2 Production of Prebiotics

Prebiotics can be obtained by three different routes, i.e., by extraction from natural (plant) sources, enzymatic synthesis, and enzymatic hydrolysis of polysaccharides. Small amounts of prebiotics occur naturally in the natural sources like plants including chicory, onion, garlic, asparagus, and artichoke; however, their large-scale production can also be achieved using various microorganisms and enzymes. Various raw materials and enzymes have been explored for the production of different prebiotics/oligosaccharides in both free and immobilized biocatalyst systems, which have been discussed in subsequent sections.

10.3 Galacto-oligosaccharides

Galacto-oligosaccharides (GOS) are a functional food ingredient consisting of β-linked galactose moieties with galactose or glucose at the reducing end. More than 30 different types of di-, tri-, and tetrasaccharides were identified as the product of various enzymatic transgalactosylation. The different unit of monomer in GOS is linked by β-(1→2), β-(1→3), β-(1→4), or β-(1→6) galactosyl moieties (Mahoney 1998; Gänzle 2011). In the commercial GOS preparations, apart from tri-, tetra-, and pentasaccharides, small amount of other carbohydrates, such as glucose-, lactose- and β-(1→3)-, or β-(1→6)-linked disaccharides, is also present (Sako et al. 1999; Splechtna et al. 2006).

10.3.1 Method of Production

Galacto-oligosaccharide can be synthesized by transgalactosylation of lactose, glucose, or galactose with lactose as galactosyl donor.

10.3.1.1 Enzymes Involved

Enzymatic production of galacto-oligosaccharide has been carried out using following enzymes:

1. β-galactosidase (EC 3.2.1.23)
2. β-glucosidases/β-glycosidases (EC 3.2.1.21)

β-galactosidase and β-glucosidases/β-glycosidases have both hydrolytic (lactose) and transgalactosylation activity.

10.3.1.2 Process

The β-galactosidase can be produced from various microbial sources like bacteria, yeast, and fungi. Among these sources, yeast has been considered as an important source of β-galactosidase from industrial point of view (Panesar et al. 2006). Apart from this, other thermostable enzymes, such as β-glucosidase/β-glycosidase, have also been used for galacto-oligosaccharide production (Akiyama et al. 2001; Choi et al. 2003). Galacto-oligosaccharide production from lactose has been carried out with microbial β-galactosidases, whole cells, permeabilized cells, as well as applying immobilized techniques (Albayrak and Yang 2002; Ladero et al. 2003; Tzortzis et al. 2005; Nakkharat et al. 2006; Nguyen et al. 2007; Sakai et al. 2008; Placier et al. 2009).

The production of GOS depends on the source of enzyme and production process. High temperature (60 °C) has been observed to favor transgalactosylation reaction and GOS yield (Cardelle-Cobas et al. 2008). The production of GOS has also been carried out by various enzymes from different microorganisms, such as thermostable β-glycosidases/β-glucosidase from *Pyrococcus furiosus* and *Sulfolobus solfataricus* (Petzelbauer et al. 2000; Boon et al. 1998; Splechtna et al. 2001) and β-galactosidase from *Sirobasidium magnum* and *Penicillium simplicissimum* (Onishi and Tanaka 1997; Cruz et al. 1999). The yield of GOS varied from 15 to 77 % by these enzymes with the lactose conversion of 45–95 % (Torres et al. 2010). GOS has been purified by continuous nanofiltration achieving a yield of 81–98 % for oligosaccharides (trisaccharide), 59–89 % for disaccharides, and only 14–18 % for the monosaccharides (Goulas et al. 2002).

Whey (by-product of dairy industry) is rich in lactose content and can be used for GOS production (López Leiva and Guzman 1995). The conversion to oligosaccharides not only depends on the reaction time but also on the initial concentration of substrate. The production of GOS has been carried out using an immobilized *Aspergillus oryzae* β-galactosidase on glutaraldehyde-treated chitosan beads in a plug reactor from whey. The maximum yield of GOS was found to be 26 % of the total saccharides on a dry weight basis for an initial concentration of lactose of 300 g L^{-1} (Sheu et al. 1998). At pilot scale membrane reactor, oligosaccharides have also been produced by hydrolysis of whey permeate (2,000 L) from Maxilact (*Kluyveromyces lactis* β-D-galactosidase). A yield of 31 % oligosaccharide was observed from whey permeate containing 20 % lactose and 0.5 % enzyme (Foda and Lopez-Leiva 2000).

Different substrates (lactose, ultrafiltration whey permeate, and recombined whey) with different enzyme concentrations (0.15–15 U mL^{-1}) have been applied for GOS synthesis. The maximum production of GOS has been observed with ultrafiltration whey permeate using an enzyme concentration of 1.5 U mL^{-1} (HelleroVá and Čurda 2009). Continuous production of GOS has also been carried out from both lactose feed solution and whey, with PVA-immobilized β-galactosidase in a packed bed reactor. A maximum GOS production of 30 % of total sugars was achieved using 40 % lactose feed solution, whereas 15 % of total sugars have been obtained in case of whey (Jovanovic-Malinovska et al. 2012).

10.4 Lactulose

Lactulose (4-*O*-β-D-galactopyranosyl-D-fructose) is a ketose disaccharide having wide range of applications in food and pharmaceutical sectors. It is composed of fructose and galactose linked through glycosidic bonds. The linkage between galactose and fructose is neither cleaved by human digestive enzymes nor absorbed in the small intestine. In the colon, lactulose stimulates the growth of *Bifidobacteria* sp. and *Lactobacillus* sp. leading to the production of a large number of short-chain fatty acids (Méndez and Olano 1979).

10.4.1 Method of Production

Lactulose can be synthesized chemically or by enzymes from various microorganisms.

10.4.1.1 Enzyme Involved

Biocatalytic production of lactulose has been carried out by using the following enzymes:

1. β-galactosidase (EC 3.2.1.23)
2. β-glycosidase (EC 3.2.1.21)

10.4.1.2 Process

Currently, for commercial utilization, lactulose is produced by alkaline isomerization of lactose. To obtain high yields, it is advantageous to use high amounts of inorganic catalysts, such as boric acid or aluminate. This approach results in expensive separation of by-products that cause difficulty in product purification and waste management (De Harr and Pluim 1991; Zokaee et al. 2002; Aider and de Halleux 2007). The problems associated with lactulose production by chemical synthesis can be solved by an enzymatic transformation process which seems to be a useful approach for clean production and easy purification of lactulose (Tang et al. 2011). Microorganisms, such as *K. lactis*, *S. solfataricus*, *Arthrobacter* sp. (β-galactosidase), and *P. furiosus* (β-glycosidases), have been applied for both lactose hydrolysis and transglycosylation leading to lactulose synthesis (Lee et al. 2004; Kim et al. 2006; Mayer et al. 2010; Tang et al. 2011).

The optimum reaction conditions for lactulose production using permeabilized cells were 40 % (w/v) lactose and 20 % (w/v) fructose at temperature 60 °C and pH of 7.0. Under these conditions, the permeabilized cells produced approximately 20 g L^{-1} lactulose (Lee et al. 2004). The use of thermostable β-galactosidase can further enhance the production of lactulose (Kim et al. 2006). The continuous enzymatic production of the lactulose through transgalactosylation has also been developed using free and immobilized (Amberlite IRA-93 or Eupergit® C) thermostable β-glycosidase (Mayer et al. 2010). Dual-enzymatic system consisting of immobilized lactase and immobilized glucose isomerase in organic-aqueous two-phase media using lactose and fructose has also been applied for the synthesis of lactulose (Xiao et al. 2010). Currently, purified β-galactosidase from Arthrobacter sp.

LAS has been used for lactulose production to reduce the nonenzymatic browning during biotransformations (Tang et al. 2011). The optimum pH and temperature for lactulose synthesis by this β-galactosidase were 6.0 and 20 °C, respectively.

Alkaline isomerization of lactose present in cheese whey ultrafiltrate permeate was carried out by the addition of boric acid (Hicks et al. 1984). However, expensive separation and purification steps were involved in this procedure. The synthesis of lactulose from dairy by-product (i.e., whey) can be carried out by using an enzyme (β-galactosidase) in the presence of fructose. The controlled enzymatic transgalactosylation of lactose in whey ultrafiltration permeate can improve the efficiency of lactulose synthesis. The factors that influenced the lactulose synthesis efficiency were enzyme source, substrate concentration, and also the ratio of lactose and fructose added to the reaction mixture (Adamczak et al. 2009; Jaindl et al. 2009).

10.5 Lactosucrose

Lactosucrose (lactosylfructoside, *O*-β-D-galactopyranosyl-(1→4)-*O*-α-D-glucopyranosyl- (1→2)-β-D-fructofuranoside) is an oligosaccharide consisting of galactose, glucose, and fructose and has several benefits. This compound is 30 % as sweet as sucrose and used as functional food ingredient. Additionally, lactosucrose also plays an important role in the refinement of sugars in the food industry (Kawase et al. 2001).

10.5.1 Method of Production

Lactosucrose, a trisaccharide, can be synthesized either by transfructosylation of lactose with sucrose or transgalactosylation of sucrose with lactose.

10.5.1.1 Enzyme Involved

The biosynthesis of lactosucrose can be obtained with lactose and sucrose as substrates by the following enzymes:

1. Levansucrase (EC 2.4.1.10)
2. Fructofuranosidase (EC 3.2.1.26)
3. β-D-galactosidase (EC 3.2.1.23)

Levansucrase and fructofuranosidase have transfructosylation activity, whereas β-D-galactosidase has both hydrolytic and transgalactosylation activity.

10.5.1.2 Process

Commercially, lactosucrose (lactosylfructoside) has been produced from sucrose and lactose by transfructosylation (Gänzle et al. 2008). The microorganisms such as

Zymomonas mobilis, *Paenibacillus polymyxa*, and *Bacillus subtilis* have been used as a source for levansucrase whereas *Bacillus circulans* and *Arthrobacter* sp. for β-D-galactosidase and fructofuranosidase, respectively (Kawase et al. 2001; Li et al. 2009; Han et al. 2009; Choi et al. 2004; Park et al. 2005). Increase in lactosucrose yields (from 29 to 43 %) was observed by a mixed enzyme system coupling the transfructosylation of lactose to glucose removal by glucose oxidase (Han et al. 2009). The selection of strain, substrate concentration, and reaction time along with other optimal conditions play an important role in the production of lactosucrose. *B. subtilis* was found to be an effective producer of levansucrase, and the maximum lactosucrose production (181 g L^{-1}) was observed at 55 °C and pH 6.0 after 10 h of incubation (Park et al. 2005). Lactosucrose has also been produced through continuous process in a packed bed reactor using mutant strain of *Sterigmatomyces elviae*, which resulted in the production of 192 g L^{-1} of lactosucrose at 50 °C and pH 6.0 with the flow rate of 1.2 mL/min (Lee et al. 2007). *Z. mobilis* strain has been applied for levansucrase, and the maximum lactosucrose conversion efficiency of 28.5 % has been observed at 23 °C and pH 7.0 with lactose monohydrate (18.0 % w/v) and sucrose (18 % w/v) as substrate. The main problem associated with the low yield of lactosucrose is the presence of other carbohydrate products. To overcome this problem, a mixed enzyme system containing a levansucrase and a glucose oxidase has also been applied, and as a result of this, the efficiency of lactose and sucrose conversion to lactosucrose increases to 43.2 % (Han et al. 2009).

10.6 Fructo-oligosaccharides

Fructo-oligosaccharides (FOS) are low caloric carbohydrate alternative sweeteners with reduced energy production of about 2 kcal g^{-1} (Molis et al. 1996; Bornet et al. 2002). They are mainly composed of 1-kestose, nystose, and 1-β-fructofuranosyl nystose (Shin et al. 2004). FOS occurs naturally or can be produced by the action of fructosyltransferase enzymes from bacteria, yeast, and fungi (Sangeetha et al. 2005a; Maugeri and Hernalsteens 2007; Hernalsteens and Maugeri 2008). These complex biomolecules are enzymatically hydrolyzed into simpler forms, such as lactate, short-chain fatty acids (acetate, propionate, and butyrate), and gas, by endogenous microbiota. They bear properties of being low caloric, non-cariogenic, and non-mutagenic; absorption of ions in gut; lowering phospholipids, triglycerides, and cholesterol level; and bifidus-stimulating functionality (Vankova et al. 2008; Bali et al. 2012). Due to all these properties and bearing "Generally Recognized as Safe (GRAS)" status by Food and Drug Administration (FDA), they are being used as functional food ingredients and worth about US $200/kg (Godshall 2007).

10.6.1 Method of Production

FOS can be synthesized chemically or produced by the action of enzymes from various microorganisms.

10.6.1.1 Enzyme Involved

Fructo-oligosaccharides can be produced in large amounts by microbial enzymes bearing fructosyltransferase activity. They can be produced enzymatically by sucrose transformation (Bali et al. 2012) which leads to the production of short-chain FOS using:

1. FFase, i.e., β-D-fructofuranosidase (EC 3.2.1.26)
2. FTase, i.e., fructosyltransferase (EC 2.4.1.9)

FFase bears both hydrolytic and transfructosylating activity, whereas FTase bears only transfructosylating activity.

10.6.1.2 Process

The fructosylating enzyme can be produced from different microbial sources, such as bacterial (*Bacillus macerans* (Park et al. 2001), *Arthrobacter* sp. (Xu et al. 2009), *Z. mobilis* (Beker et al. 2002), *Lactobacillus reuteri* (Hijum et al. 2002)), fungal (*Aspergillus japonicus*, *Aspergillus niger*, *Aspergillus sydowii*, *Aspergillus foetidus*, *A. oryzae*, *Aureobasidium pullulans*, *Penicillium citrinum*, *Penicillium frequentans*, *Fusarium oxysporum*, *Aureobasidium* sp.), and yeast (*Kluyveromyces* and *Candida* sp.) (Vranesic et al. 2002; Bali et al. 2012). Microbial production, isolation, and purification of enzyme are preferred over plant sources due to ease of production and availability in large amounts. The enzyme can be produced either intracellularly or extracellularly. The whole cell synthesis of FOS can be carried out using microorganisms bearing fructosylating activity in sucrose-based media. An economical and advantageous system of recycling *A. oryzae* CFR 202, up to six cycles with 53 % yield, was developed for the production of extracellular FTase resulting in the synthesis of FOS under submerged fermentation (Sangeetha et al. 2005b). The technique of immobilizing the enzyme or whole cells (producing FFase or FTase) on various matrices has also been applied by various researchers for enhancing the production of FOS (Cheng et al. 1996; Jung et al. 2011).

Agro-industrial waste can act as good carbon or nitrogen source for the production of enzymes useful in the synthesis of FOS, thereby, adding up to process economics. Different agro-industrial wastes, such as soybean residue and sugarcane molasses, have been explored for the microbial synthesis of enzymes (Hayashi et al. 1992; Dorta et al. 2006). Production of β-fructofuranosidase has been carried out by fermentation using *A. japonicus* and soybean residue as substrate (Hayashi et al. 1992). Among the different substrates (corncobs, coffee silverskin, and cork oak) tested, coffee silverskin was reported to be the most suitable support and nutrient source for the production of FOS using *A. japonicus* under solid-state fermentation (Mussatto and Teixeira 2010). Low-cost complex media constituting cassava peel and cassava steep liquor, under solid-state and submerged fermentation, respectively, have been explored for the production of FTase and FOS (34 %) by *Rhizopus stolonifer* LAU 07 (Lateef and Gueguim kana 2012). The production of FTase by

A. foetidus NRRL 337 using apple pomace as a substrate has been reported (Hang et al. 1995). Besides this, sugarcane molasses has also been used for the production of FOS using *A. japonicus*-FCL 119T and *A. niger* ATCC 20611 (Dorta et al. 2006).

The increase in the production of β-fructofuranosidase in both wild-type and mutant strain of *A. niger* up to 252-fold and 516-fold, respectively, was observed using wheat bran as substrate, whereas corn steep liquor improved the activity of wild-type strain up to twofold (Rajoka and Yasmeen 2005). A patented method has been reported using wheat bran as substrate for the increased production of FOS (60 % yield; w/w) using *A. niger* β-fructofuranosidase (Park and Pastore 2006). Molasses can be used as substrate for both enzyme and oligosaccharide production using microbial sources (Ghazi et al. 2006). Among the various agricultural by-products (such as cereal bran, corn by-products, sugarcane bagasse, cassava bagasse, and by-products of coffee and tea processing) employed for the production of FTase using *A. oryzae* CFR 202 by solid-state fermentation (SSF), cereal bran, rice bran, wheat bran, corn germ, spent coffee, and tea processing were observed to be most suitable as substrates (Sangeetha et al. 2004). After fermentation, the microbial cell separation is generally done using centrifugation followed by recovery of the enzyme with different precipitation methods (ethanol precipitation, ammonium sulfate precipitation, etc.). Further, analysis of FOS can be carried out by using various analytical techniques, such as HPLC, TLC, GC-MS, and NMR (Sangeetha et al. 2005a).

10.7 Inulin

Inulins are long-chain storage carbohydrates that occur naturally in small amounts in various edible vegetables, fruits, cereals, and plants, such as asparagus, chicory, onion, wheat, banana, shallot, artichokes, leek, garlic, rye, tomatoes, topinambuco, and honey (Van et al. 1995). Inulin and its polyfructans consists of GF_n molecules ($2<n<60$) with linear β-2→1-linked polyfructose chains with glucose unit at its terminal (Waterhouse and Chatterton 1993). These compounds were discovered by Rose, German scientist, from hot water extract of Elecampane (*Inula helenium*) and the term was coined by Thomson in 1818. Oligofructose (inulin hydrolysate) is composed of GF_n and F_m molecules (with $2 \leq n$ and $m \leq 10$) with DP (degree of polymerization) ≤10. It has more solubility and 30–50 % sweetness as compared to table sugar (Niness 1999).

10.7.1 Method of Production

Inulin hydrolysates can occur naturally or can be produced by enzymatic methods.

10.7.1.1 Enzyme Involved

Inulinases produce inulin hydrolysates by partial enzymatic hydrolysis of inulin (Franck 2002). There are two types of inulinases:

1. Exoinulinases (β-D-fructanfructohydrolase; EC 3.2.1.80) act on both β 2→1 and β 2→6 linkages at the nonreducing end, thereby cleaving terminal fructose residues.
2. Endoinulinases (2,1-β-D-fructanohydrolase; EC 3.2.1.7) act on specific β 2→1 internal linkages yielding inulotriose, inulotetraose, and inulopentaose.

10.7.1.2 Process

Inulin can be obtained from natural plant sources by extraction using hot water diffusion followed by purification and then drying of inulin extract to obtain pure powder (Angus et al. 2005). In spite of their natural occurrence, fructans can also be produced using microbial sources like bacteria and fungi. Inulin has been used as substrate for the production of its hydrolysates using inulinase and short-chain oligosaccharides, such as inulo-oligosaccharide, oligofructose, and fructo-oligosaccharide.

Inulinases can be produced from various bacterial strains (*Streptomyces* sp., *Pseudomonas* sp., *Bacillus* sp.), yeast strains (*Kluyveromyces* sp., *Pichia* sp., *Candida* sp.), and fungal sources (*Aspergillus* sp., *Penicillium* sp.) (Mazutti et al. 2006; Neagu and Bahrim 2011). Inulinase production has also been carried out using different agro-industrial waste, such as cassava flour, corncob, oat meal, rice straw, sugar cane bagasse, and wheat bran using *Aspergillus ochraceus* (Guimaraes et al. 2007). Sugarcane bagasse and corn steep liquor have been used for the production of inulinases using *Kluyveromyces marxianus* NRRL Y-7571 under SSF (Mazutti et al. 2006). Three exoinulinases and two endoinulinases were purified from *Aspergillus ficuum* JNSP5-06 (Chen et al. 2009).

Thermostable extracellular immobilized inulinases from *Aspergillus fumigates* have been used for the hydrolysis of inulin (Gill et al. 2006). Inulobiose and other higher oligofructosides were produced using soluble (inulobiose and DP3 oligosaccharides as product; 72 % yield) and immobilized endoinulinases (higher content of inulobiose; 83 % yield) using batch fermentation (Yun et al. 1997a). Immobilized system has also been applied for the production of inulo-oligosaccharides using immobilized enzymes or enzyme-producing whole cells. Continuous production of inulo-oligosaccharides (83 % yield) has been achieved using immobilized endoinulinases produced by *Pseudomonas* sp. and inulin as substrate (Yun et al. 1997b). Continuous production of inulo-oligosaccharides (82 % yield) has also been reported using immobilized polystyrene-bound endoinulinase and chicory juice as substrate for 28 days at 55 °C (Yun et al. 2000). The inulo-oligosaccharide with DP2 to DP4 and DP2 to DP8 was produced at 45 °C after 72 h with partially purified (pH 6.0 and 50 % yield) and purified (pH 5.0, 70 % yield) *A. ficuum* endoinulinase, respectively, using 50 g L^{-1} inulin and enzyme concentration of 10 U g^{-1} substrate

(Zhengyu et al. 2005). *Pseudomonas mucidolens* endoinulinase gene expressed on *Saccharomyces cerevisiae* cell surface resulted in hydrolysis of inulin (Jerusalem artichoke) with 2.31 U mL^{-1} of activity at temperature of 50 °C and pH of 7.0 and formation of inulotetraose (F4) as major product along with inulobiose (F2), inulotriose (F3), and inulopentaose (F5) formation (Kim et al. 2008). Similarly, *K. marxianus* CBS 6556 endoinulinase gene *INU1* which has been expressed on citric acid producing *Yarrowia lipolytica* hydrolyzed 77.9 % inulin and resulted in formation of inulin-oligosaccharides (mono-, di-, and minor tri-) at 50 °C, pH 4.5 with initial 12 % inulin concentration and inulinase activity of 181.6 U g^{-1} within 10 h of incubation (Liu et al. 2010).

10.8 Malto-oligosaccharides

Malto-oligosaccharides (MOS) are low sweeteners with properties like antistaling effect on bread, high water holding capacity, and preventing sucrose crystallization. Although α-amylases produce maltose or glucose from starch, some of the microbial α-amylases lead to the production of malto-oligosaccharides which may have their applications in MOS containing syrups (Park 1992).

10.8.1 Method of Production

10.8.1.1 Enzyme Involved

The following are enzymes which play a role in the production of malto-oligosaccharides:

1. α-Amylase (EC 3.2.1.1; 1→4-α-D-glucanohydrolase or endoamylase) results in the hydrolysis of starch by cleaving α-D-(1→4) glycosidic linkages, thereby producing specific or mixed malto-oligosaccharides.
2. α-Amylase (EC 3.2.1.2) results in the hydrolysis of starch by formation of maltose, followed by transglucosylation of maltose using α-glucosidase (EC 3.2.1.20).
3. Cyclodextrin glucanotransferase (CGTase) results in the formation of β-malto-oligosaccharides through glycosylation of bioactive tocopherols and isoflavones.

10.8.1.2 Process

Microbial amylases have been produced by *Bacillus* sp. (Nagarajan et al. 2006), *Brachybacterium* sp. (Doukyu et al. 2007), *Marinobacter* sp. (Kumar and Khare 2012), etc. for the synthesis of MOS. The transformation of maltose into isomalto-oligosaccharide using α-glucosidase from *Xanthophyllomyces dendrorhous* has been reported (Fernández-Arrojo et al. 2007). The production of α-amylases has also been carried out using agro-industrial waste/by-products, such as sugarcane molasses, cheese whey, rice husk, and wheat bran (Babu and Satyanarayana 1995; Baysal et al. 2003; Sodhi et al. 2005). *B. subtilis* has been reported to produce α-amylases on sugarcane bagasse hydrolysate (Nagarajan et al. 2006; Rajagopalan and Krishnan 2008).

Immobilized *Penicillium lilacinum* dextranase on Eupergit C resulted in isomalto-oligosaccharides production with 90 % relative activity up to 20 batch reactions (Aslana and Tanriseven 2007). A sandwich-structured enzyme membrane reactor has been used to convert maltose into isomalto-oligosaccharide with 100 % conversion and 58 % yield (Zhang et al. 2010). Transglucosidase producing *A. niger* has been reported to form various MOS (DP 2–8) from mixture of maltoheptaose and [U–^{13}C] maltose (Ota et al. 2009). The synthesis of β-malto-oligosaccharides has also been carried out from glycitein and daidzein using *Lactobacillus delbrueckii* and cyclodextrin glucanotransferase (CGTase) by sequential glycosylation (Shimoda and Hamada 2010). Similarly, β-malto-oligosaccharides as tocopherol derivatives were synthesized by glycosylation of α- and δ-tocopherols using *Klebsiella pneumoniae* and CGTase (Shimoda et al. 2009).

10.9 Xylo-oligosaccharides

Xylo-oligosaccharides (XOS) are sugar oligomers made up of xylose units and are present naturally in bamboo shoots, fruits, vegetables, milk, and honey. Xylo-oligosaccharides are nondigestible food ingredients having lower degree of polymerization (DP) and are commercially produced during the hydrolysis of xylan, the main component of the plant hemicelluloses (Brienzo et al. 2010). The composition and structure of the XOS depend upon the source and the production process. Generally, XOS are mixtures of oligosaccharides formed by xylose residues linked through β-1 → 4 linkages. The number of xylose residues involved in their formation can vary from two to ten, and depending upon the number of xylose residues, they are known as xylobiose, xylotriose, and so on. The sweetness of xylobiose is equivalent to 30 % that of sucrose and possesses no off-taste or off-odor. In addition to the health effects, XOS present interesting physicochemical properties; they are moderately sweet and stable over a wide range of pH and temperatures and have organoleptic characteristics suitable for incorporation into foods (Barreteau et al. 2006).

10.9.1 Method of Production

Xylo-oligosaccharide production from xylan containing biomass is accomplished by xylanase which hydrolyzes β-1 → 4 glycosidic linkages in xylan.

10.9.1.1 Enzyme Involved

In nature, xylanolytic enzyme systems consist of following types:

1. Endoxylanase (EC 3.2.1.8)
2. β-D-xylosidases (EC 3.2.1.37)
3. Debranching enzymes, esterases (EC 3.1.1.72)

10.9.1.2 Process

For the production of XOS, the enzyme complex should have low exoxylanase or β-xylosidase activity. Enzyme with high exoxylanase or β-xylosidase activity produces high amount of xylose which causes inhibition in XOS production (Vazquez et al. 2002). A variety of microorganisms are reported to produce endoxylanases that can degrade β-1→4 xylan in a random fashion, yielding a series of linear and branched oligosaccharide fragments. These enzymes have been widely detected in filamentous fungi, namely, *Aspergillus* sp., *Trichoderma* sp., *Penicillium* sp., and *Thermomyces lanuginosus* (Akpinar et al. 2009). Bacterial strains that are known to produce endoxylanase include *Bacillus halodurans* (Lin et al. 2011), *Cellulosimicrobium* sp. (Kim et al. 2009), *Streptomyces* sp. (Puchart and Biely 2008), and *B. subtilis* (Yuan et al. 2005). Yeasts have also been reported to produce xylanases and this group includes *Malbranchea flava* (Sharma et al. 2010), *Pseudozyma hubeiensis* (Bastawde et al. 1994), and *Pichia stipitis* (Yang et al. 2011). Most of the xylanases known to date are optimally active at temperatures below 50 °C and are active in acidic or neutral pH (Ryan et al. 2003). Investigation on novel sources of bacterial xylanase producers, which display high optimal xylanase activity and stability in more drastic conditions, is still in progress. Moreover, wide-scale industrial application of xylanase requires their cost-effective production to make the process economically viable. Annually, large quantities of lignocellulosic wastes are generated through industrial processes. There are many studies regarding the production of xylanases using waste biomass such as wheat bran, oat bran, rice straw, and corncobs.

Rice straw as an agricultural residue has the highest dry weight percentage of xylan, 24.5 %, and thus would appear to present a high potential for XOS production under optimized pretreatment processes. Other feed stocks having more than 20 % of xylan, such as beech, corn stover, bagasse, tobacco stalk, cotton stalk, sunflower stalk, wheat straw, switch grass, and big blue stem, also have great potential for XOS production. Corncob XOS were produced after enzymatic hydrolysis with a cumulative yield of 67.7 g/100 g (based on xylan in raw material), and the purity of xylo-oligosaccharides was over 70 g/100 g (Yang et al. 2005). Xylo-oligosaccharides have also been successfully isolated from hardwood xylan (Nishimura et al. 1998), delignified cottonseed residual cake (Sun et al. 2002). In another study, XOS was extracted from corncob xylan by acid hydrolysis. HPLC analysis of hydrolysate revealed increased production of XOS with time, but prolonged incubation resulted in higher amount of xylose (Samanta et al. 2012). The autohydrolysis of *Eucalyptus globulus* wood, corncobs, rice husks, and barley husks resulted in sugar oligomers, xylo-oligosaccharides, monosaccharides, acetyl and uronic acid substituents of oligomers, free acetic acid, furfural, Folin-Denis phenols, and other compounds (Parajo et al. 2004). Among different agricultural wastes, namely, tobacco stalk, cotton stalk, sunflower stalk, and wheat straw, examined for the production of xylo-oligosaccharide, the best xylan conversion into XOS was achieved with 0.25 M H_2SO_4 with 30-min reaction time followed by enzymatic hydrolysis (Akpinar et al. 2009).

10.10 Applications of Prebiotics

Prebiotics, a nondigestible oligosaccharide, have various food and non-food applications in different sectors including feed, agriculture, and pharmaceutical (Fig. 10.2). These molecules serve as substrate for probiotic bacteria, thereby, leading to selective enhancement of their growth, which in turn provides health benefits to the host. The various associated health benefits and applications of prebiotics have been discussed in subsequent sections.

10.10.1 Functional Foods

The health effects imparted by oligosaccharides make them active ingredients of "functional foods" which are similar in appearance to conventional foods that are consumed as part of a normal diet and have physiological benefits and/or reduce the

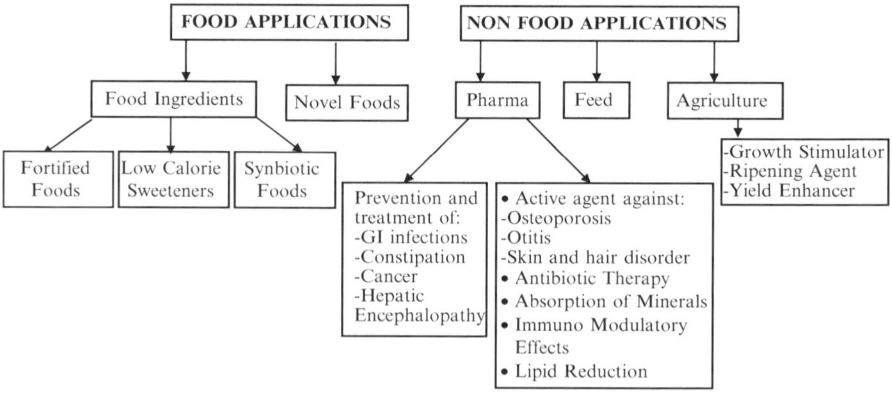

Fig. 10.2 Potential applications of prebiotics

risk of chronic disease beyond basic nutritional functions (Clydesdale 1997). As food ingredients, prebiotics have an acceptable odor and are noncarcinogenic and low caloric, allowing their utilization in anti-obesity diets (Toshio et al. 1990; Kazumitsu et al. 1997). Besides this, lactulose is also incorporated in infant formulas to stimulate health-beneficial microflora like bifidobacterium (Nagendra et al. 1995).

10.10.2 Gastroenterological Effects

Prebiotics, such as lactulose, FOS, and GOS, have laxative effects (Roberfroid 1993), with lactulose in particular being well established as a treatment for constipation (Schumann 2002). Lactulose increases the water content and volume of the stools in the bowel, making them softer and easier to pass, thereby preventing constipation.

10.10.3 Regulation of Lipid Metabolism

Nowadays, lowering of the triglyceride levels in the blood is of major concern. The deposition of cholesterol in the arterial wall may lead to atherosclerosis and coronary heart disease. Prebiotics such as galacto-oligosaccharides and xylo-oligosaccharides are reported to lower the serum cholesterol and triglyceride level, respectively, in animals (Chonan et al. 1995; Beylot 2005).

10.10.4 Absorption of Minerals

Minerals play an important role in various biosynthetic pathways, hardness of the bone, and overcoming various diseases like anemia (Heaney 1996). Mostly, calcium is stored in bone and Mg^{2+} salts are responsible for the hardness of bones (Rude 1996). The fermentation of prebiotics lowers the pH in the intestine, thereby helping in the increased absorption of various minerals, such as Mg^{2+}, Fe^{2+}, and Ca^{2+}. Therefore, prebiotics can have beneficial effect in the prevention of osteoporosis and osteopenia (Amarowicz 1999; Murosaki et al. 1999).

10.10.5 Cancer Prevention

Intake of prebiotics can help in suppressing chemically induced colorectal cancer and precancerous colon lesions (Pierre et al. 1997; Hsu et al. 2004). Short-chain fatty acid, such as butyrate, has protective action in the colon by preventing tumor growth and cell differentiation and upregulates apoptosis (Reddy 1999).

10.10.6 Immunity Enhancement

Prebiotics have influential role on the immune system, microbial composition, and metabolic product formation in the gastrointestinal tract. They provide both nonspecific (physical barrier against toxins and pathogens) and specific (gut-associated lymphoid tissue) protection and improve resistance against infection (Watzl et al. 2005). Colonic microbiota is helpful in tolerance to bacterial and dietary antigens (Roller et al. 2004). Immunological effects, such as increased level of mucosal immunoglobulin production, mesenteric lymph nodes, Peyer's patches, and altered cytokine formation and lymphocyte numbers, have been reported as an effect of prebiotic intake in the diet (Schley and Field 2002).

10.11 Market Demand of Prebiotics

Foods containing prebiotics are a growing segment in the world market due to their beneficial effects on human health (Fig. 10.3). They are known to provide specific health benefits, hence belong to a special class of foods known as functional foods. For this reason, there has been a rapid market growth in recent years all over the world. This is particularly applicable for XOS, having a selling price of 2,500 yen/kg in Japanese market, the highest among 13 different types of oligosaccharides (Stanton et al. 2001). The current largest world market for functional foods is the United States, followed by Europe and Japan, while Germany, France, United Kingdom, and the Netherlands represent the most important countries within the

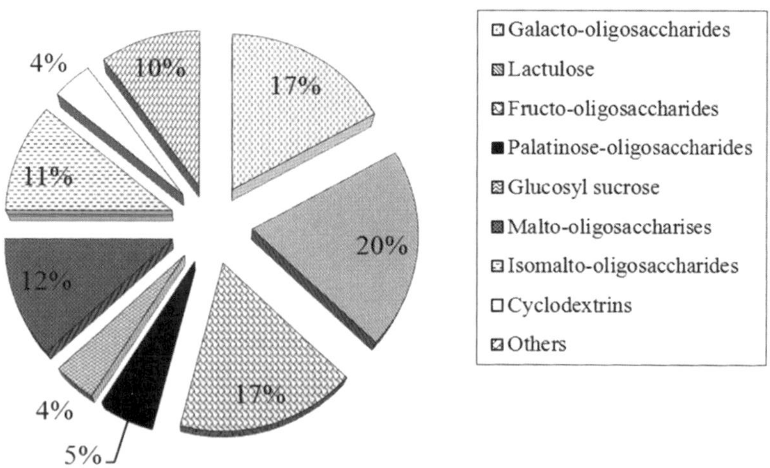

Fig. 10.3 Global production status of different prebiotics (*Source*: Singh and Singh 2010; Reproduced with permission)

functional food market in Europe. According to Global Industry Analysts (GIA) report, by 2015, the US and European prebiotic market will reach nearly $225 million and $1.12 billion, respectively, mainly impacted by prebiotic meat and snack food products (Neutraceuticals world 2010). Inulin will contribute a major portion (35 %) due to its textural resemblance with fats followed by mannan oligosaccharide (25 %) and fructan oligosaccharides (10 %) as natural sweeteners (Watson 2011). There is yet an unfulfilled potential in the world market for emerging prebiotics in terms of their production and purification that can be optimized from cellulosic biomass pretreatments.

10.12 Summary

Worldwide awareness of consumers towards diet and health has opened up new opportunities for food industries in research and development of functional foods. Foods that contain pre- and probiotics are getting special attention of consumers and are potentially exciting component of the food market. Different prebiotics can be used for the fortification of different food products for the designing of functional foods for the special target groups. Prebiotics support the growth of beneficial bacteria, thereby adding up to potential health and nutritional benefits. These compounds are associated with prevention and treatment of various chronic diseases, such as constipation, hepatic encephalopathy, and cancer. However, to improve the economics of prebiotic production, technologies based on bio-utilization of agro-industrial waste/by-products (whey, wheat and rice straw, sugarcane bagasse, etc.) need to be further strengthened.

References

Adamczak M, Charubin D, Bednarski W (2009) Influence of reaction medium composition on enzymatic synthesis of galacto-oligosaccharides and lactulose from lactose concentrates prepared from whey permeate. Chem Pap 63:111–116

Aider M, de Halleux D (2007) Isomerization of lactose and lactulose production: review. Trends Food Sci Technol 187:356–364

Akiyama K, Takase M, Horikoshi K et al (2001) Production of galactooligosaccharides from lactose using a beta-glucosidase from *Thermus sp* Z-1. Biosci Biotechnol Biochem 65:438–441

Akpinar O, Erdogan K, Bostanci S (2009) Production of xylooligosaccharides by controlled acid hydrolysis of lignocellulosic materials. Carbohydr Res 344:660–666

Albayrak N, Yang ST (2002) Production of galacto-oligosaccharides from lactose by *Aspergillus oryzae* β-galactosidase immobilized on cotton cloth. Biotechnol Bioeng 77:8–19

Amarowicz R (1999) Nutritional importance of oligosaccharides. Rocz Panstw Zakl Hig 50(1):89–95

Angus F, Smart S, Shortt C (2005) Prebiotic ingredients with emphasis on galactooligosaccharides and fructo-oligosaccharides. In: Tamime AY (ed) Probiotic dairy products. Blackwell, Oxford, UK

Aslana Y, Tanriseven A (2007) Immobilization of *Penicillium lilacinum* dextranase to produce isomaltooligosaccharides from dextran. Biochem Eng J 34(1):8–12

Babu KR, Satyanarayana T (1995) α-Amylase production by thermophilic *Bacillus coagulans* in solid state fermentation. Process Biochem 30:305–309

Bali V, Panesar PS, Bera MB (2012) Fructo-oligosaccharides: production, purification and potential applications. Crit Rev Food Sci Nutr (accepted manuscript)

Barreteau H, Delattre C, Michaud P (2006) Production of oligosaccharides as promising new food additive generation. Food Technol Biotechnol 44:323–333

Bastawde KB, Puntambekar US, Gokhale DV (1994) Optimization of cellulase free xylanase production by a novel yeast strain. J Ind Microbiol 13:220–224

Baysal Z, Uyar F, Aytekin C (2003) Solid state fermentation for production of α-amylase by a thermotolerant *Bacillus subtilis* from hot-spring water. Process Biochem 38:1665–1668

Beker M, Laukevics J, Upite D et al (2002) Fructooligosaccharide and levan producing activity of Zymomonas mobilis and extracellular levan sucrase. Process Biochem 38:701–706

Beylot M (2005) Effects of inulin-type fructans on lipid metabolism in man and in animal models. Br J Nutr 93:63–68

Boon MA, van der Oost J, de Vos WM et al (1998) Synthesis of oligosaccharides catalyzed by thermostable beta-glucosidase from *Pyrococcus furiosus*. Appl Biochem Biotechnol 75:269–278

Bornet FRJ, Brouns F, Tashiro Y et al (2002) Nutritional aspects of short-chain fructooligosaccharides: natural occurrence, chemistry, physiology and health implications. Dig Liver Dis 34(2):111–120

Brienzo M, Carvalho W, Milagres AMF (2010) Xylooligosaccharides production from alkali pretreated sugarcane bagasse using xylanase from *Thermoascus aurantiacus*. Appl Biochem Biotechnol 162:1195–1205

Cardelle-Cobas A, Villamiel M, Olano A et al (2008) Study of galacto-oligosaccharide formation from lactose using pectinex ultra SP-L. J Sci Food Agric 88:954–961

Chen HQ, Chen XM, Li Y et al (2009) Purification and characterisation of exo- and endo-inulinase from *Aspergillus ficuum* JNSP5-06. Food Chem 115:1206–1212

Cheng CY, Duan KJ, Sheu DC et al (1996) Production of fructooligosaccharides by immobilized mycelium of *Aspergillus japonicas*. J Chem Technol Biotechnol 66(2):135–138

Choi JJ, Oh EJ, Lee YJ et al (2003) Enhanced expression of the gene for beta-glycosidase of *Thermus caldophilus* GK24 and synthesis of galacto-oligosaccharides by the enzyme. Biotechnol Appl Biochem 38:131–136

Choi H-J, Kim CS, Kim P et al (2004) Lactosucrose bioconversion from lactose and sucrose by whole cells of *paenibacillus polymyxa* harboring levansucrase activity. Biotechnol Prog 20:1876–1879

Chonan O, Matsumoto K, Watanuki M (1995) Effect of galactooligosaccharides on calcium absorption and preventing bone loss in ovariectomized rats. Biosci Biotechnol Biochem 59:236–239

Clydesdale FM (1997) A proposal for the establishment of scientific criteria for health claims for functional foods. Nutr Res 55:413–423

Conway PL (2001) Prebiotics and human health: the state-of-the-art and future perspectives. Scand J Nutr 45:13–21

Cruz R, Cruz VD, Belote JG et al (1999) Production of transgalactosylated oligosaccharides (TOS) by galactosyltransferase activity from *Penicillium simplicissimum*. Bioresour Technol 70:165–171

De Harr WT, Pluim H (1991) Method of preparing lactulose. European patent 0339749

Dorta C, Cruz R, de Oliva-Neto P et al (2006) Sugarcane molasses and yeast powder used in the fructooligosaccharides production by *Aspergillus japonicus*-FCL 119T and *Aspergillus niger* ATCC 20611. J Ind Microbiol Biotechnol 33(12):1003–1009

Doukyu N, Yamagishi W, Kuwahara H et al (2007) Purification and characterization of a maltooligosaccharide-forming amylase that improves product selectivity in water-miscible organic solvents, from dimethylsulfoxide-tolerant *Brachybacterium* sp. strain LB25. Extremophiles 11(6):781–788

Fernández-Arrojo L, Marín D, De Segura AG et al (2007) Transformation of maltose into prebiotic isomaltooligosaccharides by a novel α-glucosidase from *Xanthophyllomyces dendrorhous*. Process Biochem 42(11):1530–1536

Foda MI, Lopez-Leiva M (2000) Continuous production of oligosaccharides from whey using a membrane reactor. Process Biochem 35:581–587
Franck A (2002) Technological functionality of inulin and oligofructose. Br J Nutr 87:287–291
Gänzle MG (2011) Lactose galacto-oligosaccharides. In: Fuquay JW, Fox PF, McSweeney P (eds) Encyclopedia of dairy science, 2nd edn. Elsevier, Oxford, UK
Gänzle MG, Haase G, Jelen P (2008) Lactose: crystallization, hydrolysis and value-added derivatives. Int Dairy J 18:685–694
Ghazi I, Fernandez-Arrojo L, Gomez De Segura A et al (2006) Beet sugar syrup and molasses as low-cost feedstock for the enzymatic production of fructo-oligosaccharides. J Agric Food Chem 54(8):2964–2968
Gill PK, Manhas RK, Singh P (2006) Hydrolysis of inulin by immobilized thermostable extracellular exoinulinase from *Aspergillus fumigates*. J Food Eng 76:369–375
Godshall MA (2007) Future directions for the sugar industry. http://www.spriinc.org/buton10bftpp.html
Goulas AK, Kapasakalidis PG, Sinclair HR et al (2002) Purification of oligosaccharides by nanofiltration. J Membr Sci 209:321–335
Guimaraes LHS, Terenzi HF, Polizeli ML et al (2007) Production and characterization of a thermostable extracellular β-D-fructofuranosidase produced by *Aspergillus ochraceus* with agroindustrial residues as carbon sources. Enzyme Microb Technol 42:52–57
Han W-C, Byun S-H, Kim M-H et al (2009) Production of lactosucrose from sucrose and lactose by a levansucrase from *Zymomonas mobilis*. J Microbiol Biotechnol 19:1153–1160
Hang YD, Woodams EE, Jang KY (1995) Enzymatic conversion of sucrose to ketose by fungal extracellular fructosyltransferase. Biotechnol Lett 17:295–298
Hayashi S, Matsuzaki K, Kawahara T et al (1992) Utilisation of soybean residue for the production of β-fructofuranosidase. Bioresour Technol 41(3):231–233
Heaney RP (1996) Calcium. In: Raisz LG, Rodan GA, Bilezikian JP (eds) Principle of bone biology. Academic, San Diego, CA
HelleroVá K, Čurda L (2009) Influence of type of substrate and enzyme concentration on formation of galacto-oligosaccharides. Czech J Food Sci 27:327–374
Hernalsteens S, Maugeri F (2008) Purification and characterisation of a fructosyltransferase from *Rhodotorula* sp. Appl Microbiol Biotechnol 79(4):589–596
Hicks KB, Raupp DL, Smith PW (1984) Preparation and purification of lactulose from sweet cheese whey ultrafiltrate. J Agric Food Chem 32:288–292
Hijum SV, van Geel-Schutten GH, Rahouri H et al (2002) Characterization of a novel fructosyl transferase from *Lactobacillus reuteri* that synthesizes high molecular weight inulin and inulin oligosaccharides. Appl Environ Microbiol 68:4390–4398
Hsu CK, Liao JW, Chung YC et al (2004) Xylooligosaccharides and fructooligosaccharides affect the intestinal microbiota and precancerous colonic lesion development in rats. J Nutr 134:1523–1528
Jaindl K, Schuster-Wolff-Bühring R, Fischer L et al (2009) Enzymic synthesis of prebiotic lactulose in milk and whey products. DMZ Lebensmittelindustrie und Milchwirtschaft 130:24–27
Jovanovic-Malinovska R, Fernandes P, Winkelhausen E et al (2012) Galacto-oligosaccharides synthesis from lactose and whey by β-galactosidase immobilized in PVA. Appl Biochem Biotechnol. doi:10.1007/s12010-012-9850-1
Jung KH, Bang SH, OH TK et al (2011) Industrial production of fructooligosaccharides by immobilized cells of *Aureobasidium pullulans* in a packed bed reactor. Biotechnol Lett 33(8):1621–1624
Kawase M, Pilgrim A, Araki T et al (2001) Lactosucrose production using a simulated moving bed reactor. Chem Eng Sci 56:453–458
Kazumitsu S, Boseki I, Norio S et al (1997) Production of food and drink. Japanese Patent JP 9248153
Kim Y-S, Park C-S, Oh D-K (2006) Lactulose production from lactose and fructose by a thermostable β-galactosidase from *Sulfolobus solfataricus*. Enzyme Microb Technol 39:903–908

Kim HC, Kim HJ, Choi WB et al (2008) Inulo-oligosaccharide production from inulin by *Saccharomyces cerevisiae* strain displaying cell surface endoinulases. J Microbiol Biotechnol 16(3):360–367
Kim DY, Han MKY, Lee JS et al (2009) Isolation and characterization of a cellulase-free endo-β-1, 4-xylanase produced by an invertebrate-symbiotic bacterium, *Cellulosimicrobium* sp. HY-13. Process Biochem 44:1055–1059
Kumar S, Khare SK (2012) Purification and characterization of maltooligosaccharide-forming α-amylase from moderately halophilic *Marinobacter* sp. EMB8. Bioresour Technol 116:247–251
Ladero M, Perez MT, Santos A et al (2003) Hydrolysis of lactose by free and immobilized β-galactosidase from *Thermus* sp. strain T2. Biotechnol Bioeng 81:241–252
Lateef A, Gueguim kana EB (2012) Utilization of cassava wastes in the production of fructosyltransferase by *Rhizopus stolonifer* LAU 07. Rom Biotechnol Lett 17(3):7309–7316
Lee YJ, Kim CS, Oh DK (2004) Lactulose production by β-galactosidase in permeabilized cells of *Kluyveromyces lactis*. Appl Microbiol Biotechnol 64:787–793
Lee JH, Lim JS, Park C et al (2007) Continuous production of lactosucrose by immobilized *Sterigmatomyces elviae* mutant. J Microbiol Biotechnol 17:1533–1537
Li W, Xiang X, Tang S et al (2009) Effective enzymatic synthesis of lactosucrose and its analogues by β-D-galactosidase from *Bacillus circulans*. J Agric Food Chem 57:3927–3933
Lin YS, Tsengb MJ, Lee WC (2011) Production of xylooligosaccharides using immobilized endo-xylanase of *Bacillus halodurans*. Process Biochem 46:2117–2121
Liu X-Y, Chi Z, Liu G-L et al (2010) Inulin hydrolysis and citric acid production from inulin using the surface-engineered *Yarrowia lipolytica* displaying inulinase. Metab Eng 2(5):469–476
López Leiva MHL, Guzman M (1995) Formation of oligosaccharides during enzymic hydrolysis of milk whey permeates. Process Biochem 30:757–762
Mahoney RR (1998) Galactosyl-oligosaccharide formation during lactose hydrolysis: a review. Food Chem 63:147–154
Maugeri F, Hernalsteens S (2007) Screening of yeast strains for transfructosylating activity. J Mol Catal B Enzym 49:43–49
Mayer J, Kranz B, Fischer L (2010) Continuous production of lactulose by immobilized thermostable β-glycosidase from *Pyrococcus furiosus*. J Biotechnol 145:387–393
Mazutti M, Bender JP, Treichel H et al (2006) Optimization of inulinase production by solid-state fermentation using sugarcane bagasse as substrate. Enzyme Microb Technol 39(1):56–59
Méndez A, Olano A (1979) Lactulose: a review on some chemical properties and applications in infant nutrition and medicine. Dairy Sci Abstr 41:531–535
Molis C, Flourie B, Ouarne F et al (1996) Digestion, excretion, and energy value of fructooligosaccharides in healthy humans. Am J Clin Nutr 64(3):324–328
Murosaki S, Muroyama K, Yamamoto Y et al (1999) Immunopotentiating activity of nigerooligosaccharides for the T helper 1-like immune response in mice. Biosci Biotechnol Biochem 63(2):373–378
Mussatto SI, Teixeira JA (2010) Increase in the fructooligosaccharides yield and productivity by solid-state fermentation with *Aspergillus japonicus* using agroindustrial residues as support and nutrient source. Biochem Eng J 53:154–157
Nagarajan DR, Rajagopalan G, Krishnan C (2006) Purification and characterization of a maltooligosaccharide forming α-amylase from a new *Bacillus subtilis* KCC103. Appl Microbiol Biotechnol 73:591–597
Nagendra R, Viswanatha S et al (1995) Effect of feeding milk formula containing lactulose to infants on faecal bifidobacterial flora. Nutr Res 15:15–24
Nakkharat P, Kulbe KD, Yamabhai M et al (2006) Formation of galacto-oligosaccharides during lactose hydrolysis by a novel β-galactosidase from the moderately thermophilic fungus *Talaromyces thermophilus*. Biotechnol J 1:633–638
Neagu C, Bahrim G (2011) Inulinases—a versatile tool for biotechnology. Innovat Rom Food Biotechnol 9:1–11

Neutraceuticals World (2010) Report finds significant potential in prebiotics market. http://www.nutraceuticalsworld.com/contents/view_breaking-news/2010-02-23/report-finds-significant-potential-in-prebiotics-m/
Nguyen TH, Splechtna B, Krasteva S et al (2007) Characterization and molecular cloning of a heterodimeric *β*-galactosidase from the probiotic strain *Lactobacillus acidophilus* R22. FEMS Microbiol Lett 269:136–144
Niness KR (1999) Inulin and oligofructose: what are they? J Nutr 129:1402–1406
Nishimura T, Ishihara M, Tadashi I et al (1998) Alkaline xylanases from *Bacillus mojavensis* A21: production and generation of xylooligosaccharides. Carbohydr Res 308:117–122
Onishi N, Tanaka T (1997) Purification and characterization of galacto-oligosaccharide producing *β*-galactosidase from *Sirobasidium magnum*. Lett Appl Microbiol 24:82–86
Ota M, Okamoto T, Wakabayashi H (2009) Action of transglucosidase from *Aspergillus niger* on maltoheptaose and [U–13C] maltose. Carbohydr Res 344:460–465
Panesar PS, Panesar R, Singh RS et al (2006) Microbial production, immobilization and applications of *β*-D-galactosidase. J Chem Technol Biotechnol 81:530–543
Parajo JC, Garrote G, Cruz JM et al (2004) Effects of xylooligosaccharides and sugars on the functionality of porcine myofibrillar proteins during heating and frozen storage. Trends Food Sci Technol 15:115–120
Park K (1992) Development of new carbohydrate materials. Food Sci Ind 25:73–82
Park YK, Pastore GM (2006) Process for preparing β-fructofuranosidase enzyme and a process for producing fructooligosaccharides. US Patent 7063976
Park J, Oh T, Yun JW (2001) Purification and characterization of a novel transfructosylating enzyme from *Bacillus macerans* EG-6. Process Biochem 37:471–476
Park N-H, Choi H-J, Oh D-K (2005) Lactosucrose production by various microorganisms harboring levansucrase activity. Biotechnol Lett 27:495–497
Petzelbauer I, Zeleny R, Reiter A et al (2000) Development of an ultrahigh- temperature process for the enzymatic hydrolysis of lactose: II. Oligosaccharide formation by two thermostable *β*-glycosidases. Biotechnol Bioeng 69:140–149
Pierre F, Perrin P, Champ M et al (1997) Short-chain fructo-oligosaccharides reduce the occurrence of colon tumors and develop gut-associated lymphoid tissue in Min mice. Cancer Res 57(2):225–228
Placier G, Watzlawick H, Rabiller C et al (2009) Evolved *β*-galactosidases from *Geobacillus stearothermophilus* with improved transgalactosylation yield for galacto-oligosaccharides production. Appl Environ Microbiol 75:6312–6321
Puchart V, Biely P (2008) Simultaneous production of endo-*β*-1,4-xylanase and branched xylooligosaccharides by *Thermomyces lanuginosus*. J Biotechnol 137:34–43
Rajagopalan G, Krishnan C (2008) *α*-Amylase production from catabolite derepressed *Bacillus subtilis* KCC103 utilizing sugarcane bagasse hydrolysate. Bioresour Technol 99:3044–3050
Rajoka MI, Yasmeen A (2005) Improved productivity of *β*-fructofuranosidase by a derepressed mutant of *Aspergillus niger* from conventional and non-conventional substrates. World J Microbiol Biotechnol 21:471–478
Reddy BS (1999) Possible mechanisms by which pro- and prebiotics influence colon carcinogenesis and tumor growth. J Nutr 129:1478–1482
Roberfroid MB (1993) Dietary fibre, inulin and oligofructose: a review comparing their physiological effects. Crit Rev Food Sci Nutr 33:103–148
Roberfroid MB (1997) Health benefits of non-digestible oligosaccharides. Adv Exp Med Biol 427:211–219
Roller M, Rechkemmer G, Watzl B (2004) Prebiotic inulin enriched with oligofructose in combination with the probiotics *Lactobacillus rhamnosus* and *Bifidobacterium lactis* modulates intestinal immune function in rats. J Nutr 134:153–156
Rude RK (1996) Magnesium homeostasis. In: Raisz LG, Rodan GA, Bilezikian JP (eds) Principle of bone biology. Academic, San Diego, CA
Ryan SE, Nolan K, Thompson R et al (2003) Purification and characterization of a new low molecular weight endoxylanase from *Penicillium capsulatum*. Enzyme Microb Technol 33:775–785

Sakai T, Tsuji H, Shibata S et al (2008) Repeated-batch production of galactooligosaccharides from lactose at high concentration by using alginate-immobilized cells of *Sporobolomyces singularis* YIT 10047. J Gen Appl Microbiol 54:285–293

Sako T, Matsumoto K, Tanaka R (1999) Recent progress on research and applications of non-digestible galacto-oligosaccharides. Int Dairy J 9:69–80

Samanta AK, Jayapal N, Kolte AP et al (2012) Enzymatic production of xylooligosaccharides from alkali solubilized xylan of natural grass (*Sehima nervosum*). Bioresour Technol 112:199–205

Sangeetha PT, Ramesh MN, Prapulla SG (2004) Production of fructosyl transferase by *Aspergillus oryzae* CFR 202 in solid-state fermentation using agricultural by-products. Appl Microbiol Biotechnol 65:530–537

Sangeetha PT, Ramesh MN, Prapulla SG (2005a) Recent trends in the microbial production, analysis and application of fructooligosaccharides. Trends Food Sci Technol 16:442–457

Sangeetha PT, Ramesh MN, Prapulla SG (2005b) Fructooligosaccharide production using fructosyl transferase obtained from recycling culture of *Aspergillus oryzae* CFR 202. Process Biochem 40:1085–1088

Schley PD, Field CJ (2002) The immune-enhancing effects of dietary fibres and prebiotics. Br J Nutr 87:221–230

Schumann C (2002) Medical, nutritional and technological properties of lactulose. An update. Eur J Nutr 41:17–25

Sharma M, Chadha BS, Saini HS (2010) Purification and characterization of two thermostable xylanases from *Malbranchea flava* active under alkaline conditions. Bioresour Technol 101:8834–8842

Sheu DC, Li SY et al (1998) Production of galactooligosaccharides by β-galactosidase immobilized on glutaraldehyde-treated chitosan beads. Biotechnol Tech 12:273–276

Shimoda K, Hamada H (2010) Synthesis of β-maltooligosaccharides of glycitein and daidzein and their anti-oxidant and anti-allergic activities. Molecules 15:5153–5161

Shimoda K, Akagi M, Hamada H (2009) Production of β-maltooligosaccharides of α- and δ-tocopherols by *Klebsiella pneumoniae* and cyclodextrin glucanotransferase as anti-allergic agents. Molecules 14:3106–3114

Shin HT, Baig SY, Lee SW et al (2004) Production of fructo-oligosaccharides from molasses by *Aureobasidium pullulans* cells. Bioresour Technol 93:59–62

Singh RS, Singh RP (2010) Production of fructooligosaccharides from inulin by endoinulinases and their prebiotic potential. Food Technol Biotechnol 48:435–450

Sodhi HK, Sharma K, Gupta JK et al (2005) Production of a thermostable α-amylase from *Bacillus* sp. PS-7 by solid state fermentation and its synergistic use in the hydrolysis of malt starch for alcohol production. Process Biochem 40:525–534

Splechtna B, Petzelbauer I, Baminger U et al (2001) Production of a lactose-free galacto-oligosaccharide mixture by using selective enzymatic oxidation of lactose into lactobionic acid. Enzyme Microb Technol 29:434–440

Splechtna B, Nguyen TH, Steinbock M et al (2006) Production of prebiotic galacto-oligosaccharides from lactose using β-galactosidases from *Lactobacillus reuteri*. J Agric Food Chem 54:4999–5006

Stanton C, Gardiner G, Meehan H et al (2001) Market potential for probiotics. Am J Clin Nutr 73:476–483

Sun HJ, Yoshida S, Park NH et al (2002) Enzymatic preparation of wheat bran xylooligosaccharides and their stability during pasteurization and autoclave sterilization at low pH. Carbohydr Res 337:657–661

Tang L, Li ZA, Dong XX et al (2011) Lactulose biosynthesis by β-galactosidase from a newly isolated *Arthrobacter* sp. J Ind Microbiol Biotechnol 38:471–476

Torres DPM, Goncalves MDPF, Teixeira JA et al (2010) Galacto-oligosaccharides: production, properties, applications, and significance as probiotics. Compr Rev Food Sci Food Saf 9:438–454

Toshio I, Noriyoshi I, Toshiaki K et al (1990) Production of xylobiose. Japanese Patent JP 2119790
Tzortzis G, Goulas AK, Gibson GR (2005) Synthesis of probiotic galactooligosaccharides using whole cells of a novel strain, *Bifidobacterium bifidum* NCIMB 41171. Appl Microbiol Biotechnol 68:412–416
Van LJ, Coussement P, De Leenheer L et al (1995) On the presence of inulin and oligofructose as natural ingredients in the western diet. Crit Rev Food Sci Nutr 35:525–552
Vankova K, Onderková Z, Antošová M et al (2008) Design and economics of industrial production of fructooligosaccharides. Chem Pap 62(4):3753–3781
Vazquez MJ, Garrote G, Alonso JL et al (2002) Refining of autohydrolysis liquors for manufacturing xylooligosaccharides: evaluation of operational strategies. Bioresour Technol 96:889–896
Vranesic D, Kurtanjek Z, Santos AMP et al (2002) Optimisation of inulinase production by *Kluyveromyces bulgaricus*. Food Technol Biotechnol 40(1):7–73
Waterhouse AL, Chatterton NJ (1993) Glossary of fructan terms. In: Chatterton NJ, Suzuki M (eds) Science and technology of fructans. CRC, Boca Raton, FL
Watson E (2011) Frost & Sullivan: US prebiotics market to double in five years. Prebiotics. http://www.nutraingredients-usa.com/Industry/Frost-Sullivan-US-prebiotics-market-to-double-in-five-years
Watzl B, Girrbach S, Roller M (2005) Inulin, oligofructose and immunomodulation. Br J Nutr 93:49–55
Xiao H, Ruijin Y, Wenbin Z et al (2010) Dual-enzymatic synthesis of lactulose in organic-aqueous two-phase media. Food Res Int 43:716–722
Xu ZW, Li YQ, Wang YH et al (2009) Production of β-fructofuranosidase by *Arthrobacter* sp. and its application in the modification of stevioside and rebaudioside A. Food Technol Biotechnol 47(2):137–143
Yang H, Wang H, Song X et al (2011) Production of xylooligosaccharides by xylanase from *Pichia stipitis* based on xylan preparation from triploid *Populus tomentosa*. Bioresour Technol 102:7171–7176
Yang R, Xu S, Wang Z, Yang W (2005) Aqueous extraction of corn cob xylan and production of xylooligosaccharides. LWT-Food Sci Technol 38:677–682
Yuan X, Wang J, Yao H (2005) Antioxidant activity of feruloylated oligosaccharides from wheat bran. Food Chem 90:759–764
Yun JW, Kim DH, Kim BW et al (1997a) Production of inulo-oligosaccharides from inulin by immobilized endoinulinase from *Pseudomonas* sp. J Ferment Bioeng 84:369–371
Yun JW, Kim DH, Yoon HB et al (1997b) Effect of inulin concentration on the production of inulo-oligosaccharides by soluble and immobilized endoinulinase. J Ferment Bioeng 84(4):365–368
Yun JW, Park JP, Song CH et al (2000) Continuous production of inulo-oligosaccharides from chicory juice by immobilized endoinulinase. Bioprocess Eng 22(3):189–194
Zhang L, Su Y, Zheng Y et al (2010) Sandwich-structured enzyme membrane reactor for efficient conversion of maltose into isomaltooligosaccharides. Bioresour Technol 101(23):9144–9149
Zhengyu J, Jing W, Bo J et al (2005) Production of inulooligosaccharides by endoinulinases from *Aspergillus ficuum*. Food Res Int 38:301–308
Zokaee F, Kaghazchi T, Zare A et al (2002) Isomerization of lactose to lactulose-study and comparison of three catalytic systems. Process Biochem 37:629–635

Chapter 11
Potential of Agro-residues as Sources of Bioactive Compounds

Neha Babbar and Harinder Singh Oberoi

11.1 Introduction

The agro-processing industry generates large amount of waste in the form of peels, kernels and pulp. The disposal of residues in open spaces or in municipal bins enhances environmental pollution problems (Babbar et al. 2011). The best option is, therefore, the recovery of phytochemicals/bioactive compounds from such agro-processing residues which could be used in food, cosmetic and pharmaceutical industry. Due to legislation and environmental reasons, the industry is more and more forced to find an alternative use for the residual matter. The recovery of bioactive compounds is an elegant way to reuse waste streams, while being economically interesting on the other hand.

Bioactive compounds are extra nutritional constituents that occur naturally in small quantities in plant and food products (Kris-Etherton et al. 2002). Some of the bioactive compounds are also responsible for bitterness and astringency in foods. These compounds are present in leaves, stems, roots, tubers, buds, fruits, vegetables, seeds, peels, flowers and plant-derived foods and drinks (such as tea, coffee, alcoholic beverages) and are essential for growth, maintenance and repair of the body. Most common bioactive compounds include secondary metabolites such as antibiotics, mycotoxins, alkaloids, food grade pigments, plant growth factors and phenolic compounds (Holker et al. 2004; Kris-Etherton et al. 2002). Among these, few pigments and phenolic compounds are of interest. In the past few years, many food bioactive constituents have been commercialized in the form of pharmaceutical products (pills, capsules, solutions, gels, liquors, powders, granules, etc.) that incorporate food extracts or phytochemical-enriched extracts to which a beneficial physiological function has been directly or indirectly attributed. This range of

N. Babbar • H.S. Oberoi (✉)
Central Institute of Post Harvest Engineering and Technology,
P.O. PAU, Ludhiana 141 004, Punjab, India
e-mail: hari_manu@yahoo.com

S.K. Brar et al. (eds.), *Biotransformation of Waste Biomass into High Value Biochemicals*, DOI 10.1007/978-1-4614-8005-1_11,

products cannot be truly classified as "food", and a new hybrid term between nutrients and pharmaceuticals "nutraceuticals" has been coined to designate them. Nutraceuticals are diet supplements that deliver a concentrated form of a presumed bioactive agent from a food, presented in a non-food matrix, and used with the purpose of enhancing health in dosages that exceed those that could be obtained from normal foods (Zeisel 1999). This type of health-promoting product is getting popular among health-conscious consumers, and thus, a large list of nutraceuticals containing phytochemicals is now available in the market. For example, the carotenoid lycopene, alliaceae (garlic, onion) extracts containing sulphur derivatives (i.e. alliin and allicin), glucosinolate extracts and phytosterol extracts are widely commercialized products. Some of the most common phytochemicals found in the nutraceutical market are carotenoids; polyphenols such as anthocyanins, proanthocyanidins, flavonols stilbenes, hydroxycinnamates, coumarins, ellagic acid (EA) and ellagitannins (ETs); isoflavones; and lignans.

11.2 Toxicity of Synthetic Antioxidants

In recent times, natural antioxidants have raised considerable interest among nutritionists, food manufacturers and consumers because of their presumed safety and potential therapeutic value. In food industry, synthetic antioxidants are used to prevent lipid peroxidation and oxidation of food constituents. Tert-butyl hydroxyanisole (BHA), tert-butyl hydroxytoluene (BHT), tert-butyl hydroquinone (TBHQ) and propyl gallate (PG) are the commonest synthetic phenolic antioxidants used in edible oils or lipid-based foods in order to prevent oxidative rancidity. However, the safety and toxicity of synthetic antioxidants have raised important concerns because such materials may cause liver swelling and influence liver enzyme activities and carcinogenicity (Jayaprakasha et al. 2003). Reports have also revealed that BHA and BHT could be toxic (Sherwin 1990). Moreover, the synthetic antioxidants show low solubility and moderate antioxidant activity, and therefore, strong restrictions have been placed on their applications. Hence, considerable interest has been shown in the use of natural antioxidants which are likely to have properties that can be exploited by food and pharmaceutical industry. The replacement of synthetic antioxidants by natural ones may have benefits due to health implications and functionality, such as solubility in both oil and water.

11.3 Free Radicals

A free radical can be defined as any molecular species capable of independent existence that contains an unpaired electron in an atomic orbital. The presence of an unpaired electron results in certain common properties that are shared by most radicals. Many radicals are unstable and highly reactive. They can either donate an electron to or accept an electron from other molecules, therefore behaving as oxidants or reductants (Becker et al. 2004) The most important oxygen-containing free radicals in many disease states are hydroxyl radical, superoxide anion radical, hydrogen peroxide, oxygen singlet, hypochlorite, nitric oxide radical and peroxynitrite radical. Free radicals attack important macromolecules leading to cell damage and homeostatic disruption. Targets of free radicals include all kinds of molecules in the body. Among them, lipids, nucleic acids and proteins are the major targets (Ratnam et al. 2006).

11.3.1 Production of Free Radicals

Free radical reactions are expected to produce progressive adverse changes that accumulate with age throughout the body. Such "normal" changes with age are relatively common to all. However, superimposed on this common pattern are patterns influenced by genetics and environmental differences that modulate free radical damage. These are manifested as diseases at certain ages determined by genetic and environmental factors. Cancer and atherosclerosis, two major causes of death, are salient "free radical" diseases. Cancer initiation and promotion is associated with chromosomal defects and oncogene activation. It is possible that endogenous free radical reactions, like those initiated by ionizing radiation, may result in tumour formation. The highly significant correlation between consumption of fats and oils and death rates from leukaemia and malignant neoplasia of the breast, ovaries and rectum among persons over 55 years may be a reflection of greater lipid peroxidation (Scalbert et al. 2005a, b). Studies on atherosclerosis reveal the probability that the disease may be due to free radical reactions involving diet-derived lipids in the arterial wall and serum to yield peroxides and other substances.

11.4 Antioxidants

The word "antioxidant" is increasingly popular in modern society as it gains publicity through mass media coverage of its health benefits. The dictionary definition of antioxidant is rather straightforward but with a traditional annotation (Huang et al. 2005): "a substance that opposes oxidation or inhibits reactions promoted by oxygen or peroxides, many of these substances (as the tocopherols) being used as preservatives in various products (as in fats, oils, food products, and soaps for retarding the development of rancidity, in gasoline and other petroleum products for retarding gum formation and other undesirable changes, and in rubber for retarding aging)". A more biologically relevant definition of antioxidants is "synthetic or natural substances added to products to prevent or delay their deterioration by action of oxygen in air. In biochemistry and medicine, antioxidants are enzymes or other organic substances, such as vitamin E or β-carotene, that are capable of counteracting the damaging effects of oxidation in animal tissues". The biologically relevant definition fits better to the concept of antioxidants known to the general public as people are more aware of their health than prevention of autoxidation. In food science, antioxidants have a broader scope, in that they include components that prevent fats in food from becoming rancid as well as dietary antioxidants "a substance in foods that significantly decreases the adverse effects of reactive species, such as reactive oxygen and nitrogen species, on normal physiological function in humans", as defined by the Institute of Medicine. Like the other definitions, this definition does not provide limitation on the mechanism(s) of antioxidant action. Therefore, a dietary antioxidant can (sacrificially) scavenge reactive oxygen/nitrogen species (ROS·/RNS·) to stop radical chain reactions, or it can inhibit the reactive oxidants from being formed in the first place (preventive). Dietary antioxidants often broadly include radical chain reaction inhibitors, metal chelators, oxidative enzyme inhibitors and antioxidant enzyme cofactors. A typical autoxidation, initiated by an azo compound, and the action of its inhibitors include the following elementary steps (assuming one antioxidant scavenges two radicals and oxygen is in large excess, R_2N_2 azo compound, LH substrate; AH antioxidant):

Initiation	
$R_2N_2 \rightarrow 2R\cdot + N_2$	(1)
$R\cdot + O_2 \rightarrow ROO\cdot$	(2)
$ROO\cdot + LH \rightarrow ROOH + L\cdot$	(3)
Propagation	
$L\cdot + O_2 \rightarrow LOO\cdot\cdot$	(4)
$LOO\cdot + LH \rightarrow LOOH + L\cdot$	(5)
Inhibition	
$LOO\cdot + AH \rightarrow LOOH + A\cdot$	(6)
Termination	
$A\cdot + (n-1)\ LOO\cdot \rightarrow$ Nonradical products	(7)
$LOO\cdot + LOO\cdot \rightarrow$ Nonradical products	(8)

11.4.1 Scavenging Effects of Bioactive Compounds on Free Radicals

Reactive oxygen species and nitrogen species, the so-called free radicals, are highly reactive molecules constantly produced through numerous biological reactions like mitochondrial respiratory chain and any inflammatory condition inevitably lead to an increased oxidative burden. The release of reactive oxygen species by macrophages is part of the body's defence mechanism (Hensley and Floyd 2002). The organism has different antioxidant defence mechanisms against ROS· including numerous enzymes (catalase, superoxide dismutase, glutathione reductase, glutathione peroxidase), small antioxidant molecules (uric acid, glutathione, albumin, protein–SH groups, bilirubin) and certain vitamins (ascorbic acid, α-tocopherol) as well as carotenoids (Halliwell and Gutteridge 1990) which have the capacity to neutralize free radicals acting in concert (Yeum et al. 2004). Depleted antioxidant defences can lead to oxidative stress, that is, imbalance between the rates of production and release of free radicals, increasing the likelihood of damage to other molecules. Reactive oxygen species is capable of oxidation of a variety of biomolecules, such as enzymes, proteins, DNA and lipids (Hansberg 2002). Oxidative stress is responsible for the development of chronic degenerative diseases including coronary heart disease, cancer and the degenerative processes associated with aging (Benzie 2000). The initial step for atherosclerosis development, leading to heart disease and stroke, is thought to be because of oxidized LDL. The plaques that obstruct arterial flow and cause cardiovascular disease are laid down by macrophages engorged with oxidized LDL (foam cells). Oxidized bases in DNA are potentially mutagenic and so are implicated in the process of carcinogenesis. Diabetes mellitus is associated with oxidative damage to biomolecules (Willcox et al. 2004). Therefore, diet-derived antioxidants could be important in the protection against chronic diseases (Park et al. 2003). It has been suggested that a high intake of fruits and vegetables, the main sources of antioxidants in the diet, could decrease the potential stress caused by ROS· via a number of mechanisms, including the protection of target molecules (lipids, proteins and nucleic acids) from oxidative damage, suppressing the inflammatory response and modulating vascular homeostasis (Law and Morris 1998).

11.4.2 Mode of Action of Antioxidant

The antioxidants acting in the defence systems act at different levels such as preventive, radical scavenging, repair and de novo and the fourth line of defence, i.e. the adaptation.

The first line of defence is the preventive antioxidants, which suppress the formation of free radicals. Although the precise mechanism and site of radical formation in vivo are not well elucidated yet, the metal-induced decompositions of hydroperoxides and hydrogen peroxide must be one of the important sources. To suppress such reactions, some antioxidants reduce hydroperoxides and hydrogen peroxide beforehand to alcohols and water, respectively, without generation of free radicals and some proteins sequester metal ions. Glutathione peroxidase, glutathione-s-transferase, phospholipid hydroperoxide glutathione peroxidase (PHGPX) and peroxidase are known to decompose lipid hydroperoxides to corresponding alcohols. PHGPX is unique in that it can reduce hydroperoxides of phospholipids integrated into biomembranes. Glutathione peroxidase and catalase reduce hydrogen peroxide to water.

The second line of defence is the antioxidants that scavenge the active radicals to suppress chain initiation and/or break the chain propagation reactions. Various endogenous radical-scavenging antioxidants are known: some are hydrophilic and others are lipophilic. Vitamin C, uric acid, bilirubin, albumin, and thiols are hydrophilic, radical-scavenging antioxidants, while vitamin E and ubiquinol are lipophilic radical-scavenging antioxidants. Vitamin E is accepted as the most potent radical-scavenging lipophilic antioxidant.

The third line of defence is the repair and de novo antioxidants. The proteolytic enzymes, proteases and peptidases, present in the cytosol and in the mitochondria of mammalian cells, recognize, degrade and remove oxidatively modified proteins and prevent the accumulation of oxidized proteins.

There is another important function called adaptation where the signal for the production and reactions of free radicals induces formation and transport of the appropriate antioxidant to the right site (Benzie and Strain 1996).

11.4.3 Methods for Measuring Antioxidant Activity

There are numerous published methods claiming to measure total antioxidant capacity in vitro. Ironically, the biggest problem is the lack of a validated assay that can reliably measure the antioxidant capacity of foods and biological samples. Several reviews have been published, and the opinions vary considerably. There seems to be no consensus of opinions, most probably due to the fact that the area of antioxidants is such a complex topic. On the basis of the chemical reactions involved, major antioxidant capacity assays can be roughly divided into two categories: (1) hydrogen atom transfer (HAT) reaction-based assays and (2) single electron transfer (ET) reaction-based assays. The ET-based assays involve one redox reaction with the oxidant (also as the probe for monitoring the reaction) as an indicator of the reaction endpoint. Most HAT-based assays monitor competitive reaction kinetics, and the quantitation is derived from the kinetic curves. HAT-based methods generally are composed of a synthetic free radical generator, an oxidizable molecular probe and an antioxidant. HAT- and ET-based assays are intended to measure the radical (or oxidant) scavenging capacity, instead of the preventive antioxidant capacity of a sample (Huang et al. 2005). Table 11.1 lists the major antioxidant activity assays involving the electron transfer. A desirable method for evaluating the antioxidant activity should be rapid, should be reproducible, should require small amounts of chemicals and should not be influenced by the physical properties of the matrix. Two methods commonly used to evaluate antioxidant activities are DPPH and ABTS assays, which use DPPH and ABTS as free radical generators, respectively. Methods like $ABTS^+$ (radical cation of 2,2-azinobis-3-ethylbenzothiozoline-6-sulphonate) and DPPH (2,2-diphenil-1-picrylhydrazyl) radical have been developed for determining the scavenging activity with different challengers, such as superoxide radical ($O_2\cdot$), hydroxyl (OH·), nitric oxide (NO·) and alkylperoxyl radicals. The mechanism involved in both methods is similar as the absorption spectra of the free radical changes when molecule is reduced by an antioxidant or free radical species. Some findings have suggested that the ABTS assay is better than DPPH as ABTS is soluble in water and organic solvents and it reacts relatively rapidly compared to DPPH. Trolox

Table 11.1 In vitro antioxidant activity assays

Assays involving hydrogen atom transfer reactions	
$ROO^{\cdot} + AH \rightarrow ROOH + A^{+}$	ORAC (oxygen radical absorbance capacity)
$ROO^{\cdot} + LH \rightarrow ROOH + L^{+}$	TRAP (total radical-trapping antioxidant parameter)
	Crocin bleaching assay
	IOU (inhibited oxygen uptake)
	Inhibition of linolenic oxidation
	Inhibition of LDL oxidation
Assays by electron transfer reactions	TEAC (trolox equivalent antioxidant capacity)
$M(n) + e$ (from AH) $\rightarrow AH^{++} + M(n-1)$	FRAP (ferric ion reducing antioxidant parameter)
	DPPH (diphenyl-1-picrylhydrazyl)
	Copper (II) reduction capacity
	Total phenol assay by Folin–Ciocalteu reagent
Other assays	TOSC (total oxidant scavenging capacity)
	Inhibition of Briggs–Rauscher oscillation reaction
	Chemiluminescence
	Electro-chemiluminescence

LH-substrate, AH-antioxidant

(6-hydroxy-2,5,7,8-tetramethylchroman-2-carboxylic acid, a water-soluble vitamin E analogue) equivalent antioxidant capacity (TEAC) has been used to determine the hierarchy of radical-scavenging abilities of phenolic compounds as electron or H donating agents through determination of their ability to scavenge ABTS radical/DPPH radical (Rice-Evans and Miller 1998). The assay is based on the discolouration of ABTS/DPPH by antioxidant compounds, thus reflecting the amount of ABTS/DPPH radicals that are scavenged within a fixed period of time in relation to that of trolox. Colour interference of the DPPH assay with samples that contain anthocyanins leads to under estimation of the antioxidant activity. However, this problem does not occur with the ABTS assay. Oxygen radical absorbance capacity (ORAC) is another method used to measure antioxidant capacity in vitro. This method is based on inhibition of peroxyl radical-induced oxidation initiated by thermal decomposition of azo compounds, such as 2,2-azobis (2-amidino propane) dihydrochloride (AAPH). The antioxidants react with peroxyl radicals and delay the degradation of fluorescein, a fluorescent probe. The ORAC method uses biologically relevant free radicals, integrates both time and degree of antioxidant activity into one data value, and is readily adaptable to a high-throughput assay system. The advantage of ORAC method is its ability to assay both hydrophilic and lipophilic antioxidants, which results in better measurements of total antioxidant activity (Prior et al. 2003).

The total radical-trapping antioxidant parameter (TRAP) assay has also been widely used (Ghiselli et al. 2000). These assays differ from each other in terms of substrates, probes, reaction conditions and quantification methods. Crocin bleaching assay (CBA) is a new method for determination of antioxidant capacity. In CBA, addition of hydrogen to the conjugated double bonds of crocin results in reduction of crocin and increase in the absorbance at 440 nm, which is considered as a measure of antioxidant potential. In CBA, deletion of the hydrogen atoms and/or addition of a radical to the crocin(s) results in a disruption of the conjugated double bonds of its polyene backbone, thus bleaching of the solution (Bathaie et al. 2011). The same authors concluded that CBA is a simple and useful method for determination of antioxidant potential of aqueous samples. In addition, the CBA ability to distinguish the samples that contain bilirubin or high uric acid content is helpful in clinical laboratories.

Peroxynitrite (ONOO–) scavenging capacity assay is also another useful method to determine peroxynitrous acid (ONOOH) which is a very strong oxidant. Under physiological conditions, peroxynitrite also forms an adduct with carbon dioxide dissolved in body fluid. The adduct is believed to be responsible for the oxidative damage of proteins (Squadrito and Pryor 2002).

Previous studies have reported correlations among antioxidant activities measured by different methods (Awika et al. 2005; Babbar et al. 2011). In our previous study (Babbar et al. 2011), we observed a significant correlation between DPPH and ABTS ($r^2-0.91$) for different fruit residues. In another study, the r^2 values for correlation between ORAC and ABTS and ORAC and DPPH were 0.99 and 0.98, respectively, in case of sorghum grains (Awika et al. 2003). Although, it is not possible to arrive at a single most efficient assay method which is simple, cheap, accurate and precise, we feel that the antioxidant capacity should be determined by at least two methods so that more of the types of antioxidants measured by different methods provide a reliable antioxidant profile and precise and accurate values are reported for the compound.

11.5 Bioactive Compounds as Important Antioxidants

Many of the bioactive compounds are strong antioxidants, i.e. they slow or prevent the oxidation of other chemicals. Oxidation reactions can involve the production of free radicals, which can form dangerous chain reactions. Bioactive compounds from agro-residues vary widely in chemical structure and function and are grouped accordingly (Kris-Etherton et al. 2002). Important bioactive compounds from agro-residues include phytochemicals, viz. phenolic compounds, carotenoid and tocopherols. Table 11.2 shows the important bioactive compounds present in agro-residues and their mode of action.

11.5.1 Phytochemicals

They are pronounced as "fight-o-chemicals", i.e. they fight to protect health. Phytochemicals naturally occur in fruit and vegetables that work together with vitamins, minerals and fibre to promote health benefits in many ways. Phytochemicals that are present in the diet and have been associated with health benefits include

Table 11.2 Important bioactive compounds present in agro-processing wastes and their mode of action

Compound	Structure	Forms	Solubility	Mode of action	References
Ascorbic acid	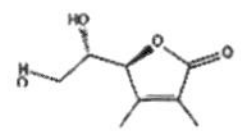	Ascorbic acid, dehydro-ascorbic acid	Hydrosoluble	Free radical-scavenging activity, enzymatic reduction due to direct electron donation from hydroxyl groups	Jacobsen et al. (1999)
Phenolic compounds	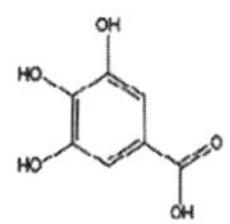	Phenolic acids, flavonoids, lignans, stilbenes	Hydrosoluble	Electron donation, metal ion chelation, ascorbic acid sparing, radical-scavenging activity	Balasundram et al. (2006)
Carotenoids	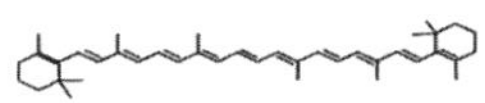	Carotenes, xanthophylls	Liposoluble	Electron donation, radical-scavenging activity	Jim and Hong-Shum (2003)
Tocopherols	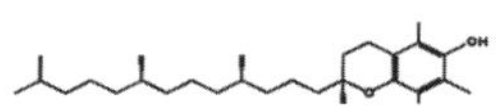	α-Tocopherol β-Tocopherol δ-Tocopherol	Liposoluble	Radical-scavenging activity	Hong-Bo et al. (2008)

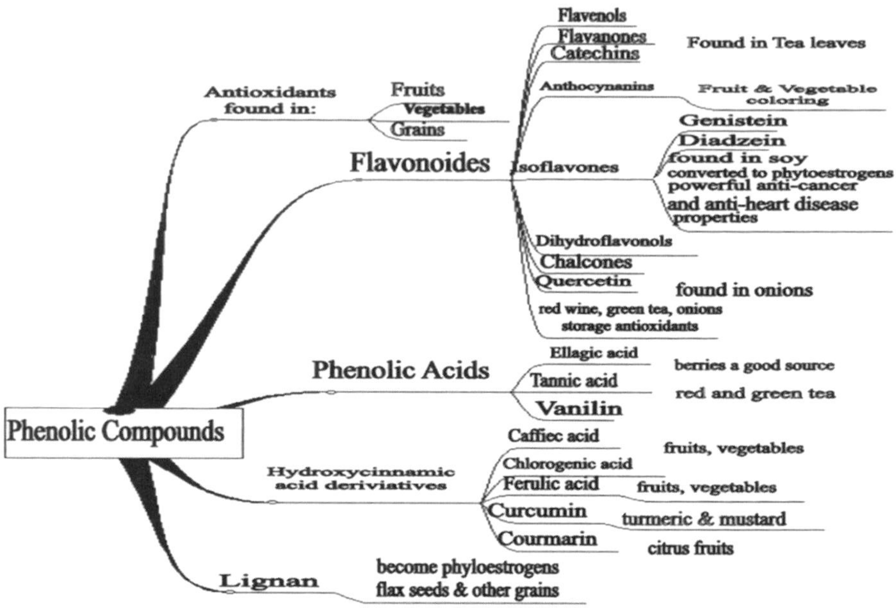

Fig. 11.1 The general breakdown of plant based phenols

glucosinolates, sulphur-containing compounds of the alliaceae, terpenoids (carotenoids, monoterpenes) and various groups of polyphenols (anthocyanins, flavones, flavan-3-ols, isoflavones, stilbenoids, ellagic acid, etc.). The bioactivity of these phytochemicals has been, to some extent, associated to their antioxidant properties, i.e. capacity to scavenge free radicals which are involved in the onset of many of chronic degenerative diseases (LDL oxidation in atheroma plaque development, DNA oxidation and cancer, oxidation and ageing, inflammation, etc.).

11.5.1.1 Phenolic Compounds

Polyphenols are secondary plant metabolites that are derived through pentose phosphate, shikimate and phenylpropanoid pathways. Phenolic compounds are found predominantly in the by-products than in the edible portions as they tend to accumulate in the dermal tissues of plant body because of their potential role in the protection against UV rays, as attractants in fruit dispersal and also as defence chemicals against pathogens. Phenolics may act as phytoalexins (Popa et al. 2008), antifeedants, attractants for pollinators, contributors to plant pigmentation, antioxidants and protective agents against UV light (Naczk and Shahidi 2006). Polyphenolic compounds are divided into several classes based on their structural diversity (Fig. 11.1). Of these, flavonoids, phenolic acids and tannins (hydrolyzable and condensed) are regarded as the main dietary phenolic compounds (D'archivio et al. 2007).

The antioxidant activity of a phenolic compound is directly proportional to its chemical structure. Following relationships between chemical structure and antioxidant activity of phenolic compounds have been encountered:

1. The antioxidant potential of phenolics depends on the number and arrangement of the hydroxyl groups (Sang et al. 2002). For phenolic compounds, the more OH groups there are in the ring, the larger is the TEAC (trolox equivalent antioxidant capacity) value (Villano et al. 2005).
2. The contribution of the 3-OH group in flavonoids is very significant (Heijnen et al. 2001). Blocking the 3-hydroxyl group in the B ring (i.e. rutin) decreases the antioxidant activity (Villano et al. 2005).
3. The presence of an ortho-dihydroxy substitution in the B ring confers higher stability to the radical structure and participates in electron delocalization and plays an important role in the antioxidant activity (Yao et al. 2004).

Phenolics have been considered powerful antioxidants in vitro (Frankel et al. 1995) and have been proved to be more potent antioxidants than vitamins E, C and carotenoids (Rice-Evans et al. 1997). The inverse relationship between fruit, vegetable intake and risk of cardiovascular and neurodegenerative diseases, cancer, diabetes and osteoporosis has partially been ascribed to phenolics (Scalbert et al. 2005a, b). Their antioxidant activity lies in their ability to donate a hydrogen or electron and their ability to delocalize the unpaired electron within the aromatic structure. They can also protect biological molecules against oxidation.

11.5.1.2 Carotenoids

Carotenoids are a diverse group of over >600 different compounds that contribute to the yellow to red colours found in many foods. Carotenoids are polyenes consisting of 3–13 conjugated double bonds and up to six carbon ring structures at one or both ends of the molecule. Carotenoids containing oxygen are known as xanthophylls (e.g. lutein and zeaxanthin) while those without oxygen are known as carotenes (e.g. lycopene and β-carotene). The carotenoids have several potential health benefits. Lutein and zeaxanthin which are found in high concentrations in the human eye have been postulated to be beneficial to age-related macular degeneration and cataracts (Stringham and Hammond 2005). Although carotenoids are generally thought to be beneficial for health, clinical trials have found that large doses of β-carotene increase the risk of lung cancer (Bendich 2004). In general, carotenoids are not strong antioxidants when added to food but relatively unstable in food systems because they are susceptible to light, oxygen and auto-oxidation (Xianquan et al. 2005).

11.5.1.3 Ascorbates

Ascorbic acid (vitamin C) is a commonly used antioxidant in many food systems for maintaining organoleptic quality. In its natural forms, ascorbic acid (e.g. L-ascorbic acid and D-isoascorbic acid) functions both as a reducing agent and as an oxygen scavenger. In addition, the metal-sequestering activity of ascorbic acid, which forms metal-ascorbate complexes that are less reactive with oxygen than with metal ions alone, provides antioxidant activity (Martell 1982). The oxygen-scavenging activity of ascorbic acid is effective in trapping both singlet oxygen and a superoxide anion, thus producing ascorbate free radicals (Zhang and Fung 1994). Ascorbate radicals so produced also react with peroxyl radicals to produce hydroperoxides (nonradical species) and the oxidized form of ascorbic acid, namely, dehydroascorbic acid. The conversion of native ascorbic acid to a salt form increases its stability and versatility in different food systems at the expense of biological activity. In muscle systems, ascorbic acid has been used to delay the formation of metmyoglobin in fresh meat products; it has also been used to prevent enzymatic browning in fresh fruits and vegetables. However, L-ascorbic acid has been shown to be ineffective as an antioxidant in poultry meat (King et al. 1995). The addition of ascorbates and sodium citrate to milk also provides protection against the loss of the lipid-soluble vitamins A and D. In clinical studies, ascorbic acid has been shown to prolong the survival of patients that have terminal cancers (Cameron and Pauling 1978). Vitamin C is one of the most popular and least toxic antioxidant of foods and has been widely used as a dietary supplement to prevent oxidative stress-mediated diseases (Gardner et al. 2002).

11.5.1.4 Tocopherols

Several lines of evidence indicate that α-tocopherol is the most effective phenolic antioxidant for reducing lipid peroxidation. The hydrophobic character of α-tocopherol enables it to be a strong antioxidant in lipid systems, where it functions as a radical scavenger and terminates the propagation of radical chain reactions by reacting with peroxyl radicals and generating unreactive phenoxyl radicals and hydroperoxide products. Dietary supplementation with vitamin E increases the plasma tocopherol concentration and the potential for associated antioxidative protection. Vitamin E is transported in plasma lipoproteins, where it can exert a positive role against peroxidative damage. In the dynamic scheme of antioxidation reactions, one mole of α-tocopherol reacts with two lipid-peroxyl radicals, yielding 14 lipid hydroperoxides and seven oxidized tocopherol molecules. Numerous studies support the contention that vitamin E is involved in antioxidant defence and chronic disease. For example, vitamin E has been shown to delay the oxidation of low-density lipoproteins (LDL), implicated to be an early step in the development of atherosclerosis. Other clinical studies have reported that α-tocopherol, although important, is not the sole factor that determines the resistance of LDL to oxidative stress (Maiorino et al. 1995). The antioxidant activity of α-tocopherol has also been shown to reduce the cytotoxic activity of lipid peroxides in tumour cells (Maiorino et al. 1995). Higher concentrations of vitamin E can also reduce transition metals and produce oxy radicals from redox reactions (Iwatsuki et al. 1995).

11.6 Agro-Processing Wastes as an Important Source of Bioactive Compounds

Agricultural and industrial residues are attractive sources of natural antioxidants (Volf and Popa 2004). Some studies have already been conducted on agricultural by-products, which could be potential sources of bioactive compounds. Lignocellulosic waste comprises of rice straw, wheat straw, sugarcane bagasse, sweet sorghum bagasse, cotton stalk, etc., which is generated in huge quantum. Lignocellulosic waste, such as rice straw, is burnt in the fields after crop harvest in many Southeast Asian countries, resulting in smoke clouds causing loss of biomass and environmental pollution. A schematic flow diagram for extraction, isolation and characterization of bioactive compounds from Agricultural residues is presented in Fig. 11.2.

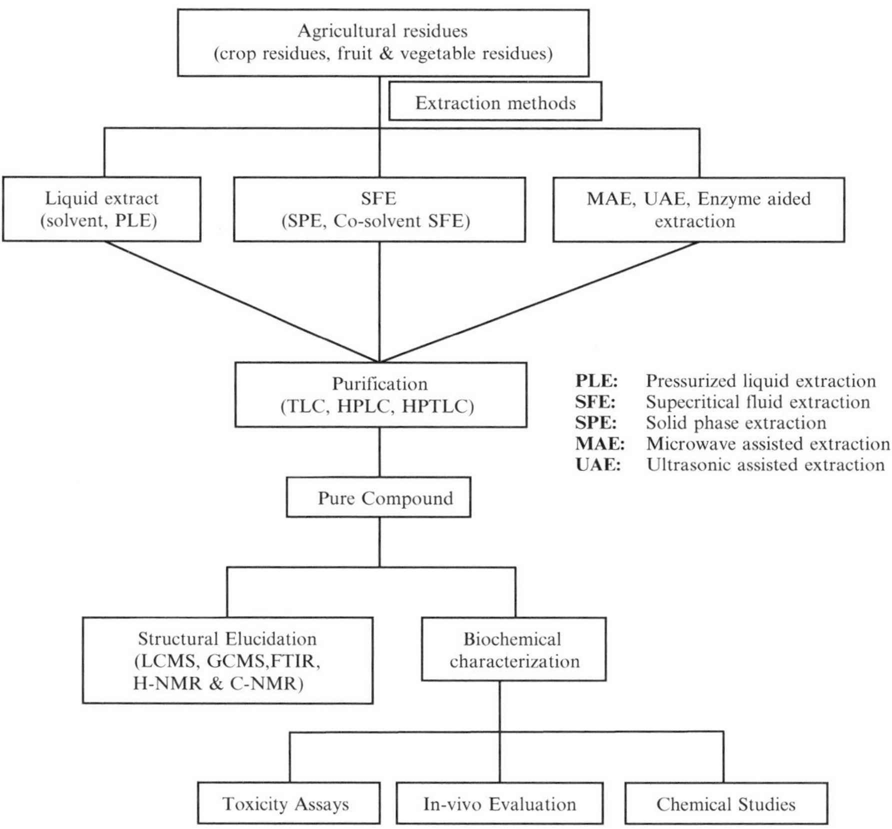

Fig. 11.2 Schematic diagram for extraction, isolation and characterization of bioactive compounds from agricultural residues

11.6.1 Lignocellulosic Waste

The agro-industrial residues mainly comprise of lignocellulosic materials and are poorly valorized or left to decay on land. Lignocellulosic materials (LCMs) are promising sources of antioxidant compounds (Domınguez et al. 2001). Lignocellulosic materials compose of cellulose, hemicellulose and lignin in varying concentrations depending on the nature of the biomass. The most important precursor of phenolic compounds from lignocellulosic waste is lignin. Lignin is a heterogeneous polymer of phenolic nature and is made up of three precursors: trans-coniferyl, trans-sinapyl and trans-*p*-coumaryl alcohols. Gymnosperm lignins show predominance of guaiacyl groups, woody angiosperms lignins contain guaiacyl–syringyl groups, and lignins from grasses contain guaiacyl–syringyl-*p*-hydroxyphenyl groups. Partial depolymerization of lignin and lignin–hemicellulose linkages occurs during the hydrolytic processing of LCM (Ando et al. 2000). Ferulic and *p*-coumaric acids (the most abundant hydroxycinnamic acids) are linked to arabinoxylans or pectins through ester bonds. In hardwoods, condensed tannins (proanthocyanidins) and hydrolyzable ellagitannins make also part of the phenolic fraction (Cadahıa et al. 2001), whereas acids (gallic, vanillic and ellagic) and aldehydes (syringaldehyde and sinapaldehyde) have also been detected (Conde et al. 1995). A selective recovery of the phenolic compounds from hydrolysates can be achieved by extraction with solvents such as ethyl acetate or diethyl ether. Ethyl acetate removes water-soluble phenolics and hemicellulose degradation products, whereas lignin–carbohydrate complexes remain in the aqueous phase (Bouchard et al. 1991). Phenolic compounds extracted from various LCMs are presented in Table 11.3.

A number of technologies are available for the mild hydrolysis of LCMs for production of various bioactive compounds (Amendola et al. 2012). Among them, the simplest one is autohydrolysis, wherein, the LCM is treated with water or steam. Related processes include the utilization of additional reagents such as mineral acids (prehydrolysis), SO_2 or oxygen (wet oxidation). Autohydrolysis of LCM is an environment-friendly process in which the hydronium ions from water autoionization and from organic acids generated in the reaction promote the hydrolytic degradation of cell wall components. A variety of compounds appear in the liquors obtained by these technologies, including sugar oligomers, monomeric sugars, sugar degradation products (furfural and hydroxymethylfurfural), organic acids (citric and malic acid coming from the cells of the biomass, formic and levulinic acid from sugar degradation products, acetic acid from acetyl groups), extractives and phenolics.

11.6.2 Fruit Residues

Fruit undergoes processing to separate the desired value product from other constituents and therefore generates large quantity of residues. These by-products or residues usually have significant value, which is generally being underutilized in

Table 11.3 Phenolic compounds from lignocellulosic waste subjected to steam explosion (SE), alkaline wet oxidation (WO) and solvent extraction

Types of lignocellulosic waste	Hydrolytic conditions	Phenolics	References
Olive whole stone/SE	236 °C, 2 min, 4.31 log Ro	Vanillic acid, syringic acid, syringaldehyde, hydroxytyrosol	Fernandez-Bolanos et al. (1999)
Olive seed husk/SE	215–229 °C, 2–3 min, 0.1 % H_2SO_4 3.69–4.29 log Ro	Vanillic acid, vanillin, syringic acid, hydroxytyrosol	Fernandez-Bolanos et al. (1999)
Wheat straw/WO	195 °C, 12 bar O_2, 10 min	Phenol, guaiacol, syringol, 4-hydroxybenzaldehude, vanillin, syringic acid	Klinke et al. (1998)
Soft woods/SE	195 °C, 2.38 min, 3.19 % SO_2	Lignans, ferulic acid, syringic	Boussaid et al. (2001)
Sugarcane bagasse/SE	205 °C, 10 min, 0 % SO_2	Benzoic acid, caffeic acid, catechol, *p*-coumaric acid, vanillic acid, protocatechuic acid	Martın et al. (2002)
Almond hulls	Solvent extraction	Chlorogenic acid, 4-O-caffeoylquinic acid	Takeoka and Dao (2002)
Buckwheat hulls	Solvent extraction	Protocatechuic acid, hyperin, rutin	Watanabe et al. (1997)

developing countries like India. Table 11.4 shows important bioactive compounds present in fruits and vegetables. Comprehensive information about the quantity of residues generated for different fruits is not available; some of the previous studies have indicated that banana peel accounts for about 30–40 % of the banana fruit weight (Oberoi et al. 2012). Kinnow peel, pulp and seeds account for more than 50 % of the total fruit weight (Kalra et al. 1989). Major wastes of mango processing industry are peels and stones, amounting for about 35–60 % of the total fruit weight (Larrauri et al. 1996).

Litchi (*Litchi chinensis Sonn.*) is a tropical fruit originating from China, with a bright red attractive pericarp surrounding a white aril (Nakasone and Paull 1998). Litchi seeds and pericarp account for about 19 and 13 %, respectively, of the fresh fruit weight. Litchi pericarp contains a large amount of anthocyanins, which are responsible for the red colour and possess antioxidant characteristics. Apple pomace is the main by-product of apple juice processing plant and has been found to be a good source of polyphenols (Chodak and Tarko 2007). Among all the known fruit residues, grape residues probably have been most extensively studied for their therapeutic potential.

Citrus fruits consist of oranges, kinnow, khatta, lime, lemon, grapefruit, malta, sweet orange, etc. Citrus by-products are major sources of phenolic compounds. The peels, in particular, are an abundant source of natural flavonoids and contain higher amount of phenolics compared to the edible portions. Flavonoids present in citrus by-products have been extensively studied for antioxidative, anticancer,

Table 11.4 Total phenolic content (TPC) of important fruit residues

Residues	TPC	References
Apple pomace, peel	52.20[a], 17.90[b], 33.00[a]	Peschel et al. (2006); Chodak and Tarko (2007); Wolfe and Liu (2003)
Strawberry waste	59.80[a]	Peschel et al. (2006)
Pear waste	18.40[a]	Peschel et al. (2006)
Litchi seed, pericarp	17.90[a], 24.60[a]	Babbar et al. (2011)
Grape seed	37.40[a], 47.30[c]	Babbar et al. (2011)
Banana peel	9–30[a]	Someya et al. (2002); Babbar et al. (2011)
Mango kernel, peel	117.00[a], 109.70[a]	Larrauri et al. (1996)
Kinnow seed, peel	3.68[a], 17.50[a]	Babbar et al. (2011)
Other citrus fruit peel	66.00–222.00[d]	Ghasemi et al. (2009)
Guava peel	58.70[a]	Lim et al. (2007)
Watermelon seeds	9.69[b]	Chodak and Tarko (2007)
Plum seeds, peels	43.68[b], 33.40[b]	Chodak and Tarko (2007)
Lemon peel	96.62[b]	Chodak and Tarko (2007)
Orange peel and seeds	10.10[a]–8.49[b], 2.12[b]	Anagnostopoulou et al. (2006); Chodak and Tarko (2007)
Star fruit residue	33.20[a]	Shui and Leong (2006)
Kiwi seeds and peel	1.02[b], 11.61[b]	Chodak and Tarko (2007)
Red grapefruit seeds and peel	2.22[b], 5.57[b]	Chodak and Tarko (2007)

[a]mg/g GAE
[b]mg/g catechin
[c]mg/g ferulic acid
[d]mg/g quercetin

antiviral and anti-inflammatory activities; effects on capillary fragility; and an observed inhibition of human platelet aggregation (Braddock 1995). Recent research suggests that citrus fruits possess another health benefit phytochemical called limonoids, which are highly oxygenated triterpenoids. Citrus limonoids appear in large amounts in citrus juice and citrus tissues as water-soluble limonoid glucosides or in seeds as water-insoluble limonoid aglycones. Limonin, nomilin and nomilinic acid are the major limonoids in citrus fruits, while both neem seeds and leaves contain the limonoid azadirachtin (Jayaprakasha et al. 2008). Currently, limonoids are under investigation for a wide variety of therapeutic effects such as antiviral, antifungal, antibacterial, antineoplastic and antimalarial (Bentley et al. 1990).

Mango (*Mangifera indica*) is one of the most important tropical fruits, and India ranks first in the production of mangoes in the world. Mango peels and stones generated in large quantities during processing of mangoes may also be used as a source of natural antioxidants. Papaya (*Carica papaya*) is a tree-like herbaceous plant, a member of the small family Caricaceae and widely cultivated for its edible fruits. Phenolic compounds from papaya waste are considered as powerful antioxidants which help in reducing oxidative stress. Table 11.4 illustrates the total phenolic content of important fruits and vegetable residues.

11.6.3 Vegetable Residues

Vegetable residues mostly contain considerable amounts of potentially interesting compounds. The olive mill wastes are also a major potential source of phenolics. The phenolic content of the olive mill waste water (OMWW) is reported to fluctuate between 1.0 and 1.8 % (Visioli and Galli 2003) depending on varietal factors and processing effects. The major components in OMWW include hydroxytyrosol, tyrosol, oleuropein and a variety of hydroxycinnamic acids (Obied et al. 2005). Besides OMWW, olive leaves are another by-product of the olive industry that has been explored as a source of phenolics (Benavente-Garcia et al. 2000). Potatoes, tomatoes, cauliflower, peas, beans, etc., leave behind a substantial amount of residues mainly in the form of peels, seeds or inedible inflorescence. Tomato (*Lycopersicon esculentum*) is a versatile vegetable that is consumed fresh as well as in the form of processed products. Tomato processing industries generate large amounts of waste in the form of seeds and peels. Tomato peel is a rich source of lycopene which has potential to be used in anticancer medicine (Wenli et al. 2001). Lycopene is effective in scavenging reactive oxygen species, such as superoxide anion, hydroxyl radical, singlet oxygen and lipid free radical (Wenli et al. 2001). Carotenoids like lycopene, β-carotene, lutein, zeaxanthin are known to be the most efficient singlet oxygen quencher in the biological systems without the production of any oxidizing products. George et al. (2004) found that the total phenolic contents of tomato seeds as catechin ranged from 188 to 465 mg/100 g on dry weight basis. Cauliflower (*Brassica oleracea* L) is an important vegetable grown all over the world. Cauliflower has the highest waste index, i.e. ratio of edible portion to nonedible portion, and thus, enormous amount of organic waste is generated which is not put to any commercial use (Oberoi et al. 2007). Total phenolic content of tomato peel, potato peel, cauliflower waste and pea pod was 21, 13.6, 9.2 and 5.4 gallic acid equivalent per gram dry weight (GAE/g-dw), respectively (Babbar et al. 2012). Phenolic compounds and vitamin C are the major antioxidants of brassica vegetables, due to their high content and high antioxidant activity. Lipid-soluble antioxidants (carotenoids and vitamin E) are responsible for up to 20 % of the brassica total antioxidant activity (Podsedek 2007). Previous studies have reported scavenging activity (SA) of 30.7–66.8 % for waste obtained from different cultivars of cauliflower (Scalzo 2005). Similarly, pea pod (*Pisum sativum*) waste is also generated in large quantities which has only limited application in the form of cattle feed. Peels are a major by-product of potato processing and a potential source of functional compounds. Pumpkin seed oil contains antioxidative components that are polar (phenolic) compounds (Fruhwirth et al. 2003). Table 11.5 presents different bioactive compounds present in various fruit and vegetable residues, while Table 11.6 gives a good idea about the composition and content of bioactive compounds present in the methanolic extracts of various plant residues as determined by HPLC.

Table 11.5 Different bioactive compounds present in various fruits and vegetable residues

Fruit	Residue	Phenolic constituent	References
Apple	Peel and pomace	Epicatechin, catechins, hydroxycinnamates, phloretin glycosides, quercetin glycosides, procyanidins, chlorogenic acid, anthocyanins	Lu and Foo (1997); Foo and Lu (1999); Wolfe and Liu (2003)
Grapes	Seed and skin	Cinnamic acid, coumaric acid, caffeic acid, ferulic acid, chlorogenic acid, neochlorogenic acid, *p*-hydroxybenzoic acid, protocatechuic acid, vanillic acid, gallic acid, proanthocyanidins, quercetin 3-O-gluuronide, quercetin, resveratrol	Shrikhande (2000); Negro et al. (2003); Maier et al. (2009)
Citrus	Peel	Hesperidin, naringin, eriocitrin, narirutin	Coll et al. (1998)
Mango	Kernel	Gallic acid, ellagic acid, gallates, gallotannins, condensed tannins	Arogba (2000); Puravankara et al. (2000)
Banana	Peel	Gallocatechin, anthocyanins, delphinidin, cyanidin, catecholamine	Someya et al. (2002)
Litchi	Pericarp, seeds	Cyanidin-3-glucosides, cyanidin-3-rutonoside, malvidin-3-glucoside, gallic acid, epicatechin-3-gallate	Lee and Wicker (1991); Duan et al. (2007)
Olive	Pomace	Phenols	Obied et al. (2005)
Tomato	Skin, pomace	Carotenoids	Strati and Oreopoulou (2011)
Potato	Peel	Gallic acid, caffeic acid, vanillic acid	Zeyada et al. (2008)
Carrot	Peel	Phenols, β-carotene	Chantaro et al. (2008)
Cucumber	Peel	Chlorophyll, pheophytin, phellandrene, caryophyllene	Zeyada et al. (2008)

11.7 Extraction of Bioactive Compounds

The extraction is one of the most important steps in sample pretreatment. During the extraction of bioactive compounds from by-products, several operational units and conditions are applied. Milling, solvent extraction and drying are the main operations that are used. The most common method for the extraction of phenolic compounds is liquid–liquid extraction (LLE), and different extraction solvents are used, i.e. ethanol, methanol or a mixture of solvents with water. Organic solvents are used for the extraction of carotenoids, and acetone resulted in the highest yield compared to ethanol, petroleum ether and hexane (Calvo 2005). Successive extractions are needed for quantitative recovery (Oreopoulou and Tzia 2007). The extracted carotenoids may be obtained as a crude pigment, after solvent evaporation at low temperature, if preceding washings are accomplished. Tocopherols and flavonoids and related compounds like coumarins, cinnamic acid derivatives, chalcones, phenolics, diterpenes and phenolic acids are isolated by solvent extraction (Oreopoulou and Tzia 2007) using nonpolar solvents (hexane, petroleum ether). Ethyl ether and ethyl acetate

Table 11.6 Composition and content of bioactive compounds in methanolic extracts of various plant materials as determined by HPLC

Plant material	Compound	Concentration (mg/g)
Tomato peel	Cis-lycopene	22.02
	Beta-carotene	6.87
	Trans-lycopene	36.49
	Lutein	1.08
	Ascorbic acid	12.27
	Quercetin	2.89
	Kaempferol	7.20
Cucumber peel	Chlorophyll	3.46
	Pheophytin	1.95
	Phellandrene	1.21
	Caryophyllene	1.49
Water melon peel	Chlorophyll	5.28
	Diosmetin	1.57
	Pheophytin	1.27
	Malvidin 3,5 diglycoside	1.23
Potato peel	Gallic acid	0.16
	Protocatechuic	1.84
	p-Hydroxybenzoic	0.26
	Caffeic acid	0.19
	Vanillic acid	0.04
	Chlorogenic acid	0.28
Orange peel	*p*-coumaric	1.02
	Ferulic acid	0.91
	Syringic acid	7.71
	Naritutin	1.21
	Naringin	3.83
	Ascorbic acid	14.9
Olive leaves	Oleuropein	71.61
	Apigenin 7-glucoside	4.10
	Rutin	0.15
	Vanillin	0.15
	Vanillic acid	1.87
	Caffeic acid	1.02
	Hydroxytyrosol	3.29

Source: Zeyada et al. (2008)

are very efficient for the recovery of flavonoid aglycones, low-molecular-weight phenolics and phenolic acids. Solvents of higher polarity (ethanol or ethanol–water mixtures) additionally can extract flavonoid glycosides and higher molecular weight phenolics, resulting in higher yields of total extracted polyphenols (Oreopoulou and Tzia 2007). It is important to point out that ethyl acetate and ethanol are unrestricted solvents permitted for the use in the preparation of food ingredients (Marriott 2010). Other modern extraction and isolation techniques as alternate techniques are also used for the extraction of bioactive compounds. These modern techniques are also known as green processes for the extraction of bioactives (Herrero et al. 2010) and include supercritical fluid extraction (SFE), pressurized liquid extraction (PLE), microwave-assisted extraction (MAE) and ultrasound-assisted extraction (UAE). General approach for extraction, isolation and characterization of bioactive compounds from agricultural residues is presented in Fig. 11.2.

11.7.1 Liquid–Liquid Extraction

The current official analytical method for extracting phenolic compounds is liquid–liquid extraction (LLE) for liquid samples. This method requires expensive and hazardous organic solvents, which are undesirable for health and disposal reasons, and they require a long time per analysis, giving rise to possible degradations. The extraction temperature usually needs to be high in order to minimize the duration of the process. For these reasons, traditional extraction sample methods have been replaced by other methodologies which are more sensitive, selective, fast and environmentally friendly (Liazid et al. 2007). Solvents, such as methanol, ethanol, propanol, acetone and ethyl acetate, and their combinations have also been used for the extraction of phenolics, often with different proportions of water. Anthocyanins are usually extracted from plant material with an acidified organic solvent, most commonly methanol. This solvent system destroys the cell membranes, simultaneously dissolves the anthocyanins, and stabilizes them. However, the acid may bring about changes in the native form of anthocyanins by breaking down their complexes with metals and co-pigments (Naczk and Shahidi 2006).

11.7.2 Solid-Phase Extraction

Solid-phase extraction (SPE) is an increasingly useful sample preparation technique. With SPE, many of the problems associated with liquid–liquid extraction, such as incomplete phase separations, low recoveries, use and disposal of large and expensive quantities of organic solvents, can be avoided, although the cost of the equipment required for SPE is higher than that for LLE. This technique is used most often to prepare liquid samples and extract semi-volatile or non-volatile analytes but can also be used with solids that are pre-extracted into solvents. They are available in a wide variety of adsorbents, and sizes so that it is necessary to select the most suitable product for each application and sample.

11.7.3 Supercritical Fluid Extraction

Usually, phenolic compounds are extracted from plant samples by SPE coupled with other techniques, such as supercritical fluid extraction (SFE). Supercritical fluid

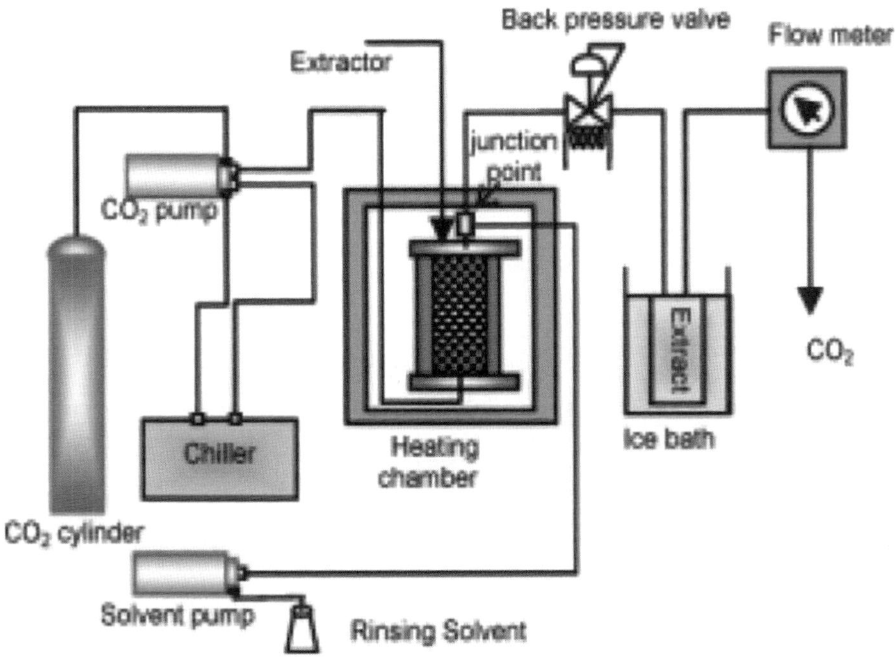

Fig. 11.3 Supercritical fluid extraction apparatus

extraction is a relatively recent technique which presents various advantages over traditional methods, such as the use of low temperatures and reduced energy consumption and high product quality due to the absence of solvents in the solute phase. However, this technique is limited to compounds of low or medium polarity. Supercritical carbon dioxide (SC-CO_2) is the most widely used solvent for SFE due to its particular characteristics, such as moderate critical conditions (31.1 °C and 73.8 MPa) and ready availability. It is also nontoxic, inflammable and chemically stable. However, SFE using CO_2 as the extracting solvent is of no use for phenolic compounds because of the low polarity of CO_2 in comparison to most phenols (Liazid et al. 2007). Generally, for this extraction procedure, several steps are followed: samples are loaded onto the sorbent of the SPE cartridge, which is inserted into the SPE/SFE extraction cell. The supercritical fluid used can be carbon dioxide, which must go through the SPE cartridge filled with the hydrolysed sample. Thus, analytes (phenolic compounds) are quantitatively trapped by a trapping solvent (e.g. methanol) at laboratory temperature (the trapping solvent is cooled naturally during the extraction by the expansion of CO_2). Finally, the extracts are evaporated to dryness, dissolved in the mobile phase and injected directly into the HPLC/ESI-MS system (Mahugo et al. 2009). Figure 11.3 shows the extraction of lycopene with supercritical CO_2. Liquid CO_2 flowing from the cylinder into the extraction vessel is compressed and controlled with a HPLC pump (slow suction and quick delivery). Liquid CO_2 is then cooled with a chiller to keep CO_2 in a liquid state. After reaching the extractor, CO_2 is transformed into a supercritical state by a heating chamber that envelops the extractor. Operating pressure is controlled by a backpressure regulator. The supercritical CO_2 flowing through the fixed bed in the extraction vessel is expanded into a collection tube immersed in an ice bath, where the extracted lycopene and the CO_2 solvent are easily separated. The amount of CO_2 consumed during the extraction period is determined by the use of a wet gas meter (Topal et al. 2006).

11.7.4 Pressurized Liquid Extraction

Pressurized liquid extraction (PLE) uses organic solvents at high pressures and temperatures above their normal boiling point. It is the newer modern method for isolation of analytes from solid samples (Klejdusa et al. 2009). In general, with PLE, a solid sample is packed into a stainless steel extraction cell and extracted with a suitable solvent under high temperatures (40–200 °C) and pressure (500–3,000 psi) for short periods of time (5–15 min). The sample extract is purged into a collection vial with the aid of a compressed gas. According to Liazid et al. (2007), PLE has been shown to be effective as a method for extracting polyphenols, while rapid methods that take only about 10 min have been developed that use high temperatures (150 °C) to accelerate the process (Liazid et al. 2007).

11.7.5 Microwave-Assisted Extraction

Microwave technology is commonly known for its use as heat treatment. For example, it is used as a heat process for commercial fruit products to achieve a fast but mild pasteurization of these products. At the same time, the use of microwaves serves to determine the stability of total polyphenol content after the treatment. According to Liazid et al. (2007), there is a clear relationship between the chemical structure and the stability of phenolic compounds that are studied under different conditions of MAE. Moreover, it has been shown that those that have a greater number of hydroxyl-type substitutes are more easily degraded under these temperature conditions (Liazid et al. 2007).

11.7.6 Ultrasound-Assisted Extraction

Ultrasonic radiation is a powerful aid in accelerating various steps of the analytical process. This energy is of great help in the pretreatment of solid samples as it facilitates and speeds up different operations including extraction of organic and inorganic compounds and homogenization. Ultrasound-assisted leaching is an effective way to extract analytes from different matrices in shorter times than with other extraction techniques (Dobias et al. 2010). Ultrasound-assisted systems have been widely used to extract capsaicinoids from hot peppers (Barbero et al. 2008). Ultrasonic extraction (USE) is considered one of the simplest extraction techniques because it is easy to perform using common laboratory equipment (i.e. ultrasonic bath). In this method, the crushed sample is mixed with the suitable solvent and placed into the ultrasonic bath, where the working temperature and extraction times are set (Klejdusa et al. 2009). The application of UAE in food-processing is of interest for facilitating the extraction of components from plant materials. The higher yield achieved in these UAE processes is of major interest from an industrial standpoint, since the technology is an add-on step to the existing process with minimum alteration, application in aqueous extraction where organic solvents can be replaced with solvents generally recognized as safe (GRAS), reduction in solvent usage and shorter extraction time.

11.7.7 Enzymatic Extraction of Bioactive Compounds

The application of enzymes for improved extraction of bioactive compounds without the use of solvents is an attractive proposition. Enzyme pretreatment of raw material normally results in a reduction in extraction time, minimizes usage of solvents and provides increased yield and quality of product (Sowbhagya and Chitra 2010). Decreased solvent use during extraction is particularly important for both regulatory and environmental reasons, providing a "greener" option than traditional nonenzymatic extraction. Enzymes have been successfully used in the extraction of various bioactive compounds. Carotenoids have been extracted from marigold flower skin with the help of enzymes (Dehghan-Shoar et al. 2011). Various enzymes, such as cellulases, pectinases and hemicellulases, are often required to disrupt the structural integrity of the plant cell wall, thereby enhancing the extraction efficiency of bioactives from plants. These enzymes hydrolyze cell wall components resulting in an increase in cell wall permeability, leading to higher extraction yields of bioactives. Enzymes have been used to increase flavonoid release from plant material while minimizing the use of solvents and heat (Kaur et al. 2010). Some of the bioactive compounds successfully extracted using the enzyme-assisted process from the plant materials are shown in Table 11.7. One example of the use of an enzyme system is in the processing of pectic polysaccharide for enhancing extraction of an antioxidant (Gan and Latiff 2010). Enzyme-aided extraction of lycopene from tomato tissues using cellulases and pectinases under optimized conditions resulted in a significant increase (20 %) in lycopene yield as against control (Choudhari and Ananthanarayan 2007). An increase in concentrations of phenolic compounds (25.90–39.72 %) and sugars (12–14 g/l) has been reported with the enzyme-assisted extraction from citrus peel and grape pomace (Kammerer et al. 2005). Application of enzymes on grape skin led to an efficient extraction of pigment (anthocyanin) during the vinification process (Munoz 2004). Extraction of lignans (secoisolariciresinol) from flax (*Linum usitatissimum*)

Table 11.7 Enzyme-assisted extraction of bioactive compounds from plants

Product	Source	Enzyme used	Yield	References
Carotenoids	Marigold flower	Viscozyme, pectinex, neutrase	97 %	Barzana et al. (2002)
Carotene	Carrot pomace	Pectinex	64 mg/kg	Stoll et al. (2003)
Lycopene	Tomato	Pancreatin	2.5-fold increase through enzyme aided extraction	Dehghan-Shoar et al. (2011)
Phenolics	Citrus peel	Cellulase	65.5 %	Li et al. (2006)
Catechins	Tea beverage	Pepsin	89–102 %	Ferruzzi and Green (2006)
Lignans (secoisolariciresinol)	Flax	Cellulase and β-glucosidase	40.75 mg/g in hulls and 15.20 mg/g in whole seeds	Renouard et al. (2010)

hulls and whole seeds was improved by using cellulase and β-glucosidase. Both enzyme preparations proved to be effective for extracting secoisolariciresinol (Renouard et al. 2010).

11.7.8 Other Extraction Methods

Use of aqueous cyclodextrins (CD) for recovery of selected bioactive phenolic compounds from grapes and their pomace was evaluated by Ratnasooriya and Rupasinghe (2012). When α, β and γ forms of CD were compared, β-CD was the most effective in recovering stilbenes, flavonols and flavan-3-ols from grape pomace (Ratnasooriya and Rupasinghe 2012).

11.8 Health Benefits of Bioactive Compounds

Numerous health benefits of bioactive compounds have been reported so far. Jang et al. (1997) reported anticancer potential of resveratrol from grape seeds. The interest in lignans and their synthetic derivatives is growing because of their potential applications in cancer chemotherapy and various other pharmacological effects (Saleem et al. 2005). Mantena et al. (2006) studied the effect of grape seed proanthocyanidins in inducing apoptosis and inhibition of metastasis in both cultured breast and colon cancer cells. Pomegranate peel extracts have been shown to retard proliferation of cells in several different human cancer cell lines (Kawaii and Lansky 2004). Pomegranate peel contains substantial amounts of polyphenols such as ellagic acid and gallic acid (Nasr et al. 1996). The presence of these polyphenols in the pomegranate peel may be responsible for antimutagenicity of peel extracts (Gil et al. 2000). Another study conducted by Jang et al. (1997) illustrated that resveratrol, a natural product derived from grapes, inhibited tumour initiation, promotion and progression. Several studies have shown that phenolic compounds reduce in vitro oxidation of low-density lipoprotein. Of them, phenolics with multiple hydroxyl groups are generally the most efficient for preventing lipid and low-density lipoproteins (LDL) oxidation and therefore reduce the risk of atherogenesis (Meyer et al. 1998). Apple pomace is the main by-product of apple juice processing plant and has been found to be a good source of polyphenols (Peschel et al. 2006). Grape seed extract has been found to decrease thiobarbituric acid reactive substances (TBARS) value and showed no effect on HDL-cholesterol while LDL-cholesterol, triglycerides and total cholesterol decreased significantly (Vigna et al. 2003). Grape seed extract has also been found to reduce chronic pancreatic problems, vomiting and pain (Banerjee and Bagchi 2001). Table 11.8 shows biological and therapeutic activities of various phenolic compounds that could be extracted from agricultural residues.

11.8.1 Therapeutic Activities of Bioactive Compounds

Different phenolic compounds act in different ways and follow different mechanisms in combating several disorders. One of the examples is the action of resveratrol, abundantly found in grape seeds in carcinogenesis. Resveratrol is able to prevent initiation phase by inhibition of carcinogen activation (R+) induction of carcinogen deactivation and subsequently blocking interaction between DNA and carcinogen (R+) [Fig. 11.4]. Resveratrol can block the action of tumour promoter and can act on tumour progression by inhibition of angiogenesis and metastatic process. Resveratrol could act on carcinogenesis by inhibiting the initiation phase which consists of the DNA alteration (mutation) of a normal cell, which is an irreversible and fast change. The initiated cell is capable to autonomous growth. The initiating event can consist of a single exposure to a carcinogenic agent, or in some cases, it may be an inherited genetic defect. The anti-initiation activity of resveratrol is linked to the suppression of the metabolic activation of carcinogens and/or the detoxifying increases via a modulation of the drug-metabolizing enzymes involved either in phase I reactions transforming a lipophilic compound into an electrophilic active carcinogen or in phase II conjugation enzyme systems converting the primary metabolite into a final hydrosoluble metabolite (Fig. 11.4).

Most of the research on bioactive molecules is still confined to screening of crude extracts. Elucidation of the structure of the active principles could be a priority area for further studies, if the extracts did show promising activity. A potential area for further studies is the use of "-omic" technologies to unravel the mechanism of action of bioactive compounds. The "-omic" technologies based on the technology platform of genomics, proteomics and metabolomics can be applied to relate

Table 11.8 Biological and therapeutic activities of bioactive phenolic compounds

Phenolic compound	Effect	References
Anthocyanin	Improvement in visual capacity	Lee et al. (2005)
	Improvement in brain function	Kang et al. (2006); Shin et al. (2006)
	Prevention of diabetes and obesity	Tsuda et al. (2003); Guo et al. (2007)
	Decrease in serum triglycerides, total cholesterol	Xia et al. (2006)
	Decrease in lipid hydroperoxides	McAnulty et al. (2005)
	Decrease in blood pressure	Herrera-Arellano et al. (2007)
Ellagic acid	Slow down tumour growth	Harris et al. (2001)
	Protection against cardiovascular diseases	Aviram et al. (2000)
	Anticancer	Aggarwal and Shishodia (2006)
Isoflavones (phytoestrogens)	Prevention of osteoporosis	Williamson-Hughes et al. (2006)
	Lowering effect on LDL and cholesterol	Jayagopal et al. (2002)
	Improvement of cognitive function	Lee et al. (2005)
	Reduction of colon cancer	Cotterchio et al. (2006)
	Modulation of immune function	Ryan-Borchers et al. (2006)
Resveratrol (3,5,4-trans-trihydroxystilbene)	Cardioprotective effect	Bertelli et al. (1995)
	Anticancer	Jang et al. (1997)
	Prevention of Ischemic damage	Bradamante et al. (2004)
Flavonones (naringenin and hesperitin and corresponding glycosides)	Cardioprotective effect	Jeon et al. (2007)
	Decrease in liver toxicity	Kaur et al. (2006)
	Protection against arthritis	Kawaguchi et al. (2006)
Proanthocyanidins	Decrease in cholesterol	Williamson and Manach (2005)
	Decrease in blood pressure	
	Maintenance of endothelial function	
	Improvement of blood circulation in legs	Christie et al. (2004)

the complex effects of the bioactives in the form of gene/protein expression profiles (Ulrich-Merzenich et al. 2007). Biotechnological approaches like microarray analysis are likely to give insights into the differential gene expression when the cells are exposed to the bioactive molecules. With the advent of bioinformatics, in silico modelling in the research on bioactive molecules could also be exploited. Such an approach is being currently applied in drug discovery wherein the bioactivity of natural products is manipulated through computer simulation.

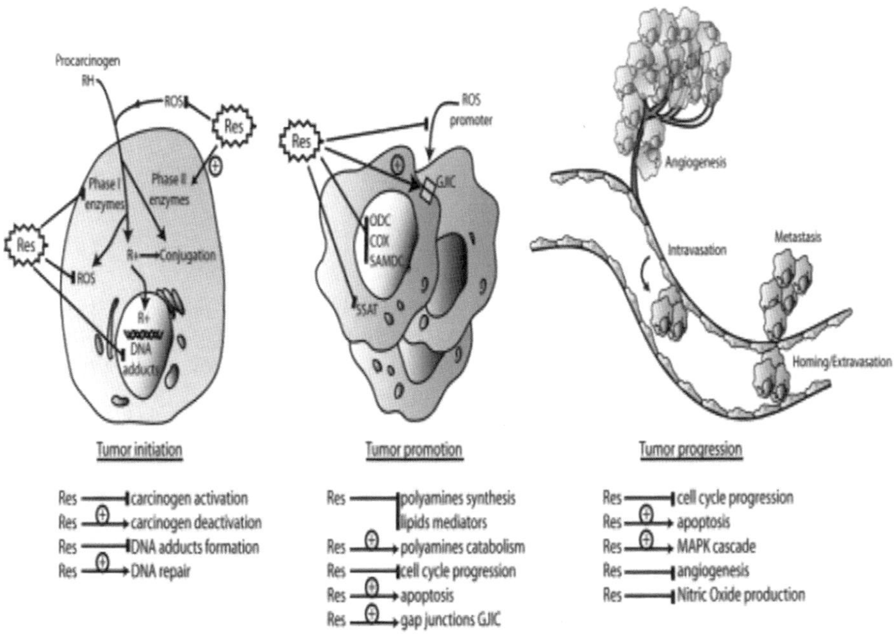

Fig. 11.4 Effect of resveratrol in carcinogenesis

In addition to the natural bioactive molecules, several studies are focusing on the production recombinant molecules. For instance, production of monoclonal antibodies and recombinant allergens, development of immunoassays, recombinant proteins with the aim of contributing to the development of diagnostic kits, have great potential for commercialization (Chu and Radhakrishnan 2008).

11.9 Conclusions

Agricultural residues though regarded as wastes have a tremendous potential as sources of bioactive compounds. Among the agricultural residues, the most important class of residues as sources of bioactive compounds are the fruit and vegetable residues which are composed of different phenolic constituents. Phenolic compounds comprise of flavonoids, phenolic acids, tannins, etc. All such compounds are known for their antioxidant potential, and some of such compounds have been successfully evaluated for their therapeutic potential. Because of the complexity in structure of different bioactive compounds, various antioxidant assays have to be employed, and the correlation between the methods used needs to be worked out in order to arrive at the antioxidant potential of the compounds. Extraction and purification of the bioactive compounds is of paramount importance for their use in food and pharmaceutical applications as the toxicity associated with the solvents has raised lot of concerns. Studies are required to select the most efficient and cost-effective method for extraction of bioactive compounds from different agricultural residues. The structural complexity of different bioactive compounds and different assays that need to be conducted for evaluation of the antioxidant potential of the bioactive compounds makes it difficult for the precise quantification of the bioactive compounds despite advancements in the field of development of sophisticated equipment. The potential of the bioactive compounds in combating disorders and diseases has been well documented; however, studies are needed to ascertain the role of each bioactive compound and elucidate the mechanism of disease control, especially for serious diseases like cancer, diabetes and cardiovascular diseases. Therefore, processes involving supercritical fluid extraction, microwave and ultrasound-assisted extractions are being evaluated. Presently, such processes are relatively expensive, but looking at the health benefits of the bioactive compounds and with technological advancements in instrumentation for development of sophisticated equipment at low cost, these processes appear quite futuristic. Further, research on elucidation of active principles and use of "-omic" technologies can lead to a significant increase in scientific knowledge on the mechanism of action of bioactive compounds, and molecular level studies are essentially needed to understand the cell-bioactive compound interactions.

References

Aggarwal BB, Shishodia S (2006) Molecular targets of dietary agents for prevention and therapy of cancer. Biochem Pharmacol 71:1397–1421

Amendola D, De Faven DM, Egues I, Serrano L, Labidi J, Spgno G (2012) Autohydrolysis and organosolv process for recovery of hemicelluloses, phenolic compounds and lignin from grape stalks. Bioresour Technol 107:267–274

Anagnostopoulou MA, Kefalas P, Papageorgiou VP, Assimopoulou AN, Boskou D (2006) Radical scavenging activity of various extracts and fractions of sweet orange peel (*Citrus sinensis*). Food Chem 94:19–25

Ando S, Sakaki T, Kokusho T, Shibata M, Uemura Y, Hatate Y (2000) Decomposition behaviour of plant biomass in hot compressed water. Ind Eng Chem Res 39:3688–3693

Arogba SS (2000) Mango (*Mangifera indica*) kernel: chromatographic analysis of the tannin, and stability study of the associated polyphenol oxidase activity. J Food Compost Anal 13: 149–156

Aviram M, Dornfeld L, Rosenblat M, Volkova N, Kaplan M, Colemann R, Hayek T, Presser D, Fuhrman B (2000) Pomegranate juice consumption reduces oxidative stress, atherogenic modifications to LDL, and platelet aggregation: studies in humans and in atherosclerotic apolipoprotein E-deficient mice. Am J Clin Nutr 71:1062–1076

Awika JM, Rooney LW, Wu X, Prior RL, Cisneros ZL (2003) Screening methods to measure antioxidant activity of sorghum (*Sorghum bicolour*) and Sorghum products. J Agric Food Chem 51:6657–6662

Awika JM, Rooney LW, Waniska RD (2005) Anthocyanins from black sorghum and their antioxidant properties. Food Chem 90:293–301

Babbar N, Oberoi HS, Uppal DS, Patil RT (2011) Total phenolic content and antioxidant capacity of extracts obtained from six important fruit residues. Food Res Int 44:391–396

Babbar N, Oberoi HS, Kaur S (2012) Influence of different solvents in extraction of phenolic compounds from vegetable residues and their evaluation as natural sources of antioxidants. J Food Sci Technol. doi:10.1007/s 13197-012-0754-4

Balasundram N, Sundram K, Samman S (2006) Phenolic compounds in plants and agri-industrial by-products: antioxidant activity, occurrence, and potential uses. Food Chem 99:191–203

Banerjee B, Bagchi D (2001) Beneficial effects of a novel IH636 grape seed proanthocyanidin extract in the treatment of chronic pancreatitis. Digestion 63:203–206

Barbero GF, Liazid A, Palma M, Barroso CG (2008) Ultrasound-assisted extraction of capsaicinoids from peppers. Talanta 75:1332–1337

Barzana E, Rubio D, Santamaria RI, Garcia-Correa O, Garcia F, Ridaura Sanz VE, Lopez-Munquia A (2002) Enzyme-mediated solvent extraction of carotenoids from marigold flower (*Tagetes erecta*). J Agric Food Chem 38:1400–1403

Bathaie SZ, Kermani FMZ, Shams A (2011) Crocin bleaching assay using purified di-gentiobiosyl crocin (α-crocin) from Iranian saffron. Iran J Basic Med Sci 14:39

Becker EM, Nissen LR, Skibsted LH (2004) Antioxidant evaluation protocols: food quality or health effects. Eur Food Res Technol 219:561–571

Benavente-Garcia O, Castillo J, Lorente J, Ortuno A, Del Rio JA (2000) Antioxidant activity of phenolics extracted from *Olea europaea* L. leaves. Food Chem 68:457–462

Bendich A (2004) From 1989 to 2001: what have we learned about the "biological actions of beta-carotene". J Nutr 134:225S–230S

Bentley MD, Rajab MS, Mendel MJ, Alford AR (1990) Limonoid model insect antifeedants. J Agric Food Chem 50:4491–4496

Benzie IF (2000) Evolution of antioxidant defence mechanisms. Eur J Nutr 39:53–61

Benzie IFF, Strain JJ (1996) The ferric reducing ability of plasma (FRAP as a measure of antioxidant power. The FRAP assay). Anal Biochem 239:70–76

Bertelli AA, Giovannini L, Giannessi D, Migliori M, Bernini W, Fregoni M, Bertelli A (1995) Antiplatelet activity of synthetic and natural resveratrol in red wine. Int J Tissue React 17:1–3

Bouchard J, Nguyen TS, Chornet E, Overend RP (1991) Analytical methodology for biomass pretreatment. Part 2: characterization of the filtrates and cumulative product distribution as a function of treatment severity. Bioresour Technol 36:121–131

Boussaid A, Cai Y, Robinson J, Gregg DJ, Nguyen Q, Saddler JN (2001) Sugar recovery and fermentability of hemicellulose hydrolysates from steam-exploded softwoods containing bark. Biotechnol Prog 17:887–892

Bradamante S, Barenghi L, Villa A (2004) Cardiovascular protective effects of resveratrol. Cardiovasc Drug Rev 22:169–188

Braddock RJ (1995) By-products of citrus fruit. Food Technol 49:74–77

Cadahıa E, Munoz L, Fernandez de Simon B, Garcıa Vallejo MC (2001) Changes in low molecular weight phenolic compounds in Spanish, French and American oak woods during natural seasoning and toasting. J Agric Food Chem 49:1790–1798

Calvo MM (2005) Lutein: a valuable ingredient of fruit and vegetables. Crit Rev Food Sci Nutr 45:671–696

Cameron E, Pauling L (1978) Supplemental ascorbate in the supportive treatment of cancer: re-evaluation of prolongation of survival times in terminal cancer patients. Proc Nat Acad Sci U S A 75:4538–4542

Chantaro P, Devahastin S, Chiewchan N (2008) Production of antioxidant high dietary fiber from carrot peel. LWT Food Sci Technol 41:1987–1994

Chodak AD, Tarko T (2007) Antioxidant properties of different fruit seeds and peels. Acta Sci Pol Technol Aliment 6:29–36

Choudhari SM, Ananthanarayan L (2007) Enzyme aided extraction of lycopene from tomato tissues. Food Chem 102:77–81

Christie S, Walker AF, Hicks SM, Abeyasekera S (2004) Flavonoid supplement improves leg health and reduces fluid retention in premenopausal women in a double-blind, placebo-controlled study. Phytomedicine 11:11–17

Chu W-L, Radhakrishnan AK (2008) Research on bioactive molecules: achievements and the way forward. IeJSME 2(Suppl 1):S21–S24

Coll MD, Coll L, Laencine J, Tomasbarberan FA (1998) Recovery of flavanons from wastes of industrially processed lemons. Z Naturforch 206:404–407

Conde E, Cadahia E, Garcia-Vallejo MC, Tomas-Barberan F (1995) Low molecular weight polyphenols in wood and bark of *Eucalyptus globulus*. Wood Fiber Sci 27:379–383

Cotterchio M, Boucher BA, Manno M, Gallinger S, Okey A, Harper P (2006) Dietary phytoestrogen intake is associated with reduced colorectal cancer risk. J Nutr 136:3046–3053

D'archivio M, Filesi CD, Benedetto R, Gargiulo R, Giovannini MR (2007) Polyphenols, dietary sources and bioavailability. Ann Chim 43:348–361

Dehghan-Shoar Z, Hardacre AK, Meerdink G, Breenan CS (2011) Lycopene extraction from extruded products containing tomato skin. Int J Food Sci Technol 46:365–371

Dobias P, Pavlíkova P, Adam M, Eisner A, Benova B, Ventura K (2010) Comparison of pressurised fluid and ultrasonic extraction methods for analysis of plant antioxidants and their antioxidant capacity. Cent Eur J Chem 8:87–95

Domınguez H, Torres JL, Nunez MJ (2001) Antioxidant phenolics as food additives from agricultural wastes. Polyphenols Actual 21:26–30

Duan X, Jiang Y, Su X, Zhang Z, Shi J (2007) Antioxidant properties of anthocyanins extracted from litchi (*Litchi chinensis Sonn.*) fruit pericarp tissues in relation to their role in the pericarp browning. Food Chem 101:1365–1371

Fernandez-Bolanos J, Felizon B, Heredia A, Guillen R, Jimenez A (1999) Characterization of the lignin obtained by alkaline delignification and of the cellulose residue from steam-exploded olive stones. Bioresour Technol 68:121–132

Ferruzzi MG, Green RJ (2006) Analysis of catechins from milk-tea beverages by enzyme assisted extraction followed by high performance liquid chromatography. Food Chem 99:484–491

Foo LY, Lu Y (1999) Isolation and identification of procyanidin in apple pomace. Food Chem 64:511–518

Frankel EN, Waterhouse AL, Teissedre PL (1995) Principal phenolic phytochemicals in selected California wines and their antioxidant activity in inhibiting oxidation of human low-density lipoproteins. J Agric Food Chem 43:890–894

Fruhwirth GO, Wenzl T, El-Toukhy R, Wagner FS, Hermetter A (2003) Fluorescence screening of antioxidants capacity in pumpkin seed oils and other natural oils. Eur J Lipid Sci Technol 105:266–274

Gan CY, Latiff AA (2010) Extraction of antioxidant pectic polysaccharide from mangosteen (*Garcinia mangostana*) rind: optimization using response surface methodology. Carbohyd Polym 83:600–607

Gardner PT, White TAC, Mc Phaid DB, Duthie GG (2002) The relative contributions of vitamin C, carotenoids and phenolics to the antioxidant potential of fruit juices. Food Chem 68:471–474

George B, Kavr C, Khurdiya D, Kapoor H (2004) Antioxidants in the tomato (*Lycopersicon esculentum*) as a function of genotype. Food Chem 84:45–51

Ghasemi K, Ghasemi Y, Ebrahimzadeh MA (2009) Antioxidant activity and flavonoid contents of 13 citrus species peels and seeds. Pak J Pharm Sci 22:277–281

Ghiselli A, Serafini M, Natella F, Scaccini C (2000) Total antioxidant capacity as a tool to assess redox status: critical review and experimental data. Free Radical Biol Med 29:1106–1114

Gil MI, Tomasbarberan FA, Hess PB, Holcroft DM, Kader AA (2000) Antioxidant activity of pomegranate juice and its relationship with phenolic composition and processing. J Agric Food Chem 48:4581–4589

Guo H, Ling W, Wang Q, Liu C, Hu Y, Xia M, Feng X, Xia X (2007) Effect of anthocyanin-rich extract from black rice (*Oryza sativa* L. *indica*) on hyperlipidemia and insulin resistance in fructose-fed rats. Plant Foods Hum Nutr 62:1–6

Halliwell B, Gutteridge JMC (1990) The antioxidants of human extracellular fluids. Arch Biochem Biophys 280:1–8

Hansberg W (2002) Biologıa de las especies de oxıgeno reactivas. In: Cea Bonilla A, del Arenal Mena IP, Riveros Rosas H, Vazquez-Contreras E (eds) Mensaje bioquımico, vol 26. Mexico, p 19–54

Harris GK, Gupta AM, Nines RG, Kresty LA, Habib SG, Frankel WL, LaPerle K, Gallaher DD, Schwartz SJ, Stoner GD (2001) Effects of lyophilized black raspberries on azoxymethane induced colon cancer and 8-hydroxy-20-deoxyguanosine levels in the Fischer 344 rat. Nutr Cancer 40:125–133

Heijnen CGM, Haenen GRMM, Van Acker FAA, Van Der Vijgh W, Bast A (2001) Flavonoids as peroxynitrite scavengers: the role of the hydroxyl groups. Toxicol In Vitro 15:3–6

Hensley K, Floyd RA (2002) Reactive oxygen species and protein oxidation in aging: a look back, a look ahead. Arch Biochem Biophys 397:377–383

Herrera-Arellano A, Miranda-Sanchez J, Avila-Castro P, HerreraAlvarez S, Jimenez-Ferrer JE, Zamilpa A, Roman-Ramos R, Ponce-Monter H, Tortoriello J (2007) Clinical effects produced by a standardized herbal medicinal product of *Hibiscus sabdariffa* on patients with hypertension. A randomized, double-blind, lisinopril controlled clinical trial. Planta Med 73:6–12

Herrero M, Plaza M, Cifuentes A, Ibanez E (2010) Green processes for extraction of bioactives from Rosemary: Chemical and functional characterization via UPLC-MS/MS and in-vitro assays. J Chromatography A 1217:2512–2520

Holker U, Höfer M, Lenz J (2004) Biotechnological advances of laboratory-scale solid-state fermentation with fungi. Appl Microbiol Biotechnol 64:175–186

Hong-Bo S, Li-Ye C, Ming-An S, Jaleel CA, Hong-Mei M (2008) Higher plant antioxidants and redox signaling under environmental stresses. C R Biol 331:433–441

Huang D, Ou B, Prior RL (2005) The chemistry behind antioxidant capacity assays. J Agric Food Chem 53:1841–1856

Iwatsuki M, Niki E, Stone D, Darley-Usmar VM (1995) A-tocopherol mediated peroxidation in the copper (II) and metmyoglobin induced oxidation of human low density lipoprotein: the influence of lipid hydroperoxides. FEBS Lett 360:271–276

Jacobsen C, Adler-Nissen J, Meyer AS (1999) Effect of ascorbic acid on iron release from the emulsifier interface and on the oxidative flavor deterioration in fish oil enriched mayonnaise. J Agric Food Chem 47:4917–4926

Jang M, Cai L, Udeani GO, Slowing KV, Thomas CF, Beecher CWW, Fong HHS, Farnsworth NR, Kinghorn AD, Mehta RG, Moon RC, Pezzuto JM (1997) Cancer chemopreventive activity of resveratrol, a natural product derived from grapes. Science 275:218–220

Jayagopal V, Albertazzi P, Kilpatrick ES, Howarth EM, Jennings PE, Hepburn DA, Atkin SL (2002) Beneficial effects of soy phytoestrogen intake in postmenopausal women with type 2 diabetes. Diabetes Care 25:1709–1714

Jayaprakasha GK, Selvi T, Sakariah KK (2003) Antibacterial and antioxidant activities of grape (*Vitis vinifera*) seed extracts. Food Res Int 36:117–122

Jayaprakasha GK, Girennavar B, Patil BS (2008) Radical scavenging activity of red grapefruits and sour orange fruit extracts in different in-vitro model systems. Bioresour Technol 99:4484–4494

Jeon SM, Kim HK, Kim HJ, Do GM, Jeong TS, Park YB, Choi MS (2007) Hypocholesterolemic and antioxidative effects of naringenin and its two metabolites in high-cholesterol fed rats. Transl Res 149:15–21

Jim S, Hong-Shum L (2003) Antioxidant. In: Jim S, Hong-Shum L (eds) Food additives data book. Blackwell Science, Iowa, IO, pp 75–118

Kalra KL, Grewal HS, Kahlon SS (1989) Bioconversion of kinnow-mandarin waste into single-cell protein. World J Microbiol Biotechnol 5:321–326

Kammerer D, Claus A, Schieber A, Carle A (2005) A novel process for the recovery of polyphenols from grape (*Vitis vinifera*) pomace. J Food Sci 70:157–163

Kang TH, Hur JY, Kim HB, Ryu JH, Kim SY (2006) Neuroprotective effects of the cyanidin-3-O-beta-D-glucopyranoside isolated from mulberry fruit against cerebral ischemia. Neurosci Lett 391:122–126

Kaur G, Tirkey N, Chopra K (2006) Beneficial effect of hesperidin on lipopolysaccharide-induced hepatotoxicity. Toxicology 226:152–160

Kaur A, Singh S, Singh RS, Schwarz WH, Puri M (2010) Hydrolysis of citrus peel naringin by recombinant a-L-rhamnosidase from *Clostridium stercorarium*. J Chem Technol Biotechnol 85:1419–1422

Kawaguchi K, Maruyama H, Kometani T, Kumazawa Y (2006) Suppression of collagen-induced arthritis by oral administration of the citrus flavonoid hesperidin. Planta Med 72:477–479

Kawaii S, Lansky EP (2004) Differentiation-promoting activity of pomegranate (*Punica granatum*) fruit extracts in HL-60 human promyelocytic leukemia cells. J Med Food 7:13–18

King AI, Uijttenboogaart TG, de Vries AW (1995) α-Tocopherol, β-carotene and ascorbic acid as antioxidants in stored poultry muscle. J Food Sci 60:1009–1012

Klejdusa B, Kopecky J, Benesova L, Vaceka J (2009) Solid-phase/supercritical-fluid extraction for liquid chromatography of phenolic compounds in freshwater microalgae and selected cyanobacterial species. J Chromatogr A 1216:763–771

Klinke HB, Schmidt AS, Thomsen AB (1998) Identification of degradation products from wheat straw in relation to pre-treatment conditions. In: Kopetz H, Weber T, Palz W, Chartier P, Ferrero GL (eds) Biomass for energy and industry. Proceedings. C.A.R.M.E.N., Rimpar, pp 484–487

Kris-Etherton PM, Hecker KD, Bonanome A, Coval SM, Binkoski AE, Hilpert KF (2002) Bioactive compounds in foods: their role in the prevention of cardiovascular disease and cancer. Am J Med 113:71S–88S

Larrauri JA, Sanchez MC, Sauracalixto F (1996) Effect of temperature on the free radical scavenging capacity of extracts from red and white grape pomace peels. J Agric Food Chem 46: 2694–2697

Law MR, Morris JK (1998) By how much does fruit and vegetable consumption reduce the risk of ischaemic heart disease. Eur J Clin Nutr 52:549–556

Lee HS, Wicker L (1991) Quantitative changes in anthocyanin pigments of lychee fruit during refrigerated storage. Food Chem 40:263–270

Lee JH, Park CH, Jung KC, Rhee HS, Yang CH (2005) Negative regulation of beta-catenin/Tcf signaling by naringenin in AGS gastric cancer cell. Biochem Biophys Res Commun 335:771–776

Li BB, Smith B, Hossain M (2006) Extraction of phenolics from citrus peels: II. Enzyme-assisted extraction method. Sep Purif Technol 48:189–196

Liazid A, Palma M, Brigui J, Barroso GC (2007) Investigation on phenolic compounds stability during microwave-assisted extraction. J Chromatogr A 1140:29–34

Lim YY, Lim TT, Tee JJ (2007) Antioxidant properties of several tropical fruits: a comparative study. Food Chem 103:1003–1008

Lu Y, Foo LY (1997) Identification and quantification of mayor polyphenols in apple pomace. Food Chem 59:187–194

Mahugo C, Sosa Z, Torres ME, Santana JJ (2009) Methodologies for the extraction of phenolic compounds from environmental samples: new approaches. Molecules 14:298–320

Maier T, Schieber T, Kammerer DR, Carle R (2009) Residues of grape (*Vitis vinifera* L.) seed oil production as a valuable source of phenolic antioxidants. Food Chem 112:551–559

Maiorino M, Zamburlini A, Roveri A, Ursini F (1995) Copper induced lipid peroxidation in liposomes, micelles, and LDL: which is the role of vitamin E? Free Radic Biol Med 18:67–74

Mantena SK, Baliga MS, Katiyar SK (2006) Grape seed proanthocyanidins induce apoptosis and inhibit metastasis of highly metastatic breast carcinoma cells. Carcinogenesis 24:1682–1691

Marriott RJ (2010) Greener chemistry preparation of traditional flavour extracts and molecules. Agro Food Industry Hi-Tech 21:46–48

Martell AE (1982) Chelates of ascorbic acid. Formation and catalytic properties. In: Seib PA, Tolbert BM (eds) Ascorbic acid: chemistry, metabolism and uses. American Chemical Society, Washington, DC, pp 153–167

Martın C, Galbe M, Nilvebrant NO, Jonsson LF (2002) Comparison of the fermentability of enzymatic hydrolyzates of sugarcane bagasse pretreated by steam explosion using different impregnating agents. Appl Biochem Biotechnol 98:699–716

McAnulty SR, McAnulty LS, Morrow JD, Khardouni D, Shooter L, Monk J, Gross S, Brown V (2005) Effect of daily fruit ingestion on angiotensin converting enzyme activity, blood pressure, and oxidative stress in chronic smokers. Free Radical Res 39:1241–1248

Meyer AS, Jepsen SM, Sorensen NS (1998) Enzymatic release of antioxidants for human low-density lipoprotein from grape pomace. J Agric Food Chem 46:2439–2446

Munoz O (2004) Effects of enzymatic treatment on anthocyanic pigments from grapes skin from Chilean wine. Food Chem 87:487–490

Naczk M, Shahidi F (2006) Phenolics in cereals, fruits and vegetables: occurrence, extraction and analysis. J Pharm Biomed Anal 41:1523–1542

Nakasone HY, Paull RE (1998) Tropical fruits. CAB International, Wallingford, pp 98–105

Nasr CB, Ayed N, Metche M (1996) Quantitative determination of the polyphenolic content of pomegranate peel. Z Naturforsch 203:374–378

Negro C, Tommasi L, Miceli A (2003) Phenolic compounds and antioxidant activity from red grape marc extracts. Bioresour Technol 87:14–41

Oberoi HS, Kalra KL, Uppal DS, Tyagi SK (2007) Effects of different drying methods of cauliflower waste on drying time, colour retention and glucoamylase production by *Aspergillus niger* NCIM 1054. Int J Food Sci Technol 42:228–234

Oberoi HS, Sandhu SK, Vadlani PV (2012) Statistical optimization of hydrolysis process for banana peels using cellulolytic and pectinolytic enzymes. Food Bioprod Process 90:257–265

Obied HK, Allen MS, Bedgood DR, Prenzler PD, Robards K, Stockmann R (2005) Bioactivity and analysis of biophenols recovered from olive mill waste. J Agric Food Chem 53:823–837

Oreopoulou V, Tzia C (2007) Utilization of plant by-products for the recovery of proteins, dietary fibers, antioxidants, and colorants. Utilization of By-Products and Treatment of Waste in the Food Industry 2:209–232

Park YK, Park E, Kim JS, Kang MH (2003) Daily grape juice consumption reduces oxidative DNA damage and plasma free radical levels in healthy Koreans. Mutat Res 529:77–86

Peschel W, Sánchez RF, Diekmann W, Plescher A, Gartzia I, Jimenez D (2006) An industrial approach in the search of natural antioxidants from vegetable and fruit wastes. Food Chem 97:137–150

Podsedek A (2007) Natural antioxidant capacity of brassica vegetables: a review. LWT Food Sci Technol 40:1–11

Popa VI, Dumitru M, Volf I, Anghel N (2008) Lignin and polyphenols as allelo chemicals. Ind Crops Prod 27:144–149

Prior RL, Hoang H, Gu L, X W, Bcchiocca M, Howard L (2003) Assays for hydrophilic and lipophilic antioxidant capacity (ORAC) of plasma and other biological and food samples. J Agric Food Chem 53:4290–4302

Puravankara D, Boghra V, Sharma RS (2000) Effect of antioxidant principles isolated from mango (*Mangifera indica* L.) seed kernels on oxidative stability of buffalo ghee (butter-fat). J Sci Food Agric 80:522–526

Ratnam DV, Ankola DD, Bharadwaj V, Sahana DK, Kumar MNVR (2006) Role of antioxidants in prophylaxis and therapy. A pharmaceutical perspective. J Control Release 13:189–207

Ratnasooriya CC, Rupasinghe HPV (2012) Extraction of phenolic compounds from grapes and their pomace using beta-cyclodextrin. Food Chem 134:625–631

Renouard S, Hano C, Corbin C, Fliniaux O, Lopez T, Montguillon J, Barakzoy E, Mesnard F, Lamblin F, Laine E (2010) Cellulase-assisted release of secoisolariciresinol from extracts of flax (*Linum usitatissimum*) hulls and whole seeds. Food Chem 122:679–687

Rice-Evans CJ, Miller NJ (1998) Structure-antioxidant activity relationships of flavonoids and isoflavonoids. In: Rice-Evans CA, Packers L (eds) Flavonoids in health and disease. Marcel Dekker, Inc, New York, NY, pp 199–219

Rice-Evans CA, Miller NJ, Papaganga G (1997) Antioxidant properties of phenolic compounds. Trends Plant Sci 4:152–159

Ryan-Borchers TA, Park JS, Chew BP, McGuire MK, Fournier LR, Beerman KA (2006) Soy isoflavones modulate immune function in healthy postmenopausal women. Am J Clin Nutr 83:1118–1125

Saleem M, Kim HJ, Ali MS, Lee YS (2005) An update on bioactive plant lignans. Nat Prod Rep 22:696–716

Sang S, Lapsley K, Jeong WS, Lachance PA, Ho CT, Rosen RTJ (2002) Antioxidative phenolic compounds isolated from almond skins (*Prunus amygdalus Batsch*). J Agric Food Chem 50:2459–2463

Scalbert A, Johnson IT, Saltmarsh M (2005a) Polyphenols: antioxidants and beyond. Am J Clin Nutr 81:215S–217S
Scalbert A, Manach C, Morand C, Remesy C (2005b) Dietary polyphenols and the prevention of diseases. Crit Rev Food Sci Nutr 45:287–306
Scalzo J (2005) Plant genotype affects total antioxidant capacity and phenolic contents in fruit. Nutrition 21:207–213
Sherwin ER (1990) Antioxidants. In: Branen AL, Davidson PM, Salminen S (eds) Food antioxidants. Marcel Dekker Inc, New York, NY
Shin WH, Park SJ, Kim EJ (2006) Protective effect of anthocyanin in middle cerebral artery occlusion and reperfusion model of cerebral ischemia in rats. Life Sci 79:130–137
Shrikhande AJ (2000) Wine by-products with health benefits. Food Res Int 33:469–474
Shui G, Leong LP (2006) Residue from star fruit as valuable source for functional food ingredients and antioxidant nutraceuticals. Food Chem 97:45–51
Someya S, Yoshiki Y, Okube K (2002) Antioxidant compounds from bananas (*Musa cavendish*). Food Chem 79:351–354
Sowbhagya HB, Chitra VN (2010) Enzyme-assisted extraction of flavorings and colorants from plant materials. Crit Rev Food Sci Nutr 50:146–161
Squadrito GL, Pryor WA (2002) Mapping the reaction of peroxynitrite with CO_2: energetics, reactive species and biological implications. Chem Res Toxicol 15:885–895
Stoll T, Schweiggert U, Schieber A, Carle R (2003) Process for the recovery of a carotene rich functional food ingredient from carrot pomace by enzymatic liquification. Innov Food Sci Emerg Technol 4:415–423
Strati IF, Oreopoulou V (2011) Effect of extraction parameters on the carotenoid recovery from tomato waste. Int J Food Sci Technol 46:23–29
Stringham JM, Hammond BR (2005) Dietary lutein and zeaxanthin: possible effects on visual function. Nutr Rev 63:59–64
Takeoka GR, Dao LT (2002) Antioxidant constituents of almond (*Prunus dulcis* Mill.) hulls. J Agric Food Chem 51:496–501
Topal U, Sasaki M, Goto M, Hayakawa K (2006) Extraction of lycopene from tomato skin with supercritical carbon dioxide: effect of operating conditions and solubility analysis. J Agric Food Chem 54:5604–5610
Tsuda T, Horio F, Uchida K, Aoki H, Osawa T (2003) Dietary cyanidin 3-O-beta-D-glucoside-rich purple corn color prevents obesity and ameliorates hyperglycemia in mice. J Nutr 133:2125–2130
Ulrich-Merzenich G, Zeitler H, Jobst D, Panek D, Vetter H, Wagner H (2007) Application of the "-Omic-" technologies in phytomedicine. Phytomedicine 14:70–82
Vigna GB, Costantini F, Alsini G, Carini M, Catapano A, Schena F, Tangerini A, Zanca R, Bombardelli E, Morazzoni P, Merzzetti A, Fellin R, Facino RM (2003) Effect of a standardized grape seed extract on low-density lipoprotein susceptibility to oxidation in heavy smokers. Metab Clin Exp 52:1250–1257
Villano D, Fernandez-Pachon MS, Troncoso AM, Garcıa-Parrilla MC (2005) Comparison of antioxidant activity of wine phenolic compounds and metabolites in vitro. Anal Chim Acta 538:391–398
Visioli F, Galli C (2003) Olives and their production waste products as sources of bioactive compounds. Curr Topics Nutraceut Res 1:85–88
Volf I, Popa VI (2004) The obtaining of active compounds with antioxidant properties from vegetable by-products: study of the extraction process of polyphenolic compounds from *Vitis* sp. wood. Rev Chim 55:707–710
Watanabe M, Ohshita Y, Tsushida T (1997) Antioxidant compounds from buckwheat (*Fagopyrum esculentum* Moench) hulls. J Agric Food Chem 45:1039–1044
Wenli Y, Yaping Z, Zhen X, Hui J, Dapu W (2001) The antioxidant properties of lycopene concentrate extracted from tomato paste. JOAC 78:697–701
Willcox JK, Ash SL, Catignani GL (2004) Antioxidants and prevention of chronic disease. Crit Rev Food Sci Nutr 44:275–295

Williamson G, Manach C (2005) Bioavailability and bioefficacy of polyphenols in humans. II. Review of 93 intervention studies. Am J Clin Nutr 81(suppl):243S–55S

Williamson-Hughes PS, Flickinger BD, Messina MJ, Empie MW (2006) Isoflavone supplements containing predominantly genistein reduce hot flash symptoms: a critical review of published articles. Menopause J North Am Menopause Soc 13:831–839

Wolfe KL, Liu RH (2003) Apple peels as a value-added food ingredient. J Agric Food Chem 51:1676–1683

Xia X, Ling W, Ma J, Xia M, Hou M, Wang Q, Zhu H, Tang Z (2006) An anthocyanin-rich extract from black rice enhances atherosclerotic plaque stabilization in apolipoprotein E-deficient mice. J Nutr 136:2220–2225

Xianquan S, Shi J, Kakuda Y, Yueming J (2005) Stability of lycopene during food processing and storage. J Med Food 8:413–422

Yao LH, Jiang YM, Shi J, Tomas-Barberan FA, Datta N, Singanusong R, Chen SS (2004) Flavonoids in food and their health benefits. Plant Foods Hum Nutr 59:113–122

Yeum KJ, Russell RM, Krinsky NI, Aldini G (2004) Biomarkers of antioxidant capacity in the hydrophilic and lipophilic compartments of human plasma. Arch Biochem Biophys 430:97–103

Zeisel SH (1999) Regulation of nutraceuticals. Science 285:1853–1855

Zeyada NN, Zeitoum MAM, Barbary OM (2008) Utilization of some vegetables and fruit waste as natural antioxidants. Alex J Food Sci Technol 5:1–11

Zhang Y, Fung LWM (1994) The roles of ascorbic acid and other antioxidants in the erythrocyte in reducing membrane nitroxide radicals. Free Radic Biol Med 16:215–222

Part IV
Pharmaceutical and Personal Care Products

Chapter 12
Biologically Active Compounds Form Seafood Processing By-Products

Se-Kwon Kim and Pradeep Dewapriya

12.1 Introduction

As in most food processing operations, seafood processing generates considerably high amount of waste as solid (carcasses, heads, viscera, skin) or liquid (blood, cleaning water). Thus, many large-scale seafood processing industries practise well-planned waste management system to avoid unnecessary environmental problems. The most common solid waste management system of seafood industries is recycling into fishmeal. However, continuously increasing seafood processing waste created wide-ranging discussion on effective utilization of fish processing waste. The necessity of an effective method to utilize the seafood waste arouses with over exploitation of marine resources. Recent advances in biotechnology and food processing have brought new paradigm to the classical way of processing waste by introducing a new avenue which generates a number of ingredients that can be used in food and pharmaceutical industries. Recent studies have identified a number of bioactive compounds from remaining waste materials. These ingredients possess a wide range of health-promoting abilities, including cure and prevention of a number of chronic diseases (Kim and Mendis 2006; Najafian and Babji 2012). Natural substances that have therapeutic values have attracted interest in diagnosis and therapy of various kinds of diseases in biomedicine since they exert a less number of side effects compared to that of synthetic origin. Moreover, awareness on added value of natural biologically active ingredients in modern society unwraps

S.-K. Kim (✉)
Marine Biochemistry Laboratory, Department of Chemistry, Pukyong National University, Busan 608-737, South Korea

Marine Bio-Process Research Center, Pukyong National University, Busan 608-737, South Korea
e-mail: sknkim@pknu.ac.kr

P. Dewapriya
Marine Biochemistry Laboratory, Department of Chemistry, Pukyong National University, Busan 608-737, South Korea

S.K. Brar et al. (eds.), *Biotransformation of Waste Biomass into High Value Biochemicals*, DOI 10.1007/978-1-4614-8005-1_12,

another growing market, nutraceutical and functional food (Myles 2003). Hence, this approach has opened up a potential way for processing seafood waste as natural health-promoting substances which have high commercial value.

12.2 Classical Way of Treating Seafood Waste

Seafood waste has been classified under the animal by-products and thus has to follow approved waste disposal methods. Under animal by-product category, seafood generates numerous waste materials that belong to several risk categories. Likewise, liquid wastes, including blood, are usually high in proteinaceous compounds and oils. These wastes have extremely high biochemical oxygen demand (BOD) and improper disposal may threaten the environment (Ababouch 2005). Therefore, practising traditional waste disposal methods have become increasingly restricted for seafood. Most of the approved seafood waste disposal methods are relatively expensive. In this regard, many large-scale processing operations tend to recover the waste disposal cost through conversion of the waste into value-added products. Most common waste-derived products are fishmeal and fish oil. Moreover, fish silage is another common waste product that is used as animal feed. In addition, converting waste into organic fertilizer is also practised (Arvanitoyannis and Kassaveti 2008). Although organic fertilizers generate little higher income than others, still all come under the category of low-value by-products. Further, increasing demand for seafood in modern society has steepened the problem by generating high amount of waste materials. Taking all into account, applying modern technological advances in seafood waste management to generate high-value ingredients would be an ideal solution.

12.3 Chitin and Chitosan from Seafood Waste

Crustacean shells and shellfish waste generated in seafood processing is one of the important waste materials. Efficient utilization of this waste material has become an environmental priority due to increased quantity as well as its slow natural degradation. The main structural component of these shells, chitin, has been identified as a potential target to be developed from these waste materials. Chitin is a long-chain polymer of *N*-acetylglucosamine (*N*-acetyl-2-deoxy-D-glucopyranose) units. This natural polymer can be easily processed into various kinds of derivatives which have range of biological activities. Chitin and its most common derivative chitosan have earned much attention as natural bioactive material with their nontoxicity, biocompatibility and biodegradability (Kim and Mendis 2006). Chitin can be simply extracted from shellfish waste with demineralization and deproteinization. The degree of deacetylation of chitin isolated in this way is around 0.1 %. The extracted chitin can be further deacetylated to a desired degree to produce chitosan.

Table 12.1 Examples of biological activities of chitin and its derivatives

Sample	Biological activity	Reference
Chitin oligosaccharide	Antioxidant	Ngo et al. (2008)
Chitin oligosaccharide	Anticancer	Huang et al. (2012)
Chitin nanoparticle	Anti-metastasis	Xu et al. (2012)
Chitosan sulphate	Anticoagulant	Fan et al. (2012)
Polysulphate chitosan	Antithrombin	Drozd et al. (2001)
Chitosan film	Antimicrobial	Dutta et al. (2009)
Chitin and chitosan	Immune enhancing	Harikrishnan et al. (2012)
Chitooligosaccharides	Anti-HIV	Artan et al. (2010)

Carboxymethyl derivative of chitin (CMC) is a water soluble form of chitin. CMC can be prepared by reacting chitin powder with monochloroacetic acid in isopropyl alcohol as a solvent using the condensation reaction (Jayakumar et al. 2010). Chitin and its derivatives possess various kinds of biological activities depending on their molecular weight, deacetylation and solubility. Hence, chitin and its derivatives have become popular in food and biomedical industries. However, applications of chitin and chitosan have faced some limitations with high viscosity and low solubility at neutral pH. Among the various means that have been explored to overcome this limitation, conversion of chitin and chitosan into their oligomers seems to be the best available option (Rinaudo 2006).

Several research articles have been published in peer-reviewed journals to reveal the biomedical and food industrial applications of chitin, chitosan and their oligomers (Table 12.1). Biodegradable, flexible and strong nature of chitin makes it possible to develop as surgical threads to be used in wound dressing. In addition, it has been shown that chitin possesses wound healing properties. Further, antibacterial and hemostasis properties of chitin that are necessary for wound healing provide additional benefits in order to cure wounds. Thus, biodegradable chitin wound dressings have added advantage over conventional dressing materials (Gupta et al. 2008). Moreover, chitin and chitosan are potent scavengers of oxidative radicals. Strength of radical scavenging activity of chitin and chitosan strongly depends on the molecular weight and degree of deacetylation. Low molecular weight chitosan oligomers with high deacetylation (90 %) have been recognized as potent scavengers. Further, it has been suggested that chitosan and its oligomers have an ability to act as scavengers of fat and cholesterol in digestive tract. It helps to reduce the low-density lipid level in liver and blood and provide hypocholesterolemic activity similar to those of dietary fibres (Prashanth and Tharanathan 2007).

Chitin and chitosan derivatives are capable of inhibiting tumour progression and cancer metastasis. In vitro and in vivo experiments have shown that low molecular weight chitosan potently inhibited cancer metastasis while improving immune functions. Thus, scientists believed that chitin and chitosan achieved the anticancer activity through immunostimulation. Further, several researchers have investigated the ability of chitosan as drug carrier. Chitin derivatives have shown fascinating ability to deliver certain drugs to the target place while maintaining its biodegradability. In the case of encapsulation, chitosan in the form of colloidal structure entraps various molecules and passes efficiently through mucosa and epithelia more. Moreover, nanoparticles made up of chitosan have also exhibited improved drug and protein delivery functions (Fuentes and Alonso 2012).

12.4 Hydrolysed Fish Proteins

Fish frames and offcuts generated from commercial fish filleting leave considerable amounts of muscle proteins which are nutritionally valuable and easily digestible with well-balanced amino acid composition (Harnedy and Fitzgerald 2012). In addition, specific degradation of remaining proteins results in biologically active protein hydrolysates and peptides which possess broad spectrum of health-promoting abilities. Since seafood is considered as prime source of proteins with high biological value, fish processing waste is an ideal and cheap source for extraction of valuable ingredients.

After removal of valuable portion, the remaining fish muscles are solubilized by means of several chemical and physical methods to obtain hydrolyzed fish proteins. Although fish protein hydrolysate (FPH) has classically been used for agricultural purposes, advanced technological developments make it possible to apply the FPH as functional ingredients in food and pharmaceuticals. Likewise, the hydrolysate is also a rich source of biologically active small peptides that have been proven for various therapeutic potentials. Hydrolysis of protein is achieved through acid-, alkali- or enzyme-mediated breakdown of parent proteins in the waste into smaller protein fractions, peptides and free amino acids. Acid hydrolysis makes the product unpalatable due to tryptophan destruction and the formation of sodium chloride after the neutralization. Alkaline hydrolysis produces some toxic compounds which are undesirable for human consumption. Among protein hydrolysis methods, enzymatic hydrolysis offers several advantages over others (Kristinsson and Rasco 2000).

Proteolytic enzymes come from several sources, such as plant (papain, ficin, bromelain), animal (trypsin, pancreatic enzymes) or microbial (Pronase, Alcalase), are employed for hydrolysis depending on the type of processing waste and the desired functionality of end product. In conventional enzymatic method, commercial enzymes are directly applied under predefined conditions such as pH, temperature, incubation time and enzyme/substrate ratio (Bhaskar et al. 2008). Fermentation approach of producing fish protein hydrolysates is not a novel concept. Proteases producing microbial strains are used as starter culture and incubated in preferred conditions to grow microbes on fish processing waste. Physicochemical as well as functional properties of enzymatic hydrolysate vary with the degree of hydrolysis which determines size of protein fractions. Over hydrolysis may impair some functional properties of food proteins or may develop off-flavours in hydrolysates (Balti et al. 2010). Further, improvement of functional properties and therapeutic value of protein hydrolysate could be obtained through a selective molecular weight cutoff. An ultrafiltration membrane system equipped with a molecular cutoff has been identified as an effective method to purify protein hydrolysates based on molecular weight of protein factions (Jeon et al. 1999). Serial enzymatic digestions in a system

Table 12.2 Bioactive fish protein hydrolysates from seafood wastes

Fish source	Bioactivity	Reference
Pacific hake	Antioxidant	Samaranayaka and Chan (2008)
Brownstripe red snapper	Antioxidant	Khantaphant et al. (2011)
Ornate threadfin bream	Antioxidant	Nalinanon et al. (2011)
Cod	ACE inhibitor	Jeon et al. (1999)
Round scad	Antioxidant	Thiansilakul et al. (2007)
Yellowfin sole	Anticoagulant	Rajapakse et al. (2005)
Blue whiting	Satiating effect	Cudennec et al. (2012a, b)
Blue whiting	Anti-proliferative	Picot et al. (2006)

of multistep recycling membrane reactor combined with ultrafiltration membrane system have been developed to produce protein hydolysates with desired molecular weights while preserving expensive photolytic enzymes (Byun and Kim 2001).

12.4.1 Bioactivities of Fish Protein Hydrolysates

The main goal of digesting protein leftover into FPH is to improve the functional properties of the original protein molecules. This improvement of functional properties accompanies with advanced health-promoting abilities due to improved functionality achieve through high amount of polar groups, solubility of hydrolysate and their bioavailability. Thus, several studies have been carried out to prove various kinds of biomedical applications of FPHs (Table 12.2). In particular, FPH possesses potent antioxidant activity which attenuate oxidative damages taking place in the body where endogenous antioxidant defence mechanism is not enough. As an economically viable product, FPHs seem good candidates to combat with production of superoxide anion (O^{2-}) and hydroxyl (OH^{1-}) radicals which are considered as causative agents for the initiation of chronic diseases such as heart disease, stroke, arteriosclerosis, diabetes and cancer (Perera and Bardessy 2011). The advantage of the product is providing protection against oxidative damage while providing an additional nutritional value. Additionally, protein hydrolysates generated from fish processing waste reduce risk of cardiovascular diseases (CVD) by triggering several key process associated with the disease, including blood clot and platelet formation, angiotensin I-converting enzyme activity and cholesterol metabolism. Moreover, a recent study (Cudennec et al. 2012a, b) has demonstrated that FPH derived from blue whiting suppresses appetite via enhanced cholecystokinin (CCK) and glucagon-like peptide-1 (GLP-1) secretion in in vitro STC-1 cells as well as in vitro experiment. The biological effects of CCK and GLP-1 stimulation lead to promising effect on body weight reduction which has gained much attention in developed countries. All these findings clearly point out that FPHs have potential of disease risk reduction. Thus, identified potential sources are currently used in production of biologically active FPHs in the form of nutraceutical and functional food.

12.5 Biologically Active Peptides

Peptides with small backbone are generated through digestion of various means, and these peptides are capable of playing an important role in metabolic regulations. Therefore, several scientific articles have highlighted that small peptides have potential to use as nutraceuticals and functional foods (Shahidi and Zhong 2008). During the investigation of biological consequence of FPHs, it was apparent that small peptides present in the hydrolysates mediate biochemical pathways to exert defined health property of the hydrolysates. Detailed studies revealed that small protein fragments containing 3–20 amino acids residues showed this potent activity, and the activity and its extent strictly depended on amino acid composition, sequence and molecular weight. Before hydrolysis, the peptides remain latent within the parent protein, and released by hydrolysis allow them to exercise hormone-like physiological effect in the body. (Himaya et al. 2012). It is well known fact that protein hydrolysis method and types of proteinase employed are crucial factors for biological activity of the peptide. The ultrafiltration membrane system has been identified as a useful tool to purify active peptides based on molecular weight. Sequential enzymatic digestion with different enzymes has been employed to achieve desired functionality of peptides (Kim et al. 2007). Biological activities of peptides derived from fish processing waste range from simple antioxidant activity to prevention and cure of serious chronic diseases such as cancer.

Several studies have reported that peptides derived from fish proteins showed antihypertensive activity by inhibiting the activity of angiotensin I-converting enzyme (ACE) which plays a vital role in regulation of blood pressure. ACE participates in the body's renin-angiotensin system by converting inactive angiotensin I into an active vasopressor angiotensin II. This conversion increased the blood pressure which triggered the function of blood vessel dilator bradykinin (Huang et al. 2005). After the discovery of critical role of ACE in blood pressure regulation, several commercial ACE inhibitors, such as captopril, enalapril, alacepril and lisinopril, were synthesized and employed to treat hypertension and heart failures. Among the natural ACE inhibitors that have been isolated from various food and natural sources, fish processing by-products derived ACE inhibitors are of great interest due to preferred sequence of peptides for the potent inhibition of ACE (Lee et al. 2010). In addition, several peptides which possess significant impact on causative agents of chronic disease have been isolated. Table 12.3 summarizes the biologically active peptides isolated from different sources of fish proteins.

12.6 Collagen and Gelatin from Fish Waste

Collagen has earned much interest as a biomaterial in medical applications due to its biodegradability and weak antigenicity. Among several collagens, Type I collagen is the most naturally abundant collagen in animal and found in skin, tendon, vascular ligature, organs and bone. Collagen is composed of three similarly sized triple helix polypeptide chains which consist of around 1,000 amino acids residues.

Table 12.3 Biological activities of peptides derived from fish muscle proteins

Fish source	Amino acid sequence	Biological activity	Reference
Alaska pollock	Leu-Pro-His-Ser-Gly-Tyr	Antioxidant	Je et al. (2005)
Croaker	Thr-Phe-Cys-Gly-Arg-His	Antioxidant	Nazeer et al. (2012)
Tuna fish	VKAGFAWTANQQLS	Antioxidant	Je et al. (2007)
Tuna fish	Gly-Asp-Leu-Gly-Lys-Thr-Thr-Thr-Val-Ser-Asn-Trp-Ser-Pro-Pro-Lys-Try-Lys-Asp-Thr-Pro	ACE inhibition	Lee et al. (2010)
Yellowfin sole	Met-Ile-Phe-Pro-Gly-Ala-Gly-Gly-Pro-Glu-leu	ACE inhibition	Jung et al. (2006a, b)
Alaska pollock	Val-Leu-Ser-Gly-Gly-Thr-Thr-Met-Ala-Met-Tyr-Thr-Leu-Val	Ca binding	Jung et al. (2006a, b)
Blue whiting	NA[a]	Cholecystokinin release	Cudennec et al. (2008)

Gelatin is structurally different form of the same macromolecules which make collagen and particularly a hydrolysed form of collagen. Common sources of collagen and gelatin, bovine hide, pig skin or chicken waste, have faced some constrains related to biological contaminants and religious issues (Aberoumand 2010). Mainly this reason sought an alternative source for collagen and gelatins for commercial purposes. Raw materials from fish waste have received considerable attention in recent years as an alternative for collagen and gelatin extraction due to its unique features. Fish skin and bones have been mainly used for collagen extraction, and three main extraction methods, neutral salt solubilization, acid solubilization and enzyme solubilization, have been used based on the characteristics of waste and end product. During extraction, triple helix structure which contributes to the unique properties of collagen should be preserved. To prepare fish gelatin, extracted which collagen is solubilized with hot water treatment by breaking down the hydrogen and covalent bonds of the triple helix, resulting in helix-to-coil transition and conversion into soluble gelatin (Guillén et al. 2011).

Several unique applications of fish by-product-derived collagen and gelatin have been reported due to enriched properties of fish collagen (Table 12.4). High hydroxyproline content of collage has resulted in reduced pain in osteoarthritis patients supplemented with collagen/gelatin hydrolysate (Moskowitz 2000). Collagen shows great advantages as a carrier molecule of drug, protein and gene through long-term maintenance of the concentration and controlled release at target sites. Moreover, collagen serves as a main scaffold for biotechnological applications. Detailed studies revealed that collagen type I, with selective removal of its telopeptides, exhibited characteristic features of bio-scaffold for bone regeneration. Experimental data confirmed that fish by-product-derived collagen and gelatin have inherent properties of collagen and can be used as an alternative to mammalian products (Senaratne et al. 2006). In addition, enzymatic hydrolysis of fish collagen and gelatin produces small peptides which have potent biological activities. Repeated Gly-Pro-Ala sequence of

Table 12.4 Biological activities of fish collagen and gelatin

Fish source	By-product	Biochemical potential	Reference
Rohu and Catla	Type I collagen	High thermal stability	Pati et al. (2010)
Nile tilapia	Collagen drink	Attenuate UVA damage	Kato et al. (2011)
Nile tilapia	Peptide	Free radical scavenging	Ngo et al. (2010)
Alaska pollock	Peptide	ACE inhibition	Byun and Kim (2001)
Pacific cod	Peptide	Antioxidant and ACE inhibition	Ngo et al. (2011)
Unicorn leatherjacket	Gelatin film	Antimicrobial	Ahmad et al. (2012)
Chum salmon	Peptide	Memory enhancement	Pei et al. (2010)

gelatin peptides has reinforced the peptide with high antioxidant and antihypertensive properties. Numerous studies have been conducted to reveal the biological activity of fish collagen- and gelatin-derived peptides.

12.7 Fish Bones as a Mineral Source

Commercial fish filleting from large as well as small fish result in huge amount of fish bones which is generally discarded as a waste. Fish frame account for approximately 10–15 % of total fish biomass. The bones are mainly composed of calcium phosphate and collagen protein with some special carbohydrates and lipids. Thus, the waste could be used as mineral source for food and biomedical industries while giving an added value to fish processing by-products. Fish bone consists of 60–70 % of inorganic substances, mainly calcium phosphate and hydroxyapatite (Toppe et al. 2007). Hydroxyapatite (HA)- and calcium phosphate-related ceramic material have earned much attention in various biomedical applications due to their close similarity to composition of natural bones. In particular, the composition [Ca_{10} $(PO_4)_6$ $(OH)_2$] and Ca/P molar ratio of HA are more similar to inorganic part of bone and teeth, and hence, this biological HA could be used as an implant material for orthopaedic and dental applications. Even though much effort has been paid to obtain synthetic hydroxyapatite, fish bone provides a cheap source for extraction of biological HA with preserved chemical characteristics and many advantages (Boutinguiza et al. 2012). Pallela et al. (2011) have developed polymer-assisted thermal calcination method to isolate micro- and nanostructured HA form tune (*Thunnus obesus*) bones. Further, findings proved that HA isolated with polymer-assisted method shows less toxicity and high biocompatibility. In addition, high level of calcium in fish bone indicates that it would be useful as a potential source to obtain calcium for dietary supplements. As most of the regular diets are calcium deficient, several calcium supplementations have been commercialized. Nevertheless, bioavailability of calcium of these products is not clearly studied. It is well known fact that small fish is a good source of balanced calcium, and fish-derived calcium is readily absorbed into human body. Thus, bones from fish processing waste can be used to produce fortified products with high biological value, and several convenient methods have been developed to soften the fish bone to convert it into an edible form.

12.8 Omega-3 Fatty Acids

Cold water oily fish, such as salmon, herring, mackerel, anchovies, and sardines, and fish oil derived from these fish have been recognized as well-balanced sources of omega-3 fatty acids, especially docosahexaenoic acid (DHA) and eicosapentaenoic acid (EPA). Due to enhanced health benefits of omega-3 fatty acids, world fish consumption has reached to a level which threatened the marine fish sources, and thus, fishing for oil extraction is not encouraged. Fish processing by-products have been identified as an ideal candidate for extraction of fish oil rich in omega-3 fatty acids while giving a positive insight into sustainable marine fisheries (Immanuel et al. 2009). The processing leftovers, such as head, skin and internal organs, are rich sources of omega-3 fatty acids, and method and conditions for the extraction are determined base of the nature of fish source. Presently, several methods, such as high-speed centrifugation, Soxhlet extraction, low-temperature solvent and supercritical fluid extraction, are employed to extract fish oil. Among them, wet reduction followed by pressing and centrifugation is the most common method used to produce fish oil from waste materials (Chantachum et al. 2000). Purification of omega-3 fatty acids from extracted fish oil is a challenge due to the presence of complex mixture of triacylglycerols and vulnerability of free fatty acids EPA and DHA to oxidize into hydroperoxides. A recent study shows that enzymatic deacidification of high-acid crude fish oils is an effective approach to extract high amount of n-3 fatty acids (Wang et al. 2012).

The inverse relationship between high level of omega-3 fatty acids present in bold and chronic disease has been reported in several studies. Findings suggest high levels of EPA and DHA in blood which in turn reduced rate of coronary heart diseases have an association with inhibition of lipid-rich atherosclerotic plaques growth, reduction in formation of thrombus, improving vascular endothelial function and lowering blood pressure (Lavie et al. 2009). Moreover, there is evidence for therapeutic value of omega-3 fatty acids. For instance, beneficial effects against diabetes mellitus, anti-inflammatory action and thereby protection against autoimmune diseases and potent activity against human carcinomas including prostate, lung, colon and breast have been reported. Basically, documented health benefits of fish oil are a result of high content of EPA and DHA, and fish processing by-product-derived oil fall under the agreement of expected fatty acid composition for biomedical treatments (Byun et al. 2008; Wu and Bechtel 2008).

12.9 Other Constituents

Internal organs, fish egg, scales, eyeball and blood have been identified as the other potential sources of high-value biological constituents which have considerable market value. Fish internal organs, especially viscera, are a rich source of digestive enzymes, proteases and lipases (Khantaphant and Benjakul 2010). As marine organisms have adopted for extreme condition prevalence in the marine environment, these enzymes have unique characteristics, including higher catalytic efficiency at low temperatures, lower sensitivity to substrate concentrations, greater stability at broader pH range and stability under high temperatures (50–60 °C) (Klomklao 2008). Owing to the special properties, these enzymes have broad applications in biomedical industry as a biocatalyst. Wound debridement, treatment for blood clots, antibiotic therapy and treatment for inflammation are classical examples of protease-based medical treatments (Seabra and Gil 2007). Furthermore, several studies reported that fish digestive enzymes are a cheap source to extract enzymes that could be used to produce biologically active peptides from different protein sources.

Fish egg derived from large-scale fish processing industries has been identified as a potential source of lectin, a naturally occurring sugar binding protein. Due to its high specificity and ability to form stable complexes with carbohydrates, it may be used as antibiotic to detect and inactivate activity of pathogens (Jung et al 2003, Jimbo et al. 2007). Hyaluronic acid is an interesting compound isolated from fish eyeball. This polymer has repeating units of *N*-acetyl-D-glucosamine and glucuronic acid and exists as cartilage in tissues. In recent years, hyaluronic acid has gained interest as an ingredient of cosmeceutical and pharmaceutical. Eyeball of certain fishes contains significant amount of hyaluronic acid which can be extracted with high purity (99.5 %) for clinical and cosmetic applications (Murado et al. 2011).

12.10 Concluding Remarks

Seafood usage has reached to a level which threatens the marine ecosystem. Several reports have highlighted that half of marine resources have been overexploited. Thus, identification and exploration of sustainable means to produce seafood have become a prioritized requirement of current seafood research. Recycling of seafood waste seems to be a positive approach to increase the utilization of existing seafood resources. However, commercial value of classical seafood by-products discourages the use of waste materials. Identification of biologically active materials and their potential application in growing fields, such as biomedical, nutraceutical and functional food, have brought a new insight to fish processing by-products. Thus, comprehensive studies on identified potential ingredients and development of commercially viable processing methods for isolation and extraction will facilitate a successful journey of fish processing by-products in the biomedical field. This seems to be an ideal approach to overcome constraints associated with conventional seafood waste management.

References

Ababouch L (2005) Fisheries and Aquaculture topics. Waste management of fish and fish products. Topics Fact Sheets. In: FAO Fisheries and Aquaculture Department. Available via FAO. http://www.fao.org/fishery/topic/12326/en. Accessed 2 August 2012

Aberoumand A (2010) Isolation and characteristics of collagen from fish waste material. World J Fish Marine Sci 2:471–474

Ahmad M, Benjakul S, Prodpran T, Agustini TW (2012) Physico-mechanical and antimicrobial properties of gelatin film from the skin of unicorn leatherjacket incorporated with essential oils. Food Hydrocolloid 28:189–199

Artan M, Karadeniz F, Karagozlu MZ, Kim MM, Kim SK (2010) Anti-HIV-1 activity of low molecular weight sulfated chitooligosaccharides. Carbohyd Res 345:656–662

Arvanitoyannis IS, Kassaveti A (2008) Fish industry waste: treatments, environmental impacts, current and potential uses. Int J Food Sci Tech 43:726–745

Balti R, Bougatef A, Ali NE, Zekri D, Barkia A, Nasri M (2010) Influence of degree of hydrolysis on functional properties and angiotensin I-converting enzyme-inhibitory activity of protein hydrolysates from cuttlefish (Sepia officinalis) by-products. J Sci Food Agr 90:2006–2014

Bhaskar N, Benila T, Radha C, Lalitha RG (2008) Optimization of enzymatic hydrolysis of visceral waste proteins of Catla (*Catla catla*) for preparing protein hydrolysate using a commercial protease. Bioresour Technol 99:335–343

Boutinguiza M, Pou J, Comesaña R, Lusquiños F, Carlos AD, León B (2012) Biological hydroxyapatite obtained from fish bones. Mat Sci Eng C 32:478–486

Byun GH, Kim SK (2001) Purification and characterization of angiotensin I converting enzyme (ACE) inhibitory peptides from Alaska pollock (*Theragra chalcogramma*) skin. Process Biochem 36:1155–1162

Byun HG, Eom TK, Jung WK, Kim SK (2008) Characterization of fish oil extracted from fish processing by-products. J Food Sci Nutri 13:7–11

Chantachum S, Benjakul S, Sriwirat N (2000) Separation and quality of fish oil from precooked and non-precooked tuna heads. Food Chem 69:289–294

Cudennec B, Peron MF, Ferry F, Duclos E, Ravallec R (2012) In vitro and in vivo evidence for a satiating effect of fish protein hydrolysate obtained from blue whiting (*Micromesistius poutassou*) muscle. J Funct Food 4:271–277

Cudennec B, Plé RR, Courois E, Peron MF (2008) Peptides from fish and crustacean by-products hydrolysates stimulate cholecystokinin release in STC-1 cells. Food Chem 111:970–975

Drozd NN, Sher AI, Makarov VA, Galbraikh LS, Vikhoreva GA, Gorbachiova IN (2001) Comparison of antithrombin activity of the polysulfate chitosan derivatives in in vivo and in vitro system. Thromb Res 102:445–455

Dutta PK, Tripathi S, Mehrotra GK, Dutta J (2009) Perspectives for chitosan based antimicrobial films in food applications. Food Chem 114:1173–1182

Fan L, Wu P, Zhang J, Gao S, Wang L, Li M, Sha M, Xie W, Nie M (2012) Synthesis and anticoagulant activity of the quaternary ammonium chitosan sulfates. Int J Biol Macromol 50:31–37

Fuentes MG, Alonso MJ (2012) Chitosan-based drug nanocarriers: where do we stand? J Control Release 161:496–504

Guillén MCG, Giménez B, Caballero MEL, Montero MP (2011) Functional and bioactive properties of collagen and gelatin from alternative sources: a review. Food Hydrocoll 25:1813–1827

Gupta B, Arora A, Saxena S, Alam MS (2008) Preparation of chitosan-polyethylene glycol coated cotton membranes for wound dressings: preparation and characterization. Polym Advan Technol 20:58–65

Harikrishnan R, Kim JS, Balasundaram C, Heo MS (2012) Dietary supplementation with chitin and chitosan on haematology and innate immune response in *Epinephelus bruneus* against *Philasterides dicentrarchi*. Ex Parasitol 131:116–124

Harnedy PA, Fitzgerald RJ (2012) Bioactive peptides from marine processing waste and shellfish: a review. J Funct Food 4:6–24

Himaya SWA, Ngo DH, Ryu B, Kim SK (2012) An active peptide purified from gastrointestinal enzyme hydrolysate of Pacific cod skin gelatin attenuates angiotensin-1 converting enzyme (ACE) activity and cellular oxidative stress. Food Chem 132:1872–1882

Huang R, Mendis E, Kim SK (2005) Improvement of ACE inhibitory activity of chitooligosaccharides (COS) by carboxyl modification. Bioorgan Med Chem 13:3649–3655

Huang R, Mendis E, Rajapakse N, Kim SK (2012) Strong electronic charge as an important factor for anticancer activity of chitooligosaccharides (COS). Life Sci 78:2399–2408

Immanuel G, Sathasivan S, Shankar VS, Peter MJP, Palavesam A (2009) Processing and characterisation of low cost balistid fish sufflamen capistratus liver oil for edible purpose. Food Chem 115:430–435

Jayakumar R, Prabaharan M, Nair SV, Tokura S, Tamura H, Selvamurugan N (2010) Novel carboxymethyl derivatives of chitin and chitosan materials and their biomedical applications. Prog Mater Sci 55:675–709

Je JY, Park PJ, Kim SK (2005) Antioxidant activity of a peptide isolated from Alaska pollock (*Theragra chalcogramma*) frame protein hydrolysate. Food Res Int 38:45–50

Je JY, Qian ZJ, Byun HG, Kim SK (2007) Purification and characterization of an antioxidant peptide obtained from tuna backbone protein by enzymatic hydrolysis. Process Biochem 42: 840–846

Jeon YJ, Byun HG, Kim SK (1999) Improvement of functional properties of cod frame protein hydrolysates using ultrafiltration membranes. Process Biochem 35:471–478

Jimbo M, Usui R, Sakai R, Muramoto K, Kamiya H (2007) Purification, cloning and characterization of egg lectins from the teleost *Tribolodon brandti*. Comp Biochem Physiol B Biochem Mol Biol 147:164–171

Jung WK, Karawita R, Heo SJ, Lee BJ, Kim SK, Jeon YJ (2006a) Recovery of a novel Ca-binding peptide from Alaska Pollock (*Theragra chalcogramma*) backbone by pepsinolytic hydrolysis. Process Biochem 41:2097–2100

Jung WK, Mendis E, Je JY, Park PJ, Son BH, Kim HC, Choi YK, Kim SK (2006b) Angiotensin I-converting enzyme inhibitory peptide from yellowfin sole (*Limanda aspera*) frame protein and its antihypertensive effect in spontaneously hypertensive rats. Food Chem 94:26–32

Jung WK, Park PJ, Kim SK (2003) Purification and characterization of a new lectin from the hard roe of skipjack tuna, *Katsuwonus pelamis*. Int J Biochem Cell Biol 35:255–265

Kato S, Matsui H, Saitoh Y, Miwa N (2011) Fish collagen-containing drink is subcutaneously absorbed and attenuates the UVA-induced tissue-integrity destruction and DNA damages in 3D-human skin tissue model. J Funct Food 3:50–55

Khantaphant S, Benjakul S (2010) Purification and characterization of trypsin from the pyloric caeca of brownstripe red snapper (*Lutjanus vitta*). Food Chem 120:658–664

Khantaphant S, Benjakul S, Ghomi MR (2011) The effects of pretreatments on antioxidative activities of protein hydrolysate from the muscle of brownstripe red snapper (*Lutjanus vitta*). LWT Food Sci Technol 44:1139–1148

Kim SK, Mendis M (2006) Bioactive compounds from marine processing byproducts – a review. Food Rese Int 39:383–393

Kim SY, Je JY, Kim SK (2007) Purification and characterization of antioxidant peptide from hoki (*Johnius belengerii*) frame protein by gastrointestinal digestion. J Nutr Biochem 18:31–38

Klomklao S (2008) Digestive proteinases from marine organisms and their applications. Songklanakarin J Sci Technol 30:37–46

Kristinsson HG, Rasco BA (2000) Fish protein hydrolysates: production, biochemical, and functional properties. Crit Rev Food Sci Nutr 40:43–81

Lavie CJ, Richard VM, Mehra MP, Ventura HO (2009) Omega-3 polyunsaturated fatty acids and cardiovascular diseases. J Am Coll Cardiol 54:585–594

Lee SH, Qian ZJ, Kim SK (2010) A novel angiotensin I converting enzyme inhibitory peptide from tuna frame protein hydrolysate and its antihypertensive effect in spontaneously hypertensive rats. Food Chem 118:96–102

Moskowitz RW (2000) Role of collagen hydrolysate in bone and joint disease. Semin Arthritis Rheum 30:87–99

Murado MA, Montemayor MI, Cabo ML, Vazquez JA, Gonzalez MP (2011) Optimization of extraction and purification process of hyaluronic acid form fish eyeball. Food Bioprod Process. doi:10.1016/j.fbp.2011.11.002

Myles DC (2003) Novel biologically active natural and unnatural products. Curr Opin Biotechnol 14:627–633

Najafian L, Babji AS (2012) A review of fish-derived antioxidant and antimicrobial peptides: their production, assessment, and applications. Peptides 33:178–185

Nalinanon S, Benjakul S, Kishimura H, Shahidi F (2011) Functionalities and antioxidant properties of protein hydrolysates from the muscle of ornate threadfin bream treated with pepsin from skipjack tuna. Food Chem 124:1354–1362

Nazeer RA, Kumara NSS, Ganesh, JR (2012) In vitro and in vivo studies on the antioxidant activity of fish peptides isolated form croaker (Otolithes ruber) muscle protein hydrolysates. Peptides 35:261–268

Ngo DH, Qian ZJ, Ryu B, Park JW, Kim SK (2010) In vitro antioxidant activity of a peptide isolated from Nile tilapia (*Oreochromis niloticus*) scale gelatin in free radical-mediated oxidative systems. J Funct Food 2:107–117

Ngo DH, Ryu B, Vo TS, Himaya SWA, Wijesekara I, Kim SK (2011) Free radical scavenging and angiotensin-I converting enzyme inhibitory peptides from Pacific cod (*Gadus macrocephalus*) skin gelatin. Int J Macromol 49:1110–1116

Ngo DN, Kim MM, Kim SK (2008) Chitin oligosaccharides inhibit oxidative stress in live cells. Carbohyd Polym 74:228–234

Pallela R, Venkatesan J, Kim SK (2011) Polymer assisted isolation of hydroxyapatite from thunnus obesus bone. Ceram Int 37:3489–3497

Pati F, Adhikari B, Dhara S (2010) Isolation and characterization of fish scale collagen of higher thermal stability. Bioresour Technol 101:3737–3742

Pei X, Yang R, Zhang Z, Gao L, Wang J, Xu Y, Zhao M, Han X, Liu Z, Li Y (2010) Marine collagen peptide isolated from Chum Salmon (*Oncorhynchus keta*) skin facilitates learning and memory in aged C57BL/6J mice. Food Chem 118:333–340

Perera MR, Bardessy N (2011) When antioxidants are bad. Nature 475:43–44

Picot L, Bordenave S, Didelot S, Arnaudin IF, Sannier F, Thorkelsson G, Berge JP, Guerard F, Chabeaud A, Piot JM (2006) Antiproliferative activity of fish protein hydrolysates on human breast cancer cell lines. Process Biochem 41:1217–1222

Prashanth HKV, Tharanathan RN (2007) Chitin/chitosan: modification and their unlimited applications potential-an overview. Trend Food Sci Tech 18:117–131

Rajapakse N, Jung WK, Mendis E, Moon SM, Kim SK (2005) A novel anticoagulant purified from fish protein hydrolysate inhibits factor XIIa and platelet aggregation. Life Sci 76:2607–2619

Rinaudo M (2006) Chitin and chitosan: properties and applications. Prog Polym Sci 31:603–632

Samaranayaka AGP, Chan ECYL (2008) Autolysis-assisted production of fish protein hydrolysates with antioxidant properties from Pacific hake (*Merluccius productus*). Food Chem 107: 768–776

Seabra JI, Gil MH (2007) Cotton gauze bandage: a support for protease immobilization for use in biomedical applications. Braz J Pharm Sci 43:535–542

Senaratne LS, Park PJ, Kim SK (2006) Isolation and characterization of collagen from brown backed toadfish (*Lagocephalus gloveri*) skin. Bioresource Technol 97:191–197

Shahidi F, Zhong Y (2008) Bioactive peptides. J AOAC Int 91:914–931

Thiansilakul Y, Benjakul S, Shahidi F (2007) Compositions, functional properties and antioxidative activity of protein hydrolysates prepared from round scad (*Decapterus maruadsi*). Food Chem 103:1385–1394

Toppe T, Albrektsen S, Hope B, Aksnes A (2007) Chemical composition, mineral content and amino acid and lipid profiles in bones from various fish species. Comp Biochem Physiol B Bichem Mol Biol 146:395–401

Wang W, Li T, Ning Z, Wang Y, Yang B, Mc Y, Yang X (2012) A process for the synthesis of PUFA-enriched triglycerides from high-acid crude fish oil. J Food Eng 109:366–371

Wu TH, Bechtel PJ (2008) Salmon by-product storage and oil extraction. Food Chem 111: 868–871

Xu Q, Guo L, Gu X, Zhang B, Hu X, Zhang J, Chen J, Wang Y, Chen C, Gao B, Kuang Y, Wang S (2012) Prevention of colorectal cancer liver metastasis by exploiting liver immunity via chitosan-TPP/nanoparticles formulated with IL-12. Biomaterials 33:3909–3918

Chapter 13
Microbial Statins

Leandro F. dos Santos, Júlio C. de Carvalho, Rosália Rubel, and Carlos Ricardo Soccol

13.1 Introduction

Statins are a class of antihypercholesterolemic (or cholesterol-lowering) drugs which act on the liver by reducing the biosynthesis of the steroid by inhibiting the activity of HMG-CoA (3-hydroxy-3-methylglutaryl-CoA) reductase, the enzyme responsible for the first step in the synthesis of cholesterol (and other biomolecules). Among statins, there are the molecules that are produced through synthetic means and those produced via fermentation-based processes with their semisynthetic derivatives. The rational production of microbial statins and discovery of new potential statin producers is stimulated by the current market value, with a market share around 60 %. Natural statins have a 2–3 billion dollar market (Bizukojc et al. 2007; Findlay et al. 2007; Morikawa et al. 2002; Murphree 2008; Rozman and Monostory 2010; Seraman et al. 2010; Vilches Ferrón et al. 2005; Weber et al. 2007), while the semisynthetic simvastatin has a 50 % market.

Among the advantages for microbial production of statins, there is the possibility of utilization of agro-industrial by-products for sustainable bioproduction and the possibility of using selected or modified microorganisms to obtain new statins. Statins also have additional biological effects, such as microvascular endothelial protection function (in the presence of hyperglycemia, thus inhibiting the early stage of diabetic microangiopathy), or regression of fatty streak lesions (Arikan et al. 2012; Bizukojc et al. 2007; Nozue et al. 2012; Seraman et al. 2010; Vilches Ferrón et al. 2005).

All bioactive secondary metabolites are common on traditional therapeutic regimens for hyperlipidemia which have been associated with increased risk of coronary

L.F. dos Santos • J.C. de Carvalho (✉) • R. Rubel • C.R. Soccol
Department of Biotechnology and Bioprocess Engineering, ACF Centro Politécnico,
Federal University of Paraná – UFPR, Rua Cel. Francisco H dos Santos, 100,
Caixa Postal 19060, 81531-980 Curitiba, PR, Brazil
e-mail: jccarvalho@ufpr.br

S.K. Brar et al. (eds.), *Biotransformation of Waste Biomass into High Value Biochemicals*,
DOI 10.1007/978-1-4614-8005-1_13, © Springer Science+Business Media New York 2013

disease, stroke, and heart attack. Hyperlipidemia is a leading cause of death in many countries, which is often triggered by hypercholesterolemia (the accumulation of cholesterol in the blood), leading to atherosclerosis (Talayero and Sacks 2011).

This chapter highlights some of the major hypolipidemic properties and aspects relevant to microbial statins, such as the mechanism of action, the market value of these drugs, the potential for new microbial statins, and the production process for the most common microbial statins. It will also present examples of well-characterized non-statin hypolipidemic agents.

13.2 Hyperlipidemia and the Processes Leading to Atherosclerosis

13.2.1 Definitions

Hyperlipidemia or dyslipidemia is a condition characterized by an increased concentration of lipids (triglycerides, cholesterol, or both) and lipoproteins (low-density lipoprotein (LDL) and very low-density lipoprotein (VLDL)) in the blood. Specific terms to increased blood concentrations of triglycerides are referred to as hypertriglyceridemia, while increased blood concentrations of cholesterol are referred to as hypercholesterolemia. The term hyperlipoproteinemia refers to increased blood concentrations of lipoproteins (Talayero and Sacks 2011).

13.2.2 Causes of Hyperlipidemia and Its Association with Atherosclerotic Processes

The mechanisms behind pathologies must be understood in order to develop better drugs. The causes underlying hyperlipidemia and atherosclerosis are discussed in this section.

13.2.2.1 Causes of Hyperlipidemia

Several causes have been reported to cause hyperlipidemia: high-fat diets, obesity, endocrine disorders (such as diabetes mellitus, hypothyroidism, or hyperadrenocorticism), and cholestasis. Among these risk factors, high-fat diets are the main cause of hyperlipidemia. Higher amounts of saturated fat, trans fat, and cholesterol intake in high-fat diet cause increased LDL levels. However, other risk factors and LDL participation have a central role in the atherosclerotic development process.

A vast number of studies confirmed the intimate and causative relationships between dyslipidemias and diabetes (Subramanian and Chait 2012). In diabetes mellitus, there is deficiency of insulin. The chylomicrons and VLDL are released into blood and should be unloaded by lipases located on the vascular endothelium of tissues. However, these enzymes cannot function to its full extent because of the insulin deficiency, since insulin can regulate lipase gene expression (Bouraoui et al. 2012). In accordance with lipase inhibition, there will be increased triglyceride levels or hypertriglyceridemia. High triglyceride levels are markers for several types of atherogenic lipoproteins, especially apo C-III (Fukui et al. 2011).

Hypercholesterolemia has been stressed as one of the common biochemical findings in primary hypothyroidism. In general, hypothyroidism is associated with hypercholesterolemia mainly due to the elevation of total cholesterol and LDL-C levels, whereas there is an increase in triglyceride levels due to decreased activity of the lipoprotein lipase. There were multiple mechanisms accounting for atherosclerosis in patients with hypothyroidism, including hypercholesterolemia, insulin resistance, and increased oxidation of LDL-C (overview of the atherogenesis below) (Arikan et al. 2012).

Comparative aspects of hyperadrenocorticism and arteriosclerosis have shown that advanced hyperadrenocorticism exhibits widespread arteriosclerosis with calcific complications. This disease shows a higher hypertriglyceridemia and hypercholesterolemia due to the associated metabolic alterations. Cortisol concentrations induce the insulin resistance and lipase activity (Ottosson et al. 1994; Wexler 1971).

Hypercholesterolemia always occurs in cholestasis, a disturbance of the secretion of bile salts by the liver, as bile is the chief pathway for the elimination of cholesterol from the body (Wagner et al. 2009).

13.2.2.2 Overview of the Atherogenesis

Atherosclerosis has been reported as a result of hyperlipidemia, and it is regarded as a lipid-induced inflammatory disease where the immune system plays a pivotal role in its initiation and progression. In general, atherogenesis begins at sites of endothelial injury caused by smoking, infection, diabetes mellitus, and hypertension and which results a nidus for monocyte and lipid/lipoprotein accumulation into the underlying arterial intima (Gu et al. 2012). Lipid and lipoprotein accumulation is due to adhesiveness and permeability of the endothelial injury. As LDL accumulates, it may be entrapped extracellularly in arteries, thus being subjected to a milieu conducive to various kinds of enzymatic and chemical modification (oxidation and glycation). Lipoproteins undergo minimal oxidation during circulation but become progressively oxidized within the arterial wall. The monocytes subsequently differentiate into macrophages and ingest modified LDL and other particles (Koga and Aikawa 2012). In a way, scavenger receptors of macrophages bind oxidized LDL but not native LDL, leading to development of cholesterol ester-engorged cells or foam cells, the precursors of atherosclerotic lesions. T cells infiltrating into the endothelial lesion may recognize antigenic signals presented by the activated macrophages and generate an immune response by inflammatory cytokines and growth factors production which stimulate smooth muscle cell migration and proliferation (Kzhyshkowska et al. 2012). The smooth muscle cells, in turn, secrete extracellular matrix components that form a fibrous cap over the underlying atherosclerotic lesion. Foam cells undergo apoptosis and release their lipid contents to form an extracellular lipid core that is contained under the fibrous cap. All this process will result in thickened intima media and atherosclerosis plaque development in arteries. Moreover, in general, atherosclerotic plaques with thin fibrous cap, large lipid cores, and numerous macrophages are most likely to rupture. Thus, once plaques rupture, the thrombogenic contents are exposed to platelets and coagulation factors in circulating blood, initiating a thrombosis. Thrombosis is a blood clot within a vessel, obstructing or stopping the flow of blood (Steinbrecher 1999; Østerud and Bjørklid 2003).

Atherosclerosis usually does not cause signs and symptoms until it severely narrows or totally blocks a vessel. Therefore, atherosclerosis represents the world's largest problem from modernity by taking 17.1 million lives a year. According to the most recent data available, one in five deaths in developing countries is due to atherosclerosis in its various forms (Ellis et al. 2010; Messner and Bernhard 2010).

Treatments for atherosclerosis may include lifestyle change, surgery, and medicines, such as antihypertensives (angiotensin-converting enzyme inhibitors, calcium channel blockers, thiazide diuretics), antiplatelets, and mainly statins against high cholesterol and LDL levels (Borshch et al. 2012).

The mechanisms discussed about the model of atherogenesis initiation outlined above are mainly based on experimental animal models with high rate of developing lesions, i.e., genetically hyperlipidemic animals and fat fed. However, there is supporting evidence that many molecules, cytokines, growth factors, scavenger receptors, etc. are expressed and produced similarly in humans, validating this whole mechanism and opening several strategies for the development of new treatments.

13.3 The Statin-Based Therapeutic Strategies

13.3.1 General Characteristics of Statins

Statins are oral inhibitors of the 3-hydroxy-3-methyl-glutaryl coenzyme A (HMG-CoA) reductase, which are well-established agents to lower cholesterol levels and prevent cardiovascular morbidity and mortality. HMG-CoA reductase is the rate-limiting enzyme that catalyzes the conversion of HMG-CoA to L-mevalonate, a key pathway for cholesterol biosynthesis. Therefore, statins are the first line of defense in drug treatment of hypercholesterolemia (Morikawa et al. 2002; Rozman and Monostory 2010; Weber et al. 2007).

13.3.2 Chemical Structure and Mode of Action: Statins

Statins have differences in their chemical structure that could translate into different pharmacological properties and pharmacokinetic parameters (bioavailability,

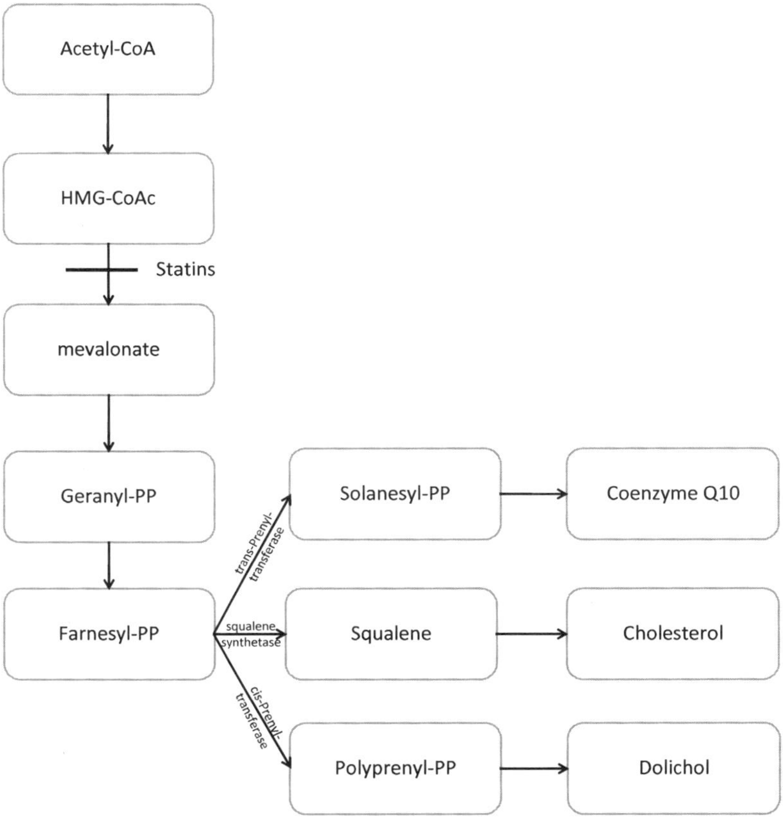

Fig. 13.1 Statins: mechanism of action

half-life, protein binding, metabolism and excretion routes, lipophilicity). All statins have a structural component which is similar to HMG portion of HMG-CoA (Fig. 13.2). Thus, statins act as competitive inhibitors of the HMG-CoA reductase. In general, competitive inhibition of HMG-CoA reductase by statins decreases the conversion of HMG-CoA to mevalonate, the rate-limiting step of cholesterol endogenous synthesis (Fig. 13.1). The reduction of downstream metabolic intermediates leads to increased expression of LDL receptor on the surface of hepatocytes and so increases uptake of LDL-C from the circulation (Campo and Carvalho 2007; Nirogi et al. 2007). Aside from their cholesterol-lowering effects, statins can reduce coenzyme Q10 and dolichol. Coenzyme Q10 is a mitochondrial coenzyme which is essential for the production of ATP and immune system while the role of the dolichol remains elusive (Cantagrel et al. 2010; Kumar et al. 2009).

Currently, seven statins are used in medical practice: lovastatin, pravastatin, simvastatin, fluvastatin, atorvastatin, rosuvastatin, and pitavastatin. From these, the last five are synthesized chemically while lovastatin and pravastatin are derived from fungal fermentation. Chemical structures of statins are shown in Fig. 13.2.

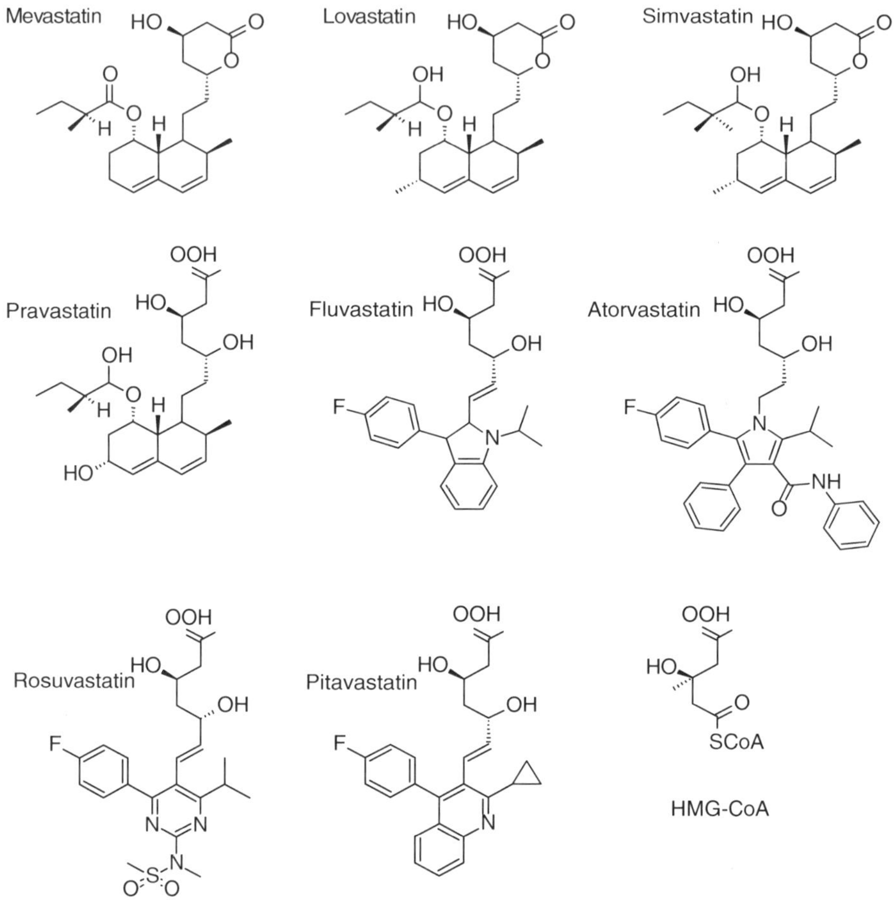

Fig. 13.2 Chemical structure of the statins and HMG-CoA, the molecule they mimic in the cholesterol synthesis inhibition

The recurrent structure in all statins is a hydroxyl carboxylic acid which mimics HMG-CoA. The lactonized forms of statins are converted to the active hydroxy acids in the liver. The efficiency with which statins are absorbed and inhibit the synthesis of cholesterol is affected by their structure, leading to a wide heterogeneity of traditional therapeutic regimens used by statins, as well as their pharmacological properties. Table 13.1 shows the therapeutic doses and half-lives for statins.

13.3.3 Other Relevant Effects of Statins

Aside from their cholesterol-lowering effects, statins are known to have a range of effects, such as vasodilatation, effects on coagulation, inflammatory response modulation, atherosclerotic plaque regression, and immunomodulatory effects.

Table 13.1 Traditional therapeutic regimens used by statins and its pharmacological properties (Betteridge 2010; Catapano 2010; Jones et al. 2003; Knopp 1999)

Statins	Therapeutic dose[a] (mg)	Elimination half-life (h)	Solubility
Fluvastatin	20–80	1–2	Lipophilic
Atorvastatin	10–80	14	Lipophilic
Rosuvastatin	5–40	19	Hydrophilic
Pitavastatin	2–4	11–18	Moderately lipophilic
Simvastatin	5–80	1–2	Lipophilic
Lovastatin	10–80	3	Lipophilic
Pravastatin	10–80	1–2	Hydrophilic

[a]Dose required for inhibition of 50 % HMG-CoA reductase

Endothelium-ameliorating effects of statin therapy were observed in patients with symptomatic heart failure. The beneficial effect of statin therapy on endothelium-dependent vasodilatation in patients was associated with coenzyme Q_{10} reductions. Therefore, statins can increase nitric oxide (NO) bioactivity, which is consistent with enhanced endothelium function, and reduce synthesis of proinflammatory proteins on the endothelial cell surface, which may reduce inflammation (Brull et al. 2001; Gajendragadkar et al. 2009; Koh et al. 2004; Strey et al. 2005).

Statins may affect the expression levels of genes involved in coagulation, such as thrombomodulin and nitric oxide syntase-3. Thrombomodulin is a transmembranous glycoprotein and plays an important role in the anticoagulant system as the receptor of thrombin leading to accelerated protein C activation (Gajendragadkar et al. 2009; Morikawa et al. 2002).

Statins have been demonstrated to inhibit superantigen-induced T cell activation, MHC class II antigen downregulation, and LFA-1 binding; reduce heart and kidney transplant rejection; reduce mortality with staphylococcal bacteremia; and reduce high-sensitive C-reactive protein in patients with coronary artery disease (Fehr et al. 2004; Weber et al. 2007).

13.3.4 Current Market Situation of Statins: The Billion Dollar Drugs

The beginning of the twenty-first century brought an economic growth reduction for most advanced countries, with an average around 2 % per year. Some economies even shrinked, and overall growth in Europe in 2012 was stagnant at 0.2 %. Despite economic losses, statin market was substantially maintained. Accounting for 6.5 % of the total pharmaceutical market share, statin drugs are the most widely sold drugs in history. To date, drug companies are earning $26 billion in annual sale (Fig. 13.3)—a single class of drugs producing more profits than Google–and are expected to continue to increase in the years ahead. The first statin, lovastatin, was launched in America in the 1980s with revenues of $200 million (Findlay et al. 2007; Murphree 2008). The statin market (Fig. 13.3) is still on the rise, although the growth is decelerating after early 2007. While new drugs are developed and the

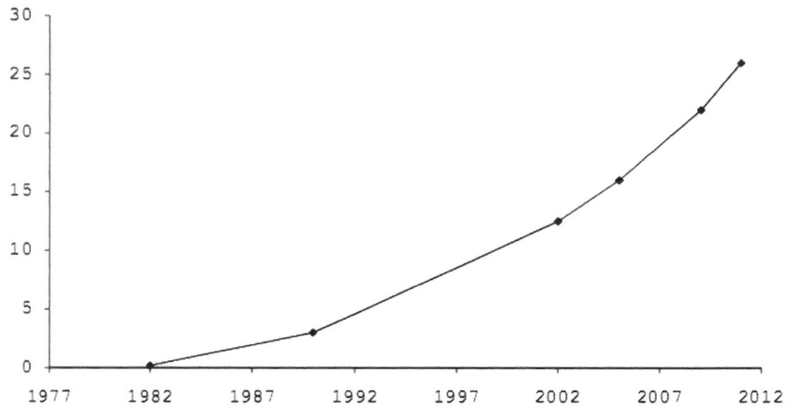

Fig. 13.3 Statin market (in billions). *Source*: Information derived by *Consumer Reports Best Buy Drugs—The statin drugs*. *The Telegraph* and *Forbes magazine*

older patents expire, statin therapies may become even more accessible, and the market may be expected to reach U$35 billion by 2023.

Statin use has increased in recent years as hyperlipidemia is being diagnosed more frequently. There is more evidence supporting the hazard of high levels of lipids in the blood, and consumers have become increasingly aware of beneficial effect of statins. New prescriptions of statins show a market share around 10 % for bio-based compounds (lovastatin and pravastatin) (Findlay et al. 2007) and a 50 % market for the semisynthetic simvastatin. This panorama is probably due to the recent expiration of patent rights for lovastatin (2001) and simvastatin (2006) and the entry of generics. The expiration of atorvastatin patents (2012) will probably bring it back as a generic statin. On the other side, with new drugs in the pipeline and an annual growth of emerging markets (such as Latin America), the microbial statin market is sure to remain huge.

13.4 Microbial Statins: Production Process and Potential for New Substances

Although most of the newer statins are synthetic, microbial statins are of interest because its production skips several synthetic steps, providing a high value-added molecule from low-cost substrates in an eco-friendly process.

ML-236B, known as compactin or mevastatin, was the first breakthrough in efforts to find a hypolipidemic agent by Akira Endo (1976). By 1980, Merck had discovered lovastatin which has been shown to be chemically very similar to mevastatin, differing by one methyl group (absent in mevastatin). Endo also had discovered the same compound, lovastatin (Brown and Goldstein 2004). Soon after the development of lovastatin, several screening efforts led to the development of similar molecules.

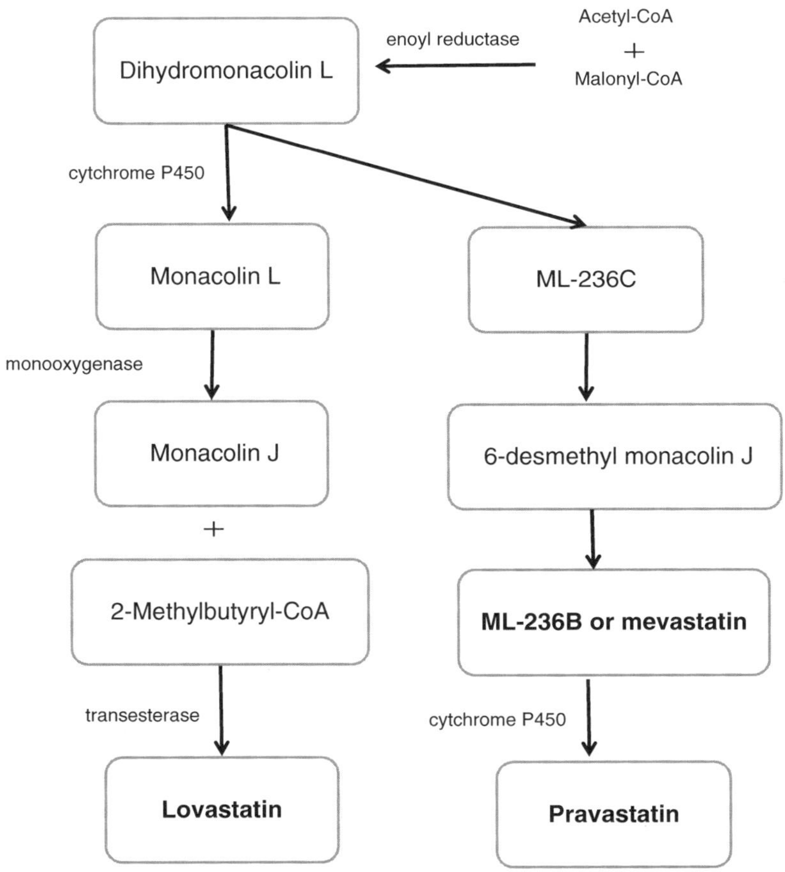

Fig. 13.4 Microbial statins biosynthetic pathway

The biosynthetic pathway for microbial statins starts with the condensation of acetyl and malonyl-CoA (Auclair et al. 2001; Barrios-gonzález and Miranda 2010; Komagata et al. 1989) and proceeds through the mevalonate pathway (Fig. 13.4).

13.4.1 Lovastatin

13.4.1.1 General

Lovastatin is a fungal secondary metabolite discovered in the seventies and introduced in the American market in the 1980s as Mevacor by Merck. It was the first statin to be approved by FDA, and it was formerly called as mevinolin or monacolin K. Lovastatin is administered as β-hydroxy lactone which over the time converts in vivo to the respective hydroxy acid form, partly similar to HMG-CoA. The hydroxyl acid form is a weak acid ($pK_a = 4.31$) with a molar mass of 422.55 while its lactone form has a higher ($pK_a = 13.5$) (Bizukojc et al. 2007; Brown and Goldstein 2004; Lisec et al. 2012; Seenivas et al. 2008; Seraman et al. 2010).

13.4.1.2 Current and Potential Uses of Lovastatin

Besides its cholesterol-lowering properties, lovastatin has been reported as a potential therapeutic agent for the treatment of various types of cancer. Recent in vitro studies have shown that lovastatin inhibits proliferation of anaplastic thyroid cancer cells through upregulation of p27 by interfering with the Rho/ROCK (serine/threonine kinase Rho kinase)-mediated pathway—this pathway has been suggested to be involved in the regulation of cancer cell motility. Other studies have shown endothelial protection function of lovastatin in the presence of hyperglycemia. Endothelial dysfunction, such as decreased endothelium-dependent vasorelaxation, plays a key role in the pathogenesis of diabetic vascular disease. Lovastatin was able to improve mesenteric responses to acetylcholine (Gajendragadkar et al. 2009; Zhong et al. 2011). Oxidative stress has been linked to the cause of many human diseases, such as heart failure, coronary artery and chronic kidney disease, and neurodegenerative disturbances. It arises from an imbalance between an excessive generation of reactive oxygen species, reactive nitrogen species, and insufficiency of antioxidant agents. Oral administration of lovastatin has been demonstrated to reduce oxidative stress and change the activities of antioxidant enzymes (Kumar et al. 2011).

13.4.1.3 Production Process

Potential Producers

Lovastatin is produced by a variety of filamentous fungi. Some of the important microbial sources are *Monascus* sp., *Penicillium* sp., and *Aspergillus* sp. Species were found to be the most significant producers of lovastatin, such as *Monascus purpureus*, *Monascus ruber*, *Aspergillus terreus*, and *Aspergillus flavipes*. Table 13.2 lists some species evaluated or developed for lovastatin production. Although several species produce low amounts of lovastatin, they are shown in order to illustrate the genera variability.

New rapid screening methods were developed to find new potential producers based on the activity of lovastatin against the yeast *Candida albicans*. In this method, the diameter of the inhibition zones (obtained on plates of *Candida albicans*) correlated linearly with the quantity of lovastatin impregnated in the paper disc (Bizukojc et al. 2007; Seraman et al. 2010; Vilches Ferrón et al. 2005). Other method for detecting lovastatin-producing strain is based on PCR for specific genes related to lovastatin synthesis, detecting suitable strains more quickly and effectively (Kim et al. 2011).

Table 13.2 Lovastatin production (in mg/L of fermented broth) by selected strains

Microorganism	Strain	mg/L	Reference
A. terreus	ATCC 20542	100	Porcel et al. (2008)
Penicillium citrinum	MTCC 1256	589	Ahmad et al. (2010)
Aspergillus terreus	Isolate	400	Szakács et al. (1998)
Aspergillus terreus	DRCC 122 (uv mutant)	2,200	Kumar et al. (2000)
Aspergillus terreus	Isolate	400	Samiee et al. (2003)
Monascus pilosus	MK-1 (a mutant strain)	725	Miyake et al. (2006)
Monascus purpureus	MTCC 369	351	Sayyad et al. (2007)
Biospora sp.	Isolate	13	Osman et al. (2011)
Cylindrocarpon radicicola	Isolate	7.1	Osman et al. (2011)
Penicillium spinulosum	Isolate	15.8	Osman et al. (2011)
Trichoderma viride	Isolate	36	Osman et al. (2011)
Mycelia sterilia	Isolate	15.3	Osman et al. (2011)
A. terreus	DSM 13596	310	Benedetti et al. (2002)

Fermentation

Submerged fermentation and solid-state fermentation (SSF) have been used for lovastatin production. Large-scale processes were developed using *Aspergillus terreus* in submerged fermentation. Enhanced strategies, such as the use of antibiotics, cultivation in fed-batch mode, and medium development, led to higher yields (Jia et al. 2010; Seenivas et al. 2008; Porcel et al. 2008).

Solid-state fermentation is a potential alternative to produce lovastatin which generates less effluent and uses less power. SSF uses various solid substrates, such as besan flour, barley, sago, and long-grain rice (all these substrates yield high lovastatin production, >110 mg/g dry substrate); mixed solids can also be used to formulate economical substrates for commercial production (Subhagar et al. 2009; Valera et al. 2005). The inocula for these processes may be either a liquid culture or a spore suspension; after inoculation, the fermenters are maintained at a temperature, pH, and aeration rate which are characteristics of each strain for several days. Typical values are 28 °C, pH 6.5, 1.5 vvm, and 7 days for *Aspergillus terreus* strains.

Culture Medium Characteristics

As with any fermentation product, the culture medium has a significant effect on the rate of production and yield of lovastatin. The type of carbon source (e.g., fructose, lactose, glycerol), nitrogen source (e.g., soybean meal, corn steep liquor, yeast extract), and the C:N mass ratio used in the medium influenced production of lovastatin and microbial biomass by *A. terreus*. The results have shown that the presence of excess carbon (slowly metabolizable carbon source) under nitrogen limitation greatly enhanced the rate of production of lovastatin. Nitrogen limitation diverts more carbon to lovastatin metabolic pathways (Casas López et al. 2003). Most liquid culture media use glucose as the main carbon source, but some authors suggest that this sugar strongly represses lovastatin synthesis (Miyake et al. 2006), which explains why continuous or fed-batch processes enhance the yield. In addition to glucose, several culture media also use starches and complex mixed sources, such as oatmeal, soybean meal, peptones, and yeast extract in the culture medium.

A variety of mineral nutrients is also added to the culture media, usually K_2HPO_4, $MgSO_4$, and ammonia, urea, or nitrates. Microelements are seldom added, being provided by the complex nutrients used.

Studies have shown that the supplementation of the culture medium with B-group vitamins enhances lovastatin synthesis by *Aspergillus terreus*. It is probable that the synthesis of lovastatin requires a high throughput of coenzymes, thus the application of its precursors in the form of B-group vitamins would give a positive effect on lovastatin production (Bizukojc et al. 2007). The impact of other supplements, such as linoleic acid, has demonstrated that micromolar concentrations of this fatty acid enhance lovastatin yield. Possibly, early supplementation of linoleic acid anticipates the production of oxylipins, thus mimicking the critical cell mass necessary for the onset of lovastatin production (Sorrentino et al. 2010). Vegetable oils stand out as a promising substrate as an additional carbon source for lovastatin production (Sripalakit et al. 2011).

Downstream Processing of Lovastatin

As lovastatin has a very low polarity, the concentration from the culture broth may be carried out by liquid-liquid extraction. However, there is a substantial portion of intracellular lovastatin (Benedetti et al. 2002), which may not be easily extracted. In addition, the molecule may be oxidized if not properly processed, leading to hard to separate impurities—antioxidants or inert atmospheres should be used. After extraction and concentration, the statin is usually purified by crystallization, although chromatography and ion exchange steps may also be used. The final compound should have a purity of at least 99.5 %. Figure 13.5 illustrates a possible lovastatin downstream process.

13.4.1.4 Simvastatin Production (Derivatization of Lovastatin)

The natural product lovastatin can be derived in its analog semisynthetic simvastatin. Substitution of the α-methylbutyrate side chain with α-dimethylbutyrate is most effective in treating hypercholesterolemia while lowering undesirable side effects. An alternative method for the simvastatin synthesis is a selective enzymatic deacylation of lovastatin. Due to the effectiveness of simvastatin, numerous multistep syntheses from lovastatin to simvastatin have been described in the patent literature. For example, a process for preparing simvastatin from lovastatin or mevinolinic acid in salt form comprises treating either starting material with cyclopropyl or butyl amine. In another process, lovastatin was hydrolyzed to acid form and then isolated in the form of amine salt like cyclopropyl or t-octylpropyl

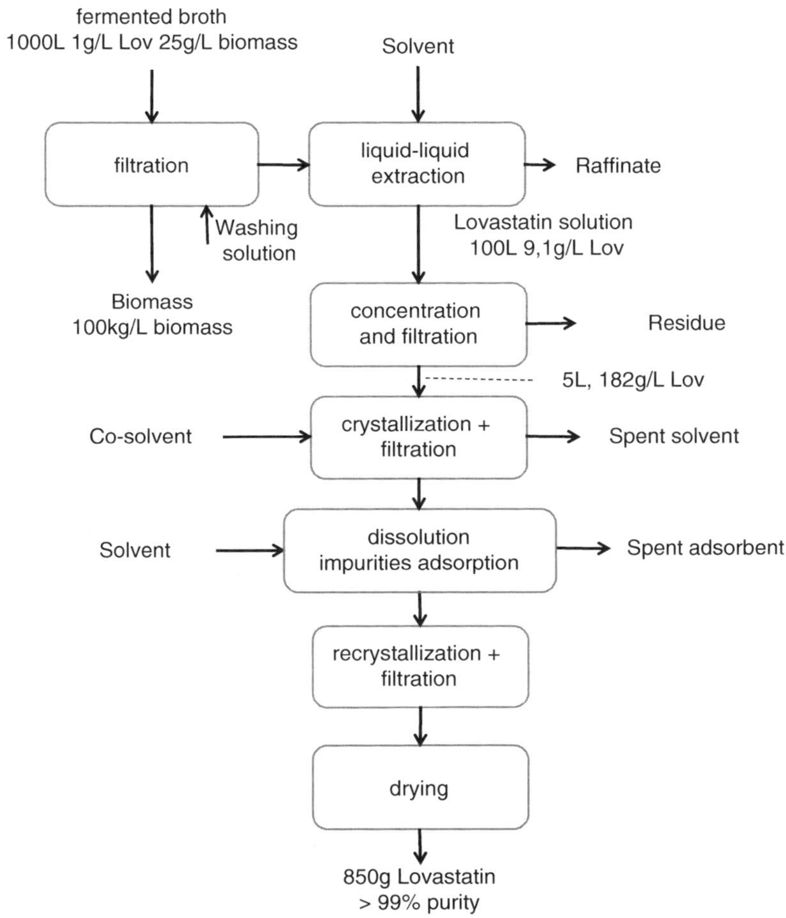

Fig. 13.5 Lovastatin downstream processing relying on crystallization operations

amines. The salts isolated were directly methylated without any protection or deprotection of hydroxyl groups. Then, simvastatin ammonium salts were converted to simvastatin by conventional methods of lactonization (Kumar et al. 1998; Vaid and Narula 2006).

13.4.2 Pravastatin

13.4.2.1 General

Pravastatin was discovered as a bioactive metabolite of mevastatin, in efforts to develop new statins. It was launched on the market in 1991 (Li 2009). Pravastatin as well as the other statins exists in two forms, lactone and open-ring hydroxy acid form (active form); besides the open ring, pravastatin has an extra hydroxyl group in comparison with lovastatin, being hydrosoluble. Pravastatin is more effective than lovastatin in moderate doses. In human plasma, pravastatin can be determined by enzyme immunoassay or in human urine by HPLC and UV detection (Darwish et al. 2009; Whigan et al. 1989).

13.4.2.2 Current and Potential Uses of Pravastatin

Neuroprotection by pravastatin in acute ischemic stroke has been demonstrated in rats even when given after stroke onset. Among the mechanisms responsible for improved neurological outcome is the inhibition the release of potentially damaging cytokines such as interleukin-6 in the early phase of cerebral ischemia (Berger et al. 2008). Other potential uses of pravastatin are in diabetic nephropathy. Diabetic nephropathy (DN) is the principal cause of end-stage renal failure in the Western world and leads to major mortality. DN is characterized by endothelial dysfunction following a variety of proinflammatory insults. Studies have provided evidence of the protective properties of pravastatin through an upregulation in endothelial constitutive nitric oxide synthase expression in diabetic group (Casey et al. 2005).

13.4.2.3 Production Process

Overview

This compound is obtained by two-step fermentation; firstly mevastatin is produced by *Penicillium citrinum* or other potential producer, such as *Streptomyces carbophilus,* and converted by hydroxylation of mevastatin to form pravastatin. The hydroxylation of mevastatin in *S. carbophilus* is catalyzed by a $CytP450_{sca}$ monooxygenase system (Sakaki 2012; Serizawa 1996). Recent advances in the molecular characterization of the $CytP450_{sca}$ and their responsiveness to mevastatin have been achieved. For example, molecular approaches for transcriptional regulation of the cytochrome $P450_{sca}$ from *S. carbophilus* by mevastatin sodium salt or cloning, characterization, and expression of the gene encoding $CytP450_{sca}$ from *S. carbophilus* involved in production of pravastatin. $CytP450_{sca}$ DNA sequences have been annotated in GenBank® (Sakaki 2012; Watanabe et al. 1995; Watanabe and Serizawa 1998). A new strain of *Streptomyces flavovirens* has been used to produce pravastatin (Gururaja et al. 2003). A *Monascus ruber* strain is capable to produce 3000 mg/L of pravastatin in a short fermentation, according to Benedetti et al. (2002). In relation to bioconversion, it has been established that actinomycetes could hydroxylate mevastatin to pravastatin. The degree of conversion by cells was 65–78 % of mevastatin added and 65–88 % of mevastatin taken up (Peng and Demain 2000).

Downstream Aspects

Pravastatin fermentation is followed by isolation and purification. A method of isolating and purifying pravastatin or its pharmaceutically salt involves a step of extracting using an organic solvent, such as ethyl acetate (Nobunari et al. 2003). Other studies have reported purification methods which use high-performance liquid chromatography (HPLC) (Haytko and Wildman 1992). Some polymorphs of pravastatin sodium have been described as obtainable from a process wherein aprotic and protic solvents are used (Pater and Wnukowski 2012). The fact that pravastatin is hydrophilic has the benefit of limiting the contamination by lipophilic compounds in the initial step of the process and the control of solvent extraction by broth acidification but may lead to partial dimerization in the concentration steps of the solvent extract.

Traces of mevastatin can still be present in the end product after lyophilization to remove solvent. Mevastatin and pravastatin are structurally closely related. Thus, purification of pravastatin is tedious but important for the production of a safe and efficient drug. Methods have been proposed for the extraction of pravastatin and the concomitant removal of impurities. It has been found that the ratio pravastatin/mevastatin has increased when the solution is added to water-immiscible solvent (e.g., isopropyl acetate, methyl isobutyl ketone) with water or an aqueous solution at a pH value ranging from 5.0 to 6.5 (Johannes et al. 2009). Mevastatin can also be highly toxic to microorganisms responsible for biotransformation, especially to mould fungi; thus mevastatin concentration must only be maintained at a low level during industrial production (Minquan et al. 2006).

Despite known pravastatin production process, still there are problems that need to be solved. During fermentation steps, there is a common problem characterized by degradation of pravastatin (e.g., hydrolysis of pravastatin) resulting in loss of product. This phenomenon also can occur with lovastatin. Therefore, use of specific nitrogen or carbon sources to avoid statin hydrolysis during fermentation, as well as deletion of genes encoding enzymes that hydrolyze statins, is necessary (Klaassen et al. 2009). During work-up procedures, unwanted loss of product occurs as result of lactonization leading to pravastatin lactone. Any breakthrough approach suppressing lactonization is therefore of great relevance in productive process. Elevated temperatures, certain pH regimes, and traces of other molecules can promote unwanted lactonization. Aprotic solvents have been appointed to suppress lactonization in pravastatin sodium downstream process (Pater and Wnukowski 2012).

13.5 Perspectives of the Non-statin Hypolipidemic Agents

High statin doses are often associated with the increased frequency of adverse effects. In addition, all statins have been observed to cause myopathy, and the risk of adverse effects on muscle increases with the use of high doses (Riphagen et al. 2012). Therefore, non-statin hypolipidemic drugs can be an alternative treatment option.

There are several promising novel therapeutic approaches for the treatment of hyperlipidemia and atherosclerosis based on non-statin hypolipidemic agents which are expected to be of great benefit for patients with adverse effects. Some of these agents may be produced using bioprocesses.

13.5.1 Niacin

Niacin, also known as vitamin B_3 or nicotinic acid, at levels higher than the required vitamin dose functions as a vitamin, favorably affecting atherogenic lipoprotein release, as well as decreases cholesterol, triglycerides, and LDL and VLDL levels and increases HDL cholesterol levels. Niacin acts on raising apoAI and lowering apoB. These apoliproteins are the main protein components of circulating HLD (apoAI) and of LDL and VLDL (apoB). However, the absolute magnitude of these lipid-modifying effects is highly dependent not only on the daily dose but also on the lipid phenotype at baseline (Chapman et al. 2010; Rozman and Monostory 2010).

13.5.2 Ezetimibe

Ezetimibe, a cholesterol-absorption inhibitor, acts on cholesterol from diet and does not affect the absorption of fat-soluble vitamins, triglycerides, or bile acids. The effect of ezetimibe on the progression of atherosclerosis remains unknown. It selectively inhibits cholesterol absorption by binding to the Niemann-Pick C1-like 1 (NPN1L1) protein. Combined therapy with ezetimibe and a statin provides an incremental reduction in LDL-C levels of 12–19 % (Rozman and Monostory 2010; Teramoto et al. 2012).

13.5.3 Cholesteryl Ester Transfer Protein (CETP) Inhibitors

CETP promotes the transfers of cholesteryl esters from antiatherogenic HDLs to proatherogenic apoB–principle protein components of circulating LDL and VLDL. Moreover, it facilitates the transport of triglycerides between the lipoproteins. Thus, CETP transfers lipids from one lipoprotein to another that results in equilibration of lipids between lipoprotein fractions. A deficiency of CETP is associated with increased HDL levels and decreased LDL levels, supporting the therapeutic potential of CETP inhibition as an approach to retarding atherogenesis (Rozman and Monostory 2010; Tzotzas et al. 2011).

13.5.4 *Fibrates*

Fibrates are generally effective in lowering elevated plasma trycerides and cholesterol. It is a class of amphipathic carboxylic acids. Fibrates can implicate five major mechanisms underlying the modulation of lipids: (a) induction of lipoprotein lipolysis, (b) induction of hepatic fatty acid uptake and reduction of hepatic triglyceride production, (c) increased removal of LDL particles, (d) reduction in neutral lipid exchange between VLDL and HDL, and (e) increase in HDL production and stimulation of reverse cholesterol transport. In general, fibrates are considered to be well tolerated, with an excellent safety profile (Rozman and Monostory 2010; Saha and Arora 2011).

13.6 Perspectives

The fermentation process for the production of statins has been generally well studied and optimized process w.r.t. process parameters and cost. It is well known that downstream processes can be considered one of the main factors involved in the final cost of the product as it can include multiple steps such as filtration, extraction, chromatography, crystallization, adsorption flow through, and drying (Straathof 2011; Winkelnkemper et al. 2011). Thus, the use of fungal biomass can be an interesting alternative in controlling lipid profile and avoid a complex downstream process (Table 13.3). There are already reports investigating the hypolipidemic effect of fungal biomasses (Santos et al. 2012). The support of this idea comes from herbal medicines which basically use plants in raw state and produce therapeutic action.

The use of statins has radically improved the therapy of coronary disease since 1990. The huge market for these molecules means that constant effort is being pursued by pharmaceutical companies and research institutes for developing new and more efficient molecules. Although semisynthetic derivatives will likely dominate the market in the near future, microorganism screening may lead to the development of new microbial

Table 13.3 Hypocholesterolemic effect of fungal biomass or its fractions on several animal models

Microorganism	Animal model	Amount/fraction	Reference
Auricularia auricula	ICR mice	Ethanol extract 150 mg/kg/day b.w.	Chen et al. (2011)
Cordyceps sinensis	ICR mice	Hot-water extract from mycelia 150 and 300 mg/kg/day	Koh et al. (2003)
Coriolus versicolor	Rats	Water extract	Hor et al. (2011)
Pleurotus ostreatus	Wistar rats	20 % biom/feed	Santos et al. (2012)
Pleurotus ostreatus	Rabbits	10 % dried fruit/feed	Bobek and Galbavy (1999)
Pleurotus ostreatus	Humans	30 g dried fruit/soup	Schneider et al. (2011)

statins. On the other side, traditional foods are being thoroughly studied in order to elucidate other cholesterol-lowering or synergistic mechanisms. This opens the path for the development of novel nutraceutical foods and dietary supplements.

References

Ahmad A, Panda BP, Mujeeb M (2010) Screening of nutrient parameters for mevastatin production by *Penicillium citrinum* MTCC 1256 under submerged fermentation using the Plackett-Burman design. J Pharm Bioallied Sci 2:44–46

Arikan S, Bahceci M, Tuzcu A et al (2012) Postprandial hyperlipidemia in overt and subclinical hypothyroidism. Eur J Intern Med 12(6):e141–e145

Auclair K, Kennedy J, Richard C, Vederas JC (2001) Conversion of cyclic nonaketides to lovastatin and compactin by a lovC deficient mutant of *Aspergillus terreus*. Bioorg Med Chem Lett 11:1527–1531

Barrios-gonzález J, Miranda RU (2010) Biotechnological production and applications of statins. Appl Microbiol Biotechnol 85:869–883

Benedetti A, Manzoni M, Nichele M, Rollini M (2002) Process for the production of pravastatin and lovastatin. EP1266967.

Berger C, Xia F, Maurer MH, Schwab S (2008) Neuroprotection by pravastatin in acute ischemic stroke in rats. Brain Res Rev 58:48–56

Betteridge J (2010) Pitavastatin – results from phase III & IV. Atheroscler Suppl 11:8–14

Bizukojc M, Pawlowska B, Ledakowicz S (2007) Supplementation of the cultivation media with B-group vitamins enhances lovastatin biosynthesis by *Aspergillus terreus*. J Biotechnol 127:258–268

Bobek P, Galbavy S (1999) Hypocholesterolemic and antiatherogenic effect of oyster mushroom (*Pleurotus ostreatus*) in rabbits. Nahrung/Food 43:339–342

Borshch VN, Andreeva ER, Kuz'min SG, Vozovikov IN (2012) New medicines and approaches to treatment of atherosclerosis. Russ J Gen Chem 82:554–563

Bouraoui L, Cruz-Garcia L, Gutiérrez J et al (2012) Regulation of lipoprotein lipase gene expression by insulin and troglitazone in rainbow trout (*Oncorhynchus mykiss*) adipocyte cells in culture. Comp Biochem Physiol 161:83–88

Brown MS, Goldstein JL (2004) A tribute to Akira Endo, discoverer of a "Penicillin" for cholesterol. Atheroscler Suppl 5:13–16

Brull DJ, Sanders J, Rumley A et al (2001) Statin therapy and the acute inflammatory response after coronary artery bypass grafting. Am J Cardiol 88:431–433

Campo VL, Carvalho I (2007) Estatinas hipolipidêmicas e novas tendências terapêuticas. Química Nova 30:425–430

Cantagrel V, Lefeber DJ, Ng BG et al (2010) SRD5A3 is required for converting polyprenol to dolichol and is mutated in a congenital glycosylation disorder. Cell 142:203–217

Casas López J, Sánchez Pérez J, Fernández Sevilla J et al (2003) Production of lovastatin by *Aspergillus terreus*: effects of the C:N ratio and the principal nutrients on growth and metabolite production. Enzyme Microb Technol 33:270–277

Casey RG, Joyce M, Roche-Nagle G et al (2005) Pravastatin modulates early diabetic nephropathy in an experimental model of diabetic renal disease. J Surg Res 123:176–181

Catapano AL (2010) Pitavastatin – pharmacological profile from early phase studies. Atheroscler Suppl 11:3–7

Chapman MJ, Redfern JS, McGovern ME, Giral P (2010) Niacin and fibrates in atherogenic dyslipidemia: pharmacotherapy to reduce cardiovascular risk. Pharmacol Ther 126: 314–345

Chen G, Luo Y-C, Ji B-P et al (2011) Hypocholesterolemic effects of *Auricularia auricula* ethanol extract in ICR mice fed a cholesterol-enriched diet. J Food Sci Technol 48:692–698
Darwish IA, Obaid ARM, Malaq HA (2009) New highly sensitive enzyme immunoassay for the determination of pravastatin in human plasma. Talanta 79:1478–1483
Ellis JT, Kilpatrick DL, Consigny P et al (2010) Therapy considerations in drug-eluting stents. Crit Rev Ther Drug Carrier Syst 22:1–25
Fehr T, Kahlert C, Fierz W et al (2004) Statin-induced immunomodulatory effects on human T cells in vivo. Atherosclerosis 175:83–90
Findlay S, Gunawardena D, Newsome-Stewart K, Skinner G (2007) The statin drugs. Consum Rep Best Buy Drugs. October 2005 to December 2006
Fukui M, Tanaka M, Toda H et al (2011) Risk factors for development of diabetes mellitus, hypertension and dyslipidemia. Diabetes Res Clin Pract 94:e15–e18
Gajendragadkar PR, Cooper DG, Walsh SR et al (2009) Novel uses for statins in surgical patients. Int J Surg (London) 7:285–290
Gu H, Tang C, Yang Y (2012) Psychological stress, immune response, and atherosclerosis. Atherosclerosis 223:69–77
Gururaja R, Goel A, Sridharan M et al (2003) Producing pravastatin sodium salt for use as antihyper-cholesterolemic agent, by fermentation under optimal fermentation parameters using new strain of Streptomyces flavovirens. WIPO WO2003027302
Haytko PN, Wildman AS (1992) Process for purification of HMG-CoA reductase inhibitors. WIPO WO/1992/016276
Hor SY, Farsi E, Yam MF et al (2011) Lipid-lowering effects of *Coriolus versicolor* extract in poloxamer 407-induced hypercholesterolaemic rats and high cholesterol-fed rats. J Med Plants Res 5:2261–2266
Jia Z, Zhang X, Zhao Y, Cao X (2010) Enhancement of lovastatin production by supplementing polyketide antibiotics to the submerged culture of *Aspergillus terreus*. Biotechnol Appl Biochem 60:2014–2025
Johannes BA, Mattheus PR, Piotr W (2009) Pravastatin extraction. WIPE WO 09728835
Jones PH, Davidson MH, Stein EA et al (2003) Comparison of the efficacy and safety of rosuvastatin versus atorvastatin, simvastatin, and pravastatin across doses (STELLAR Trial). Am J Cardiol 92:152–160
Kim JS, Youk YM, Ko HS, Kang DH (2011) PCR primer for detecting lovastatin-producing strain and a detection method using the same. Korean Patent Abstracts 1020110044613
Klaassen P, Vollebregt AWH, Van den Berg MA, Meijrink B (2009) Improved statin production. United States Patent Application 20110223640
Knopp RH (1999) Drug treatment of lipid disorders. N Engl J Med 12:498–511
Koga J, Aikawa M (2012) Crosstalk between macrophages and smooth muscle cells in atherosclerotic vascular diseases. Vascul Pharmacol 57:24–28
Koh J-H, Kim J-M, Chang U-J, Suh H-J (2003) Hypocholesterolemic effect of hot-water extract from mycelia of *Cordyceps sinensis*. Biol Pharm Bull 26:84–87
Koh KK, Son JW, Ahn JY et al (2004) Vascular effects of diet and statin in hypercholesterolemic patients. Int J Cardiol 95:185–191
Komagata D, Shimada H, Murakawa S, Endo A (1989) Biosynthesis of monacolins: conversion of monacolin L to monacolin J by a monooxygenase of *Monascus ruber*. J Antibiot 42:407–412
Kumar A, Kaur H, Devi P, Mohan V (2009) Role of coenzyme Q10 (CoQ10) in cardiac disease, hypertension and Meniere-like syndrome. Pharmacol Ther 124:259–268
Kumar MS, Jana SK, Senthil V et al (2000) Repeated fed-batch process for improving lovastatin production. Process Biochem 36:363–368
Kumar S, Srivastava N, Gomes J (2011) The effect of lovastatin on oxidative stress and antioxidant enzymes in hydrogen peroxide intoxicated rat. Food Chem Toxicol 49:898–902
Kumar Y, Thaper RK, Misra S et al (1998) Process for manufacturing simvastatin from lovastatin or mevinolinic acid. United States Patent 5763646
Kzhyshkowska J, Neyen C, Gordon S (2012) Role of macrophage scavenger receptors in atherosclerosis. Immunobiology 217:492–502

Li JJ (2009) Triumph of the heart: the story of statins. Oxford University Press, New York
Lisec B, Radež I, Žilnik LF (2012) Solvent extraction of lovastatin from a fermentation broth. Sep Purif Technol 96:187–193
Messner B, Bernhard D (2010) Cadmium and cardiovascular diseases: cell biology, pathophysiology, and epidemiological relevance. Biometals 23:811–822
Minquan M, Xiaoming J, Xiaoliang G et al (2006) The microorganism and the process for preparation of pravastatin. WIPE 04738297
Miyake T, Uchitomi K, Zhang MY et al (2006) Effects of the principal nutrients on lovastatin production by *Monascus pilosus*. Biosci Biotechnol Biochem 70:1154–1159
Morikawa S, Takabe W, Mataki C et al (2002) The effect of statins on mRNA levels of genes related to inflammation, coagulation, and vascular constriction in HUVEC. J Ateroscler Thromb 9:178–183
Murphree R (2008) Pfizer ads come clean about lipitor, but is anyone paying attention? TAC Integr Healthcare 3:45–46
Nirogi R, Mudigonda K, Kandikere V (2007) Chromatography-mass spectrometry methods for the quantitation of statins in biological samples. J Pharm Biomed Anal 44:379–387
Nobunari S, Shunshi K, Mutsuo S et al (2003) Method of purifying pravastatin. European Patent Office WO2001JP09045
Nozue T, Yamamoto S, Tohyama S et al (2012) Comparison of arterial remodeling and changes in plaque composition between patients with progression versus regression of coronary atherosclerosis during statin therapy (from the TRUTH study). Am J Cardiol 109:1247–1253
Osman M, Khattab O, Zaghlol G, Abd El-Hameed R (2011) Screening for the production of cholesterol lowering drugs (lovastatin) by some fungi. Aust J Basic Appl Sci 5:698–703
Ottosson M, Vikman-Adolfsson K, Enerbäck S et al (1994) The effects of cortisol on the regulation of lipoprotein lipase activity in human adipose tissue. J Clin Endocrinol Metab 79:820–825
Pater RM, Wnukowski P (2012) Crystalline form of pravastatine and process for the preparation thereof. WIPW WO/2012/085191
Peng Y, Demain AL (2000) Bioconversion of compactin to pravastatin by *Actinomadura* sp. ATCC 55678. J Mol Catal B: Enzym 10:151–156
Porcel EMR, López JLC, Pérez JAS, Christ Y (2008) Lovastatin production by *Aspergillus*. J Chem Technol Biotechnol 83:1236–1243
Riphagen IJ, van der Veer E, Muskiet FAJ, DeJongste MJL (2012) Myopathy during statin therapy in the daily practice of an outpatient cardiology clinic: prevalence, predictors and relation with vitamin D. Curr Med Res Opin 28:1247–1252
Rozman D, Monostory K (2010) Perspectives of the non-statin hypolipidemic agents. Pharmacol Ther 127:19–40
Saha SA, Arora RR (2011) Hyperlipidaemia and cardiovascular disease: do fibrates have a role? Curr Opin Lipidol 22:270–276
Sakaki T (2012) Practical application of cytochrome P450. Biol Pharm Bull 35:844–849
Samiee SM, Moazamil N, Haghighi S et al (2003) Screening of lovastatin production by filamentous fungi. Iran Biomed J 7:29–33
Santos LF, Zanatta AL, Soccol VT, Torres MF, Bonatto SJR, Rubel R, Soccol CR (2012) Hypolipidemic and antiatherosclerotic potential of *Pleurotus ostreatus*, cultived by submerged fermentation in the high-fat diet fed rats. Biotechnol Bioproc Eng 18(1):201–208
Sayyad SA, Panda BP, Javed S, Ali M (2007) Optimization of nutrient parameters for lovastatin production by *Monascus purpureus* MTCC 369 under submerged fermentation using response surface methodology. Appl Microbiol Biotechnol 73:1054–1058
Schneider I, Kressel G, Meyer A et al (2011) Lipid lowering effects of oyster mushroom (*Pleurotus ostreatus*) in humans. J Funct Foods 3:17–24
Seenivas A, Subhagar S, Aravindan R, Viruthagiri T (2008) Microbial production and biomedical applications of lovastatin. Indian J Pharm Sci 70:701–709
Seraman S, Rajendran A, Thangavelu V (2010) Statistical optimization of anticholesterolemic drug lovastatin production by the red mold *Monascus purpureus*. Food Bioproducts Process 88:266–276

Serizawa N (1996) Biochemical and molecular approaches for production of pravastatin, a potent cholesterol-lowering drug. Biotechnol Annu Rev 2:373–389

Sorrentino F, Roy I, Keshavarz T (2010) Impact of linoleic acid supplementation on lovastatin production in *Aspergillus terreus* cultures. Appl Microbiol Biotechnol 88:65–73

Sripalakit P, Riunkesorn J, Saraphanchotiwitthaya A (2011) Utilisation of vegetable oils in the production of lovastatin by Aspergillus terreus ATCC 20542 in submerged cultivation. Maejo Int J Sci Technol 5:231–240

Steinbrecher UP (1999) Receptors for oxidized low density lipoprotein. Biochim Biophys Acta 1436:279–298

Straathof AJJ (2011) The proportion of downstream costs in fermentative production processes. Compr Biotechnol 2:811–814

Strey CH, Young JM, Molyneux SL et al (2005) Endothelium-ameliorating effects of statin therapy and coenzyme Q10 reductions in chronic heart failure. Atherosclerosis 179:201–206

Subhagar S, Aravindan R, Viruthagiri T (2009) Response surface optimization of mixed substrate solid state fermentation for the production of lovastatin by *Monascus purpureus*. Eng Life Sci 9:303–310

Subramanian S, Chait A (2012) Hypertriglyceridemia secondary to obesity and diabetes. Biochim Biophys Acta 1821:819–825

Szakács G, Morovján G, Tegendry RP (1998) Production of lovastatin by a wild strain of *Aspergillus terreus*. Biotechnol Lett 20:411–415

Talayero BG, Sacks FM (2011) The role of triglycerides in atherosclerosis. Curr Cardiol Rep 13:544–552

Teramoto T, Kashiwagi A, Ishibashi S et al (2012) Cross-sectional survey to assess the status of lipid management in high-risk patients with dyslipidemia: clinical impact of combination therapy with Ezetimibe. Ther Res Clin Exp 73:1–15

Tzotzas T, Karras S, Gautier T et al (2011) Exploring the contribution of plasma CETP to the modulation of HDL cholesterol during niacin administration in diabetic patients with dyslipidemia. Atherosclerosis Suppl 12:184

Vaid S, Narula P (2006) Process for preparing simvastatin from lovastatin amine salts in three steps. WIPO Patent Application WO/2006/072963

Valera HR, Gomes J, Lakshmi S et al (2005) Lovastatin production by solid state fermentation using *Aspergillus flavipes*. Enzyme Microb Technol 37:521–526

Vilches Ferrón MA, Casas López JL, Sánchez Pérez JA et al (2005) Rapid screening of *Aspergillus terreus* mutants for overproduction of lovastatin. World J Microbiol Biotechnol 21:123–125

Wagner M, Zollner G, Trauner M (2009) New molecular insights into the mechanisms of cholestasis. J Hepatol 51:565–580

Watanabe I, Nara F, Serizawa N (1995) Cloning, characterization and expression of the gene encoding cytochrome P-450sca-2 from *Streptomyces carbophilus* involved in production of pravastatin, a specific HMG-CoA reductase inhibitor. Gene 163:81–85

Watanabe I, Serizawa N (1998) Molecular approaches for production of pravastatin, a HMG-CoA reductase inhibitor : transcriptional regulation of the cytochrome P450 gene from *Streptomyces carbophilus* sca by ML-236B sodium salt and phenobarbital. Gene 210:109–116

Weber MS, Steinman L, Zamvil SS (2007) Statins-treatment option for central nervous system autoimmune disease? Neurotherapeutics 4:693–700

Wexler BC (1971) Comparative aspects of hyperadrenocorticism and arteriosclerosis. Hum Pathol 2:180–181

Whigan DB, Ivashkiv E, Cohen AI (1989) Determination of pravastatin sodium and its isomeric metabolite in human urine by HPLC with UV detection. J Pharm Biomed Anal 7:907–912

Winkelnkemper T, Schuldt S, Schembecker G (2011) Systematic downstream process development for purification of baccatin III with key performance indicators. Sep Purif Technol 77:355–366

Zhong W-B, Hsu S-P, Ho P-Y et al (2011) Lovastatin inhibits proliferation of anaplastic thyroid cancer cells through up-regulation of p27 by interfering with the Rho/ROCK-mediated pathway. Biochem Pharmacol 82:1663–1672

Østerud B, Bjørklid E (2003) Role of monocytes in atherogenesis. Physiol Rev 83:1069–1112

Chapter 14
Exploring Plant and Agro-industrial Wastes for Antimicrobial Biochemicals

Sangeeta Negi

14.1 Introduction

Dating back to human history, the microbial infections have been one of leading cause of diseases. Over the year, a large number of antimicrobial compounds have been discovered which has made this world a much better place to live. Antibiotics, such as penicillin, have metamorphed the way diseases are tackled worldwide. However, indiscriminate use of antimicrobial agents, particularly antibiotics, is leading to the development of multiple drug resistance in human pathogenic microorganisms. The antibiotic-resistant bacterial strains travel very fast, making existing antibiotics ineffective. Therefore, it becomes essential to continuously look for new substances from the sources with proven antimicrobial activity.

Plant origin is the most important field for discovery of more effective and useful antimicrobial agents. Various parts of plants are used as source and template for synthesis of antimicrobial agents (Pretorius et al. 2003). In fact, plants and other natural sources provide unlimited scope for discovery of new antimicrobial agents due to chemical diversity. They are rich in secondary metabolites having antimicrobial properties, such as tannins, terpenoids, alkaloids, flavonoids, and other phenolic compounds. Plant extracts and phytochemicals as antimicrobial compounds in use for ages in crude form have been of great significance in therapeutic treatments (Kubo et al. 1993; Artizzu et al. 1995; Izzo et al. 1995; Nascimento et al. 2000). Although 250,000–500,000 species of plants are estimated to be available on the earth (Borris 1996), only less than 10 % of these are thought to be used as food and even less for medicinal purposes (Moerman 1996). In fact, major portion of plants, fruits, vegetables, and other food items are being either used as low-value items or thrown as waste.

S. Negi (✉)
Department of Biotechnology, Motilal Nehru National Institute of Technology, Allahabad, India
e-mail: sn5@mnnit.ac.in

S.K. Brar et al. (eds.), *Biotransformation of Waste Biomass into High Value Biochemicals*, DOI 10.1007/978-1-4614-8005-1_14, © Springer Science+Business Media New York 2014

As a result, huge amount of waste is being generated by almost every field, i.e., agriculture, food, and pharmaceuticals, among others. The European Union alone produced about 2 billion tonnes of wastes in the year 2006. By the year 2020, the production of waste was estimated to be 45 % higher than it was in 1995. About 30 % of the total waste was reported to be agrowaste (Piccirillo et al. 2010). There exists a huge scope for utilizing plant and agrowaste for various purposes including production of antimicrobial products. Fruit and vegetable wastes, such as pomace, skins, seeds, and other agrowaste, are rich sources of antimicrobials compounds. Therapeutic and other compounds from these natural wastes are not only eco-friendly but their recovery would also be economically attractive.

In addition to the therapeutic use, bioactive properties of plant and agrowastes have immense applications in food industry. Natural preservatives in food items have become very attractive as globalization has necessitated the transportation of high-quality food items from one corner of the world to the other (Davidson and Branen 2005). Various bioactive compounds having valuable application in food industry have been obtained from different types of agrowastes. Natural antimicrobial compounds from plants and agrowastes provide a viable and effective alternative to synthetic preservatives in the food industry.

14.2 Antimicrobial Chemistry

Antimicrobial agents are substance of chemical or biological or biochemical origin which inhibits the growth of microorganisms, such as bacteria, viruses, fungus, algae, or other parasites. Mechanism of inhibition of bacterial growth by antibacterial compounds depends on the target bacteria. Inhibition of nucleic acid synthesis, interference with cell wall synthesis, cell membrane interference, inhibition of protein synthesis, and disruption to metabolic pathway (Tenover 2006) are the common ways through which antimicrobial agents act against infectious microbes.

14.2.1 Inhibition of Nucleic Acid Synthesis

The metabolism of nucleic acids can be altered either at the DNA-dependent RNA polymerase or in the process of DNA coiling. DNA gyrase and DNA topoisomerase IV are two essential enzymes for cell growth and division. DNA gyrase controls DNA supercoiling. The enzyme binds to DNA and introduces double-strand breaks that allow the DNA to unwind. Some compounds, such as nitroimidazoles and nitrofurans, directly affect DNA (Calvo and Martinez 2009).

Antimicrobial agents, such as fluoroquinolones and rifampin, inhibit synthesis of nucleic acid. Some antimicrobials like fluoroquinolones interfere with DNA synthesis causing double-strand DNA breaks during DNA replication. Antimicrobial agents, such as rifampin, bind to DNA-dependent RNA polymerase, which blocks the synthesis of RNA and results in cell death.

14.2.2 Cell Wall and Cell Membrane Interference

The bacterial cell wall can be affected by antimicrobial compounds at different stages of either synthesis (fosfomycin, cycloserine) or transport (bacitracin, mureidomycins) of its metabolic precursors or by direct action on its structural organization (β-lactams, glycopeptides) (Calvo and Martinez 2009). Bacterial cell walls have a single layer of peptidoglycan which is a macromolecule with repeating units of N-acetylglucosamine acid (NAM) and N-acetylmuramic acid (NAG) cross-linked with amino acids between the NAM subunits. Synthesis of the peptidoglycan layer is inhibited by compounds, such as β-lactams, thus interfering with the synthesis of the bacterial cell (Tenover 2006). Antimicrobial compounds, such as vancomycin and teicoplanin, prevent the cross-linking steps required for stable cell wall synthesis by binding to the terminal D-alanine residues of the nascent peptidoglycan chain (McManus 1997). β-lactams (e.g., penicillins, cephalosporins), carbapenems, monobactams, glycopeptides, vancomycin, and teicoplanin are some of the common antibacterial compounds which inhibit bacterial cell wall synthesis.

Some antimicrobial compounds, such as polymyxins, bind to the plasma membrane of the cell and disrupt it. Polymyxin molecules diffuse through the outer membrane and cell wall of susceptible cells to the cytoplasmic membrane. They bind to the cytoplasmic membrane and disrupt and destabilize it. This causes increase in the bacterial cytoplasmic membrane permeability leading to leakage of bacterial cell contents and subsequent death of the cell. Antimicrobial compounds, such as daptomycin, irreversibly binds to the cell membrane of Gram-positive bacteria and causes rapid depolarization of the antibacterial membrane potential by directly binding to the bacterial membrane, which very often lead to death of the bacterium (Carpenter and Chambers 2004).

14.2.3 Inhibition of Protein Synthesis

A number of antimicrobial compounds, such as tetracyclines, aminoglycosides, chloramphenicol, and macrolides, inhibit protein synthesis by interfering at any of the phases of protein synthesis process, i.e., activation, initiation, binding of the transfer RNA (tRNA) amino acid complex to ribosomes, and elongation (Calvo and Martinez 2009). Protein synthesis begins with the transcription, i.e., formation of messenger RNA (mRNA) from DNA. mRNA carrying the codes for amino acids moves to the cytoplasm where these codes are translated. The ribosome moves one codon further along mRNA, releasing empty tRNA. The A site is free for the next charged tRNA, and the cycle is repeated as the ribosome moves to form polypeptide chain. Translation is terminated by stop codons recognized by release factors which help release the fully synthesized polypeptide chain from ribosomes. Translation ends with dissociation of ribosomal subunits.

Antimicrobial compounds, such as aminoglycosides, bind to the 30S subunit of bacterial ribosomes and blocks the initiation of translation and causes the misreading of mRNA. They can attach to the 30S subunit of the ribosome and prevent the 30S subunit from attaching to mRNA or presence of the aminoglycoside on the ribosome may cause misreading of the mRNA. This may either cause the insertion of the wrong amino acid into the protein or inhibit the ability of amino acids to connect with one another.

14.2.4 Inhibition of Metabolic Pathway

Certain compounds, such as trimethoprim and sulfonamides, inhibit the bacterial metabolic pathways. Both these agents act on the folic acid synthesis pathway, which is an important precursor to the synthesis of nucleic acids. Para-aminobenzoic acid (PABA) is an essential metabolite in the folic acid synthesis. Sulfonamides are structural analogs of PABA; therefore, they act as competitive inhibitors. Trimethoprim is structural analog for DHF and acts on the folic acid synthesis pathway at latter stage than sulfonamides in the pathway. Trimethoprim and sulfonamides are often used together. When used together, they produce a sequential blocking of the folic acid synthesis pathway and have a synergistic effect.

14.3 Biochemistry of Antibiotic Resistance

The site of the World Health Organization describes antimicrobial resistance as "Antimicrobial resistance (AMR) is resistance of a microorganism to an antimicrobial medicine to which it was previously sensitive. Resistant organisms (they include bacteria, viruses and some parasites) are able to withstand attack by antimicrobial medicines, such as antibiotics, antivirals, and antimalarials, so that standard treatments become ineffective and infections persist and may spread to others. AMR is a consequence of the use, particularly the misuse, of antimicrobial medicines and develops when a microorganism mutates or acquires a resistance gene." About 440,000 new cases of multidrug-resistant tuberculosis (MDR-TB) emerge annually, causing at least 150,000 deaths worldwide. Drug-resistant tuberculosis (XDR-TB) has been reported in 64 countries till mid of year 2012 (WHO 2012).

Bacterial resistance is a serious concern in the treatment of infectious disease. Effectiveness of antibacterial agent remains very good in the beginning of its introduction. However, as the use of antibacterial compound increases, pathogens develop resistance to it making the compound ineffective. In fact, pathogens acquire resistance to most of the antibacterial compounds very easily, thus creating need for new sources and more antimicrobial agents.

Bacteria develop resistance to antibacterial agents either due to their inherent characteristics, genetically encoded in its DNA, or acquire it from mutation in the host DNA or through encoded extra chromosomal materials in a horizontal transfer process. Several mechanisms have evolved which confer microbes with antimicrobial resistance. Four main mechanisms which confer antimicrobial resistance are (1) enzymatic inactivation of the antimicrobial molecules by the enzymes produced by bacteria, (2) reducing the affinity of the antimicrobial compound to the target by modifications, (3) efflux of the antibiotic from the cell through membrane-associated pumping proteins, and (4) resistance acquired through mutation or horizontal transfer.

14.3.1 Inactivation of Antimicrobial Agent

Antimicrobial molecules may be inactivated by either direct destruction or modification by enzymes synthesized by bacteria that selectively target and destroy the activity of the compound. There are three main mechanisms through which antimicrobial molecules may get deactivated: enzymatic hydrolysis, group transfer, or redox process. Many antimicrobial molecules have hydrolytically susceptible chemical bonds which are essential for their antimicrobial activity. Certain enzymes target and cleave these bonds and destroy antimicrobial activity of the compound. Bacteria produce such enzymes which require only water as a co-substrate which leads to inactivation of the antimicrobial molecules (Wright 2005). Some enzymes can induce resistance by modifying the antimicrobial molecules by chemical substitution or addition of acetyl groups to the periphery of the antibiotic molecule. This modifies the structure of the antimicrobial molecule causing poor antibiotic–bacteria interaction. Some bacteria metabolize antibiotics by oxidation and reduction mechanisms to detoxify and resist antibiotic effects.

14.3.2 Alteration of the Target

Antibacterial compounds bind to a target which is necessary to exhibit the desired antibiotic effect. However, it binds with the target only if it has affinity for it. Affinity between a target and an antimicrobial compound depends on the complementarity of the antimicrobial compound and the target. Any alteration made on the structure of the target molecules will reduce its affinity with the antibacterial compound. Such changes may occur in the target by mutation that reduces its susceptibility to antimicrobial agent (Spratt 1994). In fact, such alteration can lead to total disappearance of the affinity between the target molecule and the antibiotic compound leading to failure of antibiotic compound to bind with target, thereby causing total inhibition of antibiotic effect. Certain bacteria develop resistance to antibiotics by altering the structure of the target molecule.

14.3.3 Efflux Pumps and Reduced Outer Membrane Permeability

Certain antibiotics, such as macrolides and tetracyclines, act in the cytosol. Therefore, they have to be transported across the cell membrane before they reach their target. A minimum concentration of antimicrobial agent in the cytosol is essential to have desired antibacterial effect. In some bacteria, membrane proteins, called efflux pumps, export antibiotic molecules from the cytoplasm to the extracellular space, thus reducing the concentration of the antibiotic in the cytoplasm making the antibacterial compound less effective. Efflux pumps are drug specific but many efflux systems are multidrug transporters that are capable of expelling a wide spectrum of structurally unrelated drugs, making the bacteria multidrug resistant (Nikaido and Zgurskaya 1999; Webber and Piddock 2003).

Similarly, reduced outer membrane permeability will lead to reduced antibiotic uptake, thus reducing the effectiveness of antimicrobial compound.

14.3.4 Acquired Resistance

Above-explained mechanisms of resistance to antibacterial compound are intrinsic characteristics of the organism, but resistance to antibacterial compounds as well can be "acquired" by organism by mutation of cellular genes or through acquisition of foreign resistance genes or through combination of these two mechanisms (Dzidic et al. 2008). Two mechanisms of acquiring resistance are mutational adaptation and horizontal transfer by plasmids and the transposons (Roberts 2003).

It is now proven fact that microorganism tends to mutate in order to adapt to the environment of the antibiotic if it is exposed to inadequate doses of antibiotic for longer periods. New strain of the microorganism produced under such conditions becomes resistant to the antibiotic which was earlier effective against the microorganism in adequate doses. Antibiotic resistance occurs by nucleotide point mutations which are also able to produce a resistance phenotype (Woodford and Ellington 2007).

Horizontal transfer of resistance gene by microorganism is also a common mechanism which spreads antibiotic resistance. Microbes can acquire resistance gene from mobile genetic element, such as plasmids, integrons, and transposons. Horizontal transfer of resistance genes can take place into the recipient chromosome by recombination. Mobile genetic elements can be transmitted through transformation, transduction, or conjugation.

14.4 Common Antimicrobial Compounds from Plant and Agrowaste

Most of the isolated antimicrobial compounds from plants are secondary metabolites. These substances are normally synthesized by plants as plant defense mechanisms against microorganisms, insects, and herbivores, among others. They give odors, pigments, and flavors to plants. More than 300 natural metabolites with antimicrobial activities have been reported in the period 2000–2008 (Saleem et al. 2010). Many of these are very useful for humans as medicines and food preservatives, among others. Common antimicrobial compounds from plant and agro sources are phenolics or simple phenols, flavonoids, essential oils, coumarins, quinones, tannins, alkaloids, lignans, peptides, iridoids, and xanthones.

Polyphenols are the most common secondary metabolites of plants. These compounds are building blocks for cell wall structures and serve as defense against pathogens. Phenolic compounds from plants are well known for high antioxidant, anti-inflammatory, antiallergic, antithrombotic, and antimicrobial properties (Jayaprakasha et al. 2003; Baydar et al. 2004; Shoji et al. 2004; Alberto et al. 2006). Phenolic compounds present in fruits have antioxidant activity and protect the plant from environmental stress and fungal or bacterial infections. Waste products remaining after juice processing (peel, seeds, stems, flesh) are good sources of these ingredients. Plant and agrowaste are considered as economical and eco-friendly source of phenolic compounds.

14.4.1 Flavonoids

Flavonoids are plant compounds which contain a 2-phenylbenzopyran nucleus consisting of two benzene rings (A and B) linked through a heterocyclic pyrane ring (C). Flavonoids are one of the important secondary metabolites found in various plant species. They protect the plants from UV radiation and other environmental stresses. Flavonoids are found ample in fruits, vegetables, nuts, seeds, stems, and flowers, among others. These compounds are present in photosynthesizing cells (Cushnie and Lamb 2005; Havsteen 1983). Flavonoids present in flowers give it attractive colors for pollinators. Flavonoids in leaves protect them from fungal pathogens and UV radiation (Harborne and Baxter 1999; Harborne and Williams 2000). They have antimicrobial, antifungal, and antiviral properties. Use of various compounds of flavonoids is well known to cure human diseases.

Activity of flavonoids against pathogenic microbes is attributed to cell wall permeability and the porins in the outer membrane present in microorganisms. Activity of some flavonoids may also be due to their ability to complex with extracellular and soluble proteins and then with bacterial cell walls. Main classes of flavonoids are flavonols, flavanones, isoflavones, anthocyanidins, flavones, aurones, flavanon-3-ols flavans, flavan-3,4-diols, chalcones, and flavan-3-ols.

14.4.2 Essential Oils

Essential oils are secondary metabolites found in plants. Fragrance of the plants is carried in essential oil fraction. Their general chemical structure, called terpenes, is C10H16. Other compounds in these groups are diterpenes, triterpenes, tetraterpenes, hemiterpenes, and sesquiterpenes. Terpenoids are well known as antibacterial, antifungal, and antiviral. As per a report, about 60 % of essential oil derivatives are inhibitory to fungi, while 30 % inhibit bacteria (Saleem et al. 2010). Antibacterial mechanism of terpenes is reported to be due to membrane disruption by the lipophilic compounds.

Essential oils from various plants have shown antimicrobial activity against many pathogenic microbes (Melendez and Capriles 2006; Wannissorna et al. 2005). Essential oils from Citrus spp. are very effective antimicrobial agent against bacteria as well as fungi. Citrus essential oils are used for improving the shelf life and preservation of processed food and fruits (Lanciotti et al. 2004), skim milk, and low-fat milk. Citrus essential oils are present in fruit flavedo in great quantities (Chanthaphon et al. 2008). Antibacterial and antifungal activities of essential oils of *Mentha arvensis* and *Zingiber chrysanthum* leaves have been reported by Singh and Negi (1992a, b). There are many reports on antimicrobial activity of essential oils and plant extracts, such as rosemary, peppermint, bay, basil, tea tree, celery seed, and fennel.

14.4.3 Coumarins

Coumarins belong to benzopyrone group compounds, of which flavonoids are the other main member. Coumarin (1,2-benzopyrone) is available in a wide variety of plants. Coumarins are reported to be formed as a defense mechanism in response to traumatic injury during the wilting process or by plant diseases or through drying. Although coumarins are synthesized mainly in leaves, they accumulate on the surface of the leaves, fruits, and seeds. They are reported to protect the plants from fungal pathogens, beetles, and other terrestrials Coumarins are present in tonka bean, vanilla grass, woodruff, mullein, lavender, strawberries, apricots, cherries, cinnamon, sweet clover, bison grass, cassia, and yellow sweet clover. It was first isolated from tonka beans and has higher presence in some essential oils, particularly cinnamon bark oil and lavender oil, fruits (e.g., bilberry, cloudberry), green tea, etc. (Mirunalini and Krishnaveni 2011; Monga et al. 2012).

Several coumarins have been reported to have antimicrobial properties (Kwon et al. 1997; Kayser and Kolodziej 1997). Methanol extract from *Mitracarpus scaber* against *S. aureus* and *C. albicans* (Bisignano et al. 2000) and water extract from *Pelargonium sidoides* against *E. coli* and *Klebsiella pneumoniae* (Kayser and Kolodziej 1997) have shown antimicrobial effect.

14.4.4 Quinones

Quinones, naturally occurring pigments, are aromatic rings with two ketone substitutions. Conversion from hydroquinone to quinone and vice versa takes place through oxidation and reduction reactions. This redox potential of the particular quinone–hydroquinone pair is important in many biological systems. A number of quinones, such as anthraquinones, naphthoquinones, and benzoquinones, are widely distributed in nature. Quinone is well known to demonstrate antibacterial, antifungal, antiviral, antimicrobial, and anticancer activities (Beheshti et al. 2012).

14.4.5 Tannins

Tannins are polyphenolic substances found in bark, wood, leaves, fruits, and roots of plants. They are reported to function as chemical defenses against pathogens and herbivores (Gedir et al. 2005). They are called tannins for their ability of tanning leather. Tannins are grouped in two major structural classes: hydrolyzable tannins and condensed tannins. Condensed tannins are derived from flavonoid monomers, whereas hydrolyzable tannins are gallic acid-based multiple esters with D-glucose. Condensed tannins are more widely distributed in nature. They have been found toxic to filamentous fungi, yeasts, and bacteria. They inhibit the growth of pathogens and herbivory (Gedir et al. 2005). Tannins are natural detergent and suitable substitute for synthetic anthelmintics (Tanner et al. 1995). Mechanism of antimicrobial action of tannins may be related to their ability to inactivate microbial adhesins, enzymes, cell envelope transport proteins, etc. (Yao et al. 1992).

14.4.6 Alkaloids

Alkaloids are heterocyclic nitrogen compounds. Several alkaloids from natural sources possess potent antimicrobial properties and could be useful as antimicrobial compounds. One of the first medicinally important alkaloid was morphine isolated from *Papaver somniferum* (Saleem et al. 2010). Diterpenoid from the plants of Ranunculaceae; solamargine, a glycoalkaloid from the berries of *Solanum khasianum*; and some other alkaloids have been reported to be useful against HIV infection (Mendoza et al. 1997; Cowan 1999).

14.4.7 Peptides

Peptides are low molecular weight natural compounds and exhibit antimicrobial activity. It contains disulfide bonds. Cationic peptides are the most common peptides.

The discovery of non-ribosomally synthesized peptides present in plants and ribosomally synthesized peptides with antimicrobial properties has been very useful in fighting pathogens. Very limited resistance for the bacterial strains has made antibiotic peptides a promising source of antimicrobial drugs (Zasloff 2002). Plant peptides isolated from roots, seeds, flowers, stems, and leaves have shown activities against phytopathogens as well as against human pathogens. Many bioactive peptides have been isolated from milk-based products, eggs, meat, and fish as well as from different plant protein sources, such as soy and wheat, among others. Antimicrobial peptides act against different Gram-positive and Gram-negative bacteria (*Escherichia*, *Helicobacter*, *Listeria*, *Salmonella*, and *Staphylococcus*), yeasts, and filamentous fungi. The disruption of normal membrane permeability is understood to be partly responsible for the antibacterial mechanism of lactoferricins (Hartmann and Meisel 2007).

14.4.8 Lignan

Lignans are diphenolic compounds with a 2,3-dibenzylbutane structure. These are important plant phenolic compounds with known antimicrobial, antitumor, anti-inflammatory, and antiviral properties. Lignans are available in fiber-rich plants, such as wheat, beans, barley, oats, soybeans, lentils, garlic, asparagus, broccoli, and carrots. Flaxseed (*Linum usitatissimum*) is a very rich source of lignans. Lignans as source of plant phenolic compounds have been reported to reduce the risk of certain cancers. Lignans are one of the major classes of phytoestrogens which act as antioxidants. A lignan, (+)-lyoniresinol-3a-O-β-D-glucopyranoside, isolated from the root bark of Lycium chinense has shown antimicrobial activity against *Staphylococcus aureus* and three human-pathogenic fungi, *Candida albicans*, *Saccharomyces cerevisiae*, and *Trichosporon beigelii* (Saleem et al. 2010).

14.4.9 Xanthones

Xanthones (9H-xanthen-9-ones) are heterocyclic compounds with the dibenzo-γ-pyrone framework. Xanthone derivatives are secondary plant metabolites and have been isolated from fungi, lichens, and a few higher plants. Mangosteen (*Garcinia mangostana* Linn.) tree is a very good source of xanthones. Xanthones have been isolated from peel, whole fruit, bark, and leaves of mangosteen (VishnuPriya et al. 2010). Several studies have reported antioxidant, antitumoral, anti-inflammatory, antiallergy, antibacterial, antifungal, and antiviral activities of xanthones (Suksamrarn et al. 2006; Pedraza et al. 2008).

14.4.10 Iridoids

Iridoids are monoterpenoids, widely distributed in dicotyledonous plant families, such as the Apocynaceae, Scrophulariaceae, Diervillaceae, Lamiaceae, Loganiaceae, and Rubiaceae. Iridoids exhibit a wide range of bioactivities, such as anti-inflammatory, antioxidant, antibacterial, anticoagulant, antifungal, and antiprotozoal, among others. Aucubin and catalpol are the most spread iridoid glucosides in the *Veronica* genus. Aucubin has been reported to have shown wide pharmacological activities.

14.5 Antimicrobial Compounds from Plant and Agrowaste

Substantial amount of reusable substances, such as soluble sugars, proteins, phenols, and fibers, are present in the agro-industrial wastes which can be very valuable for the production of value-added antimicrobial products. Normally, one or two parts of plants, vegetables, fruits, cereals, etc., are used for some useful product and the rest of the parts such as peels, pomace, seed, leaves, hull, bark, and root are dumped as waste. A number of research groups are working to explore various possibilities of utilizing the plant and agrowaste as value-added products (Table 14.1). Some of the important works on exploration of these natural wastes for production of antimicrobial compounds are summed up below.

14.5.1 Antimicrobial Compounds from Fruit Waste

Huge quantities of fruits are used worldwide but their peel, seed, and pomace remain mostly unutilized, thus generating huge amount of waste. These wastes, though highly perishable and seasonal, are a serious challenge for the processing industries and pollution monitoring agencies. Suitable methods can be adopted to utilize this waste into value-added products, which can improve the overall economics of processing units and can help to reduce environmental pollution. A lot of developments have taken place in the field of utilizing fruit waste and by-products for extraction of antimicrobial compounds. Some of the important studies on utilization of fruit wastes for production of antimicrobial agents are briefed here.

14.5.1.1 Antimicrobial and Antioxidant Potential of Juice Pressing Waste

Anthocyanins, tannins, starches, saponins, polypeptides, and lectins were found in the water extract, and polyphenols, lactones, flavones, and phenons were additional

Table 14.1 Antimicrobial compounds extracted from plant and agrowaste

Type of waste	Source of waste	Compound extracted	Target microorganism	Reference
Pomace	*Fragaria ananassa* (strawberry), *Prunus cerasus* (sour cherry), *Ribes nigrum* (black currant), *Ribes rubrum* (red currant), *Rubus fruticosus* (blackberry), and *Rubus idaeus* (raspberry)	Anthocyanins, tannins, starches, saponins, polypeptides and lectins, polyphenols, lactones, flavones, and phenon	*Bacillus cereus*, *B. subtilis*, *Campylobacter jejuni*, *E. coli*, *Salmonella typhimurium*, *Serratia marcescens*, *C. albicans*, *C. krusei*, *C. glabrata*, *C. pulcherrima*, *C. parapsilosis*	Krisch et al. (2009)
Pomace	Beetroot (*Beta vulgaris* L. ssp. *Vulgaris*)	Phenolic, flavonoid betalain	*S. aureus* and *B. cereus*, *E. coli*, *P. aeruginosa*	Canadanovic et al. (2011)
Peel	*Mangifera indica* L., *Lagenaria siceraria*, *Solanum tuberosum* L., *Ananas comosus*, *Luffa acutangula* (L.), *Momordica charantia L.*, *Moringa oleifera*	–	*Staphylococcus aureus*, *Staphylococcus subflava*, *Corynebacterium rubrum*, *Salmonella typhimurium*, *Enterobacter aerogenes*, *Klebsiella pneumoniae*, *Proteus mirabilis*, *Cryptococcus luteolus*, *Candida albicans*, *Candida tropicalis*, *Candida glabrata*	Chanda et al. (2011)
Peel	*Citrus sinensis*, *Citrus limon*	Flavonoids, saponins, steroids, terpenoids, tannins, and alkaloids	*S. aureus*, *B. subtilis*, *E. coli*, *K. pneumoniae*, *S. typhi*	Ashok et al. (2011)
Peel	Kaffir lime (*Citrus hystrix* DC.)	Limonene, citronellal, and β-pinene and essential oil (β-pinene, sabinene citronellal)	*Staphylococcus aureus*, *Bacillus cereus*, *Listeria monocytogenes*, *Saccharomyces cerevisiae* var. *sake*, and *Aspergillus fumigatus*	Chanthaphon et al. (2008)
Seed kernel	Mango kernel (*Mangifera indica*)	Phenolic acids and flavonoids	*Antioxidant*	Maisuthisakul (2008)
Hull	Green tea processing, and acorn, chestnut, and persimmon hulls	Tannins	*Staphylococcus aureus*, *S. Aureus*, *E. coli*, *S. flexneri*, *L. monocytogenes*, *B. coagulans*	Sung et al. (2012)

Peel	Pomegranate	Polyphenols	*Staphylococcus aureus*, *Bacillus cereus*	Kanatt et al. (2010)
Peel, seed	Pomegranate (*Punica granatum* L.)	Polyphenols	*Penicillium italicum*, *Rhizopus stolonifer*, *Botrytis cinerea*	Tehranifara et al. (2011)
Seed	Grape	Polyphenols	*Bacillus cereus*, *Bacillus coagulans*, *Bacillus subtilis*, *S. aureus*, *E. coli*, *P. aeruginosa*	Jayaprakasha et al. (2003)
Seed	Grape	Polyphenols	*Aeromonas hydrophila*, *B. brevis*, *B. cereus*, *B. megaterium*, *B. subtilis*, *E. faecalis*, *E. coli*, *Klebsiella pneumoniae*, *L. monocytogenes*, *Mycobacterium smegmatis*, *Proteus vulgaris*, *P. aeruginosa*, and *S. aureus*	Baydar et al. (2004)
Peel	Apple	Flavonoids, polyphenols, catechin, procyanidin, caffeic acid, and chlorogenic acid	*E coli*, *S. aureus*, *P. aeruginosa*, *E. faecalis*, and *L. monocytogenes*	Escarpa and Gonzalez (1998), Vander et al. (2001), Alberto et al. (2006)
Leaves	Walnut	Caffeoylquinic acids, coumaroylquinic acids, p-coumaric acid, quercetin 3-galactoside	*Bacillus cereus*, *B. subtilis*, *Staphylococcus aureus*	Pereira et al. (2007)
Leaves	*S. macrophylla*	Phenolics, tannin, and flavonoid	*Staphylococcus aureus*, *Bacillus subtilis*, *Escherichia coli*, and *Pseudomonas aeruginosa* and, *a fungus*, *Candida albicans*	Tan et al. (2009)
Stems and leaves	Ginja cherry	Polyphenols	*Bacillus subtilis*, *Staphylococcus aureus* MSSA, *Staphylococcus aureus* MRSA, *Pseudomonas* sp., *Pseudomonas aeruginosa*, *Flavobacterium* sp., *Escherichia coli*, *Salmonella*	Piccirillo et al. (2010)
Leaves	Agave sisalana Perrine	–	*Candida albicans*	Santos et al. (2009)

(continued)

Table 14.1 (continued)

Type of waste	Source of waste	Compound extracted	Target microorganism	Reference
Leaves	*Citrus sinensis*	–	*E. coli*, *K. pneumoniae*, and *S. aureus*	Ekwenye and Edeha (2010)
Leaves	Tobacco	Rutin, flavonoids	–	Fathiazada et al. (2006)
Stalks	Beet	Azelaic acid	*S. aureus* and *L. monocytogenes*	Gollnick and Schramm (1998)
Peel	Peanut	Succinic, azelaic acids, epicatechin, caffeic acid, and coumaric acid	*S. aureus* and *L. monocytogenes*	El-Massry et al. (2009)
Seeds and marcs	Petit Verdot grape	Epicatechin caffeic and gallic acids	*S. aureus* and *L. monocytogenes*	Anastasiadi et al. (2008)
Fermentation lees	Red grapes	Gallic and ferulic acids, flavonoids	*S. aureus* and *L. monocytogenes*	Martin et al. (2012)
Bagasse	Guava	Epicatechin, quercetin, and caffeic	*S. aureus* and *L. monocytogenes*	Martin et al. (2012)
Hulls	Green tea processing and acorn, chestnut, and persimmon	Tannins	*Staphylococcus aureus*, *S. aureus*, *E. coli*, *S. flexneri*, *L. monocytogenes*, and *B. coagulans*	Sung et al. (2012)
Hull	Jojoba hull	–	*Escherichia coli*, *Staphylococcus aureus*, *Bacillus cereus*, *Listeria monocytogenes*, and *Salmonella typhimurium*	Wagdy and Taha (2012)
Mill waste	Olive oil	Phenol acids, catechol, methylcatechol, flavonoids	*Salmonella* spp., *Staphylococcus aureus*, *Clostridium botulinum*, and *Listeria monocytogenes*	Payne et al. (1989), Nychas et al. (1990), Tassou and Nychas (1995)

phytochemicals traced in the extracts of the pomace (peels, seeds, flesh) remaining after pressing the juice of *Fragaria ananassa* (strawberry), *Prunus cerasus* (sour cherry), *Ribes nigrum* (black currant), *Ribes rubrum* (red currant), *Rubus fruticosus* (blackberry), and *Rubus idaeus* (raspberry). Antimicrobial and antioxidant potential of water and methanol extracts was investigated on *Bacillus cereus*, *B. subtilis*, *Campylobacter jejuni*, *E. coli*, *Salmonella typhimurium* and *Serratia marcescens*, *C. albicans*, *C. krusei*, *C. glabrata*, *C. pulcherrima*, and *C. parapsilosis* by broth dilution method (Krisch et al. 2009). Both aqueous and methanol extracts inhibited growth of almost all bacteria. Methanol extracts had stronger inhibitory effect than water extracts.

Pomace of beetroot is also disposed as low-value feed or manure, although it is rich in phenols. During juice pressing, most of the secondary plant metabolites and dietary fiber compounds of beetroot are not transferred into the liquid phase and remain in the pomace (Will et al. 2000). Beetroot (*Beta vulgaris* L. ssp. *vulgaris*) has very high total phenolic content, i.e., in the range of 50–60 μmol/g dry weight and very good antioxidant property (Vinson et al. 1998; Kahkonen et al. 1999). Beetroot peels are reported to contain l-tryptophan, *p*-coumaric, and ferulic acids as well as cyclodopa glucoside derivatives (Kujala et al. 2001).

Phenolic content in beetroot is mainly present in the peel (50 %), crown (37 %), and flesh (13 %) (Canadanovic et al. 2011). Canadanovic et al. (2011) reported phenolic content (376.4 mg/g of dry beetroot pomace extract), flavonoid content (269.70 mg/g), and betalain (41.85 mg/g) in the ethanolic extract of the beetroot. They evaluated antibacterial activity of ethanol extract of beetroot pomace by disk diffusion and microdilution method against *Staphylococcus aureus*, *Bacillus cereus*, *Escherichia coli*, and *Pseudomonas aeruginosa*. Gram-positive bacteria *Staphylococcus aureus* and *Bacillus cereus* demonstrated higher susceptibility than Gram-negative bacteria *Escherichia coli* and *Pseudomonas aeruginosa* against beetroot extract.

14.5.1.2 Antimicrobial Activity of Citrus Fruit Peels

Citrus is one of the major fruit commercially grown all over the world. Huge quantity of wastes such as peels are generated every year, as juice yield of citrus is not even half of the total fruit mass (Manthey and Grohmann 2001). Citrus peels are rich in nutrients and contain many phytochemicals like flavanones and polymethoxylated flavones (Ahmad et al. 2006), thus can be explored for antimicrobial compounds. Therapeutic value of citrus oil as an antidiabetic, antimicrobial, antifungal, hypotensive agent, antioxidant, antibacterial, and antiviral agent has been studied by many research groups (Kumamoto et al. 1986; Caccioni et al. 1998; Hamendra and Anand 2007; Kanaze et al. 2008).

Phytochemical analysis and antimicrobial activities of peel of *Citrus limon* (lemon) and *Citrus sinensis* (sweet orange) were studied by Ashok et al. (2011). They used five different solvent extracts (ethyl acetate, acetone, ethanol, petroleum ether, and water) for both types of peels and screened their extract against five pathogenic bacteria *Staphylococcus aureus*, *Bacillus subtilis*, *Escherichia coli*, *Klebsiella pneumoniae*, and *Salmonella typhi*. They carried out phytochemical analysis of powdered plant parts and reported presence of flavonoids, saponins, tannins, alkaloids, and terpenoids in the peel extracts. Solvent extracts of *Citrus sinensis* peel and *Citrus limon* showed significant activities and presence of different phytochemicals. Acetone peel extract of *Citrus sinensis* showed highest antibacterial activity followed by the ethyl acetate peel extract of *Citrus limon*.

Chanthaphon et al. (2008) studied ethyl acetate extracts and hydrodistillated essential oils of *Citrus* spp. and *kaffir lime* peels (*Citrus hystrix* DC.) for antimicrobial activities against food-related microorganisms. Ethyl acetate extract from *kaffir lime* contained limonene (31.64 %), citronellal (25.96 %), and β-pinene (6.83 %). Essential oil obtained from hydrodistillation contained β-pinene (30.48 %), sabinene (22.75 %), and citronellal (15.66 %). They evaluated antimicrobial activities of ethyl acetate extracts from fresh peels and dried peels of tropical citrus fruits (lime), kaffir lime, and pomelo peels against pathogenic *E. coli*. Extracts from both fresh and dried limes, kaffir lime, and pomelo peels showed antibacterial activity against *S. aureus*, but the ones from fresh peels showed higher activity. Similarly, the extracts from fresh lime and kaffir lime peels showed activity against *E. coli* but no activity was observed with dried peels. They also observed that ethyl acetate extracts from all citrus peels showed better antimicrobial activities than their essential oils. The ethyl acetate extract of peels inhibited Gram-positive bacteria, yeast, and molds: *Staphylococcus aureus*, *Bacillus cereus*, *Listeria monocytogenes*, *Saccharomyces cerevisiae* var. *sake*, and *Aspergillus fumigatus*.

14.5.1.3 Antioxidant and Antimicrobial Activity of Pomegranate Peel and Seed

Pomegranate peel extract (PE) also shows excellent antioxidant and antimicrobial activity. Kanatt et al. (2010) studied the antimicrobial and antioxidant properties of pomegranate peel and seed extracts. Antimicrobial activity of pomegranate peel extract was tested against *Staphylococcus aureus*, *Bacillus cereus*, *Escherichia coli*, and *S. typhimurium*. It inhibited growth of *Staphylococcus aureus* and *Bacillus cereus* but *Escherichia coli* and *S. typhimurium* were resistant to it. Pomegranate peel extract (PE) was found very effective in scavenging hydroxyl and superoxide anion radicals. Investigators used pomegranate peel extract to enhance shelf life of chicken meat products successfully by 2–3 weeks during chilled storage.

Tehranifara et al. (2011) also investigated the antioxidant properties of peel, seed, and leaf of pomegranate (*Punica granatum* L.). They used aqueous and methanolic extraction with different concentrations (0, 500, 1,000, and 1,500 ppm) on three fungus *Penicillium italicum*, *Rhizopus stolonifer*, and *Botrytis cinerea*. Methanolic extract showed the highest inhibitory effect on the mycelia growth and spore germination. Peel and seed extracts showed more inhibitory effect than leaf extract. Antioxidant capacities of peel, seed, and leaf extracts of pomegranate were 55.3 %, 35.7 %, and 16.4 %, respectively. The phenolic content was 2.8 times higher in peel extract than leaf extract, which may be the reason for better antimicrobial activity of peel extract. These studies propose pomegranate peel and seed as source of powerful antioxidant and antifungal compounds.

14.5.1.4 Antimicrobial Effect of Apple Skins

Apples are high on phenolic compounds (Mangas et al. 1999; Podsedek et al. 2000; Shoji et al. 2003). Apple pulp and skins are reported to contain catechin, procyanidin, caffeic acid, and chlorogenic acid among other compounds. Skin of apple also contains flavonoids such as quercetin glycosides and cyanidin glycosides which are not present in pulp (Escarpa and Gonzalez 1998; Vander et al. 2001). Alberto et al. (2006) examined the antimicrobial activity of phenolic compounds extracted from the skin of two apple varieties Royal Gala and Granny Smith using the agar diffusion method. The phenolic compounds were extracted in acetone, methanol, and ethanol solvents. Total phenolic and flavonoid content was obtained with Folin–Ciocalteu reagent (Singleton and Rossi 1965). Granny Smith variety skin was found containing more polyphenols and flavonoid than Royal Gala, whereas maximum phenolics and flavonoids were obtained in acetone extract for both varieties of apples. The highest inhibitory effect of both apple varieties corresponded to extract which contained high phenolic content. Antimicrobial activities of different extracts were evaluated against *Escherichia coli, Staphylococcus aureus*, *Pseudomonas aeruginosa*, *Enterococcus faecalis*, and *Listeria monocytogenes*. Extracts of both the apple skins were reported to inhibit these microorganisms; however, extracts of Granny Smith were found more effective demonstrating a direct relationship between the phenolic content of the extracts and the antimicrobial effect. This study established the usefulness of apple skin for its antibacterial properties, thereby adding value to the already known benefits of apple to the human health.

14.5.1.5 Fruit and Vegetable Peels

Chanda et al. (2011) evaluated seven fruit and vegetable peels for their antimicrobial properties: *Mangifera indica* L. (Anacardiaceae), *Lagenaria siceraria* (Molina) Standl. (Cucurbitaceae), *Solanum tuberosum* L. (Solanaceae), *Ananas comosus* (Linnaeus) Merr. (Bromeliaceae), *Luffa acutangula* (L.) Roxb. (Cucurbitaceae), *Momordica charantia* L. (Cucurbitaceae), and *Moringa oleifera* Lam. (Moringaceae). Antimicrobial activities of hexane, chloroform, acetone, and methanol extracts of these samples were evaluated by agar well diffusion method against *Staphylococcus aureus*, *Staphylococcus subflava*, *Corynebacterium rubrum*, *Salmonella typhimurium*, *Enterobacter aerogenes*, *Klebsiella pneumoniae*, *Proteus mirabilis*, *Cryptococcus luteolus*, *Candida albicans*, *Candida tropicalis*, and *Candida glabrata*. *Mangifera indica* peel showed best and promising antimicrobial activity. Polar solvents (acetone and methanol) were found more effective than nonpolar solvents (hexane and chloroform). Activities shown by acetone extracts were the best, followed by methanol extracts. The extracts showed better antifungal activity than antibacterial activity. *C. glabrata* and *K. pneumoniae* were the most susceptible organisms. Report suggested that broad spectrum of antibacterial activity by *M. indica* peels may help to discover new chemical classes of antibiotic substances.

14.5.1.6 Antibacterial Effect of Grape Seeds

Grape seeds, which are normally a waste of winery or juice pressing, can be effectively utilized for production of antioxidants and antimicrobial compounds. Jayaprakasha et al. (2003) have reported antimicrobial activity of acetone and methanol extracts of grape seed against *Bacillus cereus*, *Bacillus coagulans*, *Bacillus subtilis*, *S. aureus*, *E. coli*, and *P. aeruginosa*. The authors reported that the extracts were free radical inhibitors and primary antioxidants that react with free radicals.

In another report on antimicrobial activity of grape seed extract, Baydar et al. (2004) revealed that the grape seed extracts in acetone/water/acetic acid and ethyl acetate/methanol/water solvents inhibited various test organisms: *Aeromonas hydrophila*, *B. brevis*, *B. cereus*, *B. megaterium*, *B. subtilis*, *E. faecalis*, *E. coli*, *Klebsiella pneumoniae*, *L. monocytogenes*, *Mycobacterium smegmatis*, *Proteus vulgaris*, *P. aeruginosa*, and *S. aureus*. The grape seed extracts were found to contain high total phenolics, which should be the reason for their strong antimicrobial activity.

14.5.2 Antimicrobial Compounds from Plant Wastes

14.5.2.1 Antimicrobial Activity of Leaf Waste

Organic and aqueous extracts of mahagony (*S. macrophylla*) seeds have been reported to possess antimicrobial, antimalaria (Soediro et al. 1990), antidiabetic, antidiarrheal, anti-inflammatory, and antitumor-promoting (Amelia et al. 1996) properties. Tan et al. (2009) studied the antimicrobial and antioxidant activities of methanol and dichloromethane extracts of *S. macrophylla* leaves, which are otherwise not considered as a valuable product. They evaluated the antimicrobial activities of the extracts against four bacteria *Staphylococcus aureus*, *Bacillus subtilis*, *Escherichia coli*, and *Pseudomonas aeruginosa* and a fungus *Candida albicans*. The methanol and the dichloromethane extracts were found more active against Gram-positive bacteria, i.e., *Staphylococcus aureus* and *Bacillus subtilis*, but showed very limited activity against *Escherichia coli* and *Pseudomonas aeruginosa*. Methanol extract also showed activity against *Candida albicans*. Considerable amount of phenolic, tannin, and flavonoid contents were found in the mahagony leaf extracts.

A study on the antimicrobial activity of extracts of the leaves and leaf waste of *Agave sisalana* Perrine, popularly known as sisal, has been carried out by Santos et al. (2009). It is a monocotyledonous plant largely grown in Brazil and Mexico. Only 5 % of the decortications of the leaves of sisal (*A. sisalana*) are reported to be used to produce useful hard fiber and the remaining 95 % being wasted as solid and juice waste. The study was carried out to evaluate the antimicrobial activity of extracts of the leaves and leaf waste. Investigators determined the antimicrobial activity by the paper disk diffusion method against bacteria and fungus both nonresistant and resistant to antibiotics. The hydroalcoholic extract obtained from leaves and from sisal waste showed significant inhibition of *C. albicans*. However, HCl and methanol extract of leave waste did not show any activity against *S. aureus*, *E. coli*, *M. luteus*, *B. cereus*, *P. aeruginosa*, and *S. choleraesuis*.

Pereira et al. (2007) used different cultivars of walnut (*Juglans regia* L.) leaves (Franquette, Mayette, Marbot, Mellanaise, and Parisienne) grown in Portugal, for evaluating their antimicrobial and antioxidant properties. Phenolic analysis of the extracts identified 3- and 5-caffeoylquinic acids, 3- and 4-p-coumaroylquinic acids, p-coumaric acid, quercetin 3-galactoside, quercetin 3-pentoside derivative, quercetin 3-arabinoside, quercetin 3-xyloside, and quercetin 3-rhamnoside in the extract. The extract was found to inhibit the growth of Gram-positive test bacteria (*Bacillus cereus*, *B. subtilis*, *Staphylococcus aureus*). However, Gram-negative test bacteria (*Pseudomonas aeruginosa*, *Escherichia coli*, *Klebsiella pneumoniae*) were resistant to the extracts.

Waste tobacco leaves have also been explored for extraction of flavonoids, mainly rutin. Tobacco plant is a rich source of medicinally useful alkaloids and flavonoids. Rutin, well-known flavonoid, is known to reduce capillary fragility, swelling, bruising, hemorrhoids, diabetic vascular disease, diabetic retinopathy, pain, tired legs, night cramps, and restless legs (Grinberg et al. 1994; Beretz and Cazenave 1988). Fathiazada et al. (2006) studied tobacco waste of tobacco factories as a source of flavonoids. They determined the content of rutin in waste and unfermented leaves of tobacco by HPLC and found the content of rutin in waste leaves of tobacco (0.6 %) less than that of unfermented leaves (1.5 %). However, waste leaves can still be an economical source of rutin.

In another report on antibacterial activity of the citrus plant leaf waste, Ekwenye and Edeha (2010) reported inhibitory effect of ethanol and aqueous extract of *Citrus sinensis* leaf against *E. coli*, *K. pneumoniae*, and *S. aureus*.

14.5.2.2 Antibacterial Activity of Other Plant Parts

Nascimento et al. (2000) studied antimicrobial activities of eight plant parts and tested antimicrobial activities of their extract against various microorganisms. The plants used for the study, plant parts used for extractions, and compounds obtained are given in Table 14.2.

Staphylococcus aureus, *Salmonella choleraesuis*, *Pseudomonas aeruginosa*, *Bacillus subtilis*, *Candida albicans*, *Proteus* spp., *K. pneumoniae*, *Shigella* spp., *Proteus* spp., *Pseudomonas aeruginosa*, *Enterobacter aerogenes*, *Escherichia*, and *Staphylococcus aureus* were used in the study. The extracts from clove and

Table 14.2 Compound extracted from plant parts

Plant	Plant part	Compound obtained
Thyme (*Thymus vulgaris* L., Lamiaceae)	Dried leaves and flowers	Essential oils (mainly thymol and carvacrol), flavonoids, tannins, and triterpenes
Rosemary (*Rosmarinus officinalis* L., Lamiaceae)	Leaf	Flavonoids, phenolic acids (caffeic, chorogenic, and rosmarinic), and essential oils (camphor and cineole)
Lemon balm (*Melissa officinalis* L., Lamiaceae)	Leaf	Essential oils, flavonoids and rosmarinic, caffeic, and chlorogenic acid
Basil (*Ocimum basilicum* L., Lamiaceae)	Leaf	Essential oils (estragol and eugenol), tannins, and flavonoids
Clove (*Syzygium aromaticum*)	Dried buds	Essential oils (eugenol), tannins, and flavonoids
Pomegranate (*Punica granatum* L., Punicaceae)	Pericarp	Ellagitannins and alkaloids
Jambolan (*Syzygium cumini*, Skeels, Myrtaceae)	Leaf	Coumarins, essential oils, flavonoids, triterpenes, and ellagitannins
Guava (*Psidium guajava* L., Myrtaceae)	Leaf	Flavonoids and tannins

jambolan were active against nine and eight test microorganisms, respectively. Other extracts were also active against at least one of the tested microorganisms.

Stems and leaves are generated as waste in the production of the cherry liquor from ginja cherry plant. Piccirillo et al. (2010) investigated the antimicrobial properties of these wastes. The ethanol extracts of stems and leaves were tested against Gram-positive and Gram-negative bacteria (*B. subtilis*, *S. aureus* MSSA, *S. aureus* MRSA, *Pseudomonas* sp., *P. aeruginosa*, *Flavobacterium* sp., *E. coli*, *Salmonella*) using the disk diffusion and the broth dilution techniques. The ethanol extracts of stems and leaves inhibited growth of Gram-positive as well as Gram-negative bacteria but it showed better antibacterial activity against Gram-positive bacteria than Gram-negative bacteria. The antimicrobial activity of ginja cherry stems and leaves waste is attributed to the presence of valuable compounds like polyphenols in the extract.

14.5.3 Antimicrobial Potential of Agro-Industrial Wastes

14.5.3.1 Extraction of Polyphenolic Compounds from Mango Seed Kernels

Huge quantities of peel and kernel by-products are generated during processing of mango. This waste has high content of phenolic compounds and saturated fatty acids. Mango kernels are good source of phospholipids, phenolic compounds, campesterol, β-sitosterols, stigmasterol, and tocopherols (Soong and Barlow 2004). Mango seed kernel enhances oxidative stability and shelf life of fresh cheese and ghee (Parmar and Sharmar 1990). Good antioxidant property of mango seed kernel is attributed to the presence of polyphenols, sesquiterpenoids, phytosterols, and microelements, such as selenium, copper, and zinc (Schiber et al. 2003). Maisuthisakul (2008) found phenolic compounds like phenolic acids and flavonoids in the extract of Thai mango (*Mangifera indica* Linn.) seed kernels.

14.5.3.2 Extraction of Tannins from Agricultural By-Products

Huge amount of agricultural wastes of tannin-containing plants are generated. Tannins, a polyphenolic compound, showed antimicrobial (Doughari et al. 2008) and antioxidant activities (Zargham and Zargham 2008). Sung et al. (2012) investigated wastes from green tea processing, acorn, chestnut, and persimmon hulls for extraction of tannin and their antibacterial and antioxidant activities. Tannin content in the ethanol, acetone, and aqueous extracts was determined. They found tannin concentrations in the extracts of chestnut hull, green tea waste, acorn hull, and persimmon hull. Tannin concentration for green tea waste was highest in ethanol extracts, whereas for chestnut hull it was highest in ethanol and acetone extracts. Antibacterial activities of various extracts were screened against *Staphylococcus aureus, E. coli, S. flexneri, L. monocytogenes*, and *B. coagulans*. Tannin extracts from green tea waste showed higher antibacterial activity than the extracts of acorn, chestnut, and persimmon hulls.

14.5.3.3 Biological Activity of Jojoba Hull Extracts

Simmondsia chinensis, better known as jojoba, is commercially cultivated in many countries all over the world. It is grown commercially for its seed oil. It is also used as food by many animals. Jojoba seed oil has medicinal properties useful for treatment of cancer, kidney disorder, obesity, sore heart, warts, and wounds (Leung and Foster 1996). Wagdy and Taha (2012) evaluated antimicrobial activity of phenolic extracts of jojoba hulls using different extracting solvents against five bacterial strains *Escherichia coli*, *Staphylococcus aureus*, *Bacillus cereus*, *Listeria monocytogenes*, and *Salmonella typhimurium*. They used five different extracts: methanol, ethanol, acetone, isopropanol, and ethyl acetate. Different extracts of jojoba hulls inhibited growth of the test microorganisms through varying degree. Maximum inhibition of *B. cereus* and *S. typhimurium* was exhibited by ethyl acetate extract. Highest inhibition of *B. cereus* was achieved with ethyl acetate. *S. aureus* was inhibited most by methanol extract. Highest inhibition of *L. monocytogenes* and *E. coli* was obtained with ethanol and acetone extract, respectively. Wagdy and Taha (2012) suggested that jojoba hull is a very promising source of bioactive compounds with very good antioxidant, antimicrobial, and anticancer properties.

14.5.3.4 Antimicrobial Compounds from Olive Oil Mill Waste

Olive fruit is a rich source of phenolic compounds, such as phenol acids, phenol alcohols, catechol, methylcatechol, phenyl alcohols (tyrosol, hydroxytyrosol), flavonoids (luteolin-7-glucoside, apigenin-7-glucoside, rutin and quercetin), and several anthocyanin pigments (cyaniding-3-glucoside and cyaniding-3-rutinoside) (Ragazzi et al. 1973; Andary et al. 1982; Romero et al. 2002). Hydroxytyrosol is one of major phenolic compound present in olive fruit with remarkable pharmacological and antioxidant activity (Visioli et al. 2004; Fernandez et al. 2006). Oleuropein is another valuable compound with certain antiviral, antibacterial, antifungal, antioxidant, and anti-inflammatory properties (Aziz et al. 1998; Visioli and Galli 2002). Phenolic compounds derived from olives have shown antibacterial activities against pathogenic bacteria, such as *Salmonella* spp., *Staphylococcus aureus*, *Clostridium botulinum*, and *Listeria monocytogenes* (Payne et al. 1989; Nychas et al. 1990; Tassou and Nychas 1995).

It is a known fact that olive oil mill waste contains polyphenols in considerable amount. Concentration of polyphenols has been reported to be so strong that phytopathogenic bacteria like *Pseudomonas syringae* and *Corynebacterium michiganense* fail to grow in it (Capasso et al. 1995). Ciafardini and Zullo (2003) studied the effect of polyphenols present in the olive oil mill wastewater (OMWW) on the crucifer seed-borne phytopathogen *Xanthomonas campestris*. They reported that polyphenols in contact with the bacterial cultures react with the protein of the bacterial cell walls and disrupt them. OMWW was able to control the seed-borne phytopathogen *Xanthomonas campestris* completely without damaging the germinability of the crucifer seeds and the metallic greenhouse structures. Therefore, Ciafardini and Zullo (2003) suggested that polyphenols from OMWW are a natural substitute for commercial corrosive chemicals like sodium hypochlorite, which is currently used to disinfect seeds and greenhouses.

14.5.3.5 Other Agrowastes

Martin et al. (2012) assessed the antimicrobial potential and chemical composition of guava bagasse (*Psidium guajava*), Cabernet Sauvignon, Pinot Noir (*Vitis vinifera*) grape marc, Isabella grape marc (*Vitis labrusca*) wastes, Petit Verdot Verdejo grape marcs, Syrah and Verdejo grape stems, Petit Verdot grape seed and red grape fermentation lees (*Vitis vinifera*), tomato bagasse (*Solanum lycopersicum*) wastes, and vegetable wastes, kale (*Brassica oleracea*), beet (*Beta vulgaris*), broccoli (*Brassica oleracea*) and turnip stems (*Brassica rapa*), carrot (*Daucus carota*) and radish leaves (*Raphanus sativus*), pumpkin (*Cucurbita* sp.) and peanut peel, and passion fruit hulls (*Passiflora edulis*) against pathogenic microorganisms of importance in food. Samples were immersed in ethanol and methanol solutions to prepare extract. Beet stalk, peanut peel, Pinot Noir grape marc, Petit Verdot grape seed and marc, red grape fermentation lees, and guava bagasse wastes showed antimicrobial activities against Gram-positive bacteria *Staphylococcus aureus* and *Listeria monocytogenes*. Methanol extract of peanut peels and ethanol extract of guava bagasse extracts showed the lowest MIC against *S. aureus* and *L. monocytogenes*. Guava bagasse extract showed the highest antimicrobial activity against *L. monocytogenes*.

14.5.4 Utilization of Agrowaste as Substrate

Agro-industrial wastes, such as wheat bran, rice husk, corn bran fenugreek straw, corn bran, and sugarcane bagasse, can be used as a substrate for production of important antimicrobial agents using various fermentation techniques (Negi and Banerjee 2009) using suitable microbes. This can make the antimicrobial compound very economical and eco-friendly. Microorganisms like endophytic microbes are rich source of novel natural compounds, and these compounds can be used for production of antimicrobial compounds using agrowaste as suitable substrate.

14.5.4.1 Sugarcane Molasses as the Source of Antimicrobial Products

Sugarcane molasses is the waste generated during sugar production and crystallization from sugarcane. Generally, it is used as a source for animal feed or biomass for ethanol production. Molasses holds the antimicrobial properties because of phenolic compounds present in it. Sugarcane molasses show strong antioxidative and tyrosinase inhibitory activities (Nakasone et al. 1996; Takara et al. 2002, 2003, 2007). Takara et al. (2007) tested the bioactive compounds present in the sugarcane molasses against two pathogenic bacteria *S. mutans* and *S. sobrinus* responsible for dental caries in human and animals.

14.5.4.2 Pine Needles as the Source of Antimicrobial Products

Chir pines, or *Pinus roxburghii*, are pines growing wild in the Himalayan range. *P. roxburghii* contains a number of phytochemicals like alkaloids, glycosides, flavonoids, saponins, tannins, and terpenoids in needles, female cones, and bark. Antibacterial and antifungal activities of essential oils of *Pinus roxburghii* stems were studied by Hassan and Amjid (2009). Major components in essential oil were α-pinene (41.9 %) followed by 3-carene (16.3 %), caryophyllene (12.3 %), p-cymene (1.9 %), terpineol (1.8 %), limonene (1.7 %), borneol acetate (1.1 %), caryophyllene oxide (1.0 %), camphene (0.9 %), and terpinyl acetate (0.8 %). Antibacterial activity of stem essential oil was observed against *Staphylococcus aureus* and *Bacillus subtilis*, while no activity was observed against *E. coli* and *Enterobacter aerogenes*. Similarly, antifungal activity of *Pinus roxburghii* essential oil was found inhibiting *Aspergillus terreus*, *Aspergillus flavus*, *Aspergillus candidus*, *Aspergillus versicolor*, *Aspergillus niger*, and *Trichoderma viride*.

14.5.4.3 Sugarcane Bagasse as the Source of Antimicrobial Products

Sugarcane bagasse is a good source for the production of coumaric acid. Ou et al. (2009) obtained coumaric acid from the sugarcane bagasse. The purified product has shown antimicrobial, anti-inflammatory, antioxidant, and free radical scavenging activities like ferulic acid.

14.5.4.4 Groundnut Shells as the Source of Antimicrobial Products

Oxytetracycline is one of the most important antibiotics produced from streptomyces, which is considered as the largest antibiotic-producing genus (Walve et al. 2001). Four agricultural wastes, namely, groundnut shells, corncob, corn pomace, and cassava peels, were used by Asagbra et al. (2005) as the substrate for the growth of *Streptomyces* spp. in solid-state fermentation. Oxytetracycline production started within three days and reached its peak on the sixth day with groundnut shells as substrate.

14.5.4.5 Production of Oxytetracycline from Cocoyam Peels

Ndubuisi Ezejiofor et al. (2012) used cocoyam peels (household kitchen waste) for production of oxytetracycline by *Streptomyces speibonae* OXS1 in solid-state fermentation. They used cocoyam peels of species *Colocasia esculenta* and *Xanthosoma esculenta* as substrate. Oxytetracycline production started on the first day of fermentation and reached its peak on third day. Higher biomass weight (140.86 g) was found for *C. esculenta* than *X. esculenta* (101.62 g) after seven days fermentation indicating higher presence of oxytetracycline in the fermented *C. esculenta*. Oxytetracycline was present in both species of the cocoyam peels.

14.6 Recent Advances

Advances in biology and chemistry along with powerful screening and isolation techniques have given the field of natural antimicrobials a big thrust. With the use of latest technologies with particular emphasis on the total synthesis and analog design, it is now possible to determine the mechanism of action of a newly discovered antibiotic and resistance mechanism of bacteria. New techniques and technologies not only help in exploration of new antimicrobial compounds but also ensure their intelligent application. Nanotechnology is one of the important emerging fields that can significantly improve the exploitation and application of antimicrobial compounds. In the recent past, there has been significant research to discover nanoparticles with novel physical, chemical, and biological properties. Nanotechnology is being used for delivery of antimicrobial phenolic compound extracts to effectively inhibit food-borne pathogens. Effectiveness of the antimicrobial compound with nanoparticle delivery compared with conventional delivery system has improved significantly. By using nanoparticle delivery system, phenolic compounds can be used as natural and better replacement of chemicals to control pathogens for commercial food safety applications.

Various nanomolecules, such as silver nanoparticles, carbon nanotubes, magnesium oxide nanoparticles, and zinc oxide nanoparticles, are being used very effectively in tackling human and food pathogens. Silver nanoparticles are nontoxic to the human body at low concentrations but have broad-spectrum antibacterial actions (Lara et al. 2010). Silver particles have strong antimicrobial effect against many drug-resistant organism such as *Staphylococcus aureus* (MRSA), *Pseudomonas aeruginosa*, *E. coli*, and *S. pyogenes*. The nano-silver materials have immense potential application due to the strong antimicrobial activity of silver against a broad spectrum of bacteria, viruses, and fungi and the low frequency of development of resistance (Egger et al. 2009).

Similarly, carbon nanotubes have applications in drug delivery and as components in medical nanodevices. They also show strong antibacterial activity like silver nanoparticles. Direct interaction of the carbon nanotubes with the bacterial cell membranes causing significant membrane damage is reported to be the reason for their bactericidal effect. Magnesium oxide nanoparticles and zinc oxide nanoparticles also have antimicrobial properties and being explored for therapeutic applications (Matthews et al. 2010).

Ravichandran et al. (2011) demonstrated the efficacy of phenolics on pathogen reduction delivered by nanoparticles by studying their effects on *Listeria monocytogenes* (*L.m.*), *Escherichia coli* O157:H7 (*E.c.*), and *Salmonella typhimurium* (*S.t.*) in brain–heart infusion broth (BHI) and meat system. Encapsulation of benzoic acid in polylactic-co-glycolic acid nanoparticles inhibited both the pathogens much better than the delivery without nanoparticles. With the advent of new nanoparticles and technology, target-specific delivery of therapeutic agents using nanocarriers has become a very promising field. Various delivery vehicles have been designed based on different nanomaterials, such as polymers, dendrimers, liposomes, nanotubes, and nanorods (Matthews et al. 2010).

14.7 Conclusion

Despite numerous advances in the field of antibiotics, bacterial and fungal infections are still major concern for pharmaceutical as well as food industry. This gets aggravated with the presence of multidrug-resistant strains. Recent advances in biology, chemistry, nanotechnology, and other associated fields, though, have improved the discoveries of novel classes of antibiotics from natural sources. Large numbers of studies are being carried out to explore the various possibilities of utilizing the plant and agro-industrial waste for production of value-added products including antimicrobial agents. There are unlimited possibilities in the field of waste utilization, which requires more push from the industry to take the benefits of the researches in the laboratories to the common man through industry. Utilization of plant and agrowaste for production of antimicrobial agents may significantly reduce the cost of medicines and food preservatives, thereby improving the health and food scenario in the society and at the same time can help to improve the environment. Although extensive research in the field of plants and agro-based antimicrobials has led to the discovery of many new natural antimicrobial agents by pharmaceutical industry, still the area of reusing the plant and agro-industrial by-products or wastes for extraction or production of antimicrobial agents needs more support from industries.

References

Ahmad MM, Salim-ur-Rehman Z, Iqbal-Anjum FM, Sultan JI (2006) Genetic variability to essential oil composition in four citrus fruit species. Pak J Bot 38(2):319–324

Alberto MR, Canavosio MAR, Manca de Nadra MC (2006) Effect of polyphenols from apple skins on human bacterial pathogens. Electron J Biotechnol 9(3)

Amelia P, Guevara AA, Hiromu S, Mutsou K, Harukuni T (1996) Anti-inflammatory, anti-mutagenicity and antitumor-promoting activities of mahogany seeds, *Swietenia macrophylla* (Meliaceae). Philip J Sci 125(4):271–278

Anastasiadi M, Chorianopoulos NG, Nychas GJE, Haroutounian SA (2008) Antilisterial activities of polyphenol-rich extracts of grapes and vinification byproducts. J Agric Food Chem 57:457–463

Andary C, Wylde R, Laffite C, Privat G, Winternitz F (1982) Structure of verbascoside and orobancoside, caffeic acid, sugar esters from *Orobanche rapum-genistae*. Phytochemistry 21:1123–1127

Artizzu N, Bonsignore L, Cottiglia F, Loy G (1995) Studies of the diuretic and antimicrobial activity of Cynodon dactylon essential oil. Fitoterapia 66:174–175

Asagbra AE, Oyewole OB, Odunfa SA (2005) Production of oxytetracycline from agricultural wastes using *streptomyces* spp. Niger Food J 2:174–181

Ashok K, Narayani M, Subanthini A, Jayakumar M (2011) Antimicrobial activity and phytochemical analysis of citrus fruit peels—utilization of fruit waste. Int J Eng Sci Technol 3(6):5414–5421

Aziz NH, Farag SE, Mousa LAA, Abo-Zaidt MA (1998) Comparative antibacterial and antifungal effects of some phenolic compounds. Microbios 93:43–54

Baydar NG, Ozkan G, Sagdic O (2004) Total phenolic contents and antibacterial activities of grapes (*Vitis vinifera* L.) extracts. Food Control 15(5):335–339

Beheshti A, Norouzi P, Ganjali MR (2012) A simple and robust model for predicting the reduction potential of quinones family electrophilicity index effect. Int J Electrochem Sci 7:4811–4821

Beretz A, Cazenave J (1988) The effect of flavonoids on blood vessel wall interactions. In: Cody V, Middleton E, Harborne JB, Beretz A (eds) Plant flavonoids in biology and medicine II: biochemical, cellular and medicinal properties. Alan R Liss, New York, pp 187–200

Bisignano G, Sanogo R, Marino A, Aquino R, D'angelo V, Germanò MP, De Pasquale R, Pizza C (2000) Antimicrobial activity of *Mitracarpus scaber* extract and isolated constituents. Lett Appl Microbiol 30:105–108

Borris RP (1996) Natural products research: perspectives from a major pharmaceutical company. J Ethnopharmacol 51:29–38

Caccioni DR, Guizzardi M, Biondi DM, Renda A, Ruberto G (1998) Relationship between volatile components of citrus fruit essential oils and antimicrobial action on *Penicillium digitatum* and *Penicillium italicum*. Int J Food Microbiol 43:73–79

Calvo J, Martinez ML (2009) Antimicrobial mechanisms of action. Enferm Infecc Microbiol Clin 27(1):44–52

Canadanovic BJM, Savatovic SS, Cetkovic GS, Vulic JJ, Djilas SM, Markov SL, Cvetkovic DD (2011) Antioxidant and antimicrobial activities of beet root pomace extracts. Czech J Food Sci 29(6):575–585

Capasso R, Evidente A, Schivo L, Orru G, Marcialis MA, Cristinzio G (1995) Antibacterial polyphenols from olive oil mill waste waters. J Appl Bacteriol 79:393–398

Carpenter CF, Chambers HF (2004) Daptomycin: another novel agent for treating infections due to drug-resistant gram-positive pathogens. Clin Infect Dis 38:994–1000

Chanda S, Baravalia Y, Kaneria M, Rakholiya K (2010) Fruit and vegetable peels — strong natural source of antimicrobics. In: Mendes-Vilas A (ed) Current research, technology and education topics in applied microbiology and applied biotechnology. Publisher: Formatex Research Center. Vol. 1, pp 444–450

Chanthaphon S, Chanthachum S, Hongpattarakere T (2008) Antimicrobial activities of essential oils and crude extracts from tropical *Citrus* spp. against food-related microorganisms. Songklanakarin J Sci Technol 30(1):125–131

Ciafardini G, Zullo BA (2003) Antimicrobial activity of oil-mill waste water polyphenols on the phytopathogen *Xanthomonas campestris* spp. Ann Microbiol 53(3):283–290

Cowan MM (1999) Plant products as antimicrobial agents. Clin Microbiol Rev 12:564–582

Cushnie TPT, Lamb AJ (2005) Antimicrobial activity of flavonoids. Int J Antimicro Ag 26:343–356

Davidson PM, Branen AL (2005) Food antimicrobials—an introduction. In: Davidson PM, Sofos JN, Branen AL (eds) Antimicrobials in food. CRC Press, Boca Raton, FL, pp 1–10

Doughari JH, El-mahmood AM, Tyoyina I (2008) Antimicrobial activity of leaf extracts of *Senna obtusifolia* (L). Afr J Pharm Pharmacol 2:7–13

Dzidic S, Suskovic J, Kos B (2008) Antibiotic resistance mechanisms in bacteria: biochemical and genetic aspects. Antibiotic resistance in bacteria. Food Technol Biotechnol 46(1):11–21

Egger S, Lehmann RP, Height MJ, Loessner MJ, Schuppler M (2009) Antimicrobial properties of a novel silver-silica nanocomposite material. Appl Environ Microbiol 75(9):2973–2976

Ekwenye UN, Edeha OV (2010) The antibacterial activity of crude leaf extract of *Citrus sinensis* (sweet orange). Int J Pharm Bio Sci 1(4):742–750

El-Massry KF, El-Ghorab AH, Shaaban HA, Shibamoto T (2009) Chemical compositions and antioxidant/antimicrobial activities of various samples prepared from *Schinus terebinthifolius* leaves cultivated in Egypt. J Agric Food Chem 57:5265–5270

Escarpa A, Gonzalez MC (1998) High-performance liquid chromatography with diode-array detection for the determination of phenolic compounds in peel and pulp from different apple varieties. J Chromatogr 823(1–2):331–337

Fathiazada F, Delazara A, Amiria R, Sarker SD (2006) Extraction of flavonoids and quantification of rutin from waste tobacco leaves. Iran J Pharm Res 3:222–227

Fernandez BJ, Rodríguez G, Rodríguez R, Guillen R, Jimenez A (2006) Potential use of olive by products; extraction of interesting organic compounds from olive oil waste. Grasas Y Aceites 57(1):95–106, Enero-Marzo

Gedir JV, Sporns P, Hudson RJ (2005) Extraction of condensed tannins from cervid feed and feces and quantification using a radial diffusion assay. J Chem Ecol 31:2761–2773

Gollnick H, Schramm M (1998) Topical therapy in acne. J Eur Acad Dermatol Venereol 11:8–12

Grinberg LN, Rachmilewitz EA, Newmark H (1994) Protective effects of rutin against hemoglobin oxidation. Biochem Pharmacol 48:643–649

Hamendra SP, Anand K (2007) Antidiabetic potential of *Citrus sinensis* and *Punica granatum* peel extracts in alloxan treated male mice. Bio Factors 31:17–24

Harborne JB, Baxter H (1999) The handbook of natural flavonoids, Vol:1 and 2. John Wiley and Sons, Chichester

Harborne JB, Williams CA (2000) Advances in flavonoid research since 1992. Phytochemistry 55:481–504

Hartmann R, Meisel H (2007) Food-derived peptides with biological activity: from research to food applications. Curr Opin Biotechnol 18:163–169

Hassan A, Amjid I (2009) Gas chromatography–mass spectrometric studies of essential oil of Pinus roxburghii stems and their antibacterial and antifungal activities. J Med Plants Res 3(9):670–673

Havsteen B (1983) Flavonoids, a class of natural products of high pharmacological potency. Biochem Pharmacol 32:1141–1148

Izzo AA, Di Carlo G, Biscardi D, Fusco R, Mascolo N, Borreli F, Capasso F, Fasulo MP, Autore G (1995) Biological screening of Italian medicinal plants for antibacterial activity. Phytother Res 9:281–286

Jayaprakasha GK, Selvi T, Sakaria KK (2003) Antibacterial and antioxidant activities of grape (*Vitis vinifera*) seed extracts. Food Res Int 36:117–122

Kahkonen MP, Hopia AI, Vuorela HJ, Rauha JP, Pihlaja K, Kujala TS, Heinonen M (1999) Antioxidant activity of plant extracts containing phenolic compounds. J Agr Food Chem 47:3954–3962

Kanatt SR, Chander R, Sharma A (2010) Antioxidant and antimicrobial activity of pomegranate peel extract improves the shelf life of chicken products. Int J Food Sci Technol 45(2):216–222

Kanaze FI, Termentzi A, Gabrieli C, Niopas I, Georgarakis M, Kokkalou E (2008) The phytochemical analysis and antioxidant activity assessment of orange peel (*Citrus sinensis*) cultivated in Greece-Crete indicates a new commercial source of hesperidin. Biomed Chromatogr 23:239–249

Kayser O, Kolodziej H (1997) Antibacterial activity of extracts and constituents of *Pelargonium sidoides* and *Pelargonium reniforme*. Planta Med 63:508–510

Krisch J, Galgoczy L, Papp T, Csaba Vagvolgyi C (2009) Antimicrobial and antioxidant potential of waste products remaining after juice pressing. J Eng tome vii (year. Fascicule 4):131–134

Kubo L, Muroi H, Himejima M (1993) Structure–antibacterial activity relationships of anacardic acids. J Agri Food Chem 41:1016–1019

Kujala T, Loponen J, Pihlaja K (2001) Betalains and phenolics in red beetroot (Beta vulgaris) peel extracts: extraction and characterisation. Zeitschrift fur Naturforschung- C 56:343–348

Kumamoto H, Matsubara Y, Iizuka Y, Okamoto K, Yokoi K (1986) Structure and hypotensive effect of flavonoid glycosides in orange (*Citrus sinensis* OSBECK) peelings. Agricult Biol Chem 50:781–783

Kwon YS, Kobayashi A, Kajiyama SI, Kawazu K, Kanzaki H, Kim CM (1997) Antimicrobial constituents of *Angelica dahurica* roots. Phytochemistry 44(5):887–889

Lanciotti R, Gianotti A, Patrignani F, Belletti N, Guerzoni EM, Gardini F (2004) Use of natural aroma compounds to improve shelf-life and safety of minimally processed fruits. Trends Food Sci Tech 15:201–208

Lara HH, Ayala-Nuñez NV, Ixtepan-Turrent L, Rodriguez-Padilla C (2010) Bactericidal effect of silver nanoparticles against multidrug-resistant bacteria. World J Microbiol Biotechnol 26:615–621

Leung AY, Foster S (1996) Encyclopaedia of common natural ingredients, 2nd edn. Wiley, New York

Maisuthisakul P (2008) Antiradical scavenging activity and polyphenolic compounds extracted from Thai mango seed kernels. As J Food Food Ag-Ind 1(02):87–96

Mangas JJ, Rodriguez R, Suarez B, Picinelli A, Dapena E (1999) Study of the phenolic profile of cider apple cultivars at maturity by multivariate techniques. J Agr Food Chem 47(10):4046–4052

Manthey A, Grohmann K (2001) Phenols in citrus peel byproducts: concentrations of hydroxycinnamates and polymethoxylated flavones in citrus peel molasses. J Agr Food Chem 49:3268–3273

Martin JGP, Porto E, Correa GB, Matias de Alencar S, Micotti da Gloria E, Cabral ISR, Maria L, de Aquino L (2012) Antimicrobial potential and chemical composition of agro-industrial wastes. J Nat Prod 5:27–36

Matthews L, Kanwar RK, Zhou S, Punj V, Kanwar JR (2010) Applications of nanomedicine in antibacterial medical therapeutics and diagnostics. Open Trop Med J 3:1–9

McManus MC (1997) Mechanisms of bacterial resistance to antimicrobial agents. Am J Health Syst Pharm 54:1420–1433

Melendez PA, Capriles VA (2006) Antibacterial properties of tropical plants from Puerto Rico. Phytomedicine 13:272–276

Mendoza L, Wilkens M, Urzua A (1997) Antimicrobial study of the resinous exudates and of diterpenoids and flavonoids isolated from some Chilean *Pseudognaphalium* (Asteraceae). J Ethnopharmacol 58:85–88

Mirunalini S, Krishnaveni M (2011) Coumarin: a plant derived polyphenol with wide biomedical applications. Int J PharmTech Res 3(3):1693–1696

Moerman DE (1996) An analysis of the food plants and drug plants of native North America. J Ethnopharmacol 52:1–22

Monga PK, Sharma D, Dubey A (2012) Comparative study of microwave and conventional synthesis and pharmacological activity of coumarins: a review. J Chem Pharm Res 4(1):822–850

Nakasone N, Takara K, Wada K, Tanaka J, Yogi J, Nakatani N (1996) Antioxidative compounds isolated from kokuto, non centrifugal cane sugar. Biosci Biotechnol Biochem 60:1714–1716

Nascimento GGF, Locatelli J, Freitas PC, Silva GL (2000) Antibacterial activity of plant extracts and phytochemicals on antibiotic resistant bacteria. Braz J Microbiol 31:247–256

Ndubuisi Ezejiofor TI, Duru CI, Asagbra AE, Ezejiofor AN, Oisakwe OE, Afonne JO, Obi E (2012) Waste to wealth production of oxytetracycline using streptomyces species from household kitchen wastes of agricultural produce. Afr J Biotechnol 11(43):10115–10124

Negi S, Banerjee R (2009) Optimization of extraction and purification of glucoamylase produced by *Aspergillus awamori* in solid-state fermentation. Biotechnol Bioproc Eng 14:60–66

Nikaido H, Zgurskaya HI (1999) Antibiotic efflux mechanisms. Curr Opin Infect Dis 12:529–536

Nychas GJE, Tassou CC, Board RG (1990) Phenolic extract from olives inhibition of *Staphylococcus aureus*. Lett Appl Microbiol 10:217–220

Ou SY, Luo YL, Huang CH, Jackson M (2009) Production of coumaric acid from sugarcane bagasse. Innov Food Sci Emerg 10:253–259

Parmar SS, Sharmar RS (1990) Effect of mango (*Mangifera indica* L.) seed kernel pre-extract on the oxidative stability of buffalo ghee. Food Chem 35:99–107

Payne K, Rico-Munoz E, Davidson PM (1989) The antimicrobial activity of phenolic compounds against *Listeria monocytogenes* and their effectiveness in a model milk system. J Food Protect 52:151–153

Pedraza CJ, Cardenas-Rodriguez N, Orozco-Ibarra M, Perez-Rojas JM (2008) Medicinal properties of mangosteen (Garcinia mangostana). Food Chem Toxicol 46:3227–3239

Pereira JA, Oliveira I, Sousa A, Valentão P, Andrade PB, Ferreira IC, Ferreres F, Bento A, Seabra R, Estevinho L (2007) Walnut (Juglans regia L.) leaves: phenolic compounds, antibacterial activity and antioxidant potential of different cultivars. Food Chem Toxicol 45(11):2287–2295

Piccirillo C, Demiray S, Franco AR, Castro PML, Pintado ME (2010) High added-value compounds with antibacterial properties from ginja cherries by-products. Waste Biomass Valorization 1(2):209–217

Podsedek A, Wilska Jeska J, Anders B, Markowski J (2000) Compositional characterization of some apple varieties. Eur Food Res Technol 210(4):268–272

Pretorius JC, Magama S, Zietsman PC (2003) Growth inhibition of plant pathogenic bacteria and fungi by extracts from selected South African plant species. S Afr J Bot 20:188–192

Ragazzi E, Veronese G, Guitto A (1973) Demethyloleuropein, a new glucoside extracted from ripe olives. Ann Chim 63:13–20

Ravichandran M, Hettiarachchy NS, Ganesh V, Ricke SC, Singh S (2011) Enhancement of antimicrobial activities of naturally occurring phenolic compounds by nanoscale delivery against *listeria monocytogenes*, *escherichia coli* O157:H7 and *salmonella* typhimurium in broth and chicken meat system. J Food Safety 31(4):462–471

Roberts MC (2003) Acquired tetracycline and/or macrolide–lincosamides–streptogramin resistance in anaerobes. Anaerobe 9(2):63–69

Romero C, Garcia P, Brenes M, Garcia A, Garrido A (2002) Phenolic compounds in natural black Spanish olive varieties. Eur Food Res Technol 215:482–496

Saleem M, Nazir M, Ali MS, Hussain H, Lee YS, Riaza N, Jabbara A (2010) Antimicrobial natural products: an update on future antibiotic drug candidates. Nat Prod Rep 27:238–254

Santos JDG, Branco A, Silva AF, Pinheiro CSR, Neto AG, Uetanabaro APT, Queiroz SROD, Osuna JTA (2009) Antimicrobial activity of *Agave sisalana*. Afr J Biotechnol 8(22):6181–6184

Schiber A, Berardini N, Carle R (2003) Identification of flavonol and xanthol glycosides from mango peels by HPLC. J Agr Food Chem 51:5006–5011

Shoji T, Akazome Y, Kanda T, Ikeda M (2004) The toxicology and safety of apple polyphenol extract. Food Chem Toxicol 42:959–967

Shoji T, Mutsuga M, Nakamura T, Kanda T, Akiyama H, Goda Y (2003) Isolation and structural elucidation of some procyanidins from apple by low-temperature NMR. J Agr Food Chem 51(13):3806–3813

Singh SP, Negi S (1992a) Antibacterial and antifungal activities of *Mentha arvensis* essential oil. Fitoterapia 63(1):76–78

Singh SP, Negi S (1992b) Antibacterial properties of essential oils from *Zingiber chrysanthum* leaves and rhizomes. Fitoterapia 63(1):73–75

Singleton VL, Rossi JA Jr (1965) Colorimetry of total phenolics with phosphomolybdic phosphotungstic acid reagents. Am J Enol Viticult 16(1):144–158

Soediro I, Padmawinata K, Wattimena JR, Rekita S (1990) Study of the active antimalarial methanolic extract of *Swietenia macrophylla* King (Meliaceae). Acta Pharm Indones 15(1):1–13

Soong Y, Barlow P (2004) Antioxidant activity and phenolic content of selected fruit seeds. Food Chem 88:411–417

Spratt BG (1994) Resistance to antibiotics mediated by target alterations. Science 264:388–393

Suksamrarn S, Komutiban O, Ratananukul P, Chimnoi N, Lartpornmatulee N, Suksamrarn A (2006) Cytotoxic prenylated xanthones from the young fruit of Garcinia mangostana. Chem Pharm Bull 54:301–305

Sung SH, Kim KH, Jeon BT, Cheong SH, Park JH, Kim DH, Kweon HJ, Moon SH (2012) Antibacterial and antioxidant activities of tannins extracted from agricultural by-products. J Med Plants Res 6(15):3072–3079

Takara K, Matsui D, Wada k, Ichiba T, Chinen L, Nakasone Y (2003) New phenolic compounds from kokuto non centrifuged cane sugar. Biosci Biotechnol Biochem 67:376–379

Takara K, Matsui D, Wada K, Ichiba T, Nakasone Y (2002) New anti-oxidative phenolic glycosides isolated from kokuto non centrifuged cane sugar. Biosci Biotechnol Biochem 60:1714–1716

Takara K, Ushijima K, Wada K, Iwasaki H, Yamashita M (2007) Phenolic compounds from sugarcane molasses possessing antibacterial activity against cariogenic bacteria. J Oleo Sci 56(11):611–614

Tan SK, Osman H, Wong KC, Boey PL, Ibrahim P (2009) Antimicrobial and antioxidant activities of Swietenia macrophylla leaf extracts. As J Food Ag-Ind 2(02):181–188

Tanner GJ, Moate PJ, Davis LH, Laby RH, Yuguang L, Larkin PJ (1995) Proanthocyanidins (condensed tannins) destabilise plant protein foams in a dose-dependent manner. Aust J Agric Res 46:1101–1109

Tassou CC, Nychas GJE (1995) Inhibition of *Salmonella enteritidis* by oleuropein in broth and in a model food system. Lett Appl Microbiol 20:120–124

Tehranifara A, Selahvarzia Y, Kharrazia M, Bakhshb VJ (2011) High potential of agro-industrial by-products of pomegranate (*Punica granatum* L.) as the powerful antifungal and antioxidant substances. Ind Crop Prod 34(3):1523–1527

Tenover FC (2006) Mechanisms of antimicrobial resistance in bacteria. Am J Med 119(6A):S3–S10

Vander SAA, Dekker M, De Jager A, Jongen WMF (2001) Activity and concentration of polyphenolic antioxidants in apple: effect of cultivar, harvest year, and storage conditions. J Agr Food Chem 49(8):3606–3613

Vinson JA, Hao Y, Su X, Zubik L (1998) Phenol antioxidant quantity and quality in foods: vegetables. J Agr Food Chem 46:3630–3634
VishnuPriya V, Mallika J, Surapaneni KM, Saraswathi P, Chandra SGSV (2010) Antimicrobial activity of pericarp extract of *garcinia mangostana linn.* Int J Pharm Sci Res 1(8):278–281
Visioli F, Galli C (2002) Biological properties of olive oil phytochemicals. Crit Rev Food Sci Nut 42:209–221
Visioli F, Grande S, Bogani P, Galli C (2004) The role of antioxidants in the Mediterranean diets: focus on cancer. Eur J Cancer Prev 13:337–343
Wagdy SM, Taha FS (2012) primary assessment of the biological activity of jojoba hull extracts. Life Sci J 9(2):244–253
Walve MG, Trekoo R, Log MM, Bhole BD (2001) How many antibiotics are produced by the genus streptomyces. Arch Microbiol 177(5):86–90
Wannissorna B, Jarikasemb S, Siriwangchaib T, Thubthimthed S (2005) Antibacterial properties of essential oils from Thai medicinal plants. Fitoterapia 76:233–236
Webber MA, Piddock LJ (2003) The importance of efflux pumps in bacterial antibiotic resistance. J Antimicrob Chemother 51:9–11
WHO (2012) Antimicrobial resistance. http://www.who.int/mediacentre/factsheets/fs194/en/index.html. Accessed 10 Aug 2012
Will F, Bauckhage K, Dietrich H (2000) Apple pomace liquefaction with pectinases and cellulases: analytical data of the corresponding juices. Eur Food Res Technol 211:291–297
Woodford N, Ellington MJ (2007) The emergence of antibiotic resistance by mutation. Eur J Clin Microbiol Infect Dis 13:5–18
Wright GD (2005) Bacterial resistance to antibiotics: enzymatic degradation and modification. Adv Drug Deliver Rev 57:451–1470
Yao XJ, Wainberg MA, Parniak MA (1992) Mechanism of inhibition of HIV-1 infection in vitro by purified extract of *Prunella vulgaris*. Virology 178:56–62
Zargham H, Zargham R (2008) Tannin extracted from Sumac inhibits vascular smooth muscle cell migration. Mcgill J Med 11:119–123
Zasloff M (2002) Antimicrobial peptides of multicellular organisms. Nature 415(6870):389–395

Chapter 15
Pharmaceutical Enzymes

Deeplina Das and Arun Goyal

15.1 Introduction

Enzymes are biomolecules that catalyze and accelerate chemical reactions. Almost all processes in a biological cell require enzymes in order to occur at desired place. Enzymes are extremely selective for their substrates, and their activity is regulated by factors like substrate concentration, pH, and temperature. Enzyme activity is determined by the quantity of substrate transformed or product formed per unit time. The reaction rate depends on a number of experimental conditions, such as temperature, pH, ionic strength, and the presence or absence of inhibitors or activators. All these attributes make the enzymes important therapeutic tools offering a diversification platform to pharmaceutical industry. For example, adenosine deaminase is highly specific towards its substrate and can be potentially used to treat severe combined immunodeficiency disease (SCID) (Aiuti 2002). Nowadays, the application of enzyme technologies to pharmaceutical research, development, and manufacturing is a growing field. Unlike common medicinal products which can temporarily solve the particular health problems, pharmaceutical enzymes address the underlying cause of health problem and the patient can achieve permanent relief.

Although pancreatic enzymes and pepsin and papain were vogue for therapeutic use even before 1940, the concept of the therapeutic enzymes is only about 40 years old. A therapeutic enzyme was described as part of replacement therapies for genetic deficiencies in the 1960s by de Duve (de Duve 1996). The industrial enzymes are required in bulk and absolute purity is not essential, whereas pharmaceutical enzymes are required in small quantities but in absolutely pure

D. Das • A. Goyal (✉)
Department of Biotechnology, Indian Institute of Technology Guwahati, Guwahati 781 039, Assam, India
e-mail: arungoyl@iitg.ernet.in

S.K. Brar et al. (eds.), *Biotransformation of Waste Biomass into High Value Biochemicals*, DOI 10.1007/978-1-4614-8005-1_15,

form. Enzymes and enzyme-generated products are administered to patients in very small dose to avoid possible side effects (Vellard 2003). The recombinant enzyme drug, Activase (recombinant human tissue plasminogen activator), used for removal of blockage of a coronary artery by a clot, was approved by the Food and Drug Administration (FDA), USA, in 1987 (Vellard 2003). After insulin in 1982, this was the second recombinant protein drug to be marketed. Adagen (pegademase bovine) was the first therapeutic enzyme considered as orphan drug approved by the FDA, USA, in 1990 under the Orphan Drug Act (Vellard 2003). The drug was particularly used for the treatment of severe combined immunodeficiency disease (SCID), which is caused by the chronic deficiency of ADA (adenosine deaminase) (Hershfield et al. 1993). The Orphan Drug Act was passed in 1983 in the United States to encourage pharmaceutical companies to develop treatments for rare medical diseases affecting only small number of people (<2,00,000). Under this act, companies that develop such a drug may sell it without competition for 7 years and may get clinical trial tax incentives. The orphan drugs can be defined as those drugs intended to treat either a rare disease or a more common disease where manufacturer cannot expect to make profits. The orphan diseases are often so rare that a physician may observe only one case a year or less (Sharma et al. 2010).

15.2 Structure of Enzyme

Enzymes have a complex globular three-dimensional structure and their activity depends upon their three-dimensional structure. A part of enzyme comprising 3–4 amino acids is called active site which is involved in substrate binding and catalysis. The shape and chemical environment of the active site facilitates the enzyme reaction. The enzymes are very specific to substrates because both enzyme and the substrate possess specific complementary geometric shapes that fit exactly into one another. Some enzymes do not need any additional components to show full activity. However, others require nonprotein molecules called cofactors for their activity. Cofactors may be inorganic ions such as Fe^{2+}, Mg^{2+}, Mn^{2+}, and Zn^{2+} and organic or metallo-organic molecules. Cofactors are classified depending upon how tightly they bind to an enzyme. Dissociable (loosely bound) cofactors are called as coenzymes and non-dissociable (tightly bound) cofactors as prosthetic groups. The inactive enzyme, without the cofactor, is called an apoenzyme, while the complete enzyme with cofactor is called holoenzyme.

15.3 Nomenclature and Classification of Enzymes

Enzymes are generally classified into six groups on the basis of the type of reactions they catalyze. Since the nomenclature was determined by the Enzyme Commission in 1961 updated in 1992, all enzymes have been assigned an "EC" number:

(a) EC 1 *Oxidoreductases*: catalyze oxidation/reduction reactions.
(b) EC 2 *Transferases*: transfer a functional group (e.g., a methyl or phosphate group).
(c) EC 3 *Hydrolases*: catalyze the hydrolysis of various bonds.
(d) EC 4 *Lyases*: cleave various bonds by means other than hydrolysis and oxidation.
(e) EC 5 *Isomerases*: catalyze isomerization changes within a single molecule.
(f) EC 6 *Ligases*: join two molecules with covalent bonds.

More than 3,000 human enzymes have been identified and named. Except for some of the originally studied enzymes like pepsin, rennin, and trypsin, most enzymes have a suffix "ase" which is attached to the name of substrate on which it acts like lactase as the enzyme breakdown disaccharide lactose to monosaccharide glucose. In some cases the type of reaction is also added to the name like DNA polymerases synthesizes the assembly of the DNA. Most of the pharmaceutical enzymes belong to the EC 3 hydrolase group, such as galsulfase. However, some pharmaceutical enzymes, such as rasburicase, belong to the oxidoreductase group.

15.4 Biopharmaceutical Enzymes

The enzymes used as drugs have two important features which distinguish them from the conventional drugs, such as:

1. Unlike drugs they bind and act on their targets with great affinity.
2. They are highly specific and act as catalyst to convert multiple target molecules to the desired products.

These two features make enzymes specific and potent drugs that can accomplish therapeutic biochemistry in the body that small molecule (synthetic active ingredient) cannot. The catalytic activity of enzymes is exploited in industrial manufacturing of drugs. Enzymes are also used as digestive aid where they are used to supplement digestive enzymes like amylase, lipase, and protease. Almost all enzyme therapies developed till date deal with the genetic disorders. Also the enzyme replacement therapy is used for relatively rare, inborn error of metabolism (Germain 2002; Chan et al. 2005). Several enzymes are also used to prevent and treat common diseases like heart attack and stroke (Longstaff et al. 2008). The enzyme collagenase has been reported for healing burn wound in children and enzyme chondroitinase ABC in the treatment of spinal cord injury (Ozcan et al. 2002; Bradbury et al. 2002). The enzymes used in pharmaceutical industry are listed in Table 15.1. Several enzymes are also used as prodrug, a drug that is administered in an inactive or significantly less active form, but once administered it is metabolized in vivo into an active metabolite through bioactivation process.

Table 15.1 Biopharmaceutical enzymes

Generic name	Treatment	Brand name(s)	Manufacturer name	Mode of action	Mode of administration	Drug Bank accession no
Asparaginase	Acute lymphoblastic leukemia	Elspar	Merck & Co., Inc., USA	Converts asparagine to aspartic acid and ammonia and facilitates production of oxaloacetate	Intravenous	DB00023
Agalsidase	Anderson-Fabry disease	Fabrazyme Replagal	Genzyme Inc. Shire Inc.	Catalyzes the hydrolysis of globotriaosylceramide (GL-3) to neutral glycolsphingolipids	Intravenous	DB00103
Alteplase	Acute ischemic stroke	Activase	Genentech Inc	Binds to fibrin in a thrombus and converts the entrapped plasminogen to plasmin	Intravenous drip	DB00009
Anistreplase	Myocardial infarction	Eminase	Wulfing Pharma GmbH	Cleaves the Arg/Val bond in plasminogen to form plasmin	Intravenous	DB00029
Pegademase bovine	Severe combined immunodeficiency disease (SCID)	Adagen	Enzon Inc.	Converts adenosine to less toxic inosine by deamination	Intramuscular	DB00061
Imiglucerase	Gaucher's disease	Cerezyme	Genzyme Inc.	Catalyzes the hydrolysis of the glycolipid, glucocerebroside, to glucose and ceramide	Intravenous	DB00053
Alglucerase	Gaucher's disease	Ceredase	Genzyme Inc Cardinal Health	Catalyzes the hydrolysis of the glycolipid, glucocerebroside, to glucose and ceramide	Intravenous drip	DB00088
Rasburicase	Hyperuricemia disease	Elitek	GlaxoSmithKline Inc.	Catalyzes enzymatic oxidation of uric acid into an inactive and soluble metabolite allantoin	Intravenous	DB00049

Dornase alfa	Cystic fibrosis	Pulmozyme Viscozyme	Genentech Inc Roche (Chile)	Endonucleolytic cleavage of extracellular DNA to 5′-phosphodinucleotide and 5′-phosphooligonucleotide end products	Respiratory (inhalation)	DB00003
Galsulfase	Mucopolysaccharidosis VI (MPS VI)	Naglazyme	BioMarin Pharmaceuticals	Catalyzes the cleavage of the sulfate ester from terminal *N*-acetylgalactosamine 4-sulfate residues of GAG chondroitin 4-sulfate and dermatan sulfate.	Intravenous drip	DB01279
Laronidase	Mucopolysaccharidosis I (MPS I)	Aldurazyme	BioMarin Pharmaceuticals	Catalyzes the hydrolysis of terminal alpha-L-iduronic acid residues of dermatan sulfate and heparin sulfate	Intravenous	DB00090
Pegaspargase	Acute lymphoblastic leukemia	Oncaspar	Ben Venue Laboratories Inc.	PEGylated L-asparagine, converts asparagine to aspartic acid and ammonia	Intravenous	DB00059
Streptokinase	Coronary artery thrombosis	Streptase	Pfizer Inc.	Helps eliminate blood clots or arterial blockages	Intravenous	DB00086
Urokinase	Pulmonary embolism	Abbokinase Kinlytic	Abbott Laboratories Microbix Biosystems Inc.	Cleaves the Arg-Val bond in plasminogen to	Intravenous	DB00013

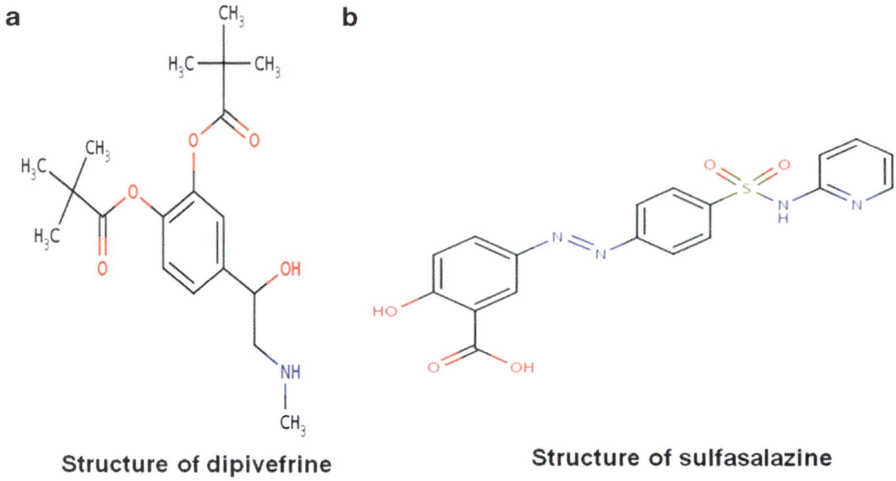

Fig. 15.1 (**a**) Structure of dipivefrine. (**b**) Structure of sulfasalazine. Available from http://www.drugbank.ca/drug

The prodrug is generally used for absorption, distribution, optimize excretion, and complete metabolism of drug (Stella et al. 1985). There are several advantages associated with the prodrugs:

1. Increased absorption and elimination of unpleasant taste.
2. Decreased toxicity and metabolic inactivation.
3. Increased chemical stability and prolonged or short-lived action.

The prodrugs can be classified in two groups:

(a) *Carrier-linked prodrugs*: It contains a group that can be easily removed enzymatically (such as an ester) to reveal the true drug. For example, dipivefrine is prodrug of epinephrine used to treat glaucoma (Fig. 15.1a). The dipivaloyl esters allow for greater corneal permeability which is hydrolyzed by corneal and aqueous humor esterases. It is available as ophthalmic solution. It causes vasoconstriction and converted to epinephrine upon penetration of the cornea (Nakamura et al. 1993).
(b) *Bioprecursor prodrugs*: These are metabolized into a new compound which may itself be active or is further metabolized into an active metabolite (e.g., amine to aldehyde to carboxylic acid). Sulfasalazine (Fig. 15.1b) is used in the treatment of ulcerative colitis, an inflammatory bowel disease, and rheumatoid arthritis (McGirt et al. 2006). Anaerobic bacteria in the lower bowel metabolically reduce sulfasalazine to the therapeutic agent 5-aminosalicylic acid. It is often well tolerated, but sometimes may cause severe depression in young males.

15.5 Application of Enzymes as Pharmaceuticals

15.5.1 Galsulfase

Galsulfase is a recombinant human *N*-acetylgalactosamine 4-sulfatase and was used in enzyme replacement therapy for the treatment of mucopolysaccharidosis VI (MPS VI) (Harmatz et al. 2004). Mucopolysaccharidosis VI (MPS VI) or Maroteaux-Lamy syndrome is a rare recessive lysosomal storage disease (LSD) resulting from a deficiency in the enzyme *N*-acetylgalactosamine 4-sulfatase. *N*-acetylgalactosamine 4-sulfatase is a soluble monomeric protein with a molecular weight of 56 kDa. The glycosylated product has an apparent molecular weight of 66 kDa on SDS-PAGE. The predicted amino acid sequence and the nucleotide sequence of the recombinant enzyme are identical to native human N-acetylgalactosamine 4-sulfatase (also known as arylsulfatase B, EC 3.1.6.12) (Bond et al. 1997). The enzyme is responsible for hydrolysis of the sulfate moiety of the glycosaminoglycan (GAG) dermatan sulfate during its stepwise degradation. Deficiency of *N*-acetylgalactosamine-4-sulfatase leads to the accumulation of its substrate, dermatan sulfate, in the lysosomes of many cell types (Neufeld and Muenzer 2001). The accumulation causes a progressive disorder with multiple organ and tissue involvement.

Recombinant-engineered galsulfase supplies *N*-acetylgalactosamine 4-sulfatase and catalyzes the cleavage of the sulfate ester from terminal *N*-acetylgalactosamine 4-sulfate residues of GAG chondroitin 4-sulfate and dermatan sulfate. Increased catabolism of GAG in turn reduces systemic dermatan sulfate accumulation, thereby reducing the primary symptoms of MPS VI. This drug was approved by the FDA, USA, in May 2005 for the treatment of patients with mucopolysaccharidosis type VI. It is the first approved product for the treatment of mucopolysaccharidosis type VI and has been granted orphan drug status (Hopwood et al. 2006). Galsulfase treatment is well tolerated although some patients developed antibodies against the enzymes and also some less serious side effects such as headache, joint pain, and eye redness were observed. Naglazyme is a formulation of galsulfase by BioMarin Pharmaceuticals Inc., USA, which is a purified human enzyme that is produced by recombinant DNA technology in a Chinese hamster ovary cell line (White et al. 2008).

15.5.2 Asparaginase

Asparaginase (E.C.5.1.1) is widely used in the treatment of childhood acute lymphoblastic leukemia (ALL). Asparaginase hydrolyzes L-asparagine to L-aspartic acid and ammonia in leukemic cells, resulting in the depletion of asparagine, inhibition of protein synthesis, cell cycle arrest in the G1 phase, and apoptosis in susceptible leukemic cell populations (McCredie et al. 2008). The asparaginases isolated from *Escherichia coli* (EcA) and *Erwinia chrysanthemi* (EcA) are useful antileukemic agents (Hill et al. 1967). In a significant number of patients with acute leukemia, particularly lymphocytic, the malignant cells depend on an exogenous source of asparagine for survival. Normal cells, however, are able to synthesize asparagine and thus are least affected by the rapid depletion produced by treatment with the enzyme asparaginase. L-asparaginase, intravenously injected into patients, is primarily distributed in the plasma (Ho et al. 1970). A study revealed that, in patients with metastatic cancer and leukemia, initial plasma levels of L-asparaginase following intravenous administration were correlated to dose. Daily administration resulted in a cumulative increase in plasma levels (Ho et al. 1970). Elspar (asparaginase) contains the enzyme L-asparaginase amidohydrolase derived from *Escherichia coli* and marketed by Merck & Co., Inc., USA. Asparaginase is also marketed by other two companies, Lundbeck Inc., USA, and Prescript Pharmaceuticals, USA.

Since asparaginase has relatively short half-life (10–0 h), PEGylated asparaginase (pegaspargase) with 10–15 times longer half-life was developed during 1970–1980. Indeed the asparaginase also has other disadvantages, like the need for frequent intramuscular injection and a very high rate of allergic reaction (Graham 2003). To overcome this problem, L-asparaginase is modified by covalently conjugating units of monomethoxy polyethylene glycol (PEG), molecular weight of 5,000, to the enzyme, forming the active ingredient PEG-L-asparaginase. Pegaspargase is more effective than asparaginase and facilitates production of oxaloacetate which is needed for general cellular metabolism (Wetzler et al. 2007). This particular drug is marketed by Ben Venue Laboratories Inc., USA, and Enzon Inc., USA.

15.5.3 Dornase Alfa

Dornase alfa is a highly purified recombinant human deoxyribonuclease I (rhDNase) used in the treatment of cystic fibrosis, a most common lethal recessive disorder in white population, and is caused by a defective cystic fibrosis transmembrane conductance regulator and a chloride channel protein, leading to improper salt balance and thick tenacious secretions (Collins 1992). Dornase alfa is produced by genetically engineered Chinese hamster ovary cells containing DNA encoding for the native human deoxyribonuclease I (DNase) (EC 3.1.21.1), an enzyme which selectively cleaves DNA. The enzyme contains 260 amino acids with an approximate molecular weight of 37 kDa (Shak et al. 1990). The primary amino acid sequence is identical to that of the native human deoxyribonuclease I. Dornase alfa hydrolyzes the DNA present in sputum/mucus of cystic fibrosis patients and reduces viscosity in the lungs, promoting improved clearance of secretions (Shah et al. 1996). The enzyme does not appear to affect sputum in the absence of an inflammatory response to infection nor does it affect the sputum of healthy individuals. Clinical trials and observation studies on the efficacy of dornase alfa (commonly known as Pulmozyme) inhalatory therapy have shown improvement in lung function and a decrease of respiratory exacerbations in patients with cystic fibrosis and moderate lung disease (Suri et al. 2001). This drug is well tolerated but some side effects such as chest pain and skin rashes have been reported (Davies et al. 1997).

15.5.4 Agalsidase

Agalsidase is used in enzyme replacement therapy for Anderson-Fabry disease, which is an X-linked defect of glycosphingolipid metabolism (El Dib and Pastores 2010). The primary driver of the disease is the accumulation of glycolipids (globotriaosylceramide [GL-3]) in a variety of cell types, including vascular endothelial cells, a range of renal cell types, cardiomyocytes, and neurons, which is caused by deficient activity of the lysosomal enzyme, alpha-galactosidase (Schaefer et al. 2009). Clinical manifestations of Fabry disease include renal failure, cardiomyopathy, and cerebrovascular accidents.

Fabrazyme (agalsidase beta) is a recombinant human α-galactosidase A enzyme (EC 3.2.1.22), responsible for the breakdown of alpha-galactosides in the lysosome (Guce et al. 2010) with the same amino acid sequence as the native enzyme. Fabrazyme (agalsidase beta) is produced by recombinant DNA technology in a Chinese hamster ovary mammalian cell expression system (Schaefer et al. 2009) and marketed by Genzyme Inc., USA. Fabrazyme is intended to provide an exogenous source of alpha-galactosidase A and to limit the accumulation of the glycolipids in the tissues. This drug also has side effects such as difficulty in breathing, choking of throat, hives, rashes, itching, and fever.

15.5.5 Therapeutic Protein Inhibitors of Elastase

Human neutrophil elastase is a trypsin-type serine protease (EC 3.4.21.37). The key physiological role of neutrophil elastase is in innate host defense. It can also participate in tissue remodeling and possesses secretagogue actions that are now recognized as important to local inflammatory responses (Chughtai and O'Riordan 2004). A number of potent reversible and irreversible inhibitors of neutrophil elastase have been developed for its potential therapeutic use:

1. Pre-Elafin: The drug pre-elafin, also known as trappin-2, is an elastase-specific inhibitor that could be an ideal candidate for the treatment of neutrophil elastase-driven lung diseases. Neutrophil elastase (EC 3.4.21.37) is a potent serine protease involved in host's defense. The inhibitory activity of pre-elafin resides in its COOH-terminal region that can be released as mature elafin (Tremblay et al. 2002).
2. Prolastin: Prolastin is an alpha-1 proteinase inhibitor marketed by Talecris Biotherapeutics C (Bayer, Germany). Its primary mechanism is inhibiting the action of the serine protease called elastase in the lungs, and its deficiency is associated with progressive ultimately fatal emphysema (Karnaukhova et al. 2006).

15.5.6 Pancrelipase

Exocrine pancreatic insufficiency is a serious condition which occurs in several diseases including chronic pancreatitis, cystic fibrosis, pancreatic cancer, and post-pancreatic surgery (Sikkens et al. 2010). Pancreatic enzymes are essential for digestion of food as it can be absorbed by the body. In the case of chronic pancreatitis or pancreatic cancer, the pancreas may not produce enough enzymes for complete digestion of food. As a result the food was not completely absorbed by the body. This phenomenon is called malabsorption (Fieker et al. 2011). The major malabsorption problems arise from incomplete fat digestion. Pancrelipase or pancreatic enzyme replacement therapy was used to treat malabsorption. It is a protein mixture isolated from porcine or bovine pancreas, sometimes called pancreatin, and contains three enzymes, amylase (to digest starchy carbohydrate), lipase (to digest fat), and protease (to digest protein). Exogenous pancrelipase reduces the amount of nitrogen and fat excreted in the stool, but the overdose of pancrelipase causes diarrhea or stomach upset (Fieker et al. 2011). The drug is marketed by the brand names of Creon, Nutrizym, Pancrease HL, or Pancrex.

15.5.7 Imiglucerase

Imiglucerase is used for the treatment of type 1 Gaucher disease, characterized by a functional deficiency in human β-glucocerebrosidase enzymatic activity (Weinreb et al. 2005). β-Glucocerebrosidase (β-D-glucosyl-*N*-acylsphingosine glucohydrolase, EC 3.2.1.45) is a lysosomal glycoprotein enzyme which catalyzes the hydrolysis of the glycolipid glucocerebroside to glucose and ceramide. Absence of the enzyme, β-glucocerebrosidase, causes accumulation of lipid glucocerebroside in tissue macrophages which become engorged which are termed as Gaucher cells. Injection of imiglucerase into Gaucher disease patients leads to elevated levels of the enzyme in serum and reduction in the accumulation of glucocerebroside which reduces anemia, thrombocytopenia, spleen and liver size, and decreased cachexia (Pastores et al. 2004).

Imiglucerase is marketed as Cerezyme by Genzyme Inc., USA, and produced by recombinant DNA technology using Chinese hamster ovary. Purified imiglucerase is a monomeric glycoprotein of 497 amino acids, containing four N-linked glycosylation sites. Cerezyme differs from placental glucocerebrosidase by one amino acid at position 495, where histidine is substituted for arginine. This drug was approved by the FDA in 1994 and more than 5,600 patients in 90 countries have been treated with Cerezyme (Weinreb et al. 2005). Cerezyme is given intravenously, and its application leads to a marked improvement in the clinical manifestation of the Gaucher disease (Pastores et al. 2004). General side effects, such as fatigue, headache, abdominal pain, fever, dizziness, chills, and backache, have been reported but only in 1.5 % patients (Ali et al. 2011).

15.5.8 Pegademase Bovine

The drug, pegademase, is bovine adenosine deaminase (EC 3.5.4.4) enzyme derived from bovine intestine. It is used for enzyme replacement therapy for treating severe combined immunodeficiency disease (SCID) associated with the deficiency of adenosine deaminase (ADA) (Polmar et al. 1975). The enzyme has been extensively PEGylated for extended serum half-life. It is a conjugate of numerous strands of monomethoxy polyethylene glycol (molecular weight 5,000), covalently attached to the enzyme adenosine deaminase (Hershfield et al. 1987). The enzyme adenosine deaminase is responsible for converting adenosine to inosine. In the absence of adenosine deaminase, the purine substrate adenosine, 2′-deoxyadenosine and its metabolites are actually toxic to lymphocytes thereby leading to diminished immune function. Pegademase converts 2′-deoxyadenosine to 2′-deoxyinosine via deamination (Pesu et al. 2005). Severe combined immunodeficiency disease (SCID) is a primary immune deficiency caused by several genetic defects in the immune system (Hershfield and Mitchell 1995). Adagen (pegademase bovine) injection was the first successful application of enzyme replacement therapy for an inherited disease (Hershfield et al. 1987) and approved in 1990 by the FDA, USA. Adagen has been used to treat nearly 150 patients with ADA-deficient SCID worldwide. The possible side effects of this drug are allergic reactions like difficulty in breathing, choking of throat, swelling of the lips, tongue, face, hives, and signs of infection such as sore throat, fever, or congestion.

15.5.9 Tissue Plasminogen Activators

Tissue plasminogen activator (tPA) (EC 3.4.21.68) is a serine protease found on endothelial cells and catalyzes the conversion of plasminogen to biologically active plasmin. It is the main enzyme for breaking down the blood clots (fibrin clots) and allows blood flow to the affected areas, thus preventing brain damage (Tsurupa and Medved 2001; Imming et al. 2006). As a protease, tPA plays a crucial role in regulating blood fibrinolysis, maintaining the homeostasis of extracellular matrix and modulating the posttranslational activation of growth factors. tPA is found not only in the blood, where its primary function is as a thrombolytic enzyme, but also in the central nervous system (CNS). It participates in a number of physiological and pathological events in the CNS, as well as the role of neuroserpin as the natural regulator of tPA's activity in these processes.

Tissue plasminogen activator (tPA) is used in clinical medicine as a thrombolytic agent to treat embolic or thrombotic diseases (Longstaff et al. 2008), like pulmonary embolism (blockage of the main artery of the lung or one of its branches by a substance that has travelled from elsewhere in the body through the bloodstream), myocardial infarction, heart attack (results from the interruption of blood supply to a part of the heart), and stroke or cerebrovascular accident (rapid loss of brain function(s) due to disturbance in the blood supply to the brain) (Del Zoppo et al. 2009). Some of the examples of tissue plasminogen activators used clinically and also approved by the FDA, USA, are mentioned below. About one third of the patients treated with intravenous thrombolytic therapy exhibited an improvement poststroke (Saver 2004).

15.5.9.1 Alteplase

Alteplase binds to fibrin-rich clots via the fibronectin finger-like domain and the kringle-2 domain. The protease domain then cleaves the Arg/Val bond in plasminogen to form plasmin. Plasmin in turn degrades the fibrin matrix of the thrombus, thereby exerting its thrombolytic action. It also produces limited conversion of plasminogen in the absence of fibrin. This drug is marketed by the name of Activase and used in all thrombolytic diseases such as acute myocardial infarction, acute ischemic stroke, and lysis of acute pulmonary emboli (Hacke et al. 2008). It is synthesized using the complementary DNA (cDNA) for natural human tissue-type plasminogen activator obtained from a human melanoma cell line. The manufacturing process involves secretion of alteplase into the culture medium by an established Chinese hamster ovary cell lines into which the cDNA for alteplase had been genetically inserted.

15.5.9.2 Reteplase

Reteplase is a recombinant non-glycosylated form of human tissue plasminogen activator, which has been modified to contain 357 of the 527 amino acids of the native human tPA (amino acids 1–3 and 176–527) and retained the activity-related kringle-2 and serine protease domains of human tPA. Retavase is considered a "third-generation" thrombolytic agent, genetically engineered to retain and delete certain portions of human tPA. Reteplase is similar to alteplase but the modifications give reteplase a longer half-life of 13–16 min. It is produced by recombinant DNA technology in *E. coli* and was marketed by the name of Retavase. The protein is isolated as inactive inclusion bodies from *E. coli*, converted into active form by an in vitro folding process and purified by chromatographic separation. Reteplase also binds fibrin with lower affinity than alteplase, improving its ability to penetrate into clots. It works best to help people with strokes caused by clots (ischemic strokes) when it is given right away after the stroke symptoms begin (Hilleman et al. 2007).

15.5.9.3 Tenecteplase

Tenecteplase is a 527 amino acid glycoprotein developed by introducing the following modifications to the complementary DNA (cDNA) for natural human tPA: a substitution of threonine 103 with asparagine and a substitution of asparagine 117 with glutamine, both within the kringle-1 domain and a tetra-alanine substitution at amino acids 296–299 in the protease domain (Ohman et al. 2005). The brand name of this drug is TNKase and is used for the treatment of myocardial infarction and lysis of intracoronary emboli. It binds to fibrin-rich clots and cleaves the Arg/Val bond in plasminogen to form plasmin (Gurbel et al. 2005). Plasmin in turn degrades the fibrin matrix of the thrombus, thereby exerting its thrombolytic action. This helps to remove blood clots and arterial blockages that cause myocardial infarction. Tenecteplase is known to have a long half-life; therefore, a single bolus injection is sufficient for the treatment (Davydov and Cheng 2001).

15.5.9.4 Anistreplase

Anistreplase is another human tissue plasminogen activator and is used to eliminate blood clots or arterial blockages that cause myocardial infarction. It cleaves the Arg/Val bond in plasminogen to form plasmin. Plasmin in turn degrades the fibrin matrix of the thrombus, thereby exerting its thrombolytic action. It was marketed by the name of Eminase by Wulfing Pharma GmbH, Germany. Eminase is a complex of Lys-plasminogen (*p*-anisoyl derivative of the primary Lys-plasminogen) and streptokinase (Lee et al. 2004). A *p*-anisoyl group is chemically conjugated to a complex of bacteria-derived streptokinase and human plasma-derived Lys-plasminogen proteins.

15.5.10 Urokinase

Urokinase (EC 3.4.21.73) is also a serine protease, which specifically cleaves the Arg-Val bond in inactive plasminogen to form active plasmin. Urokinase was originally isolated from human urine, but is present at several physiological locations, such as blood stream and the extracellular matrix. Activation of plasmin triggers a proteolysis cascade that, depending on the physiological environment, participates in thrombolysis and extracellular matrix degradation. Urokinase is used for the treatment of pulmonary embolisms (Overington et al. 2006). It is available by the name of Abbokinase and Kinlytic and administered by intravenous infusion. It has some side effects such as epistaxis, bleeding gums, and bloody or tarry stools (http://www.rxlist.com/cgi/generic2/tenecteplase.htm).

15.5.11 Streptokinase

This is another thrombolytic agent and was also used in the treatment of acute evolving transmural myocardial infarction, pulmonary embolism, deep vein thrombosis, arterial thrombosis or embolism, and occlusion of arteriovenous cannulae (Meneveau et al. 1997). Streptokinase (EC 3.4.99.0) creates an active complex to form the proteolytic enzyme plasmin. Streptokinase forms a highly specific 1:1 active enzymatic complex with plasminogen which also cleaves the Arg/Val bond in plasminogen (Mundada and Prorok 2003) and converts inactive plasminogen molecules into active plasmin (Sikri and Bardia 2007). Plasmin degrades fibrin clots as well as fibrinogen and other plasma proteins. Streptokinase is marketed under the name of Streptase. It is a purified preparation of a bacterial protein elaborated by group C (*beta*)-hemolytic *streptococci.* Streptokinase in some cases has been shown to be more effective than tissue plasminogen activator alteplase (Capstick and Henry 2005).

15.5.12 Bromelain

Bromelain is a proteolytic enzyme found in pineapple juice and stems that offers variety of health benefits. Bromelain can be taken to support healthy digestion and for treating inflammatory, cardiovascular, and skin disorders. The two main enzymes are stem bromelain (EC 3.4.22.32) and fruit bromelain (EC 3.4.22.33). Bromelain is used for treating osteoarthritis (Brien et al. 2004) and is also a promising anti-inflammatory agent (Fitzhugh et al. 2008).

15.5.13 Hyaluronidase

Hyaluronidase (EC 3.2.1.35) is used in conjunction with other drugs to speed their dispersion and delivery of medicines. Hyaluronidase hydrolyzes hyaluronic acid by splitting the glucosaminidic bond between C1 of the glucosamine moiety and C4 of glucuronic acid. This temporarily decreases the viscosity of the cellular cement and increases diffusion of injected fluids and localized transudates and exudates, facilitating their absorption (Csoka et al. 1999). It also increases the absorption rate of parenteral fluids given by hypodermoclysis and is an adjunct in subcutaneous urography for improving resorption of radiopaque agents. The drug has been branded as Hydase™ (animal-derived hyaluronidase), Vitrase (marketed by ISTA Pharmaceuticals, USA), and Amphadase (marketed by Amphastar Pharmaceuticals, USA).

15.5.14 Rasburicase

Rasburicase is derived from a cDNA code from a modified *Aspergillus flavus* strain and expressed in a modified yeast strain of *Saccharomyces cerevisiae* (Collings et al. 2010). It is recombinant urate oxidase enzyme (EC 1.7.3.3) and is used to treat lymphoid leukemia, non-Hodgkin's lymphoma, and acute myelogenous leukemia. Rasburicase converts existing uric acid to allantoin, which is 5–10 times more soluble in urine than uric acid. The oxidation of uric acid to allantoin by rasburicase produces hydrogen peroxide and carbon dioxide (Ribeiro and Pui 2003). The injection of rasburicase reduces levels of uric acid and mitigates the toxic effects of chemotherapy induced tumor lysis. This drug is marketed by the name of Elitek by Sanofi-Aventis Inc., USA. The possible side effects of this drug are vomiting, fever, nausea, abdominal pain, and diarrhea (Ho et al. 2006).

15.6 Application of Immobilized Enzymes as Pharmaceutical

Chemical immobilization of proteins and enzymes was first attempted in 1960s, and it is an emerging approach to new drug therapies. Immobilization means the enzymes with restricted mobility or rendered less motile by chemical or physical treatment. It was first prepared by loading to polymeric matrices or binding onto carrier materials. Industrial use of enzymes is greatly limited because they are relatively unstable, have a very high cost of purification, and have cumbersome process of recovery of active enzyme from reaction mixture after the completion of catalytic process. Immobilized enzymes are more stable to pH and temperature stress and less susceptible to the denaturing agents. In addition, an immobilized enzyme should have long-term stability and unaltered sensitivity and biological activity after attachment to the matrix than free enzyme when used as therapeutic purpose (Klein and Langer 1986). Immobilization has been successfully utilized for studies with such enzymes, as cytochrome P-450, UDP-glucuronosyltransferases, glutathione S-transferases, S-methyltransferases, and *N*-acetyltransferases (Dulik and Fenselaut 1998).

One of the major applications of immobilized enzymes in pharmaceutical industry is the production of 6-aminopenicillanic acid (6-APA) by deacylation of the side chain in either penicillin G or V, using penicillin acylase (penicillin amidase). Today more than 50 % of 6-APA is enzymatically produced using the immobilized route which is core of penicillin antibiotic. Penicillin amidase from *E. coli* is immobilized on cellulose triacetate fibers for producing 6-APA from penicillin G (Alvaro et al. 1990). Similarly for producing 6-APA from penicillin V, penicillin amidase is immobilized by covalent binding to Amberlite XAD-7 with glutaraldehyde through physical adsorption to bentonite or by ionic binding to DEAE-Sephadex and also by covalent binding to a copolymer of acrylamide and maleic anhydride (Arshad et al. 2007). The major reasons for its success is in obtaining a pure product, thereby minimizing the purification costs (Giordano et al. 2006). This process of the immobilized enzyme technology was also approved in India. The first industrial process for the production of 6-APA was started in 1970s by Astra, Sweden, and Riga Biochemical Plant (former USSR).

Immobilization has also been used for the production of 7-aminodeacetoxy-cephalosporanic acid, an intermediate in the production of semisynthetic cephalosporins. Conversion of 7-amino-3-deacetoxy-cephalosporanic acid (7-ADCA) to cephalexin by immobilized penicillin G acylase (IMPGA) has been investigated. It was observed that under optimized conditions, IMPGA can attain 85 % conversion of 7-ADCA to cephalexin. Furthermore, IMPGA can be reused for about ten cycles (Maladkar 1994). Production of cefazolin by immobilized cefazolin synthetase from *E. coli* as a biocatalyst has been possible. The complex of the physicochemical studies makes it possible to design a highly efficient technological process for production of cefazolin (Kurochkina and Nys 1999). Macrolide antibiotics tylosin and nikkomycin also can be produced by *Streptomycin* spp., immobilized with calcium alginate.

15.7 Production of Biopharmaceuticals at Industrial Scale

The main focus of the pharmaceutical industry is to develop such processes for producing recombinant therapeutic and diagnostic proteins of high value at large scale that utilize agricultural residue or waste as main component of the medium. Most of the enzymes are produced by submerged fermentation or by solid state fermentation at industrial scale owing to the inherent advantages such as higher yield, less energy requirement, and less cumbersome in downstream processing. The α-amylase was produced from agricultural by-products such as wheat bran, rye straw, wheat straw, and rice bran by *Bacillus cereus* MTCC 1305 by solid state fermentation (Singh et al. 2010). α-Amylases are used as a digestive aid by hydrolyzing α-(1 → 4) glycosidic linkages of polysaccharides to yield dextrin, oligosaccharides, maltose, and D-glucose. In recent years, the biopharmazymes are produced and overexpressed in *E. coli* and in mammalian cells by recombinant technology.

The pharmaceutical manufacturing of recombinant proteins most frequently employs single-cell suspension cultures in stirred-tank bioreactors of variable sizes up to 20,000 L (Wurm 2004). Advances in media and process optimization for mammalian cell culture system have already resulted in more than 100-fold improvement in yield. Monoclonal antibodies (mAb) for treating breast cancer and an immunoglobulin-TNF (tumor necrosis factor) receptor fusion protein for treating rheumatoid arthritis, are produced in suspension cell lines including suspension-adapted Chinese hamster ovary or murine myelomas in stirred-tank reactors (Chu and Robinson 2001). Further insights and subsequent targeted modifications with respect to genome-scale technologies including genomics, transcriptomics, and proteomics are expected to contribute to the development of mammalian cell-based production systems in bioreactors and are hoped to further improve protein yields and quality in the near future (Matasci et al. 2008).

15.8 Future Prospects

New strategies are continuously emerging for synthesizing and immobilization of new enzymes to enhance their role and efficiency to treat variety of diseases. In response to the need in the pharmaceutical industry for more complex, chiral molecules, fine chemical companies are adopting new manufacturing technologies to produce more efficient compounds with wide spectrum of activity. In particular, recent developments in biocatalysis combined with novel process engineering have provided improved methods for the production of valuable chemical intermediates (Huisman and Gray 2002). Enzymes with antioxidative property are still an area of intense research within the pharmaceutical industry. Superoxide dismutase which transforms the highly toxic superoxide anion to moderately toxic hydrogen peroxide has been of interest to the pharmaceutical industry for quite some time and is still under research (Veronese et al. 2002).

Several researchers are also working in the area of inhibition of human neutrophil elastase (EC 3.4.21.37) having implications in the treatment of chronic obstructive pulmonary disease (COPD) consortium, such as emphysema and chronic bronchitis. Human butyrylcholinesterase, a naturally occurring serum detoxification enzyme, acts to break down acetylcholine. It could be useful for the treatment of cocaine overdose, as demonstrated by recent results (Melov et al. 2000). Several other lysosomal storage diseases are also being investigated with enzyme replacement therapy, including Hurler's disease and Maroteaux-Lamy syndrome (Harmatz et al. 2004). The enzymes which are designed and designated as orphan drugs under investigation in the USA are listed in Table 15.2.

15.9 Conclusions

The total number of pharmaceutical enzymes in use around the world probably exceeds 3,000 not including the tens of thousands of formulations containing different combinations of these ingredients. Enzymes are important in therapeutic and in commercial processes because they accelerate specific chemical reaction to produce a useful effect or product. From the perspective of a supplier of pharmaceutically important enzymes, the success of business lies in identifying the particular technological niche for improved product. Advancements in biotechnology over the past

Table 15.2 The biopharmaceutical enzymes designated as orphan drugs under investigation by the FDA, USA

Generic name	Treatment	Brand name(s)	Year designated
Phenylalanine ammonia-lyase	Treatment of hyperphenylalaninemia	Phenylase	1995
Recombinant human α-glucosidase	Treatment of glycogen storage disease type II	Pompase1	1996
Collagenase	Treatment of Peyronie's disease	Plaquase	1996
Recombinant human acid α-glucosidase	Treatment of glycogen storage disease type II	Myozyme	1997
Papain, trypsin, and chymotrypsin	Treatment of multiple myeloma	Wobe-Mugos1	1998
PEGylated arginine deiminase	Treatment of invasive malignant melanoma	Melanoid	1999
Recombinant urate oxidase	Prophylaxis of chemotherapy induced hyperuricemia	Fasturtec	2000
Recombinant human highly phosphorylated acid α-glucosidase	Enzyme replacement therapy in patients with Pompe's disease	TBD	2000
PEG-uricase	Control of clinical consequences of hyperuricemia in patients with severe gout	Puricase	2001
Iduronate-2-sulfatase	Treatment of Hunter's syndrome	I2S	2001
Lipase, amylase, and protease	Treatment of pancreatic insufficiency	Thera CLEC-total	2002

10 years have allowed pharmaceutical companies to produce safer and cheaper enzymes with enhanced potency and specificity. More recently, identifying pharmacological activity based upon an understanding of how enzymes work at the molecular level has enabled the industry to discover many new groups of successful drugs. Along with these advances, changes in orphan drug laws and new initiatives taken by the FDA, USA, have been effective in facilitating efforts to develop enzyme drugs. India's pharmaceutical fine chemical industry is undergoing readjustment to the structural changes evolving patent regime.

References

Aiuti A (2002) Advances in gene therapy for ADA-deficient SCID. Curr Opin Mol Ther 4: 515–522

Ali MA, Saleh FM, Das K et al (2011) Gaucher disease. Mymensingh Med J 20:490–492

Alvaro G, Fernandez LR, Blanco RM et al (1990) Immobilization-stabilization of penicillin G acylase from *Escherichia coli*. Appl Biochem Biotechnol 26:181–195

Arshad R, Farooq S, Ali SS (2007) 6-Aminopenicillanic acid production by intact cells of *E. coli* containing penicillin G acylase (PGA). Pak J Biol Sci 10:3190–3194

Bond CS, Clements PR, Ashby SJ et al (1997) Structure of a human lysosomal sulfatase. Structure 5:277–289
Bradbury E, Moon L, Popat R et al (2002) Chondroitinase ABC promotes functional recovery after spinal cord injury. Nature 416:636–640
Brien S, Lewith G, Walker A (2004) Bromelain as a treatment for osteoarthritis: a review of clinical studies. eCAM 1:251–257
Capstick T, Henry MT (2005) Efficacy of thrombolytic agents in the treatment of pulmonary embolism. Eur Respir J 26:864–874
Chan B, Wara D, Bastian J et al (2005) Long-term efficacy of enzyme replacement therapy for adenosine deaminase (ADA)-deficient severe combined immunodeficiency (SCID). Clin Immunol 117:133–143
Chu L, Robinson DK (2001) Industrial choices for protein production by large-scale cell culture. Curr Opin Biotechnol 12:180–187
Chughtai B, O'Riordan TG (2004) Potential role of inhibitors of neutrophil elastase in treating diseases of the airway. J Aerosol Med 17:289–298
Collings I, Watier Y, Giffard M et al (2010) Polymorphism of microcrystalline urate oxidase from *Aspergillus flavus*. Acta Crystallogr D Biol Crystallogr 66:539–548
Collins FS (1992) Cystic fibrosis: molecular biology and therapeutic implications. Science 256:774–779
Csoka AB, Scherer SW, Stern R (1999) Expression analysis of six paralogous human hyaluronidase genes clustered on chromosomes 3p21 and 7q31. Genomics 60:356–361
Davies J, Trindade MT, Wallis C et al (1997) Retrospective review of the effects of rhDNase in children with cystic fibrosis. Pediatr Pulmonol 23:243–248
Davydov L, Cheng JWM (2001) Tenecteplase: a review. Clin Ther 23:982–997
de Duve C (1996) The significance of lysosome in pathology and medicine. Proc Inst Med Chic 26:73–76
Del Zoppo GJ, Saver JL, Jauch EC et al (2009) Expansion of the time window for treatment of acute ischemic stroke with intravenous tissue plasminogen activator: a science advisory from the American Heart Association/American Stroke Association. Stroke 40:2945–2948
Dulik DM, Fenselaut C (1998) Use of immobilized enzymes in drug metabolism studies. FASEB J 2:2235–2240
El Dib RP, Pastores GM (2010) Enzyme replacement therapy for Anderson-Fabry disease. Cochrane Database Syst Rev doi: 10.1002/14651858
Fieker A, Philpott J, Armand M (2011) Enzyme replacement therapy for pancreatic insufficiency: present and future. Clin Exp Gastroenterol 4:55–73
Fitzhugh DJ, Shan S, Dewhirst MW et al (2008) Bromelain treatment decreases neutrophil migration to sites of inflammation. Clin Immunol 128:66–74
Germain DP (2002) Fabry disease: recent advances in enzyme replacement therapy. Expert Opin Investig Drugs 11:1467–1476
Giordano RC, Ribeiro MPA, Giordano RLC (2006) Kinetics of β-lactam antibiotics synthesis by penicillin G acylase (PGA) from the viewpoint of the industrial enzymatic reactor optimization. Biotechnol Adv 24(1):27–41
Graham ML (2003) Pegaspargase: a review of clinical studies. Adv Drug Deliv Rev 55: 1293–1302
Guce AI, Clark NE, Salgado EN et al (2010) Catalytic mechanism of human alpha-galactosidase. J Biol Chem 285:3625–3632
Gurbel PA, Hayes K, Bliden KP et al (2005) The platelet-related effects of tenecteplase versus alteplase versus reteplase. Blood Coagul Fibrinolysis 16:1–7
Hacke W, Kaste M, Bluhmki E et al (2008) Thrombolysis with alteplase 3–4.5 h after acute ischemic stroke. N Engl J Med 359:1317–1329
Harmatz P, Whitley CB, Waber L et al (2004) Enzyme replacement therapy in mucopolysaccharidosis type VI (Maroteaux–Lamy syndrome). J Pediatr 144:574–580
Hershfield MS, Buckley RH, Greenberg ML et al (1987) Treatment of adenosine deaminase deficiency with polyethylene glycol-modified adenosine deaminase. N Engl J Med 316:589–596

Hershfield MS, Chaffee S, Sorensen RU (1993) Enzyme replacement therapy with polyethylene glycol-adenosine deaminase in adenosine deaminase deficiency: overview and case reports of three patients, including two now receiving gene therapy. Pediatr Res 33:S42–S48

Hershfield MS, Mitchell BS (1995) Immunodeficiency diseases caused by adenosine deaminase deficiency and purine nucleoside phosphorylase deficiency. In: Scriver CR, Beaudet AL, Sly WS, Valle D (eds) The metabolic and molecular bases of inherited disease, 7th edn. McGraw-Hill, New York, pp 1725–1768

Hill JM, Roberts J, Loeb E et al (1967) L-Asparaginase therapy for leukemia and other malignant neoplasms, remission in human leukemia. J Am Med Assoc 202:882–888

Hilleman DE, Tsikouris JP, Seals AA et al (2007) Fibrinolytic agents for the management of ST-segment elevation myocardial infarction. Pharmacotherapy 27:1558–1570

Ho DHW, Thetford BS, Carter CJK et al (1970) Clinical pharmacologic studies of L-asparaginase. Clin Pharmacol Ther 7:408–417

Ho VQ, Wetzstein GA, Patterson SG et al (2006) Abbreviated rasburicase dosing for the prevention and treatment of hyperuricemia in adults at risk for tumor lysis syndrome. Support Cancer Ther 3:178–182

Hopwood JJ, Bate G, Kirkpatrick P (2006) Galsulfase. Nat Rev Drug Discov 5:101–102

Huisman G, Gray D (2002) Towards novel processes for the fine chemical and pharmaceutical industries. Curr Opin Biotechnol 13:352–358

Imming P, Sinning C, Meyer A (2006) Drugs, their targets and the nature and number of drug targets. Nat Rev Drug Discov 5:821–834

Karnaukhova E, Ophir Y, Golding B (2006) Recombinant human alpha-1 proteinase inhibitor: towards therapeutic use. Amino Acids 30:317–332

Klein MD, Langer R (1986) Immobilized enzymes in clinical medicine: an emerging approach to new drug therapies. Trends Biotechnol 4:179–186

Kurochkina VB, Nys PS (1999) Enzymatic synthesis of beta-lactam antibiotics. I. Cefazolin. Antibiot Khimioter 44:12–16

Lee KY, Kim DI, Kim SH et al (2004) Sequential combination of intravenous recombinant tissue plasminogen activator and intra-arterial urokinase in acute ischemic stroke. Am J Neuroradiol 25:1470–1475

Longstaff C, Williams S, Thelwell C (2008) Fibrin binding and the regulation of plasminogen activators during thrombolytic therapy. Cardiovasc Hematol Agents Med Chem 6:212–223

Maladkar NK (1994) Enzymatic production of cephalexin. Enzyme Microb Technol 16:715–718

Matasci M, David L, Hacker DL, Lucia Baldi L et al (2008) Recombinant therapeutic protein production in cultivated mammalian cells: current status and future prospects. Drug Discov Today Tech 5:37–42

McCredie KB, Ho DHW, Freireich EJ (2008) L-Asparaginase for the treatment of cancer. CA Cancer J Clin 23:220–227

McGirt LY, Vasagar K, Gober LM et al (2006) Successful treatment of recalcitrant chronic idiopathic urticaria with sulfasalazine. Arch Dermatol 142:1337–1342

Melov S, Ravenscroft J, Malik S et al (2000) Extension of life-span with superoxide dismutase/catalase mimetics. Science 289:1567–1569

Meneveau N, Schiele F, Vuillemenot A et al (1997) Streptokinase vs. alteplase in massive pulmonary embolism. A randomized trial assessing right heart haemodynamics and pulmonary vascular obstruction. Eur Heart J 18:1141–1148

Mundada L, Prorok M (2003) Structure–function analysis of streptokinase amino terminus. J Biol Chem 278:24421–24427

Nakamura M, Shirasawa E, Hikida M (1993) Characterization of esterases involved in the hydrolysis of dipivefrin hydrochloride. Ophthalmic Res 25:46–51

Neufeld EF, Muenzer J (2001) The mucopolysaccharidoses. In: Scriver CR et al (eds) The metabolic and molecular bases of inherited disease, 8th edn. McGraw-Hill, New York, pp 3421–3452

Ohman EM, Van de Werf F, Antman EM et al (2005) Tenecteplase and tirofiban in ST-segment elevation acute myocardial infarction: results of a randomized trial. Am Heart J 150:79–88

Overington JP, Al-Lazikani B, Hopkins AL (2006) How many drug targets are there? Nat Rev Drug Discov 5:993–996

Ozcan C, Ergun O, Celik A et al (2002) Enzymatic debridement of burn wound with collagenase in children with partial-thickness burns. Burns 28:791–794
Pastores GM, Weinreb NJ, Aerts H et al (2004) Therapeutic goals in the treatment Gaucher disease. Semin Hematol 41:4–14
Pesu M, Candotti F, Husa M et al (2005) Jak3, severe combined immunodeficiency, and a new class of immunosuppressive drugs. Immunol Rev 203:127–142
Polmar SH, Wetzler EM, Stern RC et al (1975) Restoration of in-vitro lymphocyte responses with exogenous adenosine deaminase in a patient with severe combined immunodeficiency. Lancet 2:743–746
Ribeiro RC, Pui CH (2003) Recombinant urate oxidase for prevention of hyperuricemia and tumor lysis syndrome in lymphoid malignancies. Clin Lymphoma 3:225–232
R_XList-Theinternet drug index (2013) http://www.rxlist.com/cgi/generic2/tenecteplase.html. Accessed 21 Jan 2013
Saver JL (2004) Number needed to treat estimates incorporating effects over the entire range of clinical outcomes: novel derivation method and application to thrombolytic therapy for acute stroke. Arch Neurol 61:1066–1070
Schaefer RM, Tylki-Szymanska A, Hilz MJ (2009) Enzyme replacement therapy for Fabry disease: a systematic review of available evidence. Drugs 69:2179–2205
Shah PL, Scott SF, Knight RA et al (1996) In vivo effects of recombinant human Dnase I on sputum in patients with cystic fibrosis. Thorax 51:119–125
Shak S, Capon DJ, Hellmiss R et al (1990) Recombinant human DNase I reduces the viscosity of cystic fibrosis sputum. PNAS 87:9188–9192
Sharma A, Jacob A, Tandon M et al (2010) Orphan drug: Development trends and strategies. J Pharm Bioallied Sci 2:290–299
Sikkens EC, Cahen DL, Kuipers EJ et al (2010) Pancreatic enzyme replacement therapy in chronic pancreatitis. Best Pract Res Clin Gastroenterol 24:337–347
Sikri N, Bardia A (2007) A history of streptokinase use in acute myocardial infarction. Tex Heart Inst J 34:318–327
Singh RK, Mishra SK, Kumar N (2010) Optimization of α-amylase production on agriculture byproduct by *Bacillus cereus* MTCC 1305 using solid state fermentation. RJPBCS 1:867–876
Stella VJ, Charman WN, Naringrekar VH (1985) Prodrugs. Do they have advantages in clinical practice? Drugs 29:455–473
Suri R, Metcalfe C, Lees B et al (2001) Comparison of hypertonic saline and alternate day or daily recombinant human deoxyribonuclease in children with cystic fibrosis: a randomised trial. Lancet 358:1316–1321
Tremblay GM, Vachon E, Larouche C et al (2002) Inhibition of human neutrophil elastase-induced acute lung injury in hamsters by recombinant human pre-elafin (trappin-2). Chest 121:582–588
Tsurupa G, Medved L (2001) Identification and characterization of novel tPA- and plasminogen-binding sites within fibrinogen alpha C-domains. Biochemistry 40:801–808
Vellard M (2003) The enzyme as drug: application of enzymes as pharmaceuticals. Curr Opin Biotechnol 14:444–450
Veronese F, Calceti P, Schiavon O et al (2002) Polyethylene glycol-superoxide dismutase, a conjugate in search of exploitation. Adv Drug Deliv Rev 54:587–606
Weinreb NJ, Barranger JA, Charrow J et al (2005) Guidance on the use of miglustat for treating patients with type 1 Gaucher disease. Am J Hematol 80:223–229
Wetzler M, Sanford BL, Kurtzberg J et al (2007) Effective asparagine depletion with pegylated asparaginase results in improved outcomes in adult acute lymphoblastic leukemia: Cancer and Leukemia Group B Study 9511. Blood 109:4164–4167
White JT, Argento ML, Prince WS et al (2008) Comparison of neutralizing antibody assays for receptor binding and enzyme activity of the enzyme replacement therapeutic Naglazyme (galsulfase). AAPS J 10:439–449
Wurm FM (2004) Production of recombinant protein therapeutics in cultivated mammalian cells. Nat Biotechnol 22:1393–1398

Chapter 16
Biocosmetics

Alessandra Cristine Novak, Eduardo Bittencourt Sydney, and Carlos Ricardo Soccol

16.1 Introduction

The use of cosmetics by man dates back to the remotest times. Older cultures derived their cosmetic products from natural compounds, such as fruits, milks, vegetables, flowers, and seeds, also including mineral compounds, such as ashes and clays. Table 16.1 summarizes the main facts related to the relationship between mankind and the searching of well-being, health, and beauty.

The history of use and growth of commercial cosmetics was initially connected to religious and ornamental aspects, gaining connotation of well-being and confidence over time. This quest for beauty unleashed initially the search for active ingredients and materials that could be removed from nature and reality surrounding the human being. Thus, numerous formulae have been developed to smoothen, invigorate, and beautify skin, hair, and others. However, products and techniques are being improved and have become increasingly sophisticated. The advent of fine chemicals and the synthesis of molecules in the twentieth century and current bio- and nanotechnologies have led to the acquisition of key assets, such as hyaluronic acid, citric acid, xanthan gum, growth factors, and nanoparticulate systems.

Nowadays, a new approach in obtaining cosmetic products or raw materials for its production is the use of industrial wastes from innumerous industrial processes: physical, chemical, biological, biotechnological, and/or nanotechnological processes. This makes it possible to give a better destination to some wastes, giving them a new function as a cosmetic product.

A.C. Novak • E.B. Sydney
Department of Biotechnology and Bioprocess Engineering,
Federal University of Paraná—UFPR, Curitiba, PR, Brazil

C.R. Soccol (✉)
Department of Biotechnology and Bioprocess Engineering, Federal University of Paraná—UFPR, Av. Cel. Francisco H. dos Santos, 100., Box 19011 Zip code 81531-990, Curitiba, Parana, Brazil
e-mail: soccol@ufpr.br

S.K. Brar et al. (eds.), *Biotransformation of Waste Biomass into High Value Biochemicals*, DOI 10.1007/978-1-4614-8005-1_16, © Springer Science+Business Media New York 2014

Table 16.1 Relationship between human being, natural material, and beauty and well-being

Primitive people	Skin painting: Ornamental and religious purposes
Egypt	Eyes: Ash and dyes
	Skin: Animal and vegetable fats, beeswax, honey, milk, and aromatic oils
Ancient Greece	Visage: Blackberries and seaweed
	Lipstick: Cinnabar (mercury sulfide—pigment)
China	Nails: Pigmented polish
Rome	Skin: Donkey milk; powder to lighten skin
	Cilia and eyebrows: Coal
Middle ages	Lipstick: Saffron
	Cilia: Black soot
	Teeth: Sage (bleaching)
	Skin: Egg and Vinegar
	Prohibition of the cult of hygiene and exaltation of beauty—the disappearance of the use of cosmetics in Europe
Crusades	Resurgence of cosmetics. Growth of perfumes development
Late eighteenth century	Witchcraft in England
Twentieth century	Home production and emergence of the first cosmetic industries
1970s	Popularization of makeups
1980s	Formulas evolved for cosmetics pigmented
1990 until nowadays	New technologies and process, refined products, concerns about security, concerns about animal safety, concerns about environmental quality and preservation

Taking into consideration all these aspects, the biocosmetic industry can take advantage of the large amounts of wastes generated by innumerous industries that can provide them with large amounts of rich- and low-value materials, which can be processed by innumerous techniques with the objective of obtaining high-valued products with great biological potential and safety.

In this chapter, potential cosmetic compounds derived from industrial wastes were described as a sustainable alternative for the cosmetic industry. In Sect. 16.2, interesting compounds for the cosmetic industry such as antioxidants, ferulic acid, lycopene, sericin, and others are described to be obtained through physical and/or chemical processing of wastes originated from human activity encompassing soil exploitation. Section 16.3 explores marine processing industries' wastes as sources of important raw materials to produce biocosmetics. As in Sect. 16.2, the raw materials normally pass through a physical and/or chemical treatment in order to become suitable for the cosmetic industries. Section 16.4 presents and discusses about technologies of using microorganisms to obtain pure cosmetic ingredients through fermentation processes (and not only chemical or physical processes), which is an infinite source and avoids problems related to purity and security of those molecules obtained from animal and vegetables extractions. Special attention is given to the possibility of reuse of industrial wastewaters as source of nutrients or precursors for microbial fermentation and production of cosmetic interesting molecules, which is an innovative eco-friendly and low-cost approach that will play an important role in the next years.

16.1.1 Definition

According to the Segen's Medical Dictionary (2012), biocosmetic is a cosmetic product containing biotechnology-based product(s) or a cosmetic that has a mechanism of action based on biologic principles. In this context, biocosmetic can be defined as a product that uses, in at least one step of it production (or production of one of its raw materials), some industrial residue or waste from food, feed, or other kind of industry. When used as a substrate for cosmetic preparations, industrial residues and wastes can undergo various kinds of processing, such as chemical, physical, biological, or biotechnological, focusing on their concentration, purification, or bioactivation of molecules of interest. In order to be considered as an interesting material for the production of biocosmetics, the waste must have at least one important and desired property, such as high-antioxidant potential, high amounts of phenolic compounds, capacity to improve sensorial properties, act in the product smoothness or lightness, and being skin softener or hair shiner. Furthermore, as a common cosmetic, it must pass through tests to guarantee its physical, chemical, and microbiological safety, according to international rules and regulations.

Concerns about safety of synthetic molecules and the search for cheaper and eco-friendly molecules resulted in an increase interest in the substitution of synthetic antioxidants by natural ones, especially in cosmetics and pharmaceuticals. It has promoted research on vegetable, animal, marine, and biotechnological sources and the search for new antioxidants in new raw materials.

16.2 Products from the Earth

The human being uses many artifices for their well-being on the planet. Agriculture, livestock, fishery, poultry, textile industry, and many others are technologies in constant development that have played an important role in the evolution of the human race. Not long ago, the waste materials generated by these technologies were improperly disposed in the nature, following no rules and thus impacting directly the environment. In this context, the reuse of such wastes is a new approach in the production of new compounds, reducing their pollution potential and avoiding high costs during waste disposal (Jayathilakan et al. 2012).

Due to its composition and properties, meat, poultry, and fish processing wastes and by-products have a potential to be converted into useful high-valued products. Regulatory requirements are important as many countries restrict the use of meat by-products for reasons of food safety and quality (Jayathilakan et al. 2012).

16.2.1 Antioxidants from Agricultural Wastes

The concept of biorefinery has been used to support several research and technological developments. It comprises the concept of zero emission, or at least minimum losses in which the possibility to use one waste as raw material by another process is evaluated. Many studies have been performed based on this concept, including the valorization of agroindustrial and forestry wastes (Alonso et al. 2011) to produce bioactive molecules with high-antioxidant potential and soybean vinasse to produce lactic acid (Karp et al. 2011). Despite the problems related by Peschel et al. (2006) about the use of agricultural wastes as source of antioxidants and phenolic compounds (heterogeneity of the batches, high costs involved in drying, and some expensive steps of extraction), they seem to be an interesting source of bioactive molecules.

The cultivation of vegetables and crops results in the production of many industrial products and by-products for human and animal consumption. Consequently, it is observed that many residues and wastes are produced, which are commonly burnt or used as feed or fertilizer. Normally, vegetable materials have high levels of phenolic compounds, whose presence is also detected in its by-products and wastes. The phenolic compounds are in most cases highly related to the antioxidant capacity, which makes the agricultural wastes an interesting low-cost source of antioxidants. The properties of the bioactive molecules obtained from agricultural wastes are multiple, such as protection against free radical damage and LDL (low-density lipoproteins) oxidation, besides being safe for human and animal health (Sambanthamurthi et al. 2011).

Antioxidants are widely used to prevent deterioration of oxidizable goods, such as cosmetics, pharmaceuticals, and plastics. Polyphenols are a class of compounds with higher antioxidant activity. Other biological properties, such as anticarcinogenicity, antimutagenicity, antiallergenicity, and bleaching and anti-aging activity, have been reported for natural and synthetic antioxidants, which are some of the desirable properties for pharmaceutical and cosmetic formulae (Moure et al. 2001).

Large variety of agricultural wastes have been studied as source of antioxidant molecules, phenolic compounds, flavors, and colorants. Some examples are wastes from bergamot (Conidi et al. 2011), wastes from the *Solanaceae* family plants processing (tomato, potato, eggplant, and pepper) (Taylor and Fraser 2011), and oil palm liquor (Sambanthamurthi et al. 2011). Peschel et al. (2006) studied residues from juice production: red beet, apple, strawberry, and pear; wastes from the canning industry: tomato, artichoke, and asparagus; after harvesting wastes: chicory, endive, cucumber, and broccoli; and two medicinal herbs rich in polyphenols: golden rod herb (*Solidago virgaurea*) and woad herb (*Isatis tinctoria*). After many steps of material treatment and extraction, it was measured the total phenolic compounds of each waste extract and their antioxidant activity was determined. The best and promising materials were artichoke, apple, and tomato, considered suitable for incorporation in topical cosmetic formulations.

Camellia sinensis is a plant known for being largely used in tea preparations, as green, black, and old tea. The wastes produced in each of these processes were analyzed by Farhoosh et al. (2007) and showed highly promising antioxidant-rich extracts, with potential application in cosmetics, for both topical and ingestion formulations.

An antioxidant-rich extract, mainly composed of flavonoids, was obtained from the *Siraitia grosvenori*'s leaves by Pan et al. (2012). Taiwanese pummelo (*Citrus grandis Osbeck*) wastes were also identified as a rich source of antioxidant extracts, but also showing whitening properties and tyrosinase inhibition comparable to the kojic acid (Wu et al. 2011).

16.2.2 Ferulic Acid

Another interesting molecule that has important antioxidant activity is the phenolic compound called ferulic acid. It is a hydroxycinnamic acid found in plant cell walls which was reported to have a wide range of applications, including as cosmetic raw material due to its properties of skin whitening and antioxidant (Barberousse et al. 2008; Wang et al. 2011) and photoprotection (Graf 1992). Additionally, it can be used as feedstock for the biotechnological production of flavorings and aromatic compounds, including vanillin and vinylguaiacol (Max et al. 2010). Tilay et al. (2008) studied maize bran, rice bran, wheat bran, wheat straw, pineapple peels, and pomegranate peels for the production of ferulic acid, achieving best results with maize bran. Max et al. (2010) detected ferulic acid in prehydrolyzed solid residue of trimming vine shoots.

In cosmetic products ferulic acid can be added to sun protect products, in anti-aging treatments, as adjuvants in skin cancer treatment and in old skin care. There are industrialized products containing ferulic acid associated with vitamins for topical use (usually in concentrations between 0.1 and 0.5 %), but it can also be manipulated in manipulation pharmacies under medical prescription (in concentrations up to 5 %).

16.2.3 Resveratrol

Resveratrol, the antioxidant molecule that has gained a lot of attention in the last years, can be found in agricultural wastes. This molecule has great bioactive properties and has a global market estimation of >$30 billion according to Rayne et al. (2008). The referred authors identified grape cane waste as a potential source of trans-resveratrol and trans-viniferin, whose recovery yield is greatly influenced by the solvent used in extraction. The trans-resveratrol obtained from this waste showed economical and environmental advantages. Peanut roots were also evaluated as sources of resveratrol (Chen et al. 2002).

Another way to obtain the resveratrol is through biotechnology, using recombinant microorganisms and plant cell suspensions. The microorganisms used include *Yarrowia lipolytica*, *Lactococcus lactis*, *Aspergillus niger*, *Aspergillus oryzae*, *Saccharomyces cerevisiae*, and *Escherichia coli*, while plants species include *Arachis hypogea*, *Gossypium hirsutum L.*, and *Vitis vinifera*. Biotechnological production of resveratrol is still limited (culture conditions, optimization, and scaling-up problems), but the yields are considerable (Donnez et al. 2009).

The antioxidant activity of resveratrol occurs through inhibition of cyclooxygenase and lipoxygenase enzymes, acting as anti-inflammatory and anticoagulant. Studies show that resveratrol can induce apoptosis, programmed cell death, acting as an antiproliferative agent for some types of tumors (Zhang et al. 2012). The most common form of presentation is as supplements, providing antioxidant power; preventing wrinkles, elasticity and skin hydration, sunscreen, and anti-stain; and preventing photoaging. Some high added value cosmetic products containing resveratrol combined with other assets are on the market. Resveratrol is said to prevent formation of blemishes, sun damage, and skin lightening and also shows anti-wrinkle effect and improves the overall appearance of the skin. There are reports of products containing 1 % resveratrol that provided a significant lightening of the skin in just 14 days (Howard 2011).

There are few studies reporting the real acting benefits and transformations that resveratrol suffers in the human organism. Studies carried in pigs by Baxter (2008) demonstrated good absorbency through the stratum corneum. The same author concluded that the topical use seems to pose little risk and potentially large benefits (Baxter 2008).

16.2.4 Lycopene

Lycopene is a bioactive red colored pigment, largely used as pigment and antioxidant, belonging to carotene molecules class. It occurs in plants, such as tomato, papaya, pink grapefruit, red guava, red grapes, and watermelon (Kong et al. 2010).

Tomato processing wastes and by-products are considered as a potential source of bioactive compounds, mainly lycopene. Lycopene was detected in tomato peel and tomato peel fiber, tomato skin (Benakmoum et al. 2008), tomato paste waste (Xi 2006), tomato peels and seeds (Sandei and Leoni 2006), tomato pomace (Vagi et al. 2007), and also in pulp waste (Chiu et al. 2007).

Lycopene properties have been extensively studied. Activities against iron-induced oxidative stress damage (Matos et al. 2006), lipid peroxidation and oxidative DNA damage (Matos et al. 2000), and cardiovascular and related diseases, cancer, and diabetes (Kong et al. 2010) have also been described.

Lycopene's antioxidant properties enable it to be used as ingredient in cosmetic formulation. The natural carotenoid content in the human skin is related to the healthy diet rich in fruits and vegetables, and external factors, such as illness, UV and IR radiation of the sun, smoking, and alcohol consumption, reduce the concentration of the carotenoids in the skin (Lademann et al. 2011). In this context, topical uses of a lycopene-rich gel or cream improve the natural barrier against the premature skin aging. The challenge is to determine if the topical use has the same effects of a lycopene-rich healthy diet. Studies demonstrated that it could be a great sun-screening agent as well as an excellent antioxidant for cosmetic formulations (Maheshwari et al. 2012).

Most studied lycopene extraction method is supercritical CO_2, which causes less degradation and loss of properties and is thus considered the most effective processes to extract the lycopene (Vagi et al. 2007; Shi et al. 2009). Methods that avoid excessive heating also showed higher lycopene recovery yields (Kerkhofs et al. 2005). (Navarro-González et al. 2011) proposed the use of enzymatic or physical pretreatments prior to extraction.

Lycopene-rich extract is already commercialized (LycoMega®, Lycomplete®) in form of food supplements, but few industries have released an extract suitable for cosmetic use. An example is Hydropom, an aqueous extract compatible to creams and gels with dosage recommended between 3 and 5 %. Lycopene containing products marketing usually explores the appeal of organic, green, or animal cruelty free.

16.2.5 Olive Mill Wastes

The steps involved in the olive oil extract are crushing, pressing, and separation of oil from water. There are three different processes to extract olive oil: simple pressing (discontinuous), continuous three-phase, and continuous two-phase centrifugal olive oil extraction. The two-phase method is the most modern one, achieving higher yields and producing a low-moisture cake but generating larger amount of solid and liquid wastes (Matosa et al. 2010). The tendency of replacing discontinuous and three-phase technologies by the two-phase means that larger amounts of such wastes will be generated.

In the olive oil process, the yield is 2:3:5 olive oil to semi-solid cake to aqueous liquor, corresponding to larger production of residues. As the olive oil process requires a high heating power, these residues have been used in the energy generation within the plant (Matosa et al. 2010). Many studies have also shown the possibility to use the olive stone in many industries as plastic filling, abrasive, biosorbent, animal feed, resins, bio-oil, and activated carbon (Rodríguez et al. 2008).

The olive mill waste, rich in polyphenols, is formerly was used for long time as renewable energy source or fertilizer. It has been proposed for its use as a low-cost substrate for the production of xanthan and ethanol. The production of active compounds from this waste constitutes a viable alternative for valorizing this problematic waste (Obied et al. 2008).

Many works have shown that simple treatment can generate stable antioxidants (Capasso et al. 2006; Galanakis et al. 2010). They can play two functions in a cosmetic product: as preservative and as bioactive molecule maintaining and improving skin health. From the olive mill waste water, it is possible to produce hydroxytyrosol triacetyl from hydroxytyrosol, an abundant molecule present in this residue. Analysis has shown that hydroxytyrosol triacetyl produced through fermentation has the same activity as the one obtained through chemical path, showing advantages, such as higher stability and no toxicity (Capasso et al. 2006).

Other phenolic compounds were identified in the olive oil, including hydroxytyrosol acetate, tyrosol and catechol (Brenes et al. 2004), caffeic acid, coumaric acid, and oleuropein (Garcia-Castello et al. 2010). Concentration of such compounds reaches 600 mg/kg, which makes their recovery attractive for cosmetic application (Brenes et al. 2004).

The potential of the phenolic compounds from olive has been proved by the development of two commercial products, recommended for use against free radicals. They are composed of 3,4-dyhydroxyphenylethanol and 3,4-dihydroxyphenylglycol, compounds present in large amounts in the olive fruits, oil, and wastes (Rodríguez et al. 2007).

The olive oil itself is widely used in a variety of cosmetic products, such as shampoos, conditioners, liquid and bar soaps, creams, and lotions. The incorporation of the powder from the olive stone in cosmetics as an exfoliating agent is a new approach and is being used recently (Cosmoliva 2012; RedFlower 2012; Rodríguez et al. 2008; Mohammadi et al. 2005; Korres 2012).

16.2.6 Wastes from the Textile Industry

16.2.6.1 Sericin

Sericin is a globular protein secreted by the silkworm, *Bombyx mori,* during spinning of the cocoon. It represents 25–30 % of silk proteins and plays a fundamental role in the silk fiber acting as cement by sticking it together, providing a sticky coating in the fibroin fiber (Capara 2012; Aramwit et al. 2011). The silk fibers have many uses in a wide range of industries, such as textiles, medical, biotechnological, and cosmetic, due to its properties of water absorbency (due to its high contents of serine and aspartic acid), dying affinity, thermotolerance, luster, and insulation (Aramwit et al. 2011).

In silk manufacture of fibroin and sericin, the two major proteins secreted by the silkworm are separated through a degumming process where fibroin is recovered and sericin is discarded in the wastewater. The degumming wastewater is known for its high-oxygen demand (Fabiani et al. 1996) and if not treated correctly can result in environmental contamination. It is estimated that global production of 400,000 tons of dry cocoon, which means approximately 50,000 tons of sericin are discarded in degumming wastewater (Kim 2007; Zhang 2002).

The most suitable methods for sericin recuperation from textiles wastewaters are ultrafiltration (UF) and nanofiltration (NF) (Aramwit et al. 2011). Drying can also

Table 16.2 Biomedical applications of sericin extract from textile industry wastes

Application	Reference
Improve serum-free mammalian cell culture	Terada et al. (2005)
Enhance culture human skin fibroblast attachment and play an important role in the wound healing process	Tsubouchi et al. (2005)
Supply healing process	Ogawa et al. (2004a, b)
Enhance wound healing by sericin cream	Aramwit and Sangcakul (2007)
Suppress UVB-induced and chemical-induced acute damage and tumor promotion in mouse skin	Zhaorigetu et al. (2003)

Table 16.3 Some studies on different properties of sericin and its potential of use in cosmetic industry

Material	Characteristics	Reference
Blending protein with other materials: Polyurethane; foam	Good moisture absorbing/ deabsorbing properties	Nomura et al. (1995)
Polyvinyl alcohol (PVA); film	Highly interface of PVA/sericin complex and excellent fracture strain	Ishikawa et al. (1987)
PVA; hydrogel	Good moisture absorbing/deabsorbing properties	Yoshii et al. (2000)
Improving by the cross-linking of protein; dimethylolurea	High strength and water permeability cross-linked sericin in short time	Gimenes et al. (2007)
UV-radiation	Good mechanical properties and cytocompatibility sericin films	Kim et al. (2001)
Polyethylene glycol diglycidyl ether	Good mechanical properties and sericin can be released efficiently for promoting collagen production	Xie et al. (2007)
Genipin	Film producer and surfactant agent	Aramwit et al. (2010)

Adapted from Aramwit et al. (2012)

be used but did not show interesting results (Capara 2012). After sericin recuperation, a decrease in pH, COD, BOD_5, and total solids is observed, indicating that the extraction methods are suitable to sericin recovery and wastewater treatment. Future perspectives in this field are the use of enzymatic hydrolysis to obtain highly purified sericin.

There are large quantities of studies showing the biological activities and uses for the sericin obtained from textile wastewater. They are summarized in Tables 16.2 and 16.3. In Table 16.2, some biomedical applications of sericin extract from textile industry wastes are presented, while in Table 16.3, some studies on different properties of sericin and its potential for use in cosmetic industry are presented.

16.2.6.2 Meat and Leather Industries

Many wastes generated by the meat industry such as tendons are rich in collagen. Several studies have reported the downstream steps to obtain a hydrolyzed collagen-rich material and its use in cosmetic care preparations, such as humectant (Mokrejs et al. 2009; Cot 2004), and for the production of acylamino-carboxy surfactants (known for their favorable dermatological effects) (Mokrejs et al. 2009). Another example is the use of such collagen hydrolyzates from leather collagen proteins, edible meat product casings, and others in the production of potentially edible hydrogels for biodegradable packaging (Langmaier et al. 2008), which can be used also in cosmeceutical supplements. The collagenic biomaterials generated by using leather wastes have a wide range of applications, especially when highly purified. Collagen extracted from leather wastes can be used in formulation of gel, film, sponge, or fibers, representing economic and environmental advantages (Catalina et al. 2011).

When the leather wastes are used for protein extraction, fat is usually discarded. The transesterification of such fat produces an ester with potential use for cosmetic industry (Isler et al. 2010), completing the cycle of the total reuse of the leather wastes for cosmetics applications.

Despite being not capable to replace neither play the same function in the skin when compared with the collagen produced by the human being, the collagen extracted from different sources acts at least moisturizing and giving softness to it. The main reason of this unequal action is due to the particle size: the topical collagen cannot penetrate the skin barrier and will not supplement collagen deficiencies. The main feature of cosmetics containing collagen is the moisture added to the skin, because this molecule is capable to bind water, resulting in a conditioned and moisturized skin. So this kind of product acts decreasing water loss by the skin, improving its texture and increasing its suppleness. Manipulated formulas and industrialized products contain normally 2–10 % of collagen and are indicated for body and face skin firmness, anti-wrinkle for the eye area, streaks prevention, and others.

16.3 Products from the Sea

Currently, various metabolites such as terpenoids, tocopherol, polysaccharides, carotenes, phenolic compounds and antioxidant molecules, chitin, chitosan, and unsaturated fatty acids from different marine organisms (bacteria, micro- and macroalgae) have been researched and studied. This kind of research fits the market demand for natural cosmetic products and has shown high rates of conversion into commercial products. Major drawbacks are the sustainable production of such compounds and proof of absence of biological or chemical pollutants (Imhoff et al. 2011).

The main products from the sea with potential for use in cosmetic are collagen, the chitosan/chitin, and the polyunsaturated fatty acids (PUFA). These products have as main source the wastes generated by food industry processes involving fish, shrimps, and algae. Once large amounts of rich residues are generated, the full reuse of these materials is not only environmental friendly but also economically interesting.

16.3.1 Collagen

Collagen is the most abundant protein in vertebrates and constitutes about 25 % of vertebrate total proteins. It forms unique insoluble fibers that have high tensile strength and a right-handed triple superhelical rod consisting of almost three identical polypeptide chains (Ogawa et al. 2004a, b).

Traditionally, collagen is isolated from the skins of land-based animals, such as cow and pig. Non-denatured and denatured collagen have different applications: the first one is suitable for cosmetics, biomedical, and pharmaceutical industries, while the second finds applications in the food and biomedical industries (Ogawa et al. 2004a, b).

Despite the fact that fish collagens denature at temperatures below 30 °C and possess lower stability than the mammalian counterparts (Ogawa et al. 2004a, b), many studies have reported the extraction and purification of this molecule from wastes generated by the processing of sea animals, such as fish. Many species have been studied, such as black drum (*Pogonias cromis*), sheepshead seabream (*Archosargus probatocephalus*) (Ogawa et al. 2004a, b), yellowtail fish (Morimura et al. 2002), and others.

Downstream steps for collagen purification include molecule acid or enzymatic hydrolysis followed by steps of precipitation and re-solubilization and dialysis. Analyses of viscosity, protein, and amino acid composition are carried out to determine purity and properties of the obtained collagen. Some studies report a moderately heat-stable collagen, suggesting its use as substitute for the land-based animal ones, about which there are some concerns from consumers and manufacturers (Ogawa et al. 2004a, b).

The collagen from fish and livestock processing wastes seems to have interesting properties for the cosmetic industry: high water retention capacity, ability to repair rough skin, lack of odor, and absence of harmful effects on skin. The collagen protein hydrolyzate shows higher antiradical activity (Morimura et al. 2002), which is interesting to be used as food and cosmetic additive.

An interesting field in collagen recovery is enzymatic hydrolysis. While acidic treatment can break the structure and remove protein from the material, the enzymatic treatment can bring some improvements to the process, such as specificity and a higher-quality final material. Many microorganisms are capable of producing collagenases, such as *Aspergillus*, *Phizopus*, and *Bacillus*, and some of them can reach high yields of degradation efficiency (Morimura et al. 2002).

16.3.2 Compounds from Shrimps

The waste generated by the shrimp processing contains several bioactive compounds, such as chitin, pigments, amino acids, and fatty acids. These bioactive compounds have a wide range of applications including medical, therapies, cosmetics, paper, pulp and textile industries, biotechnology, and food applications (Kandra et al. 2012).

Shrimp heads are usually removed in peeling shed or at packaging facilities. They are a large source of protein and oils but are disposed in landfills or discarded in sea water, becoming an important concern for environmental pollution as they have higher organic load and result in unpleasant smell. About 45–48 % of the shrimp weight is discarded as waste, reinforcing its higher availability for further processing. As it still contains many interesting and valuable compounds, appropriate processing can be of great industrial interest (Kandra et al. 2012; Wang et al. 2005), and other properties, such as antimicrobial potential, have also been reported (Wang et al. 2005).

Shrimp wastes, fermented or not, have been reported as source of carotenoproteins, which, when hydrolyzed, results in a mix of amino acids, including large amount of the essential ones. It can also be an interesting source of astaxanthin protein-free, which can have important applications in cosmetic industry, as pigment and mainly antioxidant, 10 times higher than other carotenoids, such as zeaxanthin, lutein, canthaxanthin, and beta-carotene, and 500 times higher than alfa-tocopherol (Armenta and Guerrero-Legarreta 2009).

In the processing of shrimp wastes, the biotechnology again appears as an important tool to obtain most purified products. With the use of specific proteases, the proteins from shrimp waste can be hydrolyzed, and the recovery of astaxanthin and protein hydrolyzate may be feasible (Armenta and Guerrero-Legarreta 2009).

16.3.3 Polyunsaturated ω-Fatty Acids (PUFA)

Many molecules isolated from the sea products, wastes, and subproducts are being reported as having beneficial effects against heart diseases and reduced inflammatory reactions, as well as good source of polyphenol molecules, with antioxidant properties. Nowadays, these products are often used as food supplements, but have a potential to be used in the cosmetic industry (Imhoff et al. 2011). Among these important classes are the polyunsaturated fatty acids, called PUFA, which play an important role in the human and animal health.

PUFA and the products derived from them are involved in many enzymatic routes, metabolic and cellular structure functions, and prevention of degenerative diseases (such as Parkinson and Alzheimer), as well as in anti-inflammatory processes (Martin et al. 2006). The major studied PUFA include docosahexaenoic acid (DHA) and eicosapentaenoic acid (EPA) (both ω-3 fatty acids) and arachidonic acid (ARA) and gamma-linoleic acid (GLA) (both ω-6 fatty acids).

Docosahexaenoic acid (DHA) can be extracted from fish or produced by biotechnological ways. Chi et al. (2007) reported the production of DHA using a substrate for the growth of the microalga, *Schizochytrium limacinum,* using crude glycerol, a by-product from the biodiesel industry. This by-product is too costly to be purified, and its use as raw material to another process is suitable, especially for microorganisms' cultivation. This study suggests that the crude glycerol is a promising feedstock for production of DHA from heterotrophic algal culture (Chi et al. 2007).

The viscera represent one of the poorest used wastes from meat processing. In many countries, they are used for human nutrition, but the excess is still discarded without any processing. Some studies demonstrated that viscera can be an interesting source of high-valued molecules, such as PUFA. Wang et al (2009) analyzed the fatty acid profile of the yak kidneys, generally considered a waste, and found fifteen different fatty acids, including arachidonic acid. This kind of study indicates the suitability for development of possible PUFA-rich products from meat wastes.

Both eicosapentaenoic and arachidonic acids were also reported to be produced by biotechnological means. They can be produced using fungi, *Mortierella alpina*, cultivated in a mixture of crude soybean oil, a sucrose waste stream and soymeal waste stream (Cheng et al. 1999).

16.4 Biotechnological Products

According to Thakur (2006), the environment is used by humans to sustain their basic needs for survival and to provide raw materials to be converted into products and services. There is an urgent need to exploit the biotechnological innovations, using industrial wastes and by-products in the production of food, feed, energy, and also pharmaceuticals and cosmetics. There is an increasing tendency in the search for new routes and new products for the replacement of chemical and traditional technologies by biocatalysts and genetically engineered microorganisms technologies capable of using renewable raw materials without loss of quality or energy.

In this part of the chapter, various compounds, such as antimicrobials, preservatives, polysaccharides, biosurfactants, and retinoid obtained through biotechnological production from a great variety of wastes and by-products, will be discussed. At the same time, it results in waste valorization through production of high-valued cosmetic compounds, biotechnological processes from unconventional substrates avoiding costs related to waste disposal and treatment, thus being an interesting alternative for the cosmetic industry.

16.4.1 Antimicrobial and Preservative

Antimicrobial is a molecule or a group of molecules capable of limiting microorganism growth or even completely stopping it. It is a useful ingredient in cosmetic industries as preservative.

The recent discussions on the safety of parabens, a largely used preservative in the cosmetic industry, are opening new approaches for the development of new antimicrobials and preservatives. In this context, some studies (Peng et al. 2006; Nguyen et al. 2005) have contributed to elucidate the reaction mechanisms of parabens in the cell structure that confer its large activity against contaminants.

Nguyen et al. (2005) indicated that the interaction between parabens and bacteria is provided by the action in bacterial channels, collapsing the cell turgor and allowing the leakage of cytoplasmic contents. Nevertheless, side effects of parabens in animals and humans were not yet clearly identified and understood. The search for new preservative products with little or no toxicity, feasible to substitute synthetic parabens, is a field that is gaining importance. Natural parabens production was identified by Peng et al. (2006), who had isolated a bacterium capable to produce a parabens class molecule which, despite lows yields of production, was capable of preventing growth of yeasts, molds, and Gram-positive bacteria.

The innumerous studies found in the literature about preservatives and antimicrobials of biotechnological origin shows that biotechnology has a great contribution to the cosmetic industry, being capable to supply it with various classes of molecules, both as compound for the formula and as active principles.

16.4.2 Polysaccharides

The industrial applications of microbial polysaccharides are very diverse, especially in the cosmetic industry. Polysaccharides produced by bacteria, fungi, and microalgae are used in cosmetic formulations as both active ingredients and sensorial improver. Industrial wastes can be a suitable substrate for the production of microbial polysaccharides for cosmetic applications, combining economic and environmental advantages. The main polysaccharides used in cosmetics are hyaluronic acid and xanthan gum.

16.4.2.1 Hyaluronic Acid

The hyaluronic acid is a polysaccharide traditionally obtained by extraction from animals, as reported by Murado et al. (2012), who described its extraction from fish eyeball after various steps of purification, reducing the environmental impact of this residue. It is naturally present in the human body as lubricant liquid in the joints and in the eye and spread in human skin, sustaining and moisturizing it.

The internal use of the hyaluronic acid can be as a medicinal task, filling joints and in the treatment of conjunctive tissue diseases, or as a cosmetic task, used in wrinkles filling or in plastic surgery. Recent developments include capsules for ingestion and cosmetics containing fractions of hyaluronic acid. The real efficacy of this kind of products is questionable because its penetration in the skin can be tough. Many tools as nanotechnology can facilitate it; in some products, the nanoparticulated hyaluronic acid is told to penetrate to the dermis, minimizing the wrinkles and maintaining the tonus and a moisturized skin.

The main concern about the hyaluronic acid is its safety for human use due to possible viral or prion contamination as it is an animal-originated product. The steps needed to guarantee its safety enhance too much the final price and can derail its production,.

In this context, many works have been carried on biotechnological ways to produce hyaluronic acid. A strain of *Streptococcus thermophilus* was isolated from dairy products and cultured in skimmed milk. *S. thermophilus* was capable to produce hyaluronic acid with different molecular weight together with exopolysaccharides, showing potential use in medical, cosmetic, and food applications (Izawa et al. 2009).

16.4.2.2 Xanthan Gum

Xanthan gum is the second microbial polysaccharide most commercialized in the world. It was discovered in 1963 and since has been extensively studied. It has many applications in food and feed industries (bakery, beverage, dairy, dressings, pet food, syrups and toppings, relish, sauces, gravies) but also in oil industry, agricultural products, cleaners, pharmaceuticals, and coating and paper industries (Palaniraj and Jayaraman 2011).

The cosmetic application of xanthan gum includes the improvement of the flow properties of shampoos and liquid soaps, promoting also a stable, rich, and creamy lather. With its good rheological properties, it has become an excellent binder for all toothpastes, including gel and pumpable types, also improving the ribbon quality and the ease of extruding (Palaniraj and Jayaraman 2011).

Xanthomonas campestris is one of the best xanthan gum producers, and despite sucrose and glucose are recognized as the two best carbon sources for the cultivation, many other sources, considering the use of wastes, were successfully tested: hydrolyzed rice, barley and corn flour, acid whey, cheese whey, sugarcane molasses, a mixture of mannose and glucose, waste sugar beet pulp, maize, cassava starch, and peach pulp. The xanthan gum obtained using any of these materials are considered safe for human use (Palaniraj and Jayaraman 2011) and is perfectly suitable to the production of biocosmetics, aggregating value to a residue.

16.4.2.3 Other Polysaccharides and Oligosaccharides

Microorganisms isolated from extreme environments can be large producers of polysaccharides. They use polysaccharide in protection against high or low temperatures, excess of salinity, or even excess of movement.

Salipiger mucosus, isolated from halophile environments from the Mediterranean seaboard, showed the capacity of producing large amounts of exopolysaccharides (EPS) (Llamas et al. 2010). After purification, such EPS showed very low viscosity, pseudoplastic behavior, and emulsifying activity on several hydrophobic substrates. Moreover its composition showed a high amount of fucose rich, an interesting material for cosmetic products, since studies have demonstrated that fucose has action in eye-contour wrinkles and other skin benefits (Fodil-Bourahla et al. 2003).

Oligosaccharides are commonly used in cosmetics as stabilizers and bulking agents. Adachi and Vallee (2002) evaluated the skin emollient efficacy of various oligosaccharides, obtained by a controlled enzymatic depolymerization of membranous polysaccharides (agarose, carrageenan, alginate, and fucoidan) extracted from brown seaweed. The high molecular weight oligosaccharide applied to skin-protected cells from cigarette smoke, pesticides, and heavy metals, while the low molecular weight oligosaccharide showed anti-inflammatory effect. When chelated with metals, the low-weight polysaccharide presented other activities: with zinc it showed sebum regulation and antibacterial action, and with manganese and magnesium, it inhibited free radical activity and prevented Langerhans cell from UVB radiations.

Ken-ichi et al. (2002) study showed that trehalose influences the water-holding properties of lipid multilamellar structure existing in the stratum corneum.

Cyclodextrins are used to solubilize fragrance, suppress their volatility, and facilitate spraying, thus being used in deodorants to neutralize smoke, mold, cooking, stale, and pet odors emanating from clothes. They are also used in skin care products (soaps and shampoos) to stabilize antibacterial monoterpenoid (Dodziuk 2006; Patel and Goyal 2011).

16.4.3 Biosurfactants

Biosurfactant or bioemulsificant is an emulsifier molecule produced by a microorganism. It acts by reducing the superficial tension between two or more immiscible or poor miscible materials, allowing it to mixture better. It has a great structural diversity, belonging to glycolipids (Nitschke et al. 2011), lipopeptides, peptides and polymers, and fatty acids classes (Parkinson 1985). Usually, the primary step of recovery of such biomaterial results in a crude extract composed of a mixture of compounds (carbohydrate, protein, lipids, uronic acids and others, and sulfates) (Bodour et al. 2003). In many cases, this crude extract shows improved properties than the purified compound.

Generally biosurfactants' properties are compared with the synthetic surfactants' Triton and Tween, which are frequently used in cosmetic and pharmaceutical industries. These biosurfactants frequently show great potential for use in cosmetics, replacing synthetic substances in the formulation. A wide range of microorganisms are known to produce these compounds: marine bacterium (Kumar et al. 2007), yeasts (*Candida lipolytica*) (Rufino et al. 2007), *Mycobacterium*, *Nocardia*, *Rhodococcus*, *Corynebacteria*, *Pseudomonas*, *Torulopsis*, *Acinetobacter calcoaceticus*, and *Bacillus subtilis* (Parkinson 1985).

In the beginning, biosurfactants were researched solely for environmental applications, especially decontamination and bioremediation. Later, high-valued applications were researched, because of the higher cost of production. In cosmetics, they are used as emulsifiers, humectants, preservatives, and detergents. To overcome the cost production, the use of industrial wastewaters is a new interesting approach.

Due to biosurfactant's amphiphilic nature, it is necessary to provide special conditions to stimulate its production by the microorganism. In this context, oily residues can provide both a great biosurfactant production and lower costs of production.

Some oily residues used in this subject are those resulting from vegetable oil refinery (Rufino et al. 2007), sludge palm oil (Nawawi et al. 2010), soap stock from vegetable oil refining process (Benincasa et al. 2004), and distillery and dairy wastes (Makkar and Cameotra 2002).

Many advantages of biosurfactants production have been reported. The main positive aspect is the higher biodegradability as compared to synthetic surfactants, the organic load reduction of the wastewaters used as culture medium, and lower sensitivity to extreme temperatures, pH, or salinity (Parkinson 1985). Further, if produced by fermentation using cheaper renewable substrates, it is independent of fossil hydrocarbon resources (Parkinson 1985).

According to Klekner and Kosaric (1993) and Maier and Soberón-Chávez (2000), some of the products composed of surfactants among their main ingredients include insect repellents, acne pads, antidandruff products, deodorants, nail care products, and others. Biosurfactants have found a special place in the personal care market also because of their intrinsic low toxicity, excellent moisturizing properties, and skin compatibility (Brown 1991). Klekner and Kosaric (1993) described a product containing 1 mol of sophorolipid and 12 mol that had become a commercial skin moisturizer cream; patents for the use of rhamnolipids to make liposomes and emulsions (Ishigami et al. 1988) and mycolates are suited for applications in creams, pastes, sticks, and films (Lang and Philp 1998).

The review on the use and production of biosurfactants performed by Makkar and Cameotra (2002) can be used for further information in this subject.

16.4.4 Retinoids

Retinoids are lipophilic isoprenoids derived from β-carotene composed of a cyclic group and a linear chain with a hydrophilic end group (Jang et al. 2011). These compounds include retinol (vitamin A), retinal, retinyl esters, and retinoic acid and derivatives. All these molecules are known cosmetic agents and effective pharmaceuticals for skin diseases. Retinoids are chemically unstable and biologically degraded via retinoic acid. It can be produced by biotechnology, employing genetic modified microorganisms or wild strains. *E. coli* was used by Jang et al. (2011) in a two-phase culture system using a dodecane overlay to produce blocks that compound the retinoids molecules.

Retinol can be obtained also through extraction from fermented shrimp waste. An acid lactic fermentation can be used to preserve shrimp, generating a fermented waste containing a protein-rich liquor, a lipid fraction, and insoluble chitin. The lipid fraction can then be fractionated into carotenoid pigments (mainly, astaxanthin), sterols (cholesterol), and retinol (vitamin A). Nowadays, this kind of process is used in the aquaculture industries, but also has potential to be used in cosmetic industries and to be proven for its safety and absence of contaminants in the purified compound.

16.5 Future Prospects

The future of biocosmetics is promising. Every day new researches, processes, and material are being discovered, and the existing ones are frequently ameliorated. Industries and enterprises have their own R&D (research and development) departments, whose studies and developments rarely become accessible to general public. These researches are confidential as trade secret and we only have access when a new product arises in the market. It is, thus, hard to foresee a great innovative product before it launches on market, just looking for academic researches.

The use of wastes in cosmetic production allows the use of "sustainable" and "green" claims in their products, which will improve sales and generate a better acceptance by the consumers. For industrial economics, it is also a good point: the costs involved in the process are lower and the enterprise's social image is ameliorated. The production of biocosmetics aims the appearance of innovative products, satisfying the select cosmetic consumers who are always looking for novel and improved products.

The development of biocosmetic products has some drawbacks. The absence of specific laws hinders the raw materials and products to be correctly categorized. These products have to follow the same principles of absence of contamination, safety and efficacy than the normal cosmetics, but maybe other specific tests should be performed to ensure their safety. It could make the consumers feel safer enough to use the biocosmetics without discomfort and preconception. In this context, many researches are using food and pharmaceutical wastes for biocosmetic productions, once these products have to follow very restrictive healthy security parameters, including the wastes generated during their production, which makes these wastes safer than others from other sources. Another drawback for the industrial production of biocosmetics from waste is its availability: the waste has to be supplied at constant amounts and with a relatively constant composition; this can be a problem specially when the waste comes from a seasonal culture and/or varies each different batch or season.

Biorefinery concept is soaring and growing. It began to be applied first in less restrictive industries, such as biocombustibles and fertilizers, passing through food and feed industries. Now it is arriving to cosmetic and pharmaceutical ones. It was possible due to the growing development of technologies and techniques which allow wastes decontamination and assays that ensure its security. Much more can and will be done in the field of biocosmetics, a fascinating, inexhaustible, and increasing area. The diversity of wastes and microorganisms can be widely explored in the development of a variety of process and new products for health, beauty, and well-being.

References

Adachi M, Vallee R (2002) Recent research on glycolipid. New effects of oligosaccharides from brown seaweed. Fragr J 30(5):73–83

Alonso JL, Dominguez H, Garrote G, Gonzalez-Munoz MJ, Gullon B, Moure A, Santos V, Vila C, Yanez R (2011) Biorefinery processes for the integral valorization of agroindustrial and forestal wastes. CyTA J Food 9(4):282–289

Aramwit P, Sangcakul A (2007) The effects of sericin cream on wound healing in rats. Biosci Biotechnol Biochem 71:2473–2477

Aramwit P, Siritientong T, Kanokpanont S, Srichana T (2010) Formulation and characterization of silk sericin-PVA scaffold crosslinked with genipin. Int J Biol Macromol 47:668–675

Aramwit P, Siritientong T, Srichana T (2011) Potential applications of silk sericin, a natural protein from textile industry by-products. Waste Manag Res 30:217–224

Armenta RE, Guerrero-Legarreta I (2009) Amino acid profile and enhancement of the enzymatic hydrolysis of fermented shrimp carotenoproteins. Food Chem 112:310–315

Barberousse H, Roiseux O, Robert C, Paquot M, Deroanne C, Blecker C (2008) Analytical methodologies for quantification of ferulic acid and its oligomers. J Sci Food Agric 88(9):1494–1511

Baxter RA (2008) Anti-aging properties of resveratrol: review and report of a potent new antioxidant skin care formulation. J Cosmet Dermatol 7:2–7

Benakmoum A, Abbeddou S, Ammouche A, Kefalas P, Gerasopoulos D (2008) Valorization of low quality edible oil with tomato peel waste. Food Chem 110:684–690

Benincasa M, Abalos A, Oliveira I, Manresa A (2004) Chemical structure, surface properties and biological activities of the biosurfactant produced by *Pseudomonas aeruginosa* LBI from soap stock. Antonie Van Leeuwenhoek 85:1–8

Bodour AA, Drees KP, Maier RM (2003) Distribution of biosurfactant-producing bacteria in undisturbed and contaminated arid Southwestern soils. Appl Environ Microbiol 69(6):3280

Brenes M, Romero CN, Garciá A, Hidalgo FJ, Ruiz-Méndez MV (2004) Phenolic compounds in olive oils intended for refining: formation of 4-ethylphenol during olive paste storage. J Agric Food Chem 52:8177–8181

Brown MJ (1991) Biosurfactants for cosmetic applications. Int J Cosmet Sci 13:61–64

Capara G (2012) Separation of silkworm proteins in cocoon cooking wastewaters via nanofiltration: effect of solution pH on enrichment of sericin. J Memb Sci 389:509–521

Capasso R, Sannino F, Martino F, Manna C (2006) Production of triacetylhydroxytyrosol from olive mill waste waters for use as stabilized bioantioxidant. J Agric Food Chem 54:9063–9070

Catalina M, Celma P, Cot J, Manich A, Marsal A (2011) Tailor-made biomaterials from collagenic wastes. J Am Leather Chem Assoc 106(5):153–160

Chen RS, Wu PL, Chiou RYY (2002) Peanut roots as source of resveratrol. J Agric Food Chem 50(6):1665–1667

Cheng MH, Walker TH, Hulbert GJ, Raman DR (1999) Fungal production of eicosapentaenoic and arachidonic acids from industrial waste streams and crude soybean oil. Bioresource Technol 67(2):101–110

Chi Z, Pyle D, Wen Z, Frear C, Chen S (2007) A laboratory study of producing docosahexaenoic acid from biodiesel-waste glycerol by microalgal fermentation process. Biochemistry 42:1537–1545

Chiu YT, Chiu CP, Chien JT, Ho GH, Yang J, Chen BH (2007) Encapsulation of lycopene extract from tomato pulp waste with gelatin and poly(γ-glutamic acid) as carrier. J Agric Food Chem 55:5123–5130

Conidi C, Cassano A, Drioli E (2011) A membrane-based study for the recovery of polyphenols from bergamot juice. J Memb Sci 375:182–190

Cosmoliva (2012) Web site of Cosmoliva Organics. http://www.cosmoliva.com/. Accessed 25 Aug 2012

Cot J (2004) An imaginary journey to the collagen molecule for a better understanding of leather waste treatments. J Am Leather Chem Assoc 99(8):322–350

Dodziuk H (2006) Cyclodextrins and their complexes. Wiley-VCH Verlag GmbH & Co., KGaA, Weinheim

Donnez D, Jeandet P, Clément C, Courot E (2009) Bioproduction of resveratrol and stilbene derivatives by plant cells and microorganisms. Trends Biotechnol 27(12):706–713

Fabiani C, Pizzichini M, Spadoni M, Zeddita G (1996) Treatment of waste water from silk degumming processes for protein recovery and water reuse. Desalination 105(1–2):1–9

Farhoosh R, Golmovahhed GA, Khodaparast MHH (2007) Antioxidant activity of various extracts of old tea leaves and black tea wastes (*Camellia sinensis* L.). Food Chem 100(1):231–236

Farlex Segen's Medical Dictionary (2012) http://medical-dictionary.thefreedictionary.com. Accessed 2 Sep 2012

Fodil-Bourahla I, Bizbiz L, Schoevaert D, Robert AM, Robert L (2003) Effect of L-fucose and fucose-rich oligo- and polysaccharides (FROP-s) on skin aging: penetration, skin tissue production and fibrillogenesis. Biomed Pharmacother 57(5–6):209–215

Galanakis CM, Tornberg E, Gekas V (2010) Recovery and preservation of phenols from olive waste in ethanolic extracts. J Chem Technol Biotechnol 85(8):1148–1155

Garcia-Castello E, Cassano A, Criscuoli A, Conidi C, Drioli E (2010) Recovery and concentration of polyphenols from olive mill wastewaters by integrated membrane system. Water Res 44:3883–3892

Gimenes ML, Liu L, Feng X (2007) Sericin/poly (vinyl alcohol) blend membranes for pervaporation separation of ethanol/water mixtures. J Memb Sci 295:71–79

Graf E (1992) Antioxidant potential of ferulic acid. Free Radic Biol Med 13(4):435–448

Howard H (2011) Beauty spots in: skin deep. The New York Times. http://www.nytimes.com/2011/04/14/fashion/14SPOTS.html?_r=0. Accessed 22 Nov 2012

Imhoff JF, Labes A, Wiese J (2011) Bio-mining the microbial treasures of the ocean: new natural products. Biotechnol Adv 29:468–482

Ishigami Y, Gama Y, Uji Y, Masui K, Shibayama Y (1988) Japanese Patent Kokai. pp 63–77, 535

Ishikawa H, Nagura M, Tsuchiya Y (1987) Fine structure and physical properties of blend film compose of silk sericin and poly(vinyl alcohol). Sen'I Gakkaishi 43:283–287

Isler A, Sundu S, Tuter M, Karaosmanogluy F (2010) Transesterification reaction of the fat originated from solid waste of the leather industry. Waste Manag 30(12):2631–2635

Izawa N, Hanamizu T, Iizuka R, Sone T, Mizukoshi H, Kimura K, Chiba K (2009) *Streptococcus thermophilus* produces exopolysaccharides including hyaluronic acid. J Biosci Bioeng 107(2):119–123

Jang HJ, Yoon SH, Ryu HK, Kim JH, Wang CL, Kim JY, Oh DK, Kim SW (2011) Retinoid production using metabolically engineered *Escherichia coli* with a two-phase culture system. Microb Cell Fact 59(10):1–12

Jayathilakan K, Khudsia Sultana, Radhakrishna K, Bawa AS (2012) Utilization of byproducts and waste materials from meat, poultry and fish processing industries: a review. J Food Sci Technol 49(3):278–293

Kandra P, Challa MM, Jyothi HKP (2012) Efficient use of shrimp waste: present and future trends. Appl Microbiol Biotechnol 93:17–29

Karp SG, Igashiyama AG, Siqueira PF, Carvalho JC, Vandenberghe LPS (2011) Application of the biorefinery concept to produce L-lactic acid from the soybean vinasse at laboratory and pilot scale. Biores Technol 102:1765–1772

Ken-ichi A, Yusei O, Yoshio I, Minoru S. (2002) Supercritical CO2 fluid extraction of crystal water from trehalose dehydrate. Efficient production of form II (Tα) phase. Carbohydrate Research, 337(19)

Kerkhofs NS, Lister CE, Savage GP (2005) Change in color and antioxidant content of tomato cultivars following forced-air drying. Plant Foods Hum Nutr 60:117–121

Kim SJ (2007) Gas permeation through water-swollen sericin/PVA membranes. Thesis, Chemical Engineering Department, University of Waterloo, Waterloo, ON

Kim YD, Kweon H, Woo S (2001) Collecting method of silk sericin from degumming solution and characteristics of recovered sericin. Korean J Seric Sci 43:37–40

Klekner N, Kosaric N (1993) Biosurfactants: production, properties, applications, vol 48, Surfactant Science Series. Marcel Dekker, New York, pp 373–389

Kong KW, Khoo HE, Prasad KN, Ismail A, Tan CP, Rajab NF (2010) Revealing the power of the natural red pigment lycopene. Molecules 15:959–987

Korres (2012) http://www.amazon.com/Korres-Olive-Stone-Scrub/Combination/dp/B0002VXTTQ. Accessed 12 Aug 2012

Kumar AS, Mody K, Jha B (2007) Evaluation of biosurfactant/bioemulsifier production by a marine bacterium. Bull Environ Contam Toxicol 79:617–621

Lademann J, Mainke MC, Sterry W, Darvin M (2011) Carotenoids in human skin. Exp Dermatol 20:377–382

Lang S, Philp JC (1998) Surface-active lipids in Rhodococci. Antonie Van Leeuwenhoek 74:59–70

Langmaier F, Mokrejs P, Kolomaznik K, Mladek M (2008) Biodegradable packing materials from hydrolysates of collagen waste proteins. Waste Manag 28:549–556

Llamas I, Mata JA, Tallon R, Bressollier P, Urdaci MC, Quesada E, Béjar V (2010) Characterization of the exopolysaccharide produced by *Salipiger mucosus* A3T, a halophilic species belonging to the *Alphaproteobacteria*, isolated on the Spanish Mediterranean seaboard. Mar Drugs 8:2240–2251

Maheshwari R, Chauhan AK, Rani B, Singh P (2012) Lycopene's antioxidant activity in cosmetics meadow. Int Res J Pharm 3(1):46–47

Maier RM, Soberón-Chávez G (2000) Pseudomonas aeruginosa rhamnolipids: biosynthesis and potential applications. Appl Microbiol Biotechnol 54(5):625–633

Makkar RS, Cameotra SS (2002) An update on the use of unconventional substrates for biosurfactant production and their new applications. Appl Microbiol Biotechnol 58:428–434

Martin CA, Almeida VV, Ruiz MR, Visentainer JEL, Matshushita M, Souza NE, Visentainer JV (2006) Omega-3 and omega-6 polyunsaturated fatty acids: importance and occurrence in foods Rev Nutr Campinas 19(6):761–770

Matos HR, Di Mascio P, Medeiros MHG (2000) Protective effect of lycopene on lipid peroxidation and oxidative DNA damage in cell culture. Arch Biochem Biophys 383:56–59

Matos HR, Marques AS, Gomes OF, Silva AA, Heimann JC, Di Mascio P, Medeiros MHG (2006) Lycopene and ß-carotene protect in vivo iron-induced oxidative stress damage in rat prostate. Braz J Med Biol Res 39:203–210

Matosa M, Barreiro MF, Gandinia A (2010) Olive stone as a renewable source of biopolyols. Ind Crops Prod 32:7–12

Max B, Salgado JM, Cortés S, Domínguez JM (2010) Extraction of phenolic acids by alkaline hydrolysis from the solid residue obtained after prehydrolysis of trimming vine shoots. J Agric Food Chem 58:1909–1917

Mohammadi FF, Harrison JT, Czarnota A, Leonard C (2005) Nonabrasive sensory exfoliating system. Patent, National Publication Number: US 20050169868

Mokrejs P, Langmaier F, Mladek M, Janacova D, Kolomaznik K, Vasek V (2009) Extraction of collagen and gelatine from meat industry by-products for food and non food uses. Waste Manag Res 27:31–37

Morimura S, Nagata H, Uemura Y, Fahmi A, Shigematsu T, Kida K (2002) Development of an effective process for utilization of collagen from livestock and fish waste. Process Biochem 37:1403–1412

Moure A, Cruz JM, Franco D, Domingues JM, Sineiro J, Dominguez H, Nunez MJ, Parajo JC (2001) Natural antioxidant from residual sources. Food Chem 72(2):145–171

Murado MA, Montemayor MI, Cabo ML, Vazquez JA, Gonzalez MP (2012) Optimization of extraction and purification process of hyaluronic acid from fish eyeball. Food Bioprod Process 90:491–498

Navarro-González I, García-Valverde V, García-Alonso J, Periago MJ (2011) Chemical profile, functional and antioxidant properties of tomato peel fiber. Food Research International. vol 44 (5):1528–1535
Nawawi WMFW, Jamal P, Alam MZ (2010) Utilization of sludge palm oil as a novel substrate for biosurfactant production. Biores Technol 101:9241–9247
Nguyen T, Clare B, Guo W, Martinac B (2005) The effects of parabens on the mechanosensitive channels of *E. coli*. Eur Biophys J 34:389–395
Nitschke M, Costa SGVAO, Contiero J (2011) Rhamnolipids and PHAs: recent reports on Pseudomonas-derived molecules of increasing industrial interest. Process Biochem 46:621–630
Nomura M, Iwasa Y, Araya H (1995) Moisture absorbing and desorbing polyurethane foam and its production. Japan Patent Publication Number 07–292240 to Seiren Co Ltd
Obied HK, Bedgood D, Mailer R, Prenzler PD, Robards K (2008) Impact of cultivar, harvesting time, and seasonal variation on the content of biophenols in olive mill waste. J Agric Food Chem 56:8851–8858
Ogawa A, Terada S, Kanayama T, Miki M, Morikawa M, Kimura T, Yamaguchi A, Sasaki M, Yamada H (2004a) Improvement of islet culture with sericin. J Biosci Bioeng 98:217–219
Ogawa M, Portier RJ, Moody MW, Bell J, Schexnayder MA, Losso JN (2004b) Biochemical properties of bone and scale collagens isolated from the subtropical fish black drum (*Pogonias cromis*) and sheepshead seabream (*Archosargus probatocephalus*). Food Chem 88:495–501
Palaniraj A, Jayaraman V (2011) Production, recovery and applications of xanthan gum by *Xanthomonas campestris*. J Food Eng 106:1–12
Pan YM, Wei LX, Zhu ZR, Liang Y, Huang CS, Wang HS, Wang K (2012) Processing of *Siraitia grosvenori'* leaves: extraction of antioxidant substances. Biomass Bioenergy 36:419–426
Parkinson M (1985) Bio-surfactants. Biotech Adv 3:65–83
Patel S, Goyal A (2011) Functional oligosaccharides: production, properties and applications. World J Microbiol Biotechnol 27:1119–1128
Peng X, Adachi K, Chen C, Kasai H, Kanoh K, Shizuri Y, Misawa N (2006) Discovery of a marine bacterium producing 4-hydroxybenzoate and its alkyl esters, parabens. Appl Environ Microbiol 72(8):5556–5561
Peschel W, Sánchez-Rabaneda F, Diekmann W, Plescher A, Gartzía I, Jiménez D, Lamuela-Raventó SR, Buxaderas S, Codina C (2006) An industrial approach in the search of natural antioxidants from vegetable and fruit wastes. Food Chem 97:137–150
Rayne S, Karacabey E, Mazza G (2008) Grape cane waste as a source of trans-resveratrol and trans-viniferin: high-value phytochemicals with medicinal and anti-phytopathogenic applications. Ind Crops Prod 27:335–340
RedFlower (2012) Web site of Red Flower. http://www.redflower.com/. Accessed 2 Aug 2012
Rodríguez G, Rodríguez R, Fernández-Bolaños JÁ, Guillén R, Jiménez A (2007) Antioxidant activity of effluents during the purification of hydroxytyrosol and 3,4-dihydroxyphenyl glycol from olive oil waste. Eur Food Res Technol 224:733–741
Rodríguez G, Lama A, Rodríguez A, Jiménez A, Guillén R, Fernández-Bolaños J (2008) Olive stone an attractive source of bioactive and valuable compounds. Biores Technol 99:5261–5269
Rufino RD, Sarubbo LA, Campos-Takaki GM (2007) Enhancement of stability of biosurfactant produced by *Candida lipolytica* using industrial residue as substrate. World J Microbiol Biotechnol 23:729–734
Sambanthamurthi R, Tan YA, Sundram K, Abeywardena M, Sambandan TG, Rha C, Sinskey AJ, Subramaniam K, Leow SS, Hayes KC, Wahid MB (2011) Oil palm vegetation liquor: a new source of phenolic bioactives. Brit J Nutr 106(11):1655–1663
Sandei L, Leoni C (2006) Exploitation of by-products (solid wastes) from tomato processing to obtain high value antioxidants. Acta Horticult 724:249–257
Shi J, Khatri M, Xue SJ, Mittal GS, Ma Y, Li D (2009) Solubility of lycopene in supercritical CO_2 fluid as affected by temperature and pressure. Sep Purif Technol 66:322–328

Taylor MA, Fraser PD (2011) Solanesol: added value from Solanaceous waste. Phytochemistry 72:1323–1327

Terada S, Sasaki M, Yanagihara K, Yamada H (2005) Preparation of silk protein sericin as mitogenic factor for better mammalian cell culture. J Biosci Bioeng 100:667–671

Thakur IS (2006) Xenobiotics: Pollutants and their degradation-methane, benzene, pesticides, bioabsorption of metals. ENVIRONMENTAL MICROBIOLOGY. School of Environmental Sciences, Jawaharlal Nehru University. New Delhi-110 067

Tilay A, Bule M, Kishenkumar J, Annapure U (2008) Preparation of ferulic acid from agricultural wastes: its improved extraction and purification. J Agric Food Chem 56:7644–7648

Tsubouchi K, Igarashi Y, Takasu Y, Yamada H (2005) Sericin enhances attachment of cultured human skin fibroblasts. Biosci Biotech Biochem 69:403–405

Vagi E, Simandi B, Vasarhelyine KP, Daood H, Kery A, Doleschall F, Nagy B (2007) Supercritical carbon dioxide extraction of carotenoids, tocopherols and sitosterols from industrial tomato by-products. J Supercrit Fluids 40:218–226

Wang X, Du Y, Fan L, Liu H, Hu Y (2005) Chitosan-metal complexes as antimicrobial agent: synthesis, characterization and structure-activity study. Polym Bull 55:105–113

Wang Q, Wu J, Zhang S, Zhang Y, Zhang H, Fan E (2009) GC Analysis of the Fatty Acid Composition of Yak Kidney. Chromatographia 69(5–6):601

Wang QJ, Gao X, Gong H, Lin XR, Saint-Leger D, Senee J (2011) Chemical stability and degradation mechanisms of ferulic acid (F.A.) within various cosmetic formulations. J Cosm Sci 62(5):483–503

Wu S, Ng CC, Tzeng WS, Ho KC, Shyu YT (2011) Functional antioxidant and tyrosinase inhibitory properties of extracts of Taiwanese pummelo (Citrus grandis Osbeck). African J Biotechnol 10(39):7668–7674

Xi J (2006) Effect of high pressure processing on the extraction of lycopene in tomato paste waste. Chem Eng Technol 29:736–739

Xie RJ, Li M, Lu S, Sheng W and Xie Y (2007) Preparation of sericin films and its cytocompatibility. Key Eng Mat 342–343:241–244

Yoshii F, Kume T, Makuuchi K, Sato F (2000) Hydrogel composition containing silk protein. Japan Patent Publication Number 2000-169736 to Japan Atom Energy Research Institute Sato Fusamitsu

Zhang YQ (2002) Applications of natural silk protein sericin in biomaterials. Biotechnol Adv 20(2):91–100

Zhang W, Wang X, Chen T (2012) Resveratrol induces apoptosis via a Bak-mediated intrinsic pathway in human lung adenocarcinoma cells. Cell Signal 24(5):1037–1046

Zhaorigetu S, Yanaka N, Sasaki M, Watanabe H, Kato N (2003) Inhibitory effects of silk protein, sericin on UVB-induced acute damage and tumor promotion by reducing oxidative stress in the skin of hairless mouse. J Photochem Photobiol B Biol 71:11–17

Part V
Other Biochemicals

Chapter 17
Biopolymers Synthesis and Application

Emna Chaabouni, Fatma Gassara, and Satinder Kaur Brar

17.1 Introduction

Biopolymers are ecologically safe alternatives to conventional polymers because of their biodegradable nature and their origin from renewable resources (Dionisi et al. 2005). Recent discoveries in the field of bacterial polymer biosynthesis have opened up new avenues for the rational engineering of bacteria towards the production of biopolymers suitable for industrial and medical applications. Over the past decade, a better understanding of bacterial polymer biosynthesis pathways from intermediates of central metabolism has been achieved. Bioplastics and biopolysaccharides are examples of biopolymers which are synthesized by a wide variety of microorganisms. Selection of a suitable composition substrate is an important factor for the optimization of biopolymers production and properties. Since, the high cost of substrates is a major factor in biopolymers production economics, thus the usage of waste biomass could significantly reduce the total production cost (Serafim et al. 2004; Lemos et al. 2006). Therefore, the use of cheaper carbon source is required in order to reduce the high production cost of biopolymers.

This chapter highlights the recent advances of different biopolymers sources, bacterial biopolymer biosynthesis pathways, and the valorization of waste biomass to produce biopolymers, especially bioplastics and biopolysaccharides. Finally, the potential applications of these biopolymers have been discussed. These biopolymers have been selected on the basis of the characteristics of their biosynthesis pathways, the extent of knowledge available to describe their formation, and their commercial relevance and applied potential.

E. Chaabouni • F. Gassara • S.K. Brar (✉)
INRS-ETE, Université du Québec, 490, Rue de la Couronne,
Québec City, QC, Canada, G1K 9A9
e-mail: satinder.brar@ete.inrs.ca

S.K. Brar et al. (eds.), *Biotransformation of Waste Biomass into High Value Biochemicals*,
DOI 10.1007/978-1-4614-8005-1_17, © Springer Science+Business Media New York 2014

17.2 Sources of Polymers

The source of a polymer often determines its nature, common properties, and its potential applications. Depending on the source, two types of polymers are traditionally distinguished: natural polymers and synthetic polymers. However, over the past three decades, a new class of polymer resulting from blending of natural and synthetic polymers has emerged and attracted increased attention (Sionkowska 2011).

17.2.1 Natural Polymers

The term "natural polymers" refers to macromolecules that occurred naturally or synthesized by living organisms (Quélénis 2008). Even though some authors employ "biopolymers" as a synonym of "natural polymers," it's important to bear in mind that "biopolymer" is widely designated for "biodegradable polymers" or "bio-based" polymers whether natural or synthetic. Natural polymers are derived from animals, plants, and/or microbes (Cascone et al. 2001). Table 17.1 summarizes the most renowned natural polymers widely applied for the fabrication of polymeric materials and their classification. Natural polymers are formed by the juxtaposition of monomeric units linked covalently. Their characteristics are entirely determined by the physicochemical properties of the monomers and their sequence. Therefore, a precise molecular organization can lead to a rich complexity of structure and role in the organism (Rodriguez-Cabello et al. 2005).

Depending on the type of monomers, biological macromolecules can be classified into three classes: polysaccharides, proteins, and nucleic acids. Although lipids and phospholipids may associate and form very large particles, they are not considered as polymers since their association results from electrostatic or hydrophobic interactions, not from covalent ones.

17.2.1.1 Polysaccharides

Polysaccharides, also called carbohydrates, are constituted by the repetition of large number of monosaccharides, such as glucose, fructose, and galactose (Nishinari and Takahashi 2003). Monosaccharides are linked by *O*-glycosidic bonds and they are able to form either linear or branched polymers. The physical properties of polysaccharides including solubility, gelation, and surface characteristics are dictated by monomers sequence, chain shapes, and molecular weight. In the organism, polysaccharides play crucial role in membranes and intracellular communication and act as storage molecules (Malafaya et al. 2007).

Table 17.1 Biopolymers classification

Nature	Origin	Biopolymers	Characteristics	References
Polysaccharides	Plants	Starch	Composed of α-amylose and amylopectin	Malafaya et al. (2007)
			Inherent biodegradability, great abundance, and renewability	
			Brittle, extremely difficult to process, usually used with other plasticizers (water, molecular weight alcohols) or other natural polymers	
		Cellulose	Linear polysaccharide of D-glucose units	Somerville (2006)
			Considered as the most abundant organic polymer in the world	Entcheva et al. (2004)
			Readily available, has a low cost	
			Exceptional strength, water insoluble despite of their hydrophilicity, poor degradation	
		Alginates	Naturally derived polysaccharide block copolymers composed of β-D-mannuronic acid monomers and α-L-guluronic acid	Malafaya et al. (2007)
			Abundant, make reversible gelation in aqueous solution	
			Uncontrollable degradation kinetics and dissolution of gels	
		Carrageenan	Family of sulfated polysaccharides extracted from red marine algae	Mangione et al. (2005)
			Linear polymers of α-D-galactose and β-D-galactose	Bartkowiak and Hunkeler (2001)
			Forms relatively stiff and thermoreversible gels in the presence of gel-promoting salts at room temperature	
			High flexible molecules, thixotropic, high degree of protein reactivity	

(continued)

Table 17.1 (continued)

Nature	Origin	Biopolymers	Characteristics	References
	Bacterial	Chitosan	Obtained from chitin	Malafaya et al. (2007)
			Copolymers of β(1–4)glucosamine and *N*-acetyl-D-glucosamine with different degree of deacetylation	
			The second most abundant polymerized carbon found in nature	
			Biologically renewable, biodegradable, biocompatible, nonantigenic, nontoxic, and biofunctional	
		Hyaluronic acid	Naturally occurring non-sulfated glycosaminoglycan	Nishinari and Takahashi (2003)
			Linear polysaccharide with alternating disaccharide units of α-1,4-D-glucuronic acid and β-1,3-*N*-acetyl-D-glucosamine	
			Major macromolecular component of the intracellular matrix of most connective tissues	
			Enhanced viscoelastic properties	
		Bacterial cellulose	Excreted into the medium	Sutherland (1998)
			Form surface pellicule by making aggregates of microfibrils	
		Xanthan	Formed from two D-glucose, two D-mannoses, and a glucuronic acid	Donot et al. (2012)
			Has very high molecular weight ranging from 500 to 2,000 kDa	Sutherland (1998)
			Physical properties: emulsion and foam stabilization, inhibitor of crystal formation, shear thinning and viscosity control, suspending agent	
		Dextran	Branched high-molecular-weight polymer of D-glucose	Cascone et al. (2001)
			Wide range of molecular weights, several derivatives	
			Biocompatible, biodegradable	
		Gellan	Carries *O*-acetyl and glyceryl substituents on a linear polymer of 500 kDa	Sutherland (1998)
			Forms weak, elastic, and thermoreversible gels	
			Forms extensive intermolecular association and strong and brittle gels	

		Levan	Composed of D-fructofuranosyl residues Synthesized outside the cell Biodegradable , water soluble, not soluble in most organic solvents Has low intrinsic viscosity	Manandhar et al. (2009)
	Fungal	Pullulan	Linear homopolysaccharide composed of glucose Synthesized in the cytosol and then secreted Water-soluble, good physical properties making it suitable for a multitude of applications Capacity to form strong resilient films and fibers which are biodegradable, transparent, oil resistant, and impermeable to oxygen	Lazaridou et al. (2003)
		Scleroglucan	Composed of glucose residues. Glucosyl residues may be attached With varying degrees of frequency Solutions of scleroglucan are pseudoplastic. Their viscosity is not greatly affected over the temperature range 20–90 °C Its pseudoplastic behavior is not affected by various salts Degradable by different enzymes	Sutherland (1998)
Proteins	Animals	Silk	Protein polymers spun into fibers by Lepidoptera larvae such as silkworms, spiders, scorpions, mites, and flies Generally lightweight, extremely strong, and elastic Excellent mechanical properties, slow degradability, biocompatibility	Malafaya et al. (2007)
		Collagen	The major protein component of the extracellular matrix 27 types identified High mechanical strength, good biocompatibility, low antigenicity…	

(continued)

Table 17.1 (continued)

Nature	Origin	Biopolymers	Characteristics	References
		Gelatin	Natural polymer obtained by acid and alkaline processing of collagen	
			2 different types can be produced depending on the method of pretreatment	
			Flexibility to make a variety of conformations	
			Rigid chain, high molecular weight, biodegradability, biocompatibility, low antigenicity	
		Elastin	Dominant extracellular matrix protein	
			Ideal elasticity, elongation at break, resistance	
			Potential biocompatibility and controlled degradation	
	Plants	Soy protein	Abundant, renewable, inexpensive, and biodegradable	
			Tailorable degradation profile	
			Commonly mixed to other polymers to improve its properties such as mechanical properties, degradation, solubility, and hydrophilicity	

Table 17.2 Microorganisms producing biopolymers

Microbial strains	Nature of polymers	Polymers	Substrates	Polymer concentrations	References
Bacteria					
Acetobacter xylinum	Exopolysaccharides	Bacterial cellulose	Maple syrup	1.51 g/l	Zeng et al. (2011)
Gluconacetobacter medellensis			Hestrin–Schramm medium + glucose, sucrose, or fructose	4.5 g/l	Castro et al. (2012)
Xanthomonas campestris		Xanthan	Date palm juice by-products	43.35 g/l	Ben Salah et al. (2010)
Leuconostoc sp.		Dextran	Sucrose	50 % (g/g)	Shamala and Prasad (1995)
Sphingomonas paucimobilis		Gellan	Starch	35.7 g/l	Nampoothiri et al. (2003)
Halomonas sp.		Levan	Sucrose	1.8 g/l	Poli et al. (2009)
Bacillus sp.			Sucrose	70.6 g/l	
Alcaligenes latus	Polyesters	Poly(β-hydroxybutyric acid) (PHB)	Sucrose	0.4 g/g	Grothe et al. (1999)
			Starchy wastewater	55 % (g/g)	Yu (2001)
Mixed microbial communities		Polyhydroxyalkanoates (PHAs)	Glycerol	80 % (g/g)	Moralejo-Gárate et al. (2011)
Ralstonia eutropha		Poly (3-hydroxybutyric acid-co-3-hydroxyvaleric acid (Poly(3HB-co-3HV))	Glucose + propionic acid	70–80 %	Byrom (1992)
Chromobacterium violaceum		Poly 3-hydroxyvaleric acid (Poly(3HV))	Valeric acid	65–70 %	Steinbüchel and Schmack (1995)
Yeast and filamentous fungi					
Aureobasidium pullulans	Exopolysaccharides	Pullulan	Sucrose	31.3 g/l	Youssef et al. (1999)
Sclerotium rolfsii		Scleroglucan	Sugarcane juice	23.87 g/l	Survase et al. (2007)
Leuconostoc dextranicum		Glucan	Sucrose	1.0 g/l	Majumder et al. (2009)

Some natural polysaccharides, such as starch and chitin, are widely studied (Synowiecki and Al-Khateeb 2003). For example, "starch" is synthesized by plants as food reservoir. It is composed of a mixture of two polymers of α-glucose, linear amylose and a highly branched amylopectin. It is known to be one of the most promising natural polymers because of its abundance and its renewability (Tang and Alavi 2011). Similarly, "chitin" is found in cuticles of insects, crustacean shell wastes, and cell walls of fungi, and "alginate" is extracted from brown algae. All these polymers are abundant in nature and attract attention due to various applications.

17.2.1.2 Proteins

Protein-based macromolecules are the second most important class of natural polymers. They are composed by 20 essential amino acids and are linked by amide bonds. On their native environment, proteins are responsible for an impressive array of function ranging from structural to signaling and catalytic roles. Actin, tubulin, keratin, and collagen are key structural constituents, whereas hormones play a major role in intercellular signaling. The best example of catalytic proteins is enzymes. Like polysaccharide-based polymers, protein-based polymers are also regarded as promising materials due to their biocompatibility, ability of being cross-linked, and their biodegradable nature. Among the most commonly applied protein-based polymers, collagen, silk, and fibrin have well-established relevance.

17.2.1.3 Nucleic Acids

Nucleic acids include deoxyribonucleic acid (DNA) and ribonucleic acid (RNA). They are very important for all living organisms as they are the storehouse of genetic information and play a crucial role in the preservation of species. The extraction and synthesis of nucleic acids is feasible in laboratory (Raha et al. 2004). Moreover, sequencing of genomes has been made possible, and hence, nucleic acids have been applied in many medical and nanobiotechnological purposes, such as diagnostics (Dunbar 2006) and postsurgical treatments (Werk et al. 2010). Nevertheless, they aren't considered with particular interest for the conception of polymeric materials.

17.2.1.4 Other Natural Polymers

It is noted that some other "polymers" that are naturally produced by living organisms can't be simply classified as one or another of macromolecules listed above. For instance, "natural rubber" described by Puskas et al. (2006) as "the most important polymer produced by plants" and "a strategically important raw material used in many products" has such complexity that its exact structure is still undefined.

Many components are thought to be a part of "natural rubber": Nyburg (1954) detected the presence of isoprene repeat units by using X-ray diffraction studies Puskas et al. (2006) and reported that fatty acid ester may be the terminal groups and the cyclized polyisoprene sequences were also detected. Other "abnormal" chemical groups were also identified, such as aldehydes (Eng and Ong 2000), epoxides (McIntyre et al. 2001), and amines (Tanaka 1989). Natural polymers, especially proteins and polysaccharides, have many attributes required for the industrial purposes, such as biocompatibility, biodegradability, and nontoxicity. Indeed, they are regularly required as polymeric materials (e.g., for biomedical equipments) due to their resemblance with the cellular compounds, their high chemical versatility, and their relatively good biological performance. Moreover, their degradability may be controlled in some cases by cells or enzymes (Malafaya et al. 2007). Nevertheless, despite their abundance and their good properties, the use of natural polymers on an industrial scale could be problematic. First of all, their relatively weak mechanical properties and thermal instability restrict the range of shape they can be processed into. Their native structure can be damaged if the environmental conditions, such as pH and temperature, are not optimal. Furthermore, the high insolubility of some natural polymers, such as elastin, silk, and keratin, could be an obstacle for processing (Sionkowska 2011). Finally, due the cost of purification, natural polymers are more expensive than synthetic ones.

17.2.2 Synthetic Polymers

In contrary to natural polymers, synthetic ones are man-made molecules and they don't exist naturally. The development of synthetic polymers has been made possible with the advancement in knowledge and techniques related to the petrol-based chemistry. Theoretically, all kind of association between monomolecules could be imaginable if the reaction of polymerization is feasible. However, in reality, polymerization of monomers is very challenging. First of all it is primordial to conceive reactive molecules that may interact with others (Tasdelen et al. 2011). Secondly, various physicochemical parameters, such as temperature, pressure, additive products, and water removal, have to be well managed (Hvala et al. 2011). The production of synthetic polymers at an industrial scale was initially encouraged to overcome the urgent need for synthetic rubber during the World War II. Nowadays, synthetic polymers are largely present in all sectors of economic activity, such as in the form of structural materials, biomedical engineering, and wastewater treatment (Chua et al. 1997). Millions of tons of synthetic polymer latex, synthetic elastomers, and waterborne coating have been produced using synthetic polymers. Polypropylene (PP) is considered as one of the most interesting synthetic polymers in terms its mechanical properties and low cost of production. Polyethylene (PE) and polystyrene (PS) are also well-known synthetic polymers which are largely employed for plastic production.

In general, mechanical properties and thermal stability of synthetic polymers are better than those of natural macromolecules. These characteristics make them easier to use and able to be processed into a wide range of shapes. However, due to their very low degradability, toxicity (Suh and Matthew 2000), and nonrenewability of their starting materials, they can't be considered as a viable solution in the long term. For this reason, research has been focusing on the development of bio-based polymers that have the biocompatibility, biodegradability, nontoxicity, and improved mechanical and thermal stability (Sionkowska 2011). In this direction "bioartificial" polymers are very promising.

17.2.3 Bioartificial or Biosynthetic Polymers

Artificial macromolecules result from the chemical modification of natural macromolecules or from blending synthetic and natural polymers. Polymer blending has shown a massive expansion in volume and complexity over the past three decades in terms of both the scientific advancement and application as reviewed in several publications (Pingping 1997; Piza et al. 2003 ; Marsano et al. 2004; Ye et al. 2007; Wang et al. 2011). The notion of artificial product is not recent, since the first artificial polymer, nitrocellulose, was produced in 1846 by the nitration of cellulose. However, there is a regain of interest to biosynthetic polymers, especially with the growing environmental concerns and the limits of petrol-based polymers. Nowadays, the terms of "bioartificial" or "biosynthetic" polymers refer to two distinct notions:

- Polymers which are biodegradable, either from natural or synthetic sources.
- Polymers which are derived from renewable resources, such as vegetable fats and oils, corn starch, and pea starch (Hong Chua et al. 1999) using either chemical or biological processes. These polymers are not necessarily biodegradable.

The focus of current research and the main subject of this chapter is the biological production of polymers based on renewable resources with improved properties of biodegradability. The next section is giving some valuable examples of these biopolymers and their synthesis.

17.3 Synthesis of Polymers with Biological Functionality

17.3.1 Bioplastics

Bioplastics have been produced from a wide range of sources, including proteins, lipids, and polysaccharides (Avérous 2004; Hernandez-Izquierdo and Krochta 2008; Siracusa et al. 2008). They have been conceived as a solution for environmental pollution and harming wildlife caused by the nonrecyclable, nonrenewable, and nonbiodegradable nature of conventional plastics. Bioplastics are of major interest for nowadays applications. They are of many types and the promising among them are given below:

- *Biomass-based bioplastics*, such as starch-based plastics and cellulose-based plastics: Starch and cellulose are not really plastic polymers. Thus, they can't be processed into bioplastics if other synthetic polymers are not added. For example, cellulose is usually pretreated chemically before being used as a reinforcing agent in the production of polymer composites (Zhang et al. 2012). Whereas starch, usually used in small amount (6–30 %), can be melt and fluidized only in the presence of plasticizer, such as water, glycerol, and sorbitol using high temperature (Tang and Alavi 2011), the potential of these polysaccharide-based biodegradable polymers obtained from agro-resources has long been recognized. However, till date their use is limited to some applications (e.g., food industry). They are not suitable for the packaging industries, e.g., to replace conventional plastic materials. They could be an interesting alternative to overcome the limitation of the petrochemical resources in the future (Avérous 2004).
- *Bioplastics extracted from microorganisms*: Some bacteria are known to synthesize and to store polyesters (polyhydroxyalkanoates, PHAs) that exhibit mechanical properties similar to polypropylene with the added benefit of biodegradability (Barham 1990; Lenz and Marchessault 2005). PHA can be formed by the polymerization of volatile fatty acids (acetic, propionic, and butyric acids) by bacteria, such as *Alcaligenes eutrophus* from starchy wastewater (Yu 2001). *Aeromonas caviae* was also used to generate PHA from soybean oil (Kahara et al. 2004; Koller et al. 2010). Moreover, poly-3-hydroxybutyrate (PHB) has been proposed as promising microbial polyester that is comparable to some of more common petrochemical-derived thermoplastics (Grothe et al. 1999). However, a number of problems are associated with PHB, such as the fragility of some PHB-based material which increases upon aging at room temperature (Shi et al. 1996).
- PHAs could be made by more than 150 different monomers and can be in different small, medium, and large chain length PHA with extremely different properties (thermoplasticity, melting points ranging from 40 to 180 °C). To improve the physico-mechanical properties of PHAs, a large range of homo- and co-polyesters has been produced by the incorporation of different structural repeat unit types such as alkanes, alkenes, alkanols, and carboxylic acids into PHA. For example, poly(3-hydroxybutyrate-co-3-valerate) (PHBV) has been prepared with biodegradable bacterial polyester and low-cost value-added agricultural by-products, such as corn straw, soy stalk, or wheat straw (Ahankari et al. 2011).
- *Bioplastics synthesized from monomers with natural source*: Polylactic acid (PLA) is the main member of this family. PLA is an aromatic polyester derived from monomers of lactic acid which is produced by the microbial fermentation of agriculture by-products, especially carbohydrate-rich substances. PLA is considered as a sustainable alternative to petrochemical-derived products and is largely commercialized.

17.3.2 Microbial Polysaccharides

Nowadays, microbial polysaccharides are given a particular and growing interest, since they represent a potential market. Polysaccharides derived from microorganisms can be produced by bacteria, yeast, fungi, and microalgae. Comparing with plant-derived polysaccharides, microbial ones offer a wide range of advantages. First of all, production is considerably faster: few days for microbial production as compared to 3–6 months in the case of plants. Secondly, there is an opportunity to use these organisms for the nutrient removal in wastewater treatment, since they can transform industrial wastes into valuable chemicals. Moreover, utilizing microorganisms doesn't compete with arable land.

Microbes are able to synthesize a huge number of polysaccharides with remarkable physicochemical properties. Some microbial polysaccharides have well-established applications, such as thickeners, stabilizers, and gelling agents. Two classes with particular interest are exopolysaccharides (EPSs) and capsular polysaccharides (CPSs).

17.3.2.1 Exopolysaccharides

EPSs have the peculiarity to be peripheral and weakly associated to the cell. They may be released into the medium (Cescutti 2009). Structurally, EPS, as well as CPS, can be homopolymers (made up of a single type of monosaccharide) or heteropolymers (made up of several types of monosaccharides) (Donot et al. 2012). Dextran is an example of homopolysaccharides composed of units of glucose and synthesized by bacteria, such as *Leuconostoc* sp. Levan is another example of EPS, composed of units of D-fructofuranosyl and synthesized extracellularly. It is produced by bacteria, such as *Zymomonas mobilis* or *Bacillus subtilis* (Monsan et al. 2001). Among the heteropolysaccharides, xanthan is one of the most studied EPS. Rosalam and England (2006) described the composition of xanthan as follows: two D-glucose, two D-mannose, and a D-glucuronic acid.

17.3.2.2 Capsular Polysaccharides

Unlike EPS, CPSs are covalently linked to lipids or phospholipids of the cell surface and form an extracellular envelope. The capsule determines the immunological properties of some pathogenic bacteria. Both CPS and EPS are the principal constituents of biofilms. Biofilms are envelope formed around multicellular aggregates and play a major role in the interaction of these associated cells with their environment. In addition, polysaccharides can promote the adherence of microorganisms to surfaces and to each other. A notable number of CPS are produced by *Escherichia coli* and *Klebsiella pneumoniae* and confer to these bacteria their pathogenicity (Cescutti 2009).

17.4 Valorization of Biomass to Produce Biopolymers

The biopolymers are usually produced in the biological fermentation processes using renewable and sustainable agricultural feedstocks such as vegetable oils and animal fats, dairy whey, molasses, and meat-and-bone meal or other biomass components including lignocellulosics. There are also non-fermentative processes (thermochemical ones) which convert biological feedstocks to chemicals (Gray et al. 2006; Gong et al. 2008; Hames 2009; Mu et al. 2010). In addition, some of the polymers can also be produced directly in plant by industrial crops constructed via genetic modification (GM) routes (Matsumoto et al. 2009).

Till date, most of the research on the production of biopolymers is in early stage of development, while a few studies have been carried out on commercial scale (Chen 2009). Most of these polymers are manufactured via the microbial fermentation routes. A concept of a new branch of biotechnology termed white biotechnology has been developed, referring to the industrial development and implement strategies for chemical production based on biomass-derived carbon sources (Hermann and Patel 2007; Soetaert 2007). Although the technologies were developed many years ago, large-scale production of polymers from biomass was not feasible due to their expensive nature. However, in recent years, the innovations from the research sectors, particularly in biotechnology, have made some of the biological conversions competitive with the existing fossil-based processes. To mention a few, there are production of vitamins, antibiotics, and ethanol using biotechnological approaches (Hermann and Patel 2007; Soetaert 2007).

Synthetic biology (SB) has been considered as a new way of doing biotechnology. It is an emerging science and engineering field that applies engineering principles to biology. The potential benefits of SB include the development of novel medicines, renewable chemicals, and fuels (Gaisser et al. 2009). Due to the infancy of SB, a variety of definitions are circulating in the scientific community, and no consensus definition has been drawn. SB-related research has been performed in several fields, such as DNA synthesis (or synthetic genomics) (Carlson 2009; Gibson et al. 2009, 2010), engineering DNA-based biological circuits (Canton et al. 2008), minimal genome (or minimal cell) (Luisi 2007), protocells (or synthetic cells) (Bedau et al. 2009), and xenobiology (or chemical synthetic biology) (Schmidt and Pei 2011).

One of the key issues hindering the large-scale biopolymer production is the cost of sugars and fatty acids as feedstocks. This issue is an obstacle not only for the production of PHAs but also other commodity chemicals and fuels. Recently, world food prices reached a record high according to a report from the Food and Agriculture Organization (FAO) of the United Nations. The Food Price Index reached 214.7 points in December 2010 and slightly above the previous peak of 213.5 points during food crisis in 2008 (BBC News Business 2011). It has already been in debate that increasing biofuel production from starch and sugar may post further threat to the food safety. Thus, it is a key challenge to develop biological processes to harness the nonfood-based biomass. In theory, lignocellulosic biomass may be the best sustainable carbon source which could be used as substrate to produce many value-added products such as biopolymers (mainly bioplastics and polysaccharides). A great deal of work has been carried out to convert cellulosic biomass into useful products. The ideal microbes should be equipped with a couple of properties, such as the ability to hydrolyze cellulosic material effectively, with minimal requirement for preprocessing; the ability to convert the sugars released into molecules useful as liquid fuels and/or chemical industry feedstocks; self-tolerance to these molecules at a high concentration; and suitable respects for use in industrial bioreactors (French 2009). There are no super-producing microbes in nature; they can be only constructed via complex engineering, something that SB is expected to be capable of. Most of the nonedible biomass is made up by long cellulose, hemicellulose, and lignin (Lynd and Zhang 2002). With respect to their chemical composition, hemicellulose is a mixture of monomers, such as D-xylose, L-arabinose, D-mannose, and D-galactose, together with sugar derivatives, such as 4-*O*-methyl-D-glucuronic acid. Lignin is a complex chemical compound formed by polymerization of aromatic monomers. Cellulose can be digested to D-glucose by enzymatic or nonenzymatic hydrolysis. A typical enzymatic hydrolysis of cellulose involves a multiple-step process mediated by several enzymes. A handful of enzymes have been identified to degrade cellulose efficiently from different microbes (Lynd et al. 2002). Cellulosomes can be used for the degradation of cellulose and hemicellulose, and they have in fact been considered as one of nature's most elaborate and highly efficient nanomachines (Fontes and Gilbert 2010). Integration of cellulosomal components occurs via highly ordered protein–protein interactions between two proteins: cohesins and dockerins, which specifically allow the incorporation of cellulases and hemicellulases onto a molecular scaffold. Cellulosomes can be used for a range of SB applications, from clothes whitening to paper waste treatment or chemical production from lignocellulosic biomass. The first synthetic cellulosomes have already been constructed (Mitsuzawa et al. 2009). SB will play an important role in developing cellulose degradation module along with the chemical producing module where the standardization of biological parts and interdisciplinary nature of SB enable combination of multiple modules. To date, degradation of lignin and cellulose is still problematic though they can be degraded by enzymes produced by very few fungi. Using SB methods may develop lignin degradation enzymes with enhanced capability. In addition, one possible solution to make the biomass amenable to hydrolysis is the biomass derived from genetically modified plants. Such attempt has been tried by lignin-deficient genetically modified plants (Baucher et al. 2003). The same idea has been applied to develop genetically modified potatoes for industrial applications which have been approved to cultivate in EU (Ryffel 2010).

17.4.1 Valorization of Biomass to Produce Bioplastics

The major cost in the bioplastic production is the cost of the substrate (Yamane 1993). Thus, to reduce the productions cost of these bioplastics, the simplest approach is to choose renewable, inexpensive, and most readily available carbon substrates that could support both the microbial growth and bioplastic production efficiently. Microorganisms are capable of producing bioplastics such as PHA from various carbon sources ranging from inexpensive, complex waste effluents to plant oils (Fukui and Doi 1998), fatty acids (Eggink et al. 1992), and alkanes (Lageveen et al. 1988). Each year, a large amount of waste materials are discharged from agricultural and food processing industries in Canada and worldwide, and these wastes represent a potential renewable feedstock for bioplastic production. Plant oils such as soybean oil, palm oil, and corn oil are desirable carbon sources for PHA production using bacteria *Burkholderia cepacia* (Alias and Tan 2005) and *Comamonas testosteroni* (Thakor et al. 2005). These oils have been used as they are relatively cheaper than most sugars. Furthermore, bioplastics such as PHA have been produced using glycerol as substrate. Glycerol is a by-product of the palm oil refining process. It is used as a humectant as well as a plasticizer in certain applications such as oral care products, tobacco, cosmetics, food, and beverages. Glycerol is generated in large quantities as coproduct stream of biodiesel and possesses a great potential to be an attractive carbon source for PHA production by certain microorganisms (Ashby 2005; Chee et al. 2010; Zhu et al. 2010). Many studies are shown that different microorganisms such as *Burkholderia* sp. *Pseudomonas corrugata and P. oleovorans* could synthesize PHA from glycerol by under appropriate growth conditions (Ashby 2005; Chee et al. 2010; Zhu et al. 2010).

17.4.2 Valorization of Biomass to Produce Polysaccharides

To reduce the production cost of polysaccharides, many studies have been investigated to produce microbial polysaccharides using agricultural feedstocks. Currently, despite the vast number and biodiversity of the extremophilic producers of polysaccharides, namely, EPS, these nontoxic and biodegradable polymers represent only a small fraction of the current polymer market. These few marketable EPS derived from extremophiles belong only to the bacteria domain: actually no EPS produced by Archaea has a commercial application. The high production costs and the poor physicochemical properties (compared with those of industrial EPSs from plants, such as guar gum, cellulose, pectin, and starch, and from seaweed as alginate and carrageenan) make the microbial EPSs not suitable for commercial purpose. The fermentation media that account up to 30 % of the cost for a microbial fermentation usually are made of expensive nutrients, such as yeast extract, peptone, and salts. In order to maximize the cost-effectiveness of the process, recent works shifted towards using multicomponent feedstock systems, and the synthetic media were replaced by cheaper alternatives. Molasses were successfully used for fermentative production of commercial polysaccharides, such as curdlan (Lee et al. 2003), xanthan (Kalogiannis et al. 2003), dextran (Vedyashkina et al. 2005), and gellan (Banik et al. 2007). Spent malt grains, apple pomace, grape pomace, and citrus peels were used for xanthan production by solid-state fermentation (Stredansky and Conti 1999). Similarly, the olive mill wastewater was used for xanthan production (López et al. 2001). Besides the use of cheaper substrates, the reduction of production costs can be achieved by improvement of product yields by optimizing fermentation conditions or developing higher yielding strains (e.g., by mutagenesis or genetic manipulation) and by optimizing downstream processing. Moreover, the interest for the development of microbial EPSs could be related to their use in high-value market niches, such as cosmetics, pharmaceuticals, and biomedicine, where traditional polymers fail to comply with the required degree of purity or lack some of specific functional properties. In these high-value applications, quality and purity products wholly surpass the cost production and product yield issues, identifying in these interesting biopolymers suitable candidates for biotechnological applications (Kumar et al. 2007). The use of polymer in many biotechnological applications makes them very important molecules. In this context, the bacterial polymer biosynthesis pathways have been studied extensively in order to improve its yield of production.

17.5 Bacterial Polymer Biosynthesis Pathways

Biopolymers can be produced in almost all bacteria in the form of intracellular and extracellular inclusions. There are different pathways to produce biopolymers. In this chapter we have focused on the pathways of the production of bioplastics and polysaccharides. The biopolymer biosynthesis pathways using bacteria are presented in details in Fig. 17.1.

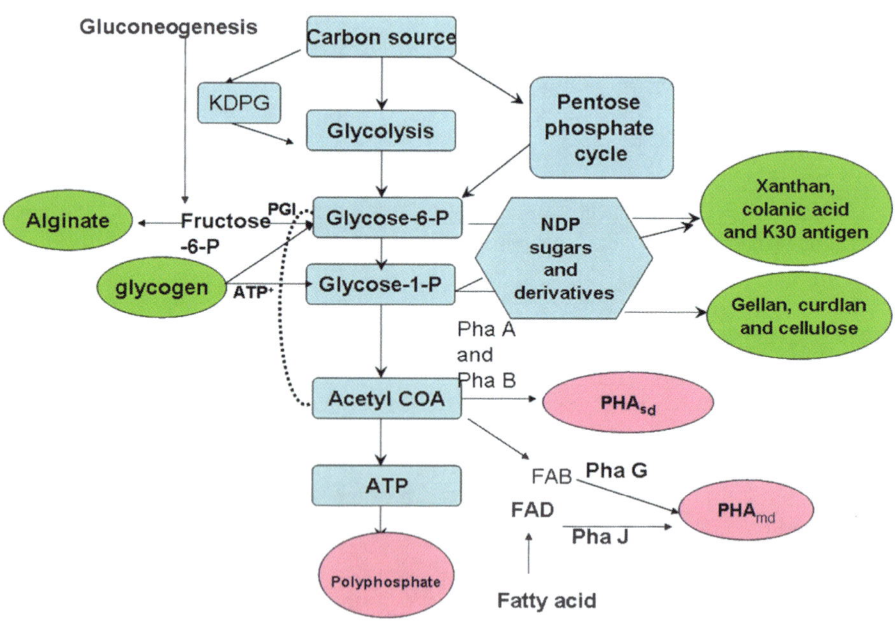

Fig. 17.1 Bacterial polymer biosynthesis pathways from intermediates of central metabolism

17.5.1 Bacterial Bioplastic Synthesis

To produce bacterial bioplastics (polyhydroxyalkanoates (PHAs); also known as bacterial bioplastics), there are three well-known natural biosynthesis pathways of PHAs (Tsuge 2002). The pathway I is most common which is found in many bacteria. It leads to generation of 3-hydroxybutyryl (3HB) monomers from acetyl CoA derived from sugars where a set of enzymes are involved, such as PhaA (3-ketothiolase which converts acetyl CoA to acetoacetyl-CoA), PhaB (NADPH-dependent acetoacetyl-CoA reductase, resulting in 3HB-CoA), and PhaC (PHA synthase, polymerizing 3HB-CoA to the final monomers). The pathways II and III are more commonly found in the genus of *Pseudomonas*. They are pathways using either sugars or fatty acids as carbon source to convert to either acetyl CoA or acyl CoA, resulting mainly in mcl-(*R*)-hydroxyacyl (3HA) monomers. The (*R*)-specific enoyl-CoA hydratase (PhaJ) and (*R*)-3-hydroxyacyl-ACP-CoA transferase (PhaG) play similar roles as PhaB in the pathway I to obtain 3HA-CoA. PHAs are natural polymers with properties similar to those of the plastics. There is a renewed interest in producing PHAs in a biological process fed with sustainable sources. Several microbes have been constructed to produce PHAs (Sandoval et al. 2007; Zhang et al. 2009; Hofer et al. 2010).

17.5.2 Bacterial Polysaccharide Biosynthesis

The polysaccharides produced by bacteria can be subdivided into the exopolysaccharides (EPSs) (e.g., xanthan, dextran, alginate, cellulose, hyaluronic acid (HA), and colonic acid), which can be either secreted or synthesized extracellularly by cell wall-anchored enzymes, the CPS (e.g., the K30 antigen), and the intracellular polysaccharide (IPS) (glycogen) (Rehm 2010). The formation of polysaccharides with such varied structures and compositions requires the recruitment of different enzymes and proteins, which is reflected in the varied organizations of the biosynthesis gene clusters (Rehm 2010). The EPS and CPS biosynthesis gene clusters are subject to extensive transcriptional regulation involving two-component signal transduction pathways, quorum sensing, alternative RnA polymerase σ-factors and anti-σ-factors, as well as integration host factor (iHF)-dependent and cyclic di-GMP-dependent processes (Leech et al. 2008). Induction of EPS biosynthesis is often correlated with establishment of the biofilm growth mode (Hall-Stoodley et al. 2004). Nucleoside diphosphate sugars (such as ADP-glucose), nucleoside diphosphate sugar acids (such as GDP-mannuronic acid), and nucleoside diphosphate sugar derivatives (such as UDP-N-acetylglucosamine) are direct precursors for bacterial polysaccharide biosynthesis. Polymer-specific biosynthesis enzymes (e.g., pyrophosphorylases and dehydrogenases) are required for the synthesis of the activated polymer precursor, which is the first committed biosynthesis step, and have been targeted by metabolic engineering to enhance polymer production (Ruffing and Chen 2006). Polymerization and secretion of EPS and CPS are often rate limiting and can substantially affect the flux of carbon towards the processive formation of high-molecular-mass exopolymers. The synthases or catalytic subunits of synthases are mostly localized in the cytoplasmic membrane and are often associated with proteins that are required for translocation of the polymer across the cytoplasmic membrane and, if needed, the outer membrane. The production of HA requires only a single protein, HA synthase (HasA), for polymerization and secretion (Brown and Pummill 2008); however, most EPSs are polymerized and secreted by membrane-spanning multiprotein complexes. These complex-mediated biosynthesis processes can be subdivided into two general pathways. One pathway is exemplified by the Wzy-dependent polymerization and secretion mechanism used in xanthan biosynthesis, which requires a lipid carrier for transfer of the repeat unit oligosaccharide across the cytoplasmic membrane. The second pathway is independent of both Wzy and a lipid carrier and is proposed for the production of EPS, such as alginate and cellulose, which are not composed of repeat units.

After discussing the bacterial biosynthesis pathways of a series of biodegradable polymers, it is interesting to elaborate on their current applications.

17.6 Applications of Polymers

A great variety of polymers are already in use at the biological interface and in many cases as their properties are either inherently biomimetic or indeed superior to their natural counterparts. The different applications of biopolymers are presented in details in Table 17.3.

17.6.1 Medical Application

Biopolymers are biodegradable materials that can be degraded and gradually absorbed and/or eliminated by the body biopolymers. They are of immense interest to biomedical technologies such as tissue engineering, bone replacement/repair, dental applications, and controlled drug delivery because these are formed of many nontoxic components and are considered to be inherently biocompatible.

Biopolymers are used in medical device applications for interaction with a biological system. These biopolymers are able to coexist with some host's reactions in a specific use (Williams et al. 1988).

Besides applications in plastics, polylactic acid (PLA)-derived bioplastics are also used extensively in biomedical applications, such as sutures, stents, dialysis devices, and drug capsules, and evaluated as a matrix for tissue engineering (Park et al. 2008).

Table 17.3 Biopolymers applications

Applications	Specific applications	Biopolymers	References
Medical application	Sutures, stents, dialysis devices, drug capsules, and evaluated as a matrix for tissue engineering	PLA-derived bioplastics	Park et al. (2008)
	Antiviral	Xanthan, sulfated dextran, and sulfated curdlan	Ghosh (2009)
	Anticancer agents	Xanthan, sulfated dextran, and sulfated curdlan	Takeuchi (2009)
	Antitumor, anti-inflammatory and immune-enhancer drugs	Fucose	Freitas (2011)
Agricultural and agro-industrial applications	Plant pots, stakes, erosion control netting, and mulch	Bioplastics	Otey and Mark (1976), McCormick and Lichatowhich (1979)
	Thickening, stabilizing, binding, and structure creation agents in foods	Bacterial EPSs	Freitas (2011)
Packaging	Packaging of food products	Polysaccharides	Nguyen (2008); Alves (2011)
	Packaging of food products	Bioplastics like PHBV, PHB	Alves (2011)
Environmental applications	Sludge dewatering	Extracellular polysaccharides	Tyagi et al. (2002); Yezza et al. (2004); Garnier et al. (2005)

Exopolysaccharides have a number of industrial applications which relate either directly or indirectly to medicine. One such application is the use of dextran and its derivatives at laboratory and industrial scale for the purification of compounds of medical interest, including pharmaceuticals, and enzymes for diagnostic purposes. Polysaccharides may also be used to encapsulate drugs for their gradual delivery, and they may be used to immobilize enzymes employed for diagnosis or for the chemical modification of pharmaceutical products (Matou et al. 2005). These applications clearly utilize the functional properties of the polysaccharides, such as their rheology or capacity for gel formation. Alternatively, the pharmacological or other biological properties of the polymers may be employed. There are essentially three types of direct application of the biological properties of exopolysaccharides to medicine (Sudip 2012). The exopolysaccharides may be used as vaccines in preference to whole microbial cells or cultures (Sudip 2012). Thus, side effects due to other cell components such as lipopolysaccharides or proteins are avoided. On the other hand, not all polysaccharides are good immunogenic propriety (Sudip 2012), nor are the exopolysaccharides necessarily the major factors in the specific disease syndromes caused by the polysaccharide-producing microbial pathogens (Arena et al. 2009).

As a result of their unique and tunable properties, several bacterial EPSs play an important role in the development of new pharmaceuticals, because not only of their ability of forming polymeric matrices but also due to their inherent biological activity. For example, xanthan, sulfated dextran, and sulfated curdlan are used as antiviral (Ghosh 2009) and anticancer agents (Takeuchi 2009). As a result of its high fucose content, FucoPol (Freitas 2011) is seen as a product with potential to be used in antitumor, anti-inflammatory, and immune-enhancer drugs.

EPS can also be used as a source of oligosaccharides and sugar monomers, which constitute added-value applications. Fucose is an example of a rare sugar that is difficult to obtain, for which supply falls short of demand in the international market. Formulations that contain fucose, as well as fucose-containing oligosaccharides, have been reported to have properties that potentiate their use, either in pharmaceuticals (e.g., as anticarcinogenic or anti-inflammatory agents) or in cosmetic products (as antiaging agents) (Kumar et al. 2007).

17.6.2 Agricultural and Agro-Industrial Applications

All main classes of polymers—that is, plastics, coating, elastomers, fibers, and water-soluble polymers—are nowadays used for controlled release of pesticides and nutrients, soil fertilization, seed coatings, and plant protection (McCormick and Lichatowhich 1979). Degradable plastics are also of serious interest as materials for crop mulching in fields or as agricultural plant containers. Their biodegradation mode (i.e., composting) is of great significance as it allows various biodegradable materials to be combined and processed into useful materials to improve the state of the soil. Bioplastics have many agricultural applications including degradable plant pots, stakes, erosion control netting, and mulch film (McCormick and Lichatowhich 1979). The companies are expecting to charge a premium for the plastic. Receiving a premium for low-value items such as disposable dinnerware or horticultural applications would be very unusual except in those political geographies that are banning nondegradable plastic items.

Bacterial EPSs have been extensively used in high-value applications, such as agro-industrial applications where they are mostly used as thickening, stabilizing, binding, and structure creation agents, as a result of their non-Newtonian behavior and high viscosity in aqueous media. When applied in food products, they must be able to maintain their properties when incorporated into formulations, in which they might experience significant variability of pH and ionic strength, along with the influence of other food components (Freitas 2011).

17.6.3 Packaging

Physical characteristics of polymers for packaging are greatly affected by their chemical structure, molecular weights, crystallinity, and processing conditions. The physical characteristics required in packaging equally depend on the items to be packed and the environments in which the packages will be stored. Products to be stored frozen for some time require specific packaging. Foods demand stricter package requirements than solid products.

Recent research has focused on developing polymeric matrices (polysaccharides) with tuned properties (e.g., transparency, barrier and mechanical properties, biocompatibility, or bioactivity) for several applications, namely, edible coatings for food products (Yasuhiro and Yoshimitsu 2010) and packaging purposes (Alves 2011).

Biologically derived polymers may be used for the production bulk packages with the same technology used for conventional materials. These data prove that they are not less important in any physical, thermal, mechanical, and barrier properties than the conventional plastics. Bioplastics do offer several other additional advantages over conventional plastics in bulk packaging. Compost derived in part from bioplastics increases the soil organic content as well as water and nutrient retention, while reducing chemical inputs and suppressing plant diseases. Moreover, starch-based bioplastics have been shown to degrade 10–20 times quicker than conventional plastics. Further, on burning traditional plastics, toxic fumes are produced which can be harmful to people's health and the environment. If any biodegradable films are burned, there is little, if any, toxic chemicals or fumes released into the air. With respect to safety, degradation tests revealed that more than 90 % of samples degraded in 10 months, according to the measurements of weight loss and CO_2 production. There are water-soluble biocomposites with solubility depending on the amount and the molecular weight and its crystallinity. Bioplastics, such as PHBV and PHB, are biodegradable in soil, river, water, seawater aerobic and anaerobic sewer sludge, and compost. For example, PHBV mineralizes in anaerobic sewer sludge to CO_2, water, and some percentage of methane to the extent of nearly 80 % in 30 days. Another example is application of a special biocomposite in making of laundry bags for hospital and other institutions, where the bag dissolves during the washing and biodegrades after disposal into sewage. Samples of bioplastic compost, obtained by mixing the test material with organic waste when compared with samples of reference compost produced only with organic waste, showed the fact when tested on plants, it did not release substances toxic for the plants and environment.

Compared to conventional plastics derived from petroleum, bio-based polymers have more diverse stereochemistry and architecture of side chains which enables research scientists a greater number of opportunities to customize the properties of the final packaging material. Thus, with these added advantages and almost similar properties of LDPE, PVC, Nylon, HDPE, and PP, we can implement bioplastics in the bulk packaging industry at the places of these petroleum-based plastics which are creating environmental pollution by its nondegradability and harmful gas emission.

17.6.4 Environmental Applications

Biopolymers, namely, exopolysaccharides, play an important role in sludge dewatering during the wastewater treatment process. Sludge dewatering is one of the most important steps in wastewater treatment and for sludge recycle. Better dewaterability of sludge leads to economical disposal and reuse, such as production of polyhydroxyalkanoates (PHAs) or any bioplastics, bricks for construction, and use as a raw material for growth of industrial microorganisms (Tyagi et al. 2002; Yezza et al. 2004). Sludge is found to be negatively charged (approximately −30 mV), and hence, cationic synthetic polymers are generally employed to neutralize the sludge charge, which facilitates the sludge settling. These polymers are expensive, and pollute environment and safety precautions must be followed during their handling (Changa et al. 2002). To overcome these existing problems, an alternative and suitable way is using biocoagulants/bioflocculants produced by bacterial strains derived from activated sludge. During the past five decades, researchers have been venturing to bioflocculate the sludge using microorganisms. Microorganisms are capable of growing in diverse environments and producing secondary metabolites during their growth. Microorganisms growing in sludge produce secondary metabolites mainly consisting of carbohydrates (extracellular polysaccharides, alginate, and chitosan), proteins (lectins), lipids (fatty acids), and DNA and RNA. It has been well documented that microbial EPS plays an important role in bioflocculation process by interacting with the sludge solids. The microbial EPS may be nonionic or may contain cationic, anionic, and/or both charges (Garnier et al. 2005).

Sludge settling ability of EPS mainly depends on the concentration and characteristics of EPS. The concentration of EPS produced by sludge is affected by sludge characteristics, EPS biochemical characteristics, operational parameters in a wastewater treatment plant, as well as type of microbial community dominating in that treatment plant. Generally, there is an optimal concentration of EPS for sludge settling (Urbain et al. 1993). EPS concentration below or above the optimal level is detrimental to sludge settling. Thus, this is also an eventual role of the biopolymers in flocculation.

17.7 Conclusions, Outlook, and Perspectives

Although plant products or plant-derived biopolymers served as commodity materials before the 1920s, petrochemically synthesized polymers (including thermoplastics, thermoset plastics, adhesives, and coatings) now dominate the commodity materials market, with an annual production estimated at 260 million tons in 2007 (Hopewell et al. 2009). This suggests a previous growth rate of approximately 9 % per annum. Owing to an increasing environmental awareness and the limitation of fossil resources, it is anticipated that renewable biopolymers will replace a substantial fraction of the market for synthetic polymers. There will also be a growing demand for bacterial production of polymers with material properties that are specifically tailored for applications in various fields of daily life. The competitive advantage of the environmentally friendly and highly process synthesis of biodegradable and, in many cases, biocompatible polymers will become increasingly attractive for industry. However, the fermentation media that account up to 30 % of the cost for a microbial fermentation usually are made of expensive nutrients, such as yeast extract, peptone, and salts. The use of renewable, inexpensive, and most readily available carbon substrates such as lignocellulosic wastes could reduce the production cost of biopolymer production process and will attract more the industries. The major problem related to the use of lignocellulosic waste for biopolymer is its complexity. Thus, lignocellulosic wastes should be fermented by microorganisms able to hydrolyze lignocellulosic material effectively and convert the sugars released into biopolymers. Bacteria remain ideal production organisms for polymers owing to the availability of genetic systems and techniques for engineering metabolic pathways, which provide an ever-growing design space for the production of biopolymers with material properties of interest. In addition, further knowledge of the molecular mechanisms of biopolymer biosynthesis could be used for the design of specific biosynthesis enzyme inhibitors that will be useful when the respective biopolymer is an important virulence factor.

However, the use of bacteria to produce biopolymers, namely, polysaccharides, is very limited. Polysaccharides recovered from plant, algae, and animal sources are still the major contributors to the overall hydrocolloid market. This is mainly because of the higher prices of bacterial polysaccharides, which are a consequence of the high value of the carbon sources commonly used and of the associated downstream costs.

Nevertheless, the research interest in bacterial production of polysaccharides is continuously growing and is focused on using low-cost substrates and improving downstream processing and on metabolic engineering that aims for production of polymers with fine-tuned properties. In fact, the main advantage of bacterial polysaccharides relies on the possibility of tailoring their chemical composition and structure, which foresees rather specific applications in pharmaceutical products, medical devices, and cosmetics. In the next few years, we can expect a significant increase in added-value products/technologies based on bacterial polysaccharides and bioplastics, especially developed for the specific market.

Acknowledgements The authors are sincerely thankful to the Natural Sciences and Engineering Research Council of Canada (Discovery Grants 355254) and INRS-ETE for financial support. The views or opinions expressed in this article are those of the authors.

References

Ahankari SS, Mohanty AK, Misra M, (2011) Mechanical behaviour of agro-residue reinforced poly(3-hydroxybutyrate-co-3-hydroxyvalerate), (PHBV) green com- posites: comparison with traditional polypropylene composites. Compos Sci Technol 71:653–657

Alias Z, Tan IKP (2005) Isolation of palm oil-utilising, polyhydroxyalkanoate (PHA)-producing bacteria by an enrichment technique. Bioresour Technol 96:1229–1234

Alves VD (2011) Characterization of biodegradable films of a new microbial polysaccharide produced using glycerol byproduct. Carbohydr Polym 83:1582–1590

Arena A, Gugliandolo C, Stassi G, Pavone B, Iannello D, Bisignano G, Maugeri TL et al (2009) An exopolysaccharide produced by *Geobacillus thermodenitrificans* strain B3-72: antiviral activity on immunocompetent cells. Immunol Lett 123:132–137

Ashby RD (2005) Synthesis of short-/medium-chain-length poly(hydroxyalkanoate) blends by mixed culture fermentation of glycerol. Biomacromolecules 6:2106–2112

Avérous L (2004) Biodegradable multiphase systems based on plasticized starch: a review. J Macromol Sci C Polym Rev 44:231–274

Banik RM, Santhiagu A, Upadhyay SN et al (2007) Optimization of nutrients for gellan gum production by *Sphingomonas paucimobilis* ATCC-31461 in molasses based medium using response surface methodology. Bioresour Technol 98:792–797

Barham PJ (1990) Physical properties of poly(hydroxybutyrate) and poly(hydroxybutyrate-co-hydroxyvalerate). In: Dawes EA (ed) Novel biodegradable microbial polymers. Kluwer, Dordrecht

Bartkowiak A, Hunkeler D (2001) Carrageenan-oligochitosan microcapsules: optimization of the formation process. Colloids Surf B Biointerfaces 21:285–298

Baucher M, Halpin C, Petit-Conil M, Boerjan W et al (2003) Lignin: genetic engineering and impact on pulping. Crit Rev Biochem Mol Biol 38:305–350

BBC News (2011) World food prices at fresh high, says UN. BBC. www.bbc.co.uk/news/business-12119539

Bedau MA, Parke EC, Tangen U, Hantsche-Tangen B et al (2009) Social and ethical checkpoints for bottom-up synthetic biology, or protocells. Syst Synth Biol 3:65–75

Ben Salah R, Chaari K, Besbes S et al (2010) Optimisation of xanthan gum production by palm date (*Phoenix dactylifera* L.) juice by-products using response surface methodology. Food Chem 121:627–633

Brown SH, Pummill PE (2008) Recombinant production of hyaluronic acid. Curr Pharm Biotechnol 9:239–241

Byrom D (1992) Production of poly-β-hydroxybutyrate: poly-β-hydroxyvalerate copolymers. FEMS Microbiol Rev 103:247–250

Canton B, Labno A, Endy D et al (2008) Refinement and standardization of synthetic biological parts and devices. Nat Biotechnol 26:787–793

Carlson R (2009) The changing economics of DNA synthesis. Nat Biotechnol 27:1091–1094
Cascone MG, Barbani N, Cristallini C et al (2001) Bioartificial polymeric materials based on polysaccharides. J Biomater Sci Polym Ed 12:267–281
Castro C, Zuluaga R, Álvarez C et al (2012) Bacterial cellulose produced by a new acid-resistant strain of Gluconacetobacter genus. Carbohydr Polym 89:1033–1037
Cescutti P (2009) Bacterial capsular polysaccharides and exopolysaccharides. In: Moran AP, Holst O, Brennan PJ, von Itzstein M (eds) Microbial glycobiology. Structures, relevance and applications. Elsevier, Amsterdam, pp 93–115
Changa LL, Bruchb MD, Griskowitzc NJ, Dentel SK et al (2002) NMR spectroscopy for determination of cationic polymer concentrations. Water Res 36:2255–2264
Chee JY, Tan Y, Samian MR, Sudesh K et al (2010) Isolation and characterization of a *Burkholderia* sp. USM (JCM15050) capable of producing polyhydroxyalkanoate (PHA) from triglycerides, fatty acids and glycerols. J Polym Environ. doi:10.1007/s10924-010-0204-1
Chen GQ (2009) A microbial polyhydroxyalkanoates (PHA) based bio- and materials industry. Chem Soc Rev 38:2434–2446
Chua H, Yu PHF, Ho LY (1997) Coupling of waste water treatment with storage polymer production. Appl Biochem Biotechnol 63–65:627–635
Chua H, Yu PHF, Ma CK (1999) Accumulation of biopolymers in activated sludge biomass. Appl Biochem Biotechnol 78:389–399
Dionisi D, Caruccia G, Petrangeli Papinia M, Riccardi C, Majone M, Carrasco F et al (2005) Olive oil mill effluents as a feedstock for production of biodegradable polymers. Water Res 39: 2076–2084
Donot F, Fontana A, Baccou JC et al (2012) Microbial exopolysaccharides: main examples of synthesis, excretion, genetics and extraction. Carbohydr Polym 87:951–962
Dunbar SA (2006) Applications of LuminexR xMAPi technology for rapid, high-through put multiplexed nucleic acid detection. Clin Chim Acta 363:71–82
Eggink G, van der Wal H, Huijberts GNM, de Waard P et al (1992) Oleic acid as a substrate for poly-3-hydroxyalkanoate formation in *Alcaligenes eutrophus* and *Pseudomonas putida*. Ind Crop Prod 1:157–163
Eng AH, Ong EL (2000) Hevea natural rubber. In: Bhowmick AK, Stephens HL (eds) Plastics engineering: handbook of elastomers. Marcel Dekker, New York
Entcheva E, Bien H, Yin LH et al (2004) Functional cardiac cell constructs on cellulose-based scaffolding. Biomaterials 25:5753–5762
Fontes CM, Gilbert HJ (2010) Cellulosomes: highly efficient nanomachines designed to deconstruct plant cell wall complex carbohydrates. Annu Rev Biochem 79:655–681
Freitas F (2011) Fucose-containing exopolysaccharide produced by the newly isolated Enterobacter strain A47 DSM 23139. Carbohydr Pol 1:159–165
French CE (2009) Synthetic biology and biomass conversion: a match made in heaven? J R Soc Interface 6:S547–S558
Fukui T, Doi Y (1998) Efficient production of polyhydroxyalkanoates from plant oils by *Alcaligenes eutrophus* and its recombinant strain. Appl Microbiol Biotechnol 49:333–336
Gaisser S, Reiss T, Lunkes A, Muller K, Bernauer H et al (2009) Making the most of synthetic biology. Strategies for synthetic biology development in Europe. EMBO Rep 10 (Suppl 1):S5–S8
Garnier C, Gorner T, Lartiges BS, Abdelouhab S, Donato P et al (2005) Characterization of activated sludge exopolymers from various origins: a combined size-exclusion chromatography and infrared microscopy study. Water Res 39:3044–3054
Ghosh T (2009) Focus on antivirally active sulphated polysaccharides: from structure–activity analysis to clinical evaluation. Glycobiology 19:2–15
Gibson DG, Young L, Chuang RY, Venter JC, Hutchison CA, Smith HO et al (2009) Enzymatic assembly of DNA molecules up to several hundred kilobases. Nat Methods 6:343–345
Gibson DG, Glass JI, Lartigue C, Noskov VN, Chuang RY, Algire MA, Benders GA, Montague MG, Ma L, Moodie MM, Merryman C, Vashee S, Krishnakumar R, Assad-Garcia N, Andrews-

Pfannkoch C, Denisova EA, Young L, Qi ZQ, Segall-Shapiro TH, Calvey CH, Parmar PP, Hutchison CA, Smith HO, Venter JC et al (2010) Creation of a bacterial cell controlled by a chemically synthesized genome. Science 329:52–56

Gong R, Zhong K, Hu Y, Chen J, Zhu G (2008) Thermochemical esterifying citric acid onto lignocellulose for enhancing methylene blue sorption capacity of rice straw. J Environ Manage 88:875–880

Gray KA, Zhao L, Emptage M (2006) Bioethanol. Curr Opin Chem Biol 10:141–146

Grothe E, Moo-Young M, Chisti Y (1999) Fermentation optimization for the production of poly (β-hydroxybutyric acid) microbial thermoplastic. Enzyme Microb Technol 25:132–141

Hall-Stoodley L, Costerton JW, Stoodley P et al (2004) Bacterial biofilms: from the natural environment to infectious diseases. Nat Rev Microbiol 2:95–108

Hames BR (2009) Biomass compositional analysis for energy applications. Methods Mol Biol 581:145–167

Hermann BG, Patel MK (2007) Today's and tomorrow's bio-based bulk chemicals from white biotechnology. Appl Biochem Biotechnol 136:361–388

Hernandez-Izquierdo VM, Krochta JM (2008) Thermoplastic processing of proteins for film formation: a review. J Food Sci 73:30–39

Hofer P, Choi YJ, Osborne MJ, Miguez CB, Vermette P, Groleau D et al (2010) Production of functionalized polyhydroxyalkanoates by genetically modified *Methylobacterium extorquens* strains. Microb Cell Fact 9:70

Hopewell J, Dvorak R, Kosior E et al (2009) Plastics recycling: challenges and opportunities. Phil Trans R Soc Lond B Biol Sci 364:2115–2126

Hvala N, Aller F, Miteva T et al (2011) Modelling, simulation and control of an industrial, semibatch, emulsion-polymerization reactor. Comput Chem Eng 35:2066–2080

Kahara P, Tsugea T, Taguchib K et al (2004) High yield production of polyhydroxyalkanoates from soybean oil by *Ralstonia eutropha* and its recombinant strain. Polym Degrad Stab 83:79–86

Kalogiannis S, Iakovidou G, Liakopoulou-Kyriakides M, Kyriakidis DA, Skaracis GN et al (2003) Optimization of xanthan gum production by *Xanthomonas campestris* grown in molasses. Process Biochem 39:249–256

Koller M, Salerno A, de Sousa M, Dias M, Reiterer A, Braunegg G et al (2010) Modern biotechnological polymer synthesis: a review. Food Technol Biotechnol 48:255–269

Kumar AS, Mody K, Jha B et al (2007) Bacterial exopolysaccharides—a perception. J Basic Microbiol 47:103–117

Lageveen RG, Huisman GW, Preusting H, Ketelaar P, Eggink G, Witholt B et al (1988) Formation of polyesters by *Pseudomonas oleovorans*: effect of substrates on formation and composition of poly-(R)-3-hydroxyalkanoates and poly-(R)-3- hydroxyalkenoates. Appl Environ Microbiol 54:2924–2932

Lazaridou A, Biliaderis CG, Kontogiorgos V (2003) Molecular weight effects on solution rheology of pullulan and mechanical properties of its films. Carbohydr Polym 52:151–166

Lee SY, Park SJ, Lee Y, Lee SH et al (2003) Economic aspects of biopolymer production. Biopolymers 10:307–337

Leech AJ, Sprinkle A, Wood L, Wozniak DJ, Ohman DE et al (2008) The NtrC family regulator AlgB, which controls alginate biosynthesis in mucoid *Pseudomonas aeruginosa*, binds directly to the algD promoter. J Bacteriol 190:581–589

Lemos PC, Serafim LS, Reis MAM et al (2006) Synthesis of Polyhydroxyalkanoates from Different Short-chain fatty acids by mixed cultures submitted to aerobic dynamic feeding. J Biotechnol 122:226–238

Lenz RW, Marchessault RH (2005) Bacterial polyesters: biosynthesis, biodegradable plastics and biotechnology. Biomacromolecules 6:1–8

López MJ, Moreno J, Ramos-Cormenzana A et al (2001) The effect of olive mill wastewaters variability on xanthan production. J Appl Microbiol 90:829–835

Luisi PL (2007) Chemical aspects of synthetic biology. Chem Biodivers 4:603–621

Lynd LR, Zhang Y (2002) Quantitative determination of cellulase concentration as distinct from cell concentration in studies of microbial cellulose utilization: analytical framework and methodological approach. Biotechnol Bioeng 77:467–475

Lynd LR, Weimer PJ, van Zyl WH, Pretorius IS et al (2002) Microbial cellulose utilization: fundamentals and biotechnology. Microbiol Mol Biol Rev 66:506–577

Majumder A, Singh A, Goyal A (2009) Application of response surface methodology for glucan production from *Leuconostoc dextranicum* and its structural characterization. Carbohydr Polym 75:150–156

Malafaya PB, Silva GA, Reis RL (2007) Natural–origin polymers as carriers and scaffolds for biomolecules and cell delivery in tissue engineering applications. Adv Drug Deliv Rev 59:207–233

Manandhar S, Vidhate S, D'Souza N (2009) Water soluble levan polysaccharide biopolymer electrospun fibers. Carbohydr Polym 78:794–798

Mangione MR, Giacomazza D, Bulone D et al (2005) K+ and Na+ effects on the gelation properties of k-carrageenan. Biophys Chem 113:129–135

Marsano E, Vicini S, Skopinska J et al (2004) Chitosan and poly(vinyl pyrrolidone): compatibility and miscibility of blends. Macromol Symp 218:251–260

Matou S, Colliec-Jouault S, Galy-Fauroux I, Ratiskol J, Sinquin C, Guezennec J, Fischer AM, Helley D et al (2005) Effect of an oversulfated exopolysaccharide on angiogenesis induced by fibroblast growth factor-2 or vascular endothelial growth factor in vitro. Biochem Pharmacol 69:751–759

Matsumoto K, Murata T, Nagao R, Nomura CT, Arai S, Arai Y, Takase K, Nakashita H, Taguchi S, Shimada H et al (2009) Production of short-chainlength/medium-chain-length polyhydroxyalkanoate (PHA) copolymer in the plastid of Arabidopsis thaliana using an engineered 3-ketoacyl-acyl carrier protein synthase III. Biomacromolecules 10:686–690

McCormick CL, Lichatowhich DK (1979) Homogeneous solution reactions of cellulose, chitin and other polysaccharides to produce controlled—activity pesticide systems. J Polm Sci Polym Lett Ed 17:479

McIntyre D, Stephens HL, Schloman WW Jr et al (2001) Guayule rubber. In: Bhowmick AK, Stephens HL (eds) Plastics engineering: handbook of elastomers. Marcel Dekker, New York

Mitsuzawa S, Kagawa H, Li Y, Chan SL, Paavola CD, Trent JD et al (2009) The rosettazyme: a synthetic cellulosome. J Biotechnol 143:139–144

Monsan P, Bozonnet S, Albenne C et al (2001) Homopolysaccharides from lactic acid bacteria. Int Dairy J 11:675–685

Moralejo-Gárate H, Mar'atusalihat E, Kleerebezem R et al (2011) Microbial community engineering for biopolymer production from glycerol. Appl Microbiol Biotechnol 92:631–639

Mu D, Seager T, Rao PS, Zhao F et al (2010) Comparative life cycle assessment of lignocellulosic ethanol production: biochemical versus thermochemical conversion. Environ Manage 46:565–578

Nampoothiri KM, Singhania RR, Sabarinath C et al (2003) Fermentative production of gellan using *Sphingomonas paucimobilis*. Process Biochem 38:1513–1519

Nguyen VT (2008) Potential of a nisin-containing bacterial cellulose film to inhibit *Listeria monocytogenes* on processed meats. Food Microbiol 25:471–478

Nishinari K, Takahashi R (2003) Interaction in polysaccharide solutions and gels. Curr Opin Colloid Interface Sci 8:396–400

Nyburg SC (1954) A statistical structure for crystalline rubber. Acta Crystallogr 7:385–392

Otey FH, Mark AM (1976) Degradable starch based agricultural mulch film. US Patent 3,949,145

Park YM, Shin BA, Oh IJ et al (2008) Poly(L-lactic acid)/polyethylenimine nanoparticles as plasmid DNA carriers. Arch Pharm Res 31:96–102

Pingping Z (1997) A new criterion of polymer–polymer miscibility detected by viscometry. Eur Polym J 33:411–414

Piza MA, Constantino CJL, Venancio EC et al (2003) Interaction mechanism of poly(o-ethoxyaniline) and collagen blends. Polymer 44:5663–5670

Poli A, Kazak H, Gürleyendağ B et al (2009) High level synthesis of levan by a novel Halomonas species growing on defined media. Carbohydr Polym 78:651–657

Puskas JE, Gautriaud E, Deffieux A et al (2006) Natural rubber biosynthesis—a living carbocationic polymerization? Prog Polym Sci 31:533–548

Quélénis A (2008) Les bioplastiques. CRCI Champagne-Ardenne http://www.veillestrategique-champagne-ardenne.fr/IMG/pdf/17bioplastiques.pdf. Accessed 10 Sept 2012

Raha T, Chattopadhyay A, Shaila MS (2004) Development of a reconstitution system for Rinderpest virus RNA synthesis in vitro. Virus Res 99:131–138

Rehm BHA (2010) Bacterial polymers: biosynthesis, modifications and applications. Rev Microbiol. doi:10.1038/nrmicro2354

Rodriguez-Cabello JC, Reguera J, Girotti A et al (2005) Developing functionality in elastin-like polymers by increasing their molecular complexity: the power of the genetic engineering approach. Prog Polym Sci 30:1119–1145

Rosalam S, England R (2006) Review of xanthan gum production from unmodified starches by *Xanthomonas campestris* sp. Enzym Microbial Technol 39:197–207

Ruffing A, Chen RR (2006) Metabolic engineering of microbes for oligosaccharide and polysaccharide synthesis. Microb Cell Fact 5:25

Ryffel GU (2010) Making the most of GM potatoes. Nat Biotechnol 28:318

Sandoval A, Arias-Barrau E, Arcos M, Naharro G, Olivera ER, Luengo JM et al (2007) Genetic and ultrastructural analysis of different mutants of *Pseudomonas putida* affected in the poly-3-hydroxy-n-alkanoate gene cluster. Environ Microbiol 9:737–751

Schmidt M, Pei L (2011) Synthetic toxicology: where engineering meets biology and toxicology. Toxicol Sci 120:204–224

Serafim LS, Lemos PC, Oliveira R,Reis MA et al (2004) Optimization of plyhydroxybutyrate production by mixed cultures submitted to aerobic dynamic feeding conditions. Botechnol Bioeng 87:145–160

Shamala TR, Prasad MS (1995) Preliminary studies on the production of high and low viscosity dextran by *Leuconostoc* spp. Process Biochem 30:237–241

Shi F, Gross RA, Rutherford DR (1996) Microbial polyester synthesis: effects of poly(ethylene glycol) on product composition, repeat unit sequence, and end group structure. Macromolecules 29:10–17

Shih IL, Chen LD, Wu JY (2010) Levan production using *Bacillus subtilis* natto cells immobilized on alginate. Carbohydr Polym 82:111–117

Sionkowska A (2011) Current research on the blends of natural and synthetic polymers as new biomaterials: review. Prog Polym Sci 36:1254–1276

Siracusa V, Rocculi P, Romani S et al (2008) Biodegradable polymers for food packaging: a review. Trends Food Sci Technol 19:634–643

Soetaert W (2007) White biotechnology: a key technology for building the biobased economy, Proc. Illmac Basel

Somerville C (2006) Cellulose synthesis in higher plants. Annu Rev Cell Dev Biol 22:53–78

Steinbüchel A, Schmack A (1995) Large scale production of poly(3-hydroxyvaleric acid) by fermentation of *Chromobacterium violaceum* technical processing and characterization of the homopolyester. J Environ Polym Degrad 3:243–258

Stredansky M, Conti E (1999) Xanthan production by solid state fermentation. Process Biochem 34:581–587

Sudip S (2012) Production and characterization of extracellular polymeric substances of Rhizobium with different carbon sources. MSc thesis

Suh JKF, Matthew HWT (2000) Application of chitosan-based polysaccharide biomaterials in cartilage tissue engineering: a review. Biomaterials 21:2589–2598

Survase SA, Saudagar PS, Singhal RS (2007) Use of complex media for the production of scleroglucan by *Sclerotium rolfsii* MTCC 2156. Bioresource Technol 98:1509–1512

Sutherland IW (1998) Novel and established applications of microbial polysaccharides. Trends Biotechnol 16:41–46

Synowiecki J, Al-Khateeb NA (2003) Production, properties, and some new applications of chitin and its derivatives. Crit Rev Food Sci Nutr 43:145–171
Takeuchi A (2009) Oral administration of xanthan gum enhances antitumor activity through Toll-like receptor 4. Int Immunopharmacol 9:1562–1567
Tanaka Y (1989) Structure and biosynthesis mechanism of natural polyisoprene. Prog Polym Sci 14:339–371
Tang X, Alavi S (2011) Recent advances in starch, polyvinyl alcohol based polymer blends, nanocomposites and their biodegradability. Carbohydr Polym 85:7–16
Tasdelen MA, Kahveci MU, Yagci Y (2011) Telechelic polymers by living and controlled/living polymerization methods. Prog Polym Sci 36:455–567
Thakor N, Trivedi U, Patel KC et al (2005) Biosynthesis of medium chain length poly(3-hydroxyalkanoates) (mcl-PHAs) by *Comamonas testosteroni* during cultivation on vegetable oils. Bioresource Technol 96:1843–1850
Tsuge T (2002) Metabolic improvements and use of inexpensive carbon sources in microbial production of polyhydroxyalkanoates. J Biosci Bioeng 94:579–584
Tyagi RD, Sikati-Foko V, Barnabe S, Vidyarthi AS, Valéro JR, Surampalli RY et al (2002) Simultaneous production of biopesticide and alkaline proteases by *Bacillus thuringiensis* using sewage sludge as a raw material. Water Sci Technol 46:247–254
Urbain V, Block JC, Manem J et al (1993) Bioflocculation in activated sludge: an analytical approach. Water Res 27:829–838
Vedyashkina TA, Revin VV, Gogotov IN (2005) Optimizing the conditions of dextran synthesis by the bacterium *Leuconostoc mesenteroides* grown in a molasses-containing medium. Appl Biochem Microbiol 41:361–364
Wang X, Sang L, Luo D et al (2011) From collagen–chitosan blends to three-dimensional scaffolds: the influences of chitosan on collagen nanofibrillar structure and mechanical property. Colloids Surf B Biointerfaces 82:233–240
Werk D, Wengel J, Wengel SL et al (2010) Application of small interfering RNAs modified by unlocked nucleic acid (UNA)to inhibit the heart-pathogenic coxsackievirus B3. FEBS Lett 584:591–598
Williams DL, Sherwood ER, Browder LW, McNamee RB, Jones EL, Di Luzio NR et al (1988) The role of complement in glucan-induced protection against septic. Circ Shock 25:53–60
Yamane T (1993) Yield of poly-D(−)-3-hydroxybutyrate from various carbon sources: a theoretical study. Biotechnol Bioeng 41:165–170
Yasuhiro M, Yoshimitsu K (2010) Development of a wound dressing composed of hyaluronic acid sponge containing arginine and epidermal growth factor. J Biomat Sci Polym Ed 21:715–726
Ye Y, Dan W, Zeng R et al (2007) Miscibility studies on the blends of collagen/chitosan by dilute solution viscometry. Eur Polym J 43:2066–2071
Yezza A, Tyagi RD, Valero JR, Surampalli RY, Smith JC et al (2004) Scale-up of biopesticides production process using wastewater sludge as a raw material. J Ind Microbiol Biotechnol 31:545–552
Youssef F, Roukas T, Biliaderis CG (1999) Pullulan production by a non-pigmented strain of *Aureobasidium pullulans* using batch and fed-batch culture. Process Biochem 34:355–366
Yu J (2001) Production of PHA from starchy wastewater via organic acids. J Biotechnol 86:105–112
Zeng X, Small DP, Wan W (2011) Statistical optimization of culture conditions for bacterial cellulose production by *Acetobacter xylinum* BPR 2001 from maple syrup. Carbohydr Polym 85:506–513
Zhang X, Luo R, Wang Z, Deng Y, Chen GQ et al (2009) Application of (R)-3-hydroxyalkanoate methyl esters derived from microbial polyhydroxyalkanoates as novel biofuels. Biomacromolecules 10:707–711
Zhang X, Wub X, Gaoa D et al (2012) Bulk cellulose plastic materials from processing cellulose powder using back pressure-equal channel angular pressing. Carbohydr Polym 87:2470–2476
Zhu C, Nomura CT, Perrotta JA, Stipanovic AJ, Nakas JP et al (2010) Production and characterization of poly-3-hydroxybutyrate from biodiesel-glycerol by *Burkholderia cepacia* ATCC 17759. Biotechnol Prog 26:424–430

Chapter 18
Exploitation of Agro-Industrial Wastes to Produce Low-Cost Microbial Surfactants

Partap Bir Singh and Harvinder Singh Saini

18.1 Introduction

Surfactants are amphipathic molecules having characteristic property of reducing interfacial tension between two immiscible liquids which helps to improve their miscibility. Due to these properties, use of synthetic surfactants as emulsifiers, detergents, solubilizers, and foaming and wetting agents became widespread for household and industrial processes has increased their demand over 300 % within the last decade (Van Bogaert et al. 2007). Majority of the surfactants and emulsifiers are being synthesized chemically from petrochemicals. This results in additional pressure on already depleting stocks of petroleum products. Additionally, due to extensive use and release of synthetic surfactants in the environment, issues relating to human health and environment safety have arisen over the last years. The commonly used synthetic surfactants are persistent in environment as these are composed of branched chain fatty acids which cannot be degraded by majority of microbes. Moreover, synthetic colorant and optical brighteners present in commercial preparations of synthetic surfactants are known to cause allergic reactions and reproductive and developmental problems in mammals and fish. In light of these ecological issues, extensive research is being carried out for production of biologically safe surface-active molecules from non-petrochemical sources.

Biosurfactants, the secondary metabolites of microbial origin, are biodegradable and environmentally safer alternative to synthetic surfactants. Additionally, they have ability to retain their surface-active properties at wide range of temperature, pH, and salinity which is suitable for their use in harsh operating condition during their applications in environment management, laundry, mining, and metallurgical industries. The microbial surfactants due to their better biocompatibility as compared to synthetic surfactants could be used safely in cosmetics, in therapeutics, and as

P.B. Singh • H.S. Saini (✉)
Department of Microbiology, Guru Nanak Dev University, Amritsar, India
e-mail: sainihs@yahoo.com

S.K. Brar et al. (eds.), *Biotransformation of Waste Biomass into High Value Biochemicals*,
DOI 10.1007/978-1-4614-8005-1_18,

functional food ingredients in food products for human consumption. The microbial species belonging to different genera *Pseudomonas* sp., *Bacillus* sp., *Acinetobacter* sp., *Agrobacterium* sp., *Rhodococcus* sp., *Arthrobacter* sp., *Nocardia* sp., *Corynebacterium* sp., *Mycobacterium* sp., and *Candida* sp. have been reported to produce biosurfactants. These biomolecules based on their diverse structural compositions can be classified as glycolipids, phospholipids, lipopolysaccharides, lipoproteins/lipopeptides, and hydroxylated/cross-linked fatty acids. This vast diversity of functional groups lacking in synthetic surfactants provides a wide range of surface-active properties to biosurfactants. However, in spite of such advantages, large-scale application of these biomolecules is still limited because the overall production cost of biosurfactant is three to ten times higher than synthetic surfactants (Mukherjee et al. 2006). Thus, the use of biosurfactants is not economical for applications which require their bulk quantities, such as bioremediation of polluted sites, laundry, and enhanced oil recovery. Presently, the use of biosurfactants is limited only to certain high-value applications, such as cosmetics, diagnostic kits, and stabilization of nano-emulsions, as high cost of these products could offset the overall cost of biosurfactant production.

The three basic factors primarily responsible for high cost of biosurfactant production are (1) cost of raw materials, (2) low yield of biosurfactant by microorganisms, and (3) cost of downstream processing. It is estimated that cost of initial raw materials account for almost 50 % of the overall cost of biosurfactant production (Rodrigues et al. 2006). Thus, lowering the cost of growth medium could significantly lower the overall cost of biosurfactant production. The large amount of renewable agricultural waste and industrial by-products rich in organic matter is generally disposed in the environment. Asokan et al. (2005) reported release of 350 million tons of wastes from agricultural sources in India which includes sugarcane bagasse, paddy and wheat straw, coconut husk, and vegetable wastes among others. The utilization of such agricultural wastes as raw material for production of high-value biochemicals not only would reduce the overall cost of production but also may reduce the risk related to their discharge in the environment. Similarly, industrial by-products/wastes, such as molasses, soy molasses, potato process effluent, cassava wastewater, orange fruit peelings, frying oils, oil refinery wastes, olive oil mill effluent, peanut oil cake, soap stock, whey, distillery waste, and used motor lubricant oil, are other potential nutrient supplements which could support microbial biomass/product formation at laboratory/pilot scale (Sengupta 2002). However, production of microbial surfactant at commercial level requires a lot of basic studies for developing nutritionally balanced medium formulation as certain by-products/wastes are available only in specific regions of the world which limits their generalized use for process development. The cost may further be reduced by developing protocols to produce more than one value-added product in the same process or harnessing secondary products.

This chapter presents an account of reports regarding evaluation of suitable low-cost renewable agro-industrial wastes as nutrient sources for production of microbial surfactants. The literature survey indicated that statistical-based process optimization studies and efficient bioreactor designing approaches could provide a suitable road map for cost effective production of such biomolecules at commercial scale.

18.2 Biosurfactants: Microbial Origin

The water-soluble compounds are preferred source of nutrition for microbial growth. The nonaqueous phase soluble hydrocarbons, although are rich source of carbon and energy for microbial growth, are not readily bioavailable to microorganisms. Some of the microbial populations present in such ecosystem produce surface-active compounds which help in emulsification of these nonaqueous phase hydrocarbons to aqueous phase. The microemulsions of hydrocarbons thus formed increase their availability to microbes which can use them as growth substrates. This ability of these microorganisms provides them a competitive edge over non-biosurfactant-producing microbes to thrive in oligotrophic environment. The microorganisms belonging mainly to *Pseudomonas* spp., *Rhodococcus* spp., *Arthrobacter* spp., *Bacillus* spp., etc. are being explored for production of surface-active molecules for diverse applications.

18.2.1 Classification and Chemical Composition

Biosurfactants are natural surface-active secondary metabolites of microbial origin which are known for their diverse chemical structures (Fig. 18.1). The surface-active compounds of microbial origin are primarily classified into two major classes: biosurfactants (low-molecular-weight compounds) and bioemulsifiers (high-molecular-weight compounds). Microbial surfactants are amphipathic biomolecules having both hydrophobic and hydrophilic domains. The hydrophilic domain can be a carbohydrate, phosphate, cyclic peptide, amino acid, carboxylic acid, or alcohol. Most of the biosurfactants are either anionic or neutral while some are cationic, having amine groups. The hydrophobic moiety is frequently a hydrocarbon chain comprised of a long-chain fatty acids, hydroxy fatty acids, or α-alkyl-β-hydroxy fatty acids. These wide ranges of structural variations in biosurfactants provide them characteristic surface-active properties, ranging from emulsifiers to biosurfactants, which could be exploited for multifarious applications. The broad grouping of structurally diverse biosurfactants and their microbial origin are summarized in Table 18.1.

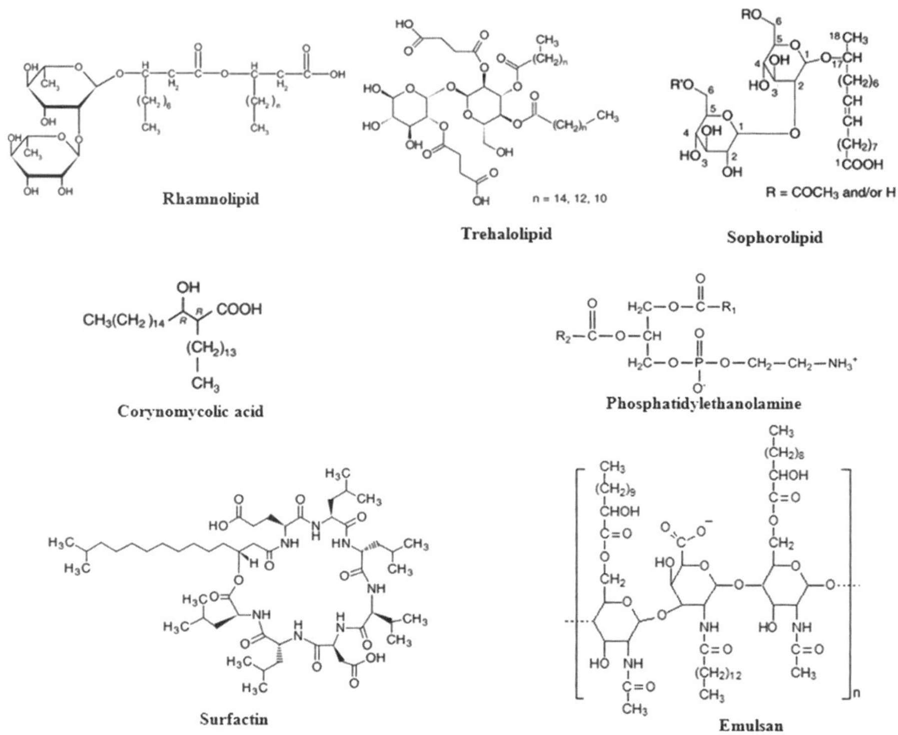

Fig. 18.1 Structural variations in different types of microbial surfactants

18.3 Comparative Account of Biosafety and Surface-Active Properties of Biosurfactants and Synthetic Surfactants

18.3.1 Biodegradability and Toxicity

Presently, majority of surfactants available in the market are products/by-products of petrochemical industries. The common nonionic synthetic surfactants include ethylene, ethoxylates, sorbitan ester, and propylene oxide copolymers while ionic synthetic surfactants include fatty acids, ester sulfates or sulfonates, and quaternary ammonium salts. These surfactants are being used in various sectors, such as laundry and cleaning (54 %), textiles, leather and paper (13 %), cosmetics and pharmaceuticals (10 %), chemical processes (10 %), food industry (3 %), and agriculture (2 %) (Rahman and Gakpe 2008). According to a recent estimate, the worldwide production of synthetic surfactants is increasing over ten million tons each year which is putting immense pressure on already depleting sources of petrochemicals (Van Bogaert et al. 2007). The products based on these synthetic surfactants generally contain artificial colorants and aminotriazine or

Table 18.1 Classification of biosurfactants produced by microorganisms

Biosurfactants			
Group	Class	Microorganisms	References
Glycolipids	Rhamnolipids, sophorolipids, trehalose lipids, mannosylerythritol lipids	*Pseudomonas* spp., *Rhodococcus* spp., *Arthrobacter* spp., *Candida* spp.	Mata-Sandoval et al. (1999), Mercade et al. (1996), Duvnjak et al. (1982), Sobrinho et al. (2008)
Phospholipids	Phosphatidylethanolamine	*Rhodococcus* spp., *Acinetobacter* sp.	Mercade et al. (1996), Kappeli and Finnerty (1979)
Lipopeptides	Surfactin, iturin, fengycin, viscosin, lichenysin, serrawettin	*Bacillus subtilis, B. amyloliquefaciens, Pseudomonas fluorescens, B. licheniformis, Serratia marcescens*	Vedaraman and Venkatesh (2011), Arima et al. (1968), Wang and Mulligan (2004), McInerney et al. (1990)
Fatty acids/neutral lipids	Corynomycolic acids	*Corynebacterium lepus*	Rosenberg and Ron (1999)
Polymeric surfactants	Emulsan, alasan, liposan, lipomanan	*Acinetobacter calcoaceticus, A. radioresistens, C. lipolytica, C. tropicalis*	Banat et al. (2000), Cirigliano and Carman (1985)
Particulate compounds	Vesicles, whole microbial cells	*Acinetobacter calcoaceticus, Pseudomonas marginilis, P. maltophilia, cyanobacteria*	Kappeli and Finnerty (1979)

stilbene-based nonbiodegradable optical brighteners which can cause skin/eye irritations and allergic reactions in mammals and fishes (RPA 2006). Moreover, diethanolamine-based surfactants can chemically react with natural nitrogen oxides in the atmosphere to form carcinogenic nitrosamines (IARC 2000). Some synthetic surfactants contain inorganic phosphates, such as aluminosilicates which are responsible for excessive algal growth resulting in eutrophication of water bodies. Thus, excessive use and accumulation of synthetic surfactants in the environment is not biologically safe.

However, microbial surfactants being of biological origin are easily biodegradable and less toxic than their synthetic counterparts. Lima et al. (2011) reported 42.5–73.4 % degradation of five microbial surfactants produced by *Flavobacterium* sp., two *Bacillus* sp., *Dietzia maris,* and *Arthrobacter oxydans* which is significantly higher as compared to 24.8 % achieved of synthetic surfactant sodium dodecylsulfate (SDS) by mixed cultures of the soil after 7 days of incubation. The hemolytic activity of lipopeptide produced by *Bacillus subtilis* ATCC6633 is lower than that of synthetic surfactants cetyltrimethylammonium bromide (CTAB), tetradecyltrimethylammonium bromide (TTAB), benzalkonium chloride (BC), and SDS (Noudeh et al. 2005). Das and Mukherjee (2005) reported that a glycolipid biosurfactant produced by *P. aeruginosa* is safe for human consumption as it has no detrimental effect on the human heart, lung, liver, and kidney tissues. Poremba et al. (1991) reported that the lethal concentration causing 50 % mortality (LC_{50}) of synthetic anionic surfactant (Corexit) against *Photobacterium phosphoreum* was ten times lower than that of rhamnolipids indicating that rhamnolipid being ten times less toxic than synthetic surfactant.

18.3.2 Surface Activity

The addition of surfactant in a solution up to a concentration known as the critical micelle concentration (CMC) reduces the surface/interfacial tension to a level where no further reduction takes place with addition of surfactant. The surfactant monomer aggregate at CMC due to weak hydrophobic and van der Waals interactions and form micelles to entrap hydrophobic compounds in the core resulting in their emulsification in aqueous phase. The CMC of biosurfactants is generally 10–40 times lower than synthetic surfactants, thus biosurfactant at low concentration can achieve higher emulsification activity. These low concentrations are safe and economical in different industrial and biotechnological applications (Desai and Banat 1997). The high-molecular-weight surface-active biomolecules which do not efficiently decrease the surface tension are generally good emulsifiers and used in cosmetics and food industries to form stable emulsions of oil in water (Muthusamy et al. 2008). These studies showed that microbial surfactants with diverse chemical composition, low toxicity, and biodegradability have distinct functional properties which can be exploited for specialized applications in various commercial sectors as summarized in Table 18.2.

Table 18.2 Commercial applications of biosurfactants

Commercial sectors	Biosurfactants	Functional properties	Applications	References
Agricultural	Fengycins, rhamnolipid	Antifungal, competition, hypovirulence, detergents	Control pathogenic fungi of agricultural crops Facilitation of biocontrol mechanisms of microbes Prevent the caking of certain fertilizer during storage	Makkar and Rockne (2003), Stanghellini and Miller (1997)
Bioremediation	Rhamnolipids, sophorolipids, phospholipids, emulsan, alasan	Solubilizers, emulsifiers, demulsifiers	Solubilization and biodegradation of: Crude petroleum hydrocarbons Pesticides Polyaromatic compounds at contaminated sites	Providenti et al. (1995), Millioli et al. (2009), Singh et al. (2009), Sharma et al. (2009)
Food	Rhamnolipids, sophorolipids, emulsan, surfactin, iturin	Suspension agent, wetting agent, thickener, lubricating agent, antifungal	For cleaning sanitizing Improve removal of pesticides Solubilize flavor oils, control consistency Crystallization of sugar, reduce processing time Reduction of the mycoflora in stored grains of corn, peanuts, and cottonseeds	Muthusamy et al. (2008), Nitschke and Costa (2007)
Petroleum processes	Surfactin, lichenysin, rhamnolipid, trehalose, cellobiose lipid	Emulsifiers	Microbially enhanced oil recovery (MEOR) by lowering interfacial tension at the oil-rock interface and reduction in oil viscosity	Banat (1995), Jack et al. (1998), Yakimov et al. (1997)

(continued)

Table 18.2 (continued)

Commercial sectors	Biosurfactants	Functional properties	Applications	References
Therapeutics and biomedical	Rhamnolipid, surfactin, iturin, fengycins, trehalose, mannosylerythritol lipids	Antimicrobial, anti-adhesive, antitumor, chelating agents	Antimicrobial activity against *Mycobacterium tuberculosis,* several bacterial, and yeast strains Antitumor activity against Ehrlich's ascites carcinoma cells Antiviral activity against HSV and influenza virus Chelating properties explain the membrane-disrupting effect of lipopeptides	Gerard et al. (1997), Vollenbroich et al. (1997), Kitamoto et al. (1993)
Personal care	Rhamnolipids, sophorolipids, mannosylerythritol lipids	Antibacterial, antioxidant, moisturizers, wetting agent, emulsifier, dispersant	Lotions, body washes, hair products, lip color, acne treatment, deodorants, skin smoothing, antiwrinkle products	Kosaric (1992), Brown (1991), Singh et al. (2007)
Mining	Rhamnolipids, sophorolipids, trehalose, surfactin	Emulsifying agent	Catalyze the fracturing of limestone and lowers the energy required for cleaving the micro-fractures into smaller particles Stabilization of coal slurries for transportation of coal	Rosenberg et al. (1988)
Others	Rhamnolipid, surfactin, sophorolipids	Emulsifying agent and dispersant	Suspension made in biosurfactant solution gives better spreadability and improved mixing properties in: Pulp and paper Textiles Uranium ore processing Paint	Horowitz and Currie (1990), McInerney et al. (1990),

18.4 Substrates for Biosurfactants Production

18.4.1 Conventional Substrates for Microbial Surfactants Production

The carbon source used in growth medium primarily affects the quantity and quality of biosurfactants produced by microorganisms (Sen 1997; Abouseoud et al. 2008). Different microbes have been reported to use wide range of carbon sources, such as alkanes, pyruvate, fructose, succinate, citrate, glycerol, mannitol, lactose, *n-paraffin*, hexadecane, and glucose for production of biosurfactants (Robert et al. 1989; Abu-Ruwaida et al. 1991; Pruthi and Cameotra 2003). The commonly used organic and inorganic nitrogen supplements for biosurfactants production includes urea, beef extract, yeast extract, casein hydrolysate, ammonium nitrate, potassium nitrate, sodium nitrate, and ammonium chloride (Abouseoud et al. 2008; Aparna et al. 2012). Although, use of these conventional carbon and nitrogen sources is suitable for lab scale studies but is not economically viable to support production at commercial scale. Thus, there is a need to explore possibilities of using various low-cost agro-industrial waste or by-products as nutrient supplements in fermentation media to lower the cost of biosurfactant production.

18.4.2 Low-Cost Substrates: Cheap and Value-Added Alternatives

Farming and agricultural production worldwide support basic food requirements of mankind, and processing of agricultural raw materials in industries produces a considerable amount of solid waste and by-products. The processing of wheat, rice, corn, sorghum, and barley grains in food industries releases large amounts of wastewater rich in carbohydrates in the environment. Thus, studies have been carried out to explore possibilities of using low-cost and renewable agro-industrial wastes including distillery wastes, plant oils, oil wastes, starchy substances, and whey as substrates for cost effective production of biosurfactant (Deleu and Paquot 2004; Ferreira 2008; Montoneri et al. 2009). The agricultural residues of various crops and industrial wastes reported as substrates for biosurfactant production are presented in Table 18.3.

18.4.2.1 Starchy Substrates

The carbohydrate-enriched crops, such as corn, wheat, rice, cassava, and potatoes, are being grown worldwide. Starch obtained from these crops is a major industrial

Table 18.3 Agro-industrial wastes utilized for biosurfactant production

Agro-industrial wastes/by-products		Biosurfactants	References
Starchy waste substrates	Peel and stalks of sweet potato Bran and straw of rice Corn steep liquor Wastewater from the processing of cereals and pulses Cassava waste water	Surfactin, rhamnolipids, glycolipids, polyhydroxyalkanoates	Fox and Bala (2000), Noah et al. (2005), Nitschke and Pastore (2006)
Vegetable oils	Sunflower, soybean, rapeseed, babassu, olive, palm, coconut, corn oils, etc.	Rhamnolipids, glycolipids, sophorolipid, mannosylerythritol	Trummler et al. (2003), Coimbra et al. (2009), Camargo-de-Morais et al. (2003)
Oil processing waste	Fried oils, olive oil mill effluents, soybean refinery waste, palm oil mill waste, lubricating oil waste	Sophorolipids, surfactin, rhamnolipids, glycolipids	Pandey et al. (2000), Makkar and Cameotra (2002), Lima et al. (2009), Onbasli and Aslim (2009), Krieger et al. (2010)
Dairy industry wastes	Buttermilk Lactose/cheese whey	Sophorolipids, rhamnolipids, glycolipids	Deshpande and Daniels (1995), Dubey and Juwarkar (2001), Siddhartha et al. (2009)
Sugar industry wastes	Molasses Spent wash from alcohol distilleries Crude glycerol of biodiesel processing	Sophorolipids, surfactin, rhamnolipid, glycolipids	Daniel et al. (1998), Makkar and Cameotra (1999), Dubey and Juwarkar (2004), Onbasli and Aslim (2009)

product used as a thickener in food industries for making puddings, custards, soups, noodles, pastas, etc. and as a nonfood product in paper and clothing industries. Potato is a rich source of starch (9.1–22.6 g/100 g), protein (0.8–4.2 g/100 g), micronutrients (iron, potassium, magnesium, phosphorus, and calcium), and vitamin C (Lutaladio and Castaldi 2009). The potato peel containing about 40 % dietary fibers and 28–51 % starch depending upon the peeling process is usually discarded as waste (Camire et al. 1997). Moreover, it is estimated that 41 % of the potato crop cannot be utilized for making consumable products due to its overproduction/spoilage and thus used in animal feed and alcohol production (Natu et al. 1991; Fox and Bala 2000).

Fox and Bala (2000) utilized waste from potato processing industry to produce biosurfactant by *B. subtilis* ATCC 21332 in shake flask studies. The production of biosurfactant reduced the surface tension of growth medium to 28.3 mN/m from 71.3 mN/m with critical micellar concentration of 0.10 g/l. Similarly, Das and Mukherjee (2007) reported biosurfactant (lipopeptides) production by two *Bacillus subtilis* strains DM-03 and DM-04 using powder of potato peels as substrate in solid state fermentation (SSF) and submerged fermentation (SmF). *Bacillus subtilis* DM-03 supported 67.0 mg/gram dry substrate (mg/gds) to 80 mg/gds in SSF and SmF systems, respectively. Zhu et al. (2012) reported production of 50.01 mg/gds lipopeptides by *Bacillus amyloliquefaciens* XZ-173 by utilizing starch-rich rice straw (3.67 g) and soybean flour (5.58 g) as substrates in SSF after 2 days of incubation.

Cassava is cultivated as an annual crop in tropical and subtropical regions for its edible starch-rich tuberous roots. The cassava wastewater is a rich source of protein, fructose, glucose, maltose, nitrate, phosphorus, and minerals (Nitschke and Pastore 2006). Nitschke and Pastore (2003, 2004, 2006) used cassava wastewater for surfactin production by *Bacillus subtilis* ATCC 21332 and *B. subtilis* LB5a. The yield of surfactin was 2.2 g/l and 3.0 g/l respectively during shake flask studies in the fermentation medium. While, Barros et al. (2008) reported production of a biosurfactant by *Bacillus subtilis* LB5a in a 40 l pilot scale batch bioreactor by using cassava wastewater. The foam was collected simultaneously by fractionation which was precipitated to yield 2.4 g/l of biosurfactant during the fermentative process. The microbial surfactant supported good surface activity with reduction in surface tension of growth medium to 27 mN/m and the CMC of 11 mg/l. Further, Costa et al. (2009) reported use of cassava wastewater as a substrate for the production of rhamnolipids by *P. aeruginosa* in lab scale shake flask studies. The rhamnolipid production was 660 mg/l with reduction in surface tension of growth medium up to 30 mN/m and CMC of 26.5 mg/l.

Corn steep liquor, a by-product obtained after separation of starch by wet milling of corns, is generally used as a nutritive source in animal feed as it contains protein (47 %), lactic acid (26 %), dextrose (2.5 %), nitrogen (7.5 %), and fat (0.4 %). It can be potentially used as a low-cost nutrient for microbial growth (White and Johnson 2003). Sobrinho et al. (2008) reported a low-cost medium containing 5.0 % (w/v) groundnut oil refinery residue and 2.5 % (w/v) corn steep liquor as substrates for biosurfactant (4.5 g/l) production by yeast *Candida sphaerica* (UCP 0995). Recently, Luna et al. (2013) reported 9 g/l biosurfactant production by using same

Table 18.4 Fatty acid composition of different vegetable oils

	Fatty acids (%)					
Vegetable oils	Lauric acids (C12:0)	Palmitic acid (C16:0)	Myristic acids (C14:0)	Stearic acids (C18:0)	Oleic acid (C18:1)	Linoleic acid (C18:2)
Sunflower oil	–	5.9	–	4.5	19.5	65.7
Safflower oil	–	4.29	–	1.92	14.36	74.6
Soya-bean oil	–	10.3	0.1	3.8	22.8	56.0
Coconut oil	44.6	8.2	16.8	2.8	5.8	1.8
Rice bran oil	–	16.9	0.7	1.6	39.1	33.4
Olive oil		10.93		1.98	72.3	9.21

strain of *C. sphaerica* (UCP 0995) in medium supplemented with 9.0 % corn steep liquor and 9.0 % groundnut oil refinery residue. The biosurfactant with CMC of 0.25 mg/ml lowers the surface tension of water from 70 to 25 mN/m. Thus, these studies indicated the potential of wastes/by-products containing starch/others micronutrients as low-cost substrates for economical production of biosurfactants.

18.4.2.2 Oils and Oil Wastes as Substrates

Vegetable Oils

The production of oil and fats around the world from animal and plant sources was more than three million tons, and out of that 75 % of the oil is obtained from plants (Haba et al. 2000). According to a recent report, the total worldwide production of vegetable oils in 2008–2009 was around 133 million tons (USDA 2009). The vegetable oils are comprised of varied ratios of different fatty acids such as palmitic acid (C16:0), linoleic acid (C18:2), oleic acid (C18:1), lauric acids (C12:0), stearic acids (C18:0), and myristic acids (C14:0) as described in Table 18.4.

Mercade et al. (1988) reported production of 7,10 dihydroxy-8(E)-octadecanoic acid by *Pseudomonas* strain 42A2 by utilizing by-product of vegetable oils mixtures containing 98 % (w/w) of oleic acid. Kitamoto et al. (1993) reported use of soybean oil as substrate for *Candida antarctica* T-34 which supported production of mannosylerythritol lipids (MEL-A and B) having antimicrobial properties against gram-positive bacteria. Vollbrecht et al. (1999) used oleic acid-rich oils, rapeseed oil, and sunflower oil individually as substrates for *Tsukamurella* sp. DSM 44370 and obtained better growth and glycolipid production than that obtained in synthetic media. The sunflower oil supported overall yield of 5 g/l of glycolipids GL1, GL2, and GL3 with reduction in surface tension of water from 72 mN/m to 35 mN/m, 23 mN/m and 24 mN/m respectively. CMC value of glycolipids was 10 mg/l. Similarly, Casas and Garcia-Ochoa (1999) used sunflower oil as a substrate for *Candida bombicola* and reported 120 g/l production of sophorolipid after 8 days of incubation. Mata-Sandoval et al. (1999) reported improved production of rhamnolipid (100–165 mg/g substrate) by *Pseudomonas aeruginosa* UG2 using corn oil as a sole carbon source in the medium instead of succinic acid and glucose as these supported only 12–36 mg rhamnolipid per gram substrate. Rau et al. (2001) reported economical production of sophorolipid with overall yield of 57 g/l/day in feed-batch and 76 g/l/day in continuous mode using rapeseed oil along with glucose as carbon source. Trummler et al. (2003) obtained overall yield of 45 g/l biosurfactant mixtures of mono- and dirhamnolipids by using rapeseed oil as substrate for *Pseudomonas* sp. DSM 2874. Camargo-de-Morais et al. (2003) used 1 % (v/v) olive oil as carbon source for *Penicillium citrinum* which supported production of a glycolipid. Pekin and Vardar-Sukan (2005) reported capability of *Candida bombicola* ATCC 22214 to produce 400 g/l of sophorolipids in a 3 l bioreactor using Turkish corn oil, glucose, and cheap market honey as carbon sources. Thaniyavarn et al. (2006) studied the production of biosurfactant by using 2 % (v/v) of different vegetable oils as a carbon sources in the growth medium for *Pseudomonas aeruginosa* A41. The maximum biosurfactant yield of 6.58 g/l was obtained with olive oil followed by 2.91 g/l with palm oil and 2.93 g/l with coconut oil. Thaniyavarn et al. (2008) further reported use of 4 % (v/v) soybean oil as carbon source for production of sophorolipid by a thermotolerant strain *Pichia anomala* PY1. The biosurfactant lowered the surface tension of growth medium to 28 mN/m and possesses crude oil displacement property of 69.43 cm^2 for a potential application in extraction of crude oil from oil wells. Coimbra et al. (2009) reported production of biosurfactant by six *Candida* strains utilizing mixture of oil-rich substrates including groundnut, oil refinery residue, and soybean oil with corn steep liquor and glucose. The biosurfactant produced was able to remove 90 % of the hydrophobic contaminants from sand. Daverey and Pakshirajan (2009) used soybean, sunflower, and olive oil in combination with sugarcane molasses as a low-cost medium and reported production of 24 g/l sophorolipids from the yeast *Candida bombicola*. In subsequent fermentor (3 l) studies using 50 g/l of sugarcane molasses with 50 g/l of soybean oil as the substrates under optimal conditions the authors achieved 60 g/l of sophorolipids production (Daverey and Pakshirajan 2010).

Waste Vegetable Oil

Vegetable oil wastes generated from food industries, oil refineries, soap industries, and kitchen are generally disposed in sewers or drains which causes pollution and blockages in waterways. Thus, oil waste residues, such as olive oil mill effluents (OOME), soybean oil refinery wastes, and fried oils, are being exploited as low-cost substrates for the production of rhamnolipid, sophorolipid, mannosylerythritol, and lipopeptides (Trummler et al. 2003; Rahman et al. 2002; Pekin and Vardar-Sukan 2005).

Olive oil mill effluents, a water-soluble fraction released after extraction of oil, contain toxic polyphenols and are not suitable for human consumption but can be used as a nutrient-rich substrate for microbes to produce biochemicals of commercial importance. OOME contains sugars (20–80 g/l), nitrogen (12–24 g/l), organic acids (5–15 g/l), and residual oil (0.3–5 g/l) and was reported for first time in 1993 as a substrate for production of rhamnolipids (0.058 g/g OOME) by *Pseudomonas* sp. JAMM (Mercade et al. 1993). The biosurfactant produced was able to lower the surface tension of growth medium from 40 mN/m to about 30 mN/m.

The used frying oils of sunflower, canola, soybean, olive, rice bran, corns, etc. are being released at the large scale in the environment. The restaurants in the United States are producing waste frying oil at an average rate of 100 billion liters per week (Shah et al. 2007). The nutritional value of frying oils varies depending upon food products fried and number of times it has been reused. The used frying oils have 30 % higher polar hydrocarbons (triacylglycerol oligomers, triacylglycerol dimmers, oxidized triacylglycerol monomers, and diacylglycerols) than the fresh oil and can be exploited as a low-cost substrate for production of microbial products (Marmesat et al. 2007). There is a considerable difference in the composition and chain lengths of fatty acids in fresh and fried oils. Haba et al. (2000) reported that used olive and sunflower oils have 22.52 % more myristic acid and lauric acid as compared to the standard unused oils. They reported production of biosurfactants by *Pseudomonas* spp., *Bacillus* spp., *Rhodococcus* sp., *Acinetobacter calcoaceticus,* and *Candida* spp. in medium supplemented with 2 % (v/v) used olive oil and sunflower oil as carbon source. All the cultures lowered the surface tension of growth medium to 40 mN/m from 62 mN/m. *Pseudomonas aeruginosa* strain 47T2 produced 2.7 g/l of rhamnolipid using waste frying sunflower and olive oil as substrates (Haba et al. 2000). Abalos et al. (2001) reported production of new rhamnolipids (9.5 g/l) by *Pseudomonas aeruginosa* AT110 using soybean oil refinery wastes, and the CMC of biosurfactant was 122 mg/l. Shah et al. (2007) studied biosurfactant production by *C. bombicola* in fed-batch fermentations using restaurant oil waste and reported 34 g/l of sophorolipids production. *Vedaraman and Venkatesh* (2011) reported production of surfactin, a cyclic lipopeptides (CLP) by *Bacillus subtilis* MTCC 2423 in submerged batch cultivation using waste frying sunflower oil and rice bran oil. The overall yield of biosurfactant was 14.9 g/kg and 11.0 g/kg of frying sunflower oil and rice bran oil respectively. *Wadekar* et al. (2012) reported production of rhamnolipid mainly dirhamnolipids by *Pseudomonas aeruginosa* (ATCC 10145) using waste frying oil. The yield of biosurfactant was 2.8 g/l and 7.5 g/l respectively before and after activated earth treatment of waste frying oil. The acid activated earth upon mixing with 50 g of heated (90 °C) waste frying oil adsorbed the peroxide, hydroperoxides, and low-molecular-weight aldehydes and ketones from the oil. The treatment increased the consumption of linoleic acid from 58 to 75 % by *P. aeruginosa* which indicated that presence of peroxides in frying oil might have restricted the consumption of linoleic acid and decreased the yield of rhamnolipids.

Spent Lubricating Oils

The worldwide estimated use of lubricating oils, such as hydraulic oil, motor oil, transmission oil, brake fluids, crankcase oil, and gear box oil generated from different industries, was 42 million tons in 2010 (IETC 2012). The lube oil is composed of synthetic base oils (90 %), such as polyalphaolefin (PAO), polyalkylene glycols, phosphate esters, silicate esters, and synthetic esters, and variety of oil additives (10 %), such as emulsifiers/demulsifiers, friction modifiers, and corrosion inhibitors (Bartels 2005). The difference between fresh and used lube oil is that polyaromatic hydrocarbons (PAHs) are generally undetectable in unused oil while these are abundant in all used lubricating oils. Moreover, concentrations of alkanes are higher in used oils by 2–3 folds than fresh lube oils. The used lube oil is either disposed of by burning on-site in permitted hazardous waste incinerators or used as fuel in industrial furnaces and boilers. However, only 38 % is recycled or properly managed and remaining is a potential source of pollution in the environment. Mercade et al. (1996) reported the screening and isolation of microorganisms utilizing waste lubricating oil as sole carbon source from hydrocarbon-contaminated soil for production of biosurfactants. The biosurfactant produced by *Rhodococcus* sp. and *Bacillus* sp. was characterized as trehalose, glycolipids, and lipopeptide. Thavasi et al. (2008) used a mixture of waste motor lubricant oil and low-costing peanut oil cake a by-product obtained from peanut oil manufacturing as substrates for glycolipid biosynthesis by *Bacillus megaterium, Azotobacter chroococcum,* and *Corynebacterium kutscheri.*

18.4.2.3 Waste of Dairy Industries

The wastes from animal origin, such as whey from dairy industries, animal fat, and tallow from meat processing industries, are generally being discarded through the effluent treatment systems. These wastes being rich in proteins, amino acids, lipids, vitamins, and minerals can be utilized as medium supplements to support microbial growth and biosurfactant production (Deshpande and Daniels 1995; Dubey and Juwarkar 2001).

Daniel et al. (1998) reported two-stage fed-batch (3 l) cultivation process using deproteinized whey and rapeseed oil as substrates for production of sophorolipids by *Candida bombicola* ATCC 22214 and *Cryptococcus curvatus* ATCC 20509. The first stage involved consumption of lactic acid in deproteinized whey (1.5 l) by *Candida bombicola* ATCC 22214 in stirred fed-batch. Thereafter cells of *Candida bombicola* in the medium were homogenized at high pressure to release crude cell debris and single cell-oil in the medium. This medium was then supplemented with rapeseed oil (100 g/l) and inoculated with *Cryptococcus curvatus* ATCC 20509 for production of sophorolipids. This two-stage fed-batch system yielded 422 g/l sophorolipids. Dubey and Juwarkar (2001) used mixture of distillery and whey wastes as substrate for production of biosurfactant by *Pseudomonas aeruginosa* strain BS2 which supported overall yield of 0.92 g/l during the stationary growth phase. The biosurfactant with CMC of 0.028 mg/ml reduced the surface tension of water from 72 to 27 mN/m. Further, a decrease in chemical oxygen demand (COD) by 81.0 % and 87.0 % of distillery and whey wastes respectively was observed indicating ability of cells to assimilate these waste as carbon and nitrogen sources. *Daverey* and

Pakshirajan (2010) used deproteinized whey and glucose as substrate for sophorolipids production by *Candida bombicola*. The supplements supported up to 33.3 g/l of biosurfactant production by yeast in bioreactor. The CMC of biosurfactant was 27.17 mg/l and able to reduce surface tension of water down to 34.18 mN/m. The biosurfactant was able to solubilize nonaqueous phase liquids (*n*-hexane, sunflower oil, and olive oil) indicating its potential application in bioremediation of hydrocarbon polluted sites. *Jahanshah* et al. (2013) used whey to enrich microbial diversity present in municipal waste compost and isolated two biosurfactant producers viz. *Bacillus* sp. and *Streptomyces* sp. The biosurfactant produced by these isolates resulted in >50 % removal of heavy metals (lead, nickel, chromium, and cadmium) during bioremediation. The anionic biosurfactants form a complex with positively charged metal ions to remove them from the negatively charged humus and organic matters of compost matrix. Thus, removal of heavy metals using crude biosurfactant decreased the bio-toxicity of polluted site and enhanced the decomposition rate of organic matter by 42.3 %.

18.4.2.4 Waste of Sugar Industries

The sugar industries release molasses as a by-product after crystallization of the sugar from liquid extracts of sugarcane or sugar beet as further extraction of sugar from molasses is not economical (Maneerat 2005). Sugarcane molasses is a complex mixture of 48–56 % total sugar (mainly sucrose), 9–12 % nonsugar organic matter, 2–4 % protein, 1.5–5 % potassium, 0.4–0.8 % calcium, 0.06 % magnesium, 0.6–2.0 % phosphorus, 1.0–3.0 mg/kg biotin, 15–55 mg/kg pantothenic acid, and 2,500–6,000 mg/kg inositol, and 1.8 mg/kg can be an effective growth medium supplement to support microbial growth (Makkar and Cameotra 1997). Soy molasses, a by-product of soybean oil processing, is rich in fermentable carbohydrate (30 % w/v) mainly glucose, arabinose, sucrose, raffinose, and stachyose. Due to increase in demand of soy protein-based foods and drinks, soy molasses is available as by-product and can be used as a medium supplement for supporting microbial growth (Deak and Johnson 2006). The cost of soy molasses is approximately 20 % lower than glucose, thus reducing the overall cost of biosurfactant production (Lynd et al. 1999).

Solaiman et al. (2004) used 333 g/l soy molasses equivalent to 100 g/l of carbohydrates and 90 ml/l oleic acid as substrates to produce 21 g/l of pure sophorolipids by *Candida bombicola*. The yield of sophorolipids increased to 55 g/l in 12 l benchtop fermentor with 4 l of working volume using same fermentative medium (Solaiman et al. 2007). Rashedi et al. (2005) used sugar beet molasses as carbon source to support rhamnolipid production by *Pseudomonas aeruginosa*. The yield of rhamnolipid per gram biomass was 0.003 g, 0.009 g, 0.053 g, 0.041 g, and 0.213 g in medium supplemented with 2 %, 4 %, 6 %, 8 %, and 10 % (v/v) molasses, respectively. The rhamnolipid was able to form stable emulsions with aromatics, *n*-alkanes, olive oil, and crude oil. Rodrigues et al. (2006) used sugarcane molasses and cheese whey as alternative substrates to synthetic media for *Lactococcus lactis* 53 and *Streptococcus thermophilus*. The authors observed 1.2–1.5 time increase in biosurfactant production per gram cell dry weight and 75 % reduction in overall cost of production process at lab scale. Raza et al. (2007) reported gamma ray-induced *Pseudomonas aeruginosa* EBN-8 mutant utilizing clarified blackstrap molasses as a sole carbon and energy source for production of rhamnolipid-based surfactant with a yield of 1.45 g/l after 96 h of incubation. Moldes et al. (2007) reported ability of *Lactobacillus pentosus* to produce 0.71 g of biosurfactant per gram of biomass while growing in medium supplemented with hemicellulosic sugar hydrolyzates derived from trimming vine shoots as carbon source. Moldes et al. (2011) used biosurfactant for bioremediation of soil contaminated with octane by mobilizing the target contaminant from the soil surface to make it bioavailable for the microbial population. They reported 62.5 % degradation of octane by *Lactobacillus pentosus* in presence of biosurfactant as compared to 24 % achieved in absence of biosurfactant after 15 days of treatment. Abdel-Mawgoud et al. (2008) reported 1.12 g/l of surfactin production by *Bacillus subtilis* BS5 in medium having of 16 % (w/v) molasses and 5 g/l $NaNO_3$. Joshi et al. (2008) reported lipopeptide biosurfactant production based upon decrease surface tension (29 mN/m) of growth medium by *Bacillus licheniformis* K51, *B. subtilis* 20B, *B. subtilis* R1, and *Bacillus* strain HS3 using 5.0–7.0 % (w/v) of molasses as a sole source of nutrition at 45 °C. *Onbasli* and *Aslim* (2009) reported rhamnolipid biosurfactant production by *Pseudomonas luteola* B17 and *Pseudomonas putida* B12n using 1–5 % (w/v) sugar beet molasses supplements to growth medium.

Glycerol, a trihydroxy alcohol, syrup is abundant in nature and can be utilized by many microorganism as a carbon source (Syldatk et al. 1985). The glycerol is obtained economically from renewable resources by hydrolysis of animal fat and vegetable oil and also as a coproduct of biodiesel production as every 100 kg of biodiesel production generates 10 kg of glycerol. The depleting natural fuel resources, and increasing price and demand of fuel have increased production of biodiesel with average annual growth rate of 38 % during 2005–2010 (REN21 2011). There are reports regarding use of glycerol obtained from bio-refineries without pretreatment or purification to support enhanced production of biosurfactants. Ashby et al. (2005) used biodiesel coproduct stream (BCS) containing crude glycerol, free fatty acids (FAA), and fatty acyl methyl esters (FAME) as a feedstock for synthesis of sophorolipids (SL) by *Candida bombicola* with an overall yield of 60 g/l which was significantly higher than 9 g/l obtained using pure glycerol. This may be due to easy availability of FAA and FAME and other micronutrients in BCS required for SL synthesis. Samadi et al. 2007 isolated *Brevibacterium* sp. strain S-34 capable of utilizing glycerol as sole carbon source and supported yield of 2.4 g/l glycolipid after 72 h of incubation. Dos Santos et al. (2010) reported two strains of *Pseudomonas* spp. capable of using vegetable oils and glycerol as alternative low-cost carbon sources supporting production of biosurfactant as evident by decrease in surface tension and mineral oil emulsification efficiency of cell-free culture broths ranging from 50 to 59 %. Sousa et al. (2011) used coproduct of biodiesel production as carbon and energy source for *Pseudomonas aeruginosa* MSIC02 and reported the production of rhamnolipids with overall yield of 1.2 g/l. The biosurfactant obtained was reported to possess emulsifying properties with emulsification index (IE_{24}) = 65 % and form stable emulsions with mineral and vegetable oils.

18.5 Process Optimization

It is evident from the presented data that microorganisms due to their metabolic diversity could utilize diverse type of nutrients for their growth. However, there is a need to induce the desired biochemical pathway for improving the production of biological molecules of commercial interest. The maximal expression of these metabolic pathways can be achieved by optimizing the physicochemical conditions and medium supplements to improve the yield of biosurfactant. The process optimization methods involve classical one-variable-at-a-time approach and statistics-based response surface methodology (RSM). This approach can help to achieve maximum production levels by understanding the cumulative effect of key variables on the overall yield of biosurfactant.

18.5.1 Optimization of Medium and Physicochemical Conditions

The production of biosurfactants is significantly influenced by variety of substrates which may induce or repress the biosurfactant production. The water-insoluble substrates (*n*-alkanes) or hydrocarbons generally induce biosurfactants (lipoprotein) production under natural conditions (Radwan and Sorkhoh 1993). Although, glucose is a preferred source of biosurfactant production; however, glucose could repress synthesis of biosurfactants by catabolic repression as reported in case of *Arthrobacter paraffineus* where no biosurfactant production was observed when hexadecane was replaced with glucose (*Duvnjak* et al. 1982). Thus, yield and activity of end product is largely affected by variations in carbon, nitrogen, and phosphorous sources and their concentrations (Guerra-Santos et al. 1986).

Abdel-Mawgoud et al. (2008) observed that fermentation conditions and medium components (carbon, nitrogen, and minerals) significantly improved surfactin production efficiency of *Bacillus subtilis* BS5 in mineral salts medium. The medium supplemented with glucose (20 g/l) and $NaNO_3$ (5 g/l) supported yield of 0.35 g/l biosurfactant yield. The authors reported threefold increase in biosurfactant yield to 1.12 g/l by replacing glucose with 160 ml/l molasses under similar physicochemical conditions of pH (6.5–6.8) at 30 °C. Mutalik et al. (2008) optimized the production of biosurfactant from *Rhodococcus* spp. MTCC 2574 using one-factor-at-a-time approach which suggested mannitol, yeast extract, and meat peptone as suitable carbon and nitrogen sources and *n*-hexadecane as the inducer for biosurfactant production. Seven hydrocarbons were evaluated for the enhancement of biosurfactant production, and *n*-hexadecane gave maximum EI_{24} of 59.2 % and selected as an inducer. The concentrations of these media components were further optimized by using central composite rotatable design (CCRD) of RSM. The optimized concentrations of mannitol (1.6 g/l), yeast extract (6.92 g/l), meat peptone (19.65 g/l), and *n*-hexadecane (63.8 g/l) increased the yield of biosurfactant by 3.4-fold from 3.2 to 10.9 g/l.

Reis et al. (2004) studied production of biosurfactant by *Bacillus subtilis* ATCC 6633 using commercial sugar, glycerol, mannitol, and soybean oil. The cells grown in medium supplemented with 10 g/l commercial sugar achieved maximum reduction in surface tension (28.7 mN/m) after 48 h of incubation with CMC of 78.6 mg/l. Thavasi et al. (2011) optimized the concentrations of waste motor lubricant oil and peanut oil cake for production of glycolipopeptide biosurfactant in 3 l fermentor. The maximum biosurfactant production of 6.4 mg ml/l was achieved using 2 % (w/v) waste motor lubricant oil and peanut oil cake in mineral medium (pH 8) using 8 mg/l of dissolved oxygen (DO) at 38 °C in a run of 120 h. The biosurfactant was able to emulsify crude oil, waste motor lubricant oil, kerosene, diesel, peanut oil, xylene, naphthalene, and anthracene. Similarly, Ghribi and Ellouze-Chaabouni (2011) optimized the concentrations of orange peels and soya bean as growth substrates for improving biosurfactant production efficiency of *Bacillus subtilis* SPB1 by employing central composite design. The optimal medium composed of orange peels (15.0 g/l), soya bean (15 g/l), and diluted sea water (200 ml/l) increased the biosurfactant production to 4.45 g/l from 2.39 g/l obtained from un-optimized medium using orange peels (15.0 g/l), soya bean (45 g/l), and diluted sea water (40 %). In this study, biosurfactant was efficiently used as a biocontrol agent against olive moth *Prays oleae*.

Jamal et al. (2012) studied effect of different combinations of sludge palm oil (SPO), sucrose, and glucose to improve biosurfactant (phospholipids) production by *Klebsiella pneumoniae* WMF02 in SmF using one-factor-at-a-time optimization approach. The optimized medium containing 5 g/l sucrose and 4 % (v/v) SPO lowered the surface tension of growth medium to 25.70 mN/m as compared to 36.2 mN/m achieved in un-optimized medium. The studies carried out in our lab improved the overall biosurfactant production by lab isolate *P. aeruginosa* ChlD by 2.5 times from 3.2 to 7.2 g/l when grown in medium supplemented with molasses (2.0 %) and corn steep liquor (2.5 %) as carbon and nitrogen sources as compared to that obtained by cells grown in presence of glucose (2 % w/v) and tryptone (0.25 % w/v) supplemented medium.

18.5.2 Bioreactor Designing

The implementation of these successful lab/pilot scale studies at commercial level is possible only by designing suitable bioreactor configurations capable of supporting high product yield over a specific period of time. However, production of biosurfactant at bioreactor level confronts certain operative problems. The excessive formation of foam by increasing the levels of aeration and agitation is the basic problem while lowering the rate of aeration significantly affects the growth of microbes and resulting biosurfactant production. Further, maintaining the pH and temperature at large scale is not cost effective. Thus there is a need to design efficient bioreactor to support enhanced production at fairly low price. There are only few studies available in literature regarding scale-up of production from shake flask to bioreactor level at pilot scale promising the enhanced production for commercial applications.

Mao-Sung et al. (2006) developed an innovative batch bioreactor to lower the severe foaming raised during surfactin production by *Bacillus subtilis* ATCC 21332. The bioreactor was integrated with a foam collector and biosurfactant precipitation unit. In addition, activated carbon was used as a solid carrier in the fermentation broth to increase cell mass concentration and surfactin yield. The bioreactor allowed stable surfactin fermentation at the rate of 190 mg/l/h using 6.45 g/l glucose under intensive foaming conditions. The controlled foaming allowed aeration of 1.5 vvm and agitation at 300 rpm without addition of antifoam agents.

Daverey and Pakshirajan (2009) developed a 3 l fermentor containing medium supplemented with 50 g/l sugarcane molasses and 50 g/l soybean oil and inoculated with 5 % (v/v) yeast *Candida bombicola*. The temperature was maintained at 30 °C with agitation of 200 rpm. These optimum conditions yielded 47 g/l sophorolipids without controlling pH of the growth medium which reduces to 3.5 during the process. However, the yield increased to 60 g/l in the fermentor under when pH was maintained constantly at 6.0.

Chtioui et al. (2012) reported a horizontal glass bioreactor equipped with stainless steel rotating discs mounted on a central shaft for production of lipopeptides fengycin and surfactin by *Bacillus subtilis* ATCC 21332. The cells were immobilized on the surfaces of rotating discs partially submerged in 1.2 l of medium. The aeration was applied on the exposed surface of the rotating blades rotating at 30 rotations per minute which maintained a non-foaming fermentation process. Although, the maximal yield of fengycin and surfactin were only 838 mg/l and 212 mg/l respectively after 72 h, but further modification in this foam free process could advance the scale-up of the fermenters for production of biosurfactants.

18.5.3 Downstream Processing

The extraction of end product in its pure form from a complex growth medium increases the complexity and cost of the overall process (Smyth et al. 2010). The use of biosurfactants for human consumption as antimicrobials or food ingredients demands high purity which involves several complex extraction and purification steps. The use of less complex substrates and development of effective downstream protocols may decrease the number of steps required to obtain the pure end product. Thus, development of cost-effective downstream processing will be a step forward to achieve commercialization of these biomolecules. The most frequently used downstream processes involve uses of solvent extraction, acid precipitation, ammonium sulfate precipitation, crystallization, centrifugation, adsorption, and foam fractionation (Mukherjee et al. 2006; Chen et al. 2008; Neto et al. 2009).

The drawbacks regarding use of volatile organic solvents and chemical substances in large quantity are their high cost, air pollution and health-related hazards. The use of solvent systems chloroform:ethanol (2:1) and chloroform:methanol

(65:15) is reported for extraction of rhamnolipid from cell free supernatant and extraction of surfactin from acid precipitated dried pellet respectively (Nitschke and Pastore 2006; Costa et al. 2006). Whereas, ethyl acetate, dichloromethane and methanol are being used as individual solvents to dissolve acid-precipitated lyophilized lipopeptide biosurfactant (Smyth et al. 2010; Joshi et al. 2008; Das et al. 2008). Thus, development of bioreactor with integrated foam collector and use of different solid activated carriers to immobilize biomass and increase surfactant yield is the need of the day for successful and environmentally safe production of biosurfactant at commercial scale.

18.6 Conclusion

The increased production of petroleum-based synthetic surfactants and their reckless use for various applications is of major environmental concerns as it had adversely affected the microflora of natural habitats and regeneration ability of the polluted sites. Thus, there is a need to replace such toxic compounds with environmentally safe and bio-based surfactants. The microbial surfactants due to their diverse range of applications, biodegradable properties, ability to maintain activity at wide range of temperature, and pH are emerging as most desirable biomolecules of future. The economics of biosurfactant is essential for its commercialization. The exploitation of abundantly available cheap agro-industrial by-products/wastes as alternative substrates for biosurfactant production and prudent optimization of process parameters and medium components could increase the yield and lower the production cost. Further, development of innovative bioreactor designs and integrated downstream processing are desirable to support large-scale production and economical harvesting of biosurfactants at commercial level.

References

Abalos A, Pinaso A, Infante MR, Casals M, Garcia F, Manresa A (2001) Physiochemical and antimicrobial activities of new rhamnolipids by *Pseudomonas aeruginosa* AT10 from soybean oil refinery wastes. Langmuir 17:1367–1371

Abdel-Mawgoud A, Aboulwafa M, Hassouna N (2008) Optimization of surfactin production by *Bacillus subtilis* isolate BS5. Appl Biochem Biotechnol 150:305–325

Abouseoud M, Yataghene A, Amrane A, Maachi R (2008) Biosurfactant production by free and alginate entrapped cells of *Pseudomonas fluorescens*. J Ind Microbiol Biotechnol 35:1303–1308

Abu-Ruwaida AS, Banat M, Haditirto SS, Kadri A (1991) Isolation of biosurfactant producing bacteria product characterization and evaluation. Acta Biotech 11:315–324

Aparna A, Srinikethan G, Smitha H (2012) Production and characterization of biosurfactant produced by a novel *Pseudomonas* sp. 2B. Colloid Surf B 95:23–29

Arima K, Kakinuma A, Tamura G (1968) Surfactin, a crystalline peptide surfactant produced by *Bacillus subtilis*: isolation, characterisation and its inhibition of fibrin clot fomation. Biochem Biophys Res Commun 31:488–494

Ashby RD, Nuñez A, Solaiman DKY, Foglia TA (2005) Sophorolipid biosynthesis from a biodiesel co-production stream. J Am Oil Chem Soc 82:625–630

Asokan P, Saxena M, Asolekar SR (2005) Coal combustion residues environmental implications and recycling potentials. Resour Conserv Recy 43:239–262

Banat IM (1995) Biosurfactants production and possible uses in microbial enhanced oil-recovery and oil pollution remediation – a review. Bioresour Technol 51:1–12

Banat IM, Makkar RS, Cameotra SS (2000) Potential commercial applications of microbial surfactants. Appl Microbiol Biotechnol 53:495–508

Barros FF, Ponezi AN, Pastore GM (2008) Production of biosurfactant by *Bacillus subtilis* LB5a on a pilot scale using cassava wastewater as substrate. J Ind Microbiol Biotechnol 35:1071–1078

Bartels T (2005) Lubricants and lubrication, in Ullmann's encyclopedia of industrial chemistry. Wiley, Weinheim. doi:10.1002/14356007.a15_423

Brown MJ (1991) Biosurfactants for cosmetic applications. Int J Cosmet Sci 13:61–64

Camargo-de-Morais M, Ramos SAF, Pimentel M, De Morais M Jr, Lima Filho J (2003) Production of an extracellular polysaccharide with emulsifier properties by *Penicillium citrinum*. World J Microbiol Biotechnol 19:191–194

Camire ME, Violette D, Dougherty MP, McLaughlin MA (1997) Potato peels dietary fiber composition: effects of peeling and extrusion cooking processes. J Agric Food Chem 45:1404–1408

Casas J, Garcia-Ochoa F (1999) Sophorolipid production by *Candida bombicola*: medium composition and culture methods. J Biosci Bioeng 88:488–494

Chen HL, Chen Y-S, Juang R-S (2008) Recovery of surfactin from fermentation broths by a hybrid salting-out and membrane filtration process. Sep Purif Technol 59:244–252

Chtioui O, Dimitrov K, Gancel F, Dhulster P, Nikov I (2012) Rotating discs bioreactor, a new tool for lipopeptides production. Process Biochem 47:2020–2024

Cirigliano MC, Carman GM (1985) Purification and characterization of liposan, a bioemulsifier from candida lipolytica. Appl Environ Microbiol 50:846–850

Coimbra CD, Rufino RD, Luna JM, Sarubbo LA (2009) Studies of the cell surface properties of *Candida* species and relation with the production of biosurfactants for environmental applications. Curr Microbiol 58:245–249

Costa LG (2006) Current issues in organophosphate toxicology. Clinica Chimica Acta 366:1–13

Costa SG, Lépine F, Milot S, Déziel E, Nitschke M, Contiero J (2009) Cassava wastewater as a substrate for the simultaneous production of rhamnolipids and polyhydroxyalkanoates by *Pseudomonas aeruginosa*. J Ind Microbiol Biotechnol 36:1063–1072

Daniel HJ, Reuss M, Syldatk C (1998) Production of sophorolipids in high concentration from deproteinized whey and rapeseed oil in a two stage fed batch process using *Candida bombicola* ATCC 22214 and *Cryptococcus curvatus* ATCC 20509. Biotechnol Lett 20:1153–1156

Das K, Mukherjee AK (2005) Characterization of biochemical properties and biological activities of biosurfactants produced by *Pseudomonas aeruginosa* mucoid and non-mucoid strains isolated from hydrocarbon-contaminated soil samples. Appl Microbiol Biotechnol 69:192–199

Das K, Mukherjee AK (2007) Comparison of lipopeptide biosurfactants production by *Bacillus subtilis* strains in submerged and solid state fermentation systems using a cheap carbon source: some industrial applications of biosurfactants. Pro Biochem 42:1191–1199

Das P, Mukherjee S, Sen R (2008) Antimicrobial potential of a lipopeptide biosurfactant derived from a marine *B. circulans*. J Appl Microbiol 104:1675–1684

Daverey A, Pakshirajan K (2009) Production, characterization, and properties of sophorolipids from the yeast *Candida bombicola* using a low-cost fermentative medium. Appl Biochem Biotechnol 158:663–674

Daverey A, Pakshirajan K (2010) Kinetics of growth and enhanced sophorolipids production by *Candida bombicola* using a low-cost fermentative medium. Appl Biochem Biotechnol 160:2090–2101

Deak N, Johnson L (2006) Functional properties of protein ingredients prepared from high-sucrose/low-stachyose soybeans. JOCOS 83:811–818

Deleu M, Paquot M (2004) From renewable vegetables resources to micro-organisms: new trends in surfactants. Comptes Rendus Chimie 7:641–646

Desai JD, Banat IM (1997) Microbial production of surfactants and their commercial potential. Microbiol Mol Biol Rev 61:47–64

Deshpande M, Daniels L (1995) Evaluation of sophorolipid biosurfactant production by *Candida bombicola* using animal fat. Bioresour Technol 54:143–150

Dos santos SC, Fernandez LG, Rossi-alva JC, De abreu roque MR (2010) Evaluation of substrates from renewable sources in biosurfactant production by Pseudomonas strains. Afr J Biotechnol 9:5074–5711

Dubey K, Juwarkar A (2001) Distillery and curd whey wastes as viable alternative sources for biosurfactant production. World J Microbiol Biotechnol 17:61–69

Dubey K, Juwarkar A (2004) Determination of genetic basis for biosurfactant production in distillery and curd whey wastes utilizing *Pseudomonas aeruginosa* strain BS2. Ind J Biotechnol 3:74–81

Duvnjak Z, Cooper DG, Kosaric N (1982) Production of surfactant by *Arthrobacter paraffineus* ATCC19558. Biotechnol Bioeng 24:165–175

Ferreira NL (2008) Industrial exploitation of renewable resources: from ethanol production to bio products development. J Soc Biol 202:191–199

Fox SL, Bala GA (2000) Production of surfactant from Bacillus subtilis ATCC 21332 using potato substrates. Bioresour Technol 75:235–240

Gerard J, Lloyd R, Barsby T, Haden P, Kelly MT, Andersen RJ (1997) Massetolides A-H, antimycobacterial cyclic depsipeptides produced by two pseudomonads isolated from marine habitats. J Nat Prod 60:223–229

Ghribi D, Ellouze-Chaabouni S (2011) Enhancement of Bacillus subtilis lipopeptide biosurfactants production through optimization of medium composition and adequate control of aeration. Biotechnol Res Int 2011:653654. doi:10.4061/2011/653654

Guerra-Santos L, Kappeli O, Fiechter A (1986) Dependence of *Pseudomonas aeruginosa* continuous culture biosurfactant production on nutritional and environmental factors. Appl Microbiol Biotechnol 24:443–448

Haba E, Espuny M, Busquets M, Manresa A (2000) Screening and production of rhamnolipids by *Pseudomonas aeruginosa* 47T2 NCIB 40044 from waste frying oils. J Appl Microbiol 88:379–387

Horowitz S, Currie JK (1990) Novel dispersants of silicon and aluminum nitride. J Dispersion Sci Technol 11:637–659

IARC (2000) Some industrial chemicals. Monographs on the evaluation of carcinogenic risks to humans. World Health Organization (WHO) international agency for research on cancer (IARC) 77:A-1–31

International Advisory Board of the International Environmental Technology Centre (IETC) Eighth meeting Osaka, Japan, 7 November 2012. UNEP(DTIE)/IETC/IAB.8/3

Jack TR (1998) Microbially enhanced oil recovery. Biorecovery 1:59–73

Jahanshah G, Nahvi I, Zarkesh-Esfahani SH, Ghanavati H, Khodaverdi H, Barani M (2013) Enhancing compost quality by using whey-grown biosurfactant-producing bacteria as inocula. Annal Microbiol 63:91–100. doi:10.1007/s13213-012-0448-1

Jamal P, Nawawi WMFW, Alam MZ (2012) Optimum medium components for biosurfactant production by *Klebsiella pneumonia* WMF02 utilizing sludge palm oil as a substrate. Aust J Basic Appl Sci 6:100–108

Joshi S, Bharucha C, Desai AJ (2008) Production of biosurfactant and antifungal compound by fermented food isolate *Bacillus subtilis* 20B. Bioresour Technol 99:4603–4608

Käppeli O, Finnerty WR (1979) Partition of alkane by an extracellular vesicle derived from hexadecane- grown *Acinetobacter*. J Bacteriol 140:707–712

Kitamoto D, Yanagishita H, Shinbo T, Nakane T, Kamisawa C, Nakahara T (1993) Surface active properties and antimicrobial activities of mannosylerythritol lipids as biosurfactants produced by *Candida antarctica*. J Biotechnol 29:91–96

Kosaric N (1992) Biosurfactants in industry. Pure Appl Chem 64:1731–1737

Krieger N, Doumit C, David AM (2010) Production of microbial biosurfactants by solid-state cultivation. Adv Exp Medic Bio 672:203–210

Lima de CJB, Ribeiro EJ, Sérvulo EFC, Resende MM, Cardoso VL (2009) Biosurfactant production by *Pseudomonas aeruginosa* grown in residual soybean oil. Appl Biochem Biotechnol 152:156–168

Lima TMS, Procópio LC, Brandão FD, Carvalho AMX, Tótola MR, Borges AC (2011) Biodegradability of bacterial surfactants. Biodegradation 22:585–592

Luna JM, Rufino RD, Sarubbo LA, Campos-Takaki GM (2013) Characterisation, surface properties and biological activity of a biosurfactant produced from industrial waste by *Candida sphaerica* UCP0995 for application in the petroleum industry. Colloids Surf B Biointerfaces 102:202–209

Lutaladio NB, Castaldi L (2009) Potato: The hidden treasure. J Food Compos Anal 22:491–493

Lynd LR, Wyman CE, Gerngross TU (1999) Biocommodity engineering. Biotechnol Prog 15:777–793

Makkar RS, Cameotra SS (1997) Utilization of molasses for biosurfactant production by two *Bacillus* strains at thermophilic conditions. J Am Oil Chem Soc 74:887–889

Makkar RS, Cameotra SS (1999) Biosurfactant production by microorganisms on unconventional carbon sources. J Surf Detergents 2:237–241

Makkar RS, Cameotra SS (2002) An update on use of unconventional substrates for biosurfactants production and their new applications. Appl Microbiol Biotechnol 58:428–434

Makkar RS, Rockne KJ (2003) Comparison of synthetic surfactants and biosurfactants in enhancing biodegradation of polycyclic aromatic hydrocarbons. Environ Toxicol Chem 22:2280–2292

Maneerat S (2005) Production of biosurfactants using substrates from renewable resources. Songklanakarin J Sci Techno 27:675–683

Mao-Sung Y, Yu-Hong W, Jo-Shu C (2006) Bioreactor design for enhanced carrier-assisted surfactin production with *Bacillus subtilis*. Process Biochem 41:1799–1805

Marmesat S, Rodrigues E, Velasco J, Dobarganes C (2007) Quality of used frying fats and oils: comparison of rapid tests based on chemical and physical oil properties. Int J Food Sci Technol 42:601–608

Mata-Sandoval JC, Karns J, Torrents A (1999) High-performance liquid chromatography method for the characterization of rhamnolipid mixtures produced by *Pseudomonas aeruginosa* UG2 on corn oil. J Chromat 864:211–220

McInerney MJ, Javaheri M, Nagle DP (1990) Properties of the biosurfactant produced by *Bacillus licheniformis* strain JF-2. J Indust Microbiol 5:95–102

Mercade E, Robert M, Espuny MJ, Bosch MP, Manreesa MA, Parra JL, Guinea J (1988) New Surfactant Isolated from *Pseudomonas* sp. 42A2. J Am Oil Chem Soc 65:1915–1916

Mercadé M, Monleón L, de Andrés C (1996) Screening and selection of surfactant-producing bacteria from waste lubricating oil. J Appl Bacteriol 81:161–166

Mercade ME, Manresa MA, Robert M, Espuny MJ, de Andres C, Guinea J (1993) Olive oil mill effluent (OOME). New substrate for biosurfactant production. Bioresour Technol 43:1–6

Millioli VS, Servulol ELC, Sobrald LGS, de Carvalho DD (2009) Bioremediation of crude oil-bearing soil: evaluating the effect of rhamnolipid addition to soil toxicity and to crude oil biodegradation efficiency. Global Nest J 11:181–188

Moldes AB, Paradelo R, Rubinos D, Devesa-Rey R, Cruz JM, Barral MT (2011) Ex situ treatment of hydrocarbon-contaminated soil using biosurfactants from *Lactobacillus pentosus*. J Agric Food Chem 59:9443–9447

Moldes AB, Torrado AM, Barral MT, Dominguez JM (2007) Evaluation of biosurfactant production from various agricultural residues by *Lactobacillus pentosus*. J Agric Food Chem 55:4481–4486

Montoneri E, Savarino P, Bottigliengo S, Boffa V, Prevot AB, Fabbri D, Pramauro E (2009) Biomass wastes as renewable source of energy and chemicals for the industry with friendly environmental impact. Fresenius Environ Bull 18:219–223

Mukherjee S, Das P, Sen R (2006) Towards commercial production of microbial surfactants. Trends Biotechnol 24:509–515

Mutalik SR, Vaidya BK, Joshi RM, Desai KM, Nene SN (2008) Use of response surface optimization for the production of biosurfactant from *Rhodococcus* spp. MTCC 2574. Bioresource Technol 99:7875–7880

Muthusamy K, Gopalakrishnan S, Ravi TK, Sivachidambaram P (2008) Biosurfactants: properties, commercial production and application. Curr Sci 94:736–774

Natu RB, Mazza G, Jadhav SJ (1991) Waste utilization. Potato: production, processing, and products. CRC Press, Boca Raton, FL, pp 175–201

Neto DC, Meira JA, Tiburtius E, Zamora PP, Bugay C, Mitchell DA, Krieger N (2009) Production of rhamnolipids in solid-state cultivation: characterization, downstream processing and application in the cleaning of contaminated soils. Biotechnol J 4:748–755

Nitschke M, Costa S (2007) Biosurfactants in food industry. Trends Food Sci Technol 18:252–259

Nitschke M, Pastore G (2003) Cassava flour wastewater as a substrate for biosurfactant production. Appl Biochem Biotechnol 106:295–302

Nitschke M, Pastore GM (2004) Biosurfactant production by *Bacillus subtilis* using cassava-processing effluent. Appl Biochem Biotechnol 112:163–172

Nitschke M, Pastore GM (2006) Production and properties of a surfactant obtained from *Bacillus subtilis* grown on cassava wastewater. Bioresource Technol 97:336–341

Noah KS (2005) Surfactin production from potato process effluent by *Bacillus subtilis* in a chemostat. Appl Biochem Biotechnol 122:465–474

Noudeh GD, Housaindokht M, Bazzaz BSF (2005) Isolation, characterization, and investigation of surface and hemolytic activities of a lipopeptide biosurfactant produced by *Bacillus subtilis* ATCC 6633. J Microbiol 43:272–276

Onbasli D, Aslim B (2009) Determination of rhamnolipid biosurfactant production in molasses by some *Pseudomonas* spp. New Biotechnol 25:S255

Pandey A, Soccol CR, Mitchell D (2000) New developments in solid state fermentation: I-bioprocesses and products. Proc Biochem 35:1153–1169

Pekin G, Vardar-Sukan FNK (2005) Production of sophorolipids from *Candida bombicola* ATCC 22214 using turkish corn oil and honey. Eng Life Sci 5:357–362

Poremba K, Gunkel W, Lang S, Wagner F (1991) Marine biosurfactants, III. Toxicity testing with marine micro-organisms and comparison with synthetic surfactants. Z Naturforsch 46:210–221

Providenti MA, Flemming CA, Lee H, Trevors JT (1995) Effect of addition of rhamnolipid biosurfactant or rhamnolipid-producing *Pseudomonas aeruginosa* on phenanthrene mineralization in soil slurries. Fed Eur Microbiol Soc Microbiol Ecol 17:15–26

Pruthi V, Cameotra SS (2003) Effect of nutrients on optimal production of biosurfactants by *Pseudomonas putida*—a Gujarat oil field isolate. J Surfact Deterg 6:65–68

Radwan S, Sorkhoh N (1993) Lipids of *n*-alkane-utilizing micro-organisms and their application potential. Adv Appl Microbiol 39:29–90

Rahman KS, Rahman TJ, McClean S, Marchant R, Banat IM (2002) Rhamnolipid biosurfactant production by strains of *Pseudomonas aeruginosa* using lowcost raw materials. Biotechnol Prog 18:1277–1281

Rahman PKSM, Gakpe E (2008) Production, characterisation and applications of biosurfactants-review. Biotechnology 7:360–370

Rashedi H, Assadi MM, Bonakdarpour B, Jamshidi E (2005) Environmental importance of rhamnolipid production from molasses as a carbon source. Int J Environ Sci Technol 2:59–62

Rau U, Hammen S, Heckmann R, Wray V, Lang S (2001) Sophorolipids: a source for novel compounds. Ind Crop Prod 13:85–92

Raza ZA, Khan MS, Khalid ZM (2007) Physicochemical and surface-active properties of biosurfactant produced using molasses by a *Pseudomonas aeruginosa* mutant. J Environ Sci Health A Tox Hazard Subst Environ Eng 42:73–80

Reis FA, Servulo EF, De Franca FP (2004) Lipopeptide surfactant production by *Bacillus subtilis* grown on low-cost raw materials. Appl Biochem Biotechnol 115:899–912

REN21: Renewable Energy Policy Network for the 21st Century (2011) Global status report. www.ren21.net

Robert M, Mercadé ME, Bosch MP, Parra JL, Espuny MJ, Manresa MA, Guinea J (1989) Effect of the carbon source on biosurfactant production by *P. aeruginosa* 44T1. Biotechnol Lett 11:871–874

Rodrigues LR, Teixeira JA, Oliveira R (2006) Low-cost fermentative medium for biosurfactant production by probiotic bacteria. Biochemical Eng J 32:135–142

Rosenberg E, Ron EZ (1999) High and low molecular weight mass microbial surfactants. Appl Microbiol Biotechnol 52:154–162

Rosenberg E, Rubinovitz C, Gottlieb A, Rosenhak S, Ron EZ (1988) Production of biodispersan by *Acinetobacter calcoaceticus* A2. Appl Environ Microbiol 54:317–322

RPA (2006) Non-surfactant organic ingredients and zeolite-based detergents. Final report prepared for the European Commission. Risk & Policy Analysts Limited (RPA)

Samadi N, Abadian N, Akhavan A, Fazeli MR, Tahzibi A, Jamalifar H (2007) Biosurfactant production by the strain isolated from contaminated soil. J Biol Sci 7:1266–1269

Sen R (1997) Response surface optimization of the critical media components for the production of surfactin. J Chem Tech Biotechnol 68:263–270

Sengupta J (2002) Recycling of agro-industrial wastes for manufacturing of building materials and components in India. An over view. Civil Eng Constr Rev 15:23–33

Shah V, Jurjevic M, Badia D (2007) Utilization of restaurant waste oil as a precursor for sophorolipid production. Biotechnol Prog 23:512–515

Sharma S, Singh PB, Raj M, Chadha BS, Saini HS (2009) Aqueous phase partitioning of hexachlorocyclohexane (HCH) isomers by biosurfactant produced by *Pseudomonas aeruginosa* WH-2. J Haz Mat 171:1178–1182

Siddhartha GV, Costa AO, Lépine F, Milot S, Déziel E, Nitschke M (2009) Cassava wastewater as a substrate for the simultaneous production of rhamnolipids and polyhydroxyalkanoates by *Pseudomonas aeruginosa*. J Indust Microbiol Biotechnol 36:1063–1072

Singh A, Van Hamme JD, Ward OP (2007) Surfactants in microbiology and biotechnology. Part 2. Application aspects. Biotechnol Adv 25:99–121

Singh PB, Sharma S, Saini HS, Chadha BS (2009) Biosurfactant production by *Pseudomonas* sp. and its role in aqueous phase partitioning and biodegradation of chlorpyrifos. Lett Appl Microbiol 49:378–383

Smyth TJP, Perfumo A, Marchant R, Banat IM (2010) Isolation and analysis of low molecular weight microbial glycolipids, Handbook of Hydrocarbon and Lipid Microbiology. Springer, Berlin, pp 3705–3723

Sobrinho HBS, Rufino RD, Luna JM, Salgueiro AA, Campos-Takaki GM, Leite LFC, Sarubbo LA (2008) Utilization of two agro-industrial by-products for the production of a surfactant by *Candida sphaerica* UCP0995. Process Biochem 43:912–917

Solaiman D, Ashby R, Zerkowski J, Foglia T (2007) Simplified soy molasses-based medium for reduced-cost production of sophorolipids by *Candida bombicola*. Biotechnol Lett 29:1341–1347

Solaiman DKY, Ashby RD, Nun˜ez A, Foglia A (2004) Production of sophorolipids by *Candida bombicola* grown on soy molasses as substrate. Biotechnol Lett 26:1241–1245

Sousa JR, Correia JAC, Almeida JGL, Rodrigues S, Pessoa ODL, Melo VMM, Gonçalves LRB (2011) Evaluation of a co-product of biodiesel production as carbon source in the production of biosurfactant by *P. aeruginosa* MSIC02. Proc Biochem 46:1831–1839

Stanghellini ME, Miller RM (1997) Biosurfactants: their identity and potential efficacy in the biological control of zoosporic plant pathogens. Plant Disease 81:4–12

Syldatk C, Lang S, Wagner F (1985) Chemical and physical characterization of four interfacial active rhamnolipids from *Pseudomonas* sp. DSM 2874 grown on *n*-alkanes. Z Naturforsch 40:51–60

Thaniyavarn J, Chianguthai T, Sangvanich P, Roongsawang N, Washio K, Morikawa M, Thaniyavarn S (2008) Production of sophorolipid biosurfactant by *Pichia anomala*. Biosci Biotechnol Biochem 72:2061–2068

Thaniyavarn J, Chongchin A, Wanitsuksombut N, Thaniyavarn S, Pinphanichakarn P, Leepipatpiboon N, Morikawa M, Kanaya S (2006) Biosurfactant production by *Pseudomonas aeruginosa* A41 using palm oil as carbon source. J Gen Appl Microbiol 52:215–222

Thavasi R, Jayalakshmi S, Balasubramanian T, Banat IM (2008) Production and characterization of a glycolipid biosurfactant from *Bacillus megaterium* using economically cheaper sources. World J Microbiol Biotechnol 24:917–925

Thavasi R, Jayalakshmi S, Banat IM (2011) Application of biosurfactant produced from peanut oil cake by *Lactobacillus delbrueckii* in biodegradation of crude oil. Bioresour Technol 102:3366–3372

Trummler K, Effenberger F, Syldatk C (2003) An integrated microbial/enzymatic process for production of rhamnolipids and L-(+)-rhamnose from rapeseed oil with *Pseudomonas* sp. DSM 2874. Eur J Lipid Sci Technol 105:563–571

USDA (2009) Oilseeds: world market and trade. FOP 1–09:01–12 http://www.fas.usda.gov/oilseeds/circular/2009/January/Oilseedsfull0109.pdf

Van Bogaert INA, Saerens K, De Muynck C, Develter D, Soetaert W, Vandamme EJ (2007) Microbial production and application of sophorolipids. Appl Microbiol Biotechnol 76:23–34

Vedaraman N, Venkatesh N (2011) Production of surfactin by *Bacillus subtilis* MTCC 2423 from waste frying oils. Braz J Chem Eng 28:0104–6632

Vollbrecht E, Rau U, Lang S (1999) Microbial conversion of vegetable oils into surface-active di-, tri-, and tetrasaccharide lipids (biosurfactants) by the bacterial strain *Tsukamurella* sp. Lipid - Fett 101:389–394

Vollenbroich D, Özel M, Vater J, R Kamp M, Pauli G (1997) Mechanism of inactivation of enveloped viruses by the biosurfactant surfactin from *Bacillus subtilis*. Biologicals 25:289–297

Wadekar SD, Kale SB, Lali AM, Bhowmick DN, Pratap AP (2012) Utilization of sweetwater as a cost-effective carbon source for sophorolipids production by *Starmerella bombicola* (ATCC 22214). Prep Biochem Biotechnol 42:125–142

Wang S, Mulligan CN (2004) Surfactant foam technology in remediation of contaminated soil. Chemosphere 57:1079–1089

White P, Johnson LA (eds) (2003) Corn: chemistry and technology, 2nd edn. American Association of Cereal Chemists, St. Paul, MN

Yakimov M, Amro M, Bock M (1997) The potential of *Bacillus licheniformis* strains for in situ enhanced oil recovery. J Petro Sci Eng 18:147–160

Zhu Z, Zhang G, Luo Y, Ran W, Qirong S (2012) Production of lipopeptides by *Bacillus amyloliquefaciens* XZ-173 in solid state fermentation using soybean flour and rice straw as the substrate. Bioresource Technol 112:254–260

Chapter 19
C3–C4 Platform Chemicals Bioproduction Using Biomass

Emna Chaabouni, Saurabh Jyoti Sarma, Fatma Gassara, and Satinder Kaur Brar

19.1 Introduction

The term "platform chemicals" broadly indicates the group of chemicals that can be used as building block for manufacturing of structurally related valuable chemicals, such as polymers. For instance, lactic acid could be considered as the platform chemical for the production of polylactide (Jang et al. 2012c). At present, most of the commercial platform chemicals are produced from petroleum-based products. However, fossil-derived resources are nonrenewable and their amount is increasingly depleting (Bechthold et al. 2008). Additionally, application of these materials has a number of environmental concerns. In this context, the concept of biorefinery has been developed (Jiang et al. 2011). Broadly, biorefinery is a manufacturing concept where industrially important chemicals, including biofuels and their precursors, are produced by environment-friendly microbial processes using biomass/organic material. Thus, biorefinery-based platform chemicals may become an industrially important sustainable alternative for commercially available petrochemical process-based platform chemicals. In fact, in recent years, considerable importance has been given on renewable resources-based platform chemicals (Bechthold et al. 2008; Jang et al. 2012a, b, c) for the production of fuels, chemicals, and industrial materials (Sels et al. 2013). Owing to its high availability and renewable nature, lignocellulose is an attractive substrate of bioproduction of platform chemicals (Jäger and Büchs 2012). In a report from the US Department of Energy, top 15 value-added platform chemicals that could be produced from starch, cellulose, or hemicellulose have been listed, and the list includes important C2–C4

E. Chaabouni • S.J. Sarma • F. Gassara • S.K. Brar (✉)
INRS-ETE, Université du Québec, 490, Rue de la Couronne, Québec City, QC, Canada G1K 9A9
e-mail: satinder.brar@ete.inrs.ca

S.K. Brar et al. (eds.), *Biotransformation of Waste Biomass into High Value Biochemicals*, DOI 10.1007/978-1-4614-8005-1_19, © Springer Science+Business Media New York 2014

platform chemicals, such as glycerol, succinic acid, aspartic acid, and malic acid (Jäger and Büchs 2012). Likewise, a number of reports could be found where biomass/organic waste materials, such as cheese-processing industry wastewater (Hwang and Hansen 1997), corn steep liquor/corn fiber hydrolysate (Zhu et al. 2002), *Jerusalem artichoke* hydrolysate (Huang et al. 2011), wood hydrolysate (Kim et al. 2004), and straw hydrolysates (Zheng et al. 2009), have been used for sustainable production of C2–C4 platform chemicals. Thus, bioproduction of platform chemicals has the potential to be integrated as a profitable waste-utilization strategy. Being a microbial/biochemical process, bioproduction of platform chemicals has certain concerns, such as product inhibition. Fortunately, recent technologies such as in situ product removal have been successfully implemented for improved process performance. Thus, the purpose of the present chapter is to discuss recent developments in as well as potential implementation of platform chemical bioproduction.

19.2 Platform Chemicals and Their Applications in Industry

There are various types of platform chemicals which are produced by industries for various purposes. Today with the growth of industrialization, there has been a substantial growth in the demand for industrial platform chemicals for various applications. Platform chemicals have been used in various forms including as detergents, polishes, cleaning agents, varnishes, adhesives, solvents, bioenergy, dyes used for photocopying, and so forth. Platform chemical applications are presented in details in Table 19.1. Most of the platform chemicals shown are produced from fossil oil; however, some of them could be replaced by bio-based chemicals at this time (Murphy 2011; Carbonell et al. 2012; Curran and Alper 2012; Hong and Nielsen 2012; Ye and Bhatia 2012). The bio-based platform chemical market includes all the chemicals obtained from renewable feedstock such as agricultural raw materials, agricultural waste products or biomass, and microorganisms (Hatti-Kaul et al. 2007; Jang et al. 2012b). Platform chemicals are estimated to reach a market size of US$ 3.5 billion in 2014 from US$ 1.9 billion in 2009 at an optimistic CAGR of 12.6 % from 2009 to 2014. Platform chemicals play an important role in the renewable chemical market since their multiple functional groups can be converted to families of highly useful chemicals. Immense market opportunity lies for the renewable chemical market in Japan and the rapidly developing economies of India, China, and Russia, which largely consume chemicals from petrochemical feedstock. Governmental support and initiatives to encourage the use of renewable chemicals is expected to provide the necessary boost to the market in these economies. Strategic alliances and new technology developments are a few of the most popular strategies being deployed by the market players in this segment to gain competitive edge and also to build upon their market presence by knowledge and resource sharing. The market is driven by the increasing demand from the food-packaging industry, biodegradable and compostable plastics, and other consumer

Table 19.1 Applications of platform chemicals in industry

Classification	Platform chemical	Applications	References
C3 chemicals	Propionic acid	Food and feed preservatives	Boyaval and Corre (1995)
	Lactic acid	Bioplastic food industry as a preservative in the textile and pharmaceutical industries and in the chemical industry as a raw material	Sauer et al. (2008); Akerberg and Zacchi (2000)
	3-Hydroxypropionic acids	Manufacture of various plastics, coatings, adhesives, elastomers, as well as floor polishes, and paints	Ashok et al. (2011)
	1-Propanol	Esterifying reagents	Hanai et al. (2007)
	Isopropanol	Esterifying reagents	Hanai et al. (2007)
C4 chemicals	Butyric acid	of food, cosmetic, pharmaceutical, and fine chemical manufacturing	Dwidar et al. (2012)
	Succinic Acid	Platform chemical for specialized polyester raw material for nylon precursor (adipic acid)	McKinlay et al. (2007); Song and Lee (2006)
	1-Butanol	Biofuel	Kirschner (2006); Lee et al. (2008)
	Isobutanol	Cosmetics, such as nail polish solvent, rubber, and fuel ingredients	Atsumi et al. (2008)

products. Platform chemicals possess chemical moieties that serve as linkages during polymerization or chemical reactions. For example, diamines and dicarboxylic acids serve as building block compounds by offering amine and carboxylic groups, respectively, to form amide bonds during the polymerization stages of polyamide production (Jang et al. 2012c). Dicarboxylic acids (oxalic, malonic, succinic, glucaric, adipic, fumaric, and malic acids), diamines (ethylenediamine, cadaverine, and putrescine), diols (ethylene glycol, propanediols, and butanediols), and aldehydes (formaldehyde) are the most common building block compounds used in condensation polymerization reactions (Odian 2004). Another building block chemical, ethylene, has a carbon–carbon double bond that serves as a p bond to make an s-bond linkage to an adjacent ethylene molecule during the polymerization stage of PE production (Jang et al. 2012c). Building block compounds containing carbon-to-carbon double bonds (e.g., ethylene, isobutylene, acrylonitrile, vinyl chloride, styrene, methyl methacrylate, vinyl acetate, and isoprene) are typically used for addition polymerization (Odian 2004). Some platform chemicals, such as monocarboxylic acids (acetic, propionic, butyric, pentanoic, and hexanoic acids) and monools (ethanol, propanol, *n*-/isobutanol, and pentanols) are used as precursors to form polymers, followed by chemical reactions that convert them into building block compounds. Propionic acid has many application areas, including the production of cellulose fibers, herbicides, perfumes, and pharmaceuticals (Boyaval and Corre 1995). In industry, propionic acid is mainly produced through petrochemical routes, using ethylene as a starting material (Hohenschutz et al. 1974).

19.3 Production of C3 Chemicals

In this part of this book chapter, we concentrate on the bio-based production of representative C3 platform chemicals, including carboxylic acid (propionic acid), hydroxy acids (lactic and 3-hydroxypropionic acids), and alcohols (1-propanol and isopropanol) using renewable resources. The general pathway of platform chemical from biomass is presented in details in Fig. 19.1, and microorganisms and substrates used to produce platform chemicals are presented in details in Table 19.2.

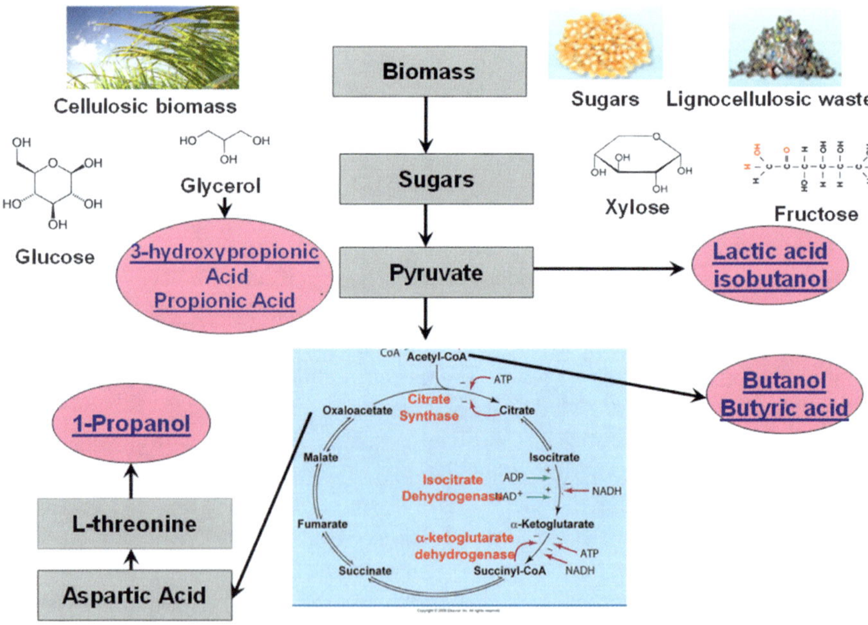

Fig. 19.1 Platform chemical production from biomass

Table 19.2 Biotechnological production of platform chemicals using microorganisms

Classification	Platform chemical	Microorganisms	Substrate, type of fermentation	Yield	Productivity	References
C3 chemicals	Propionic acid	*P. acidipropionici*	Glycerol, fed-batch fermentation	0.56	0.035	Zhang and Yang (2009)
	Lactic acid	*Sporolactobacillus* *E. coli*	Glucose, fed-batch fermentation	0.86–0.93	3.5–3.8	Wang et al. (2011), Zhu et al. (2007)
	3-Hydroxypropionic acids	*K. pneumonia*	Glycerol, fed-batch fermentation		0.01	Ashok et al. (2011)
	1-Propanol	*E. coli*	Glucose, flask culture		0.04	Atsumi and Liao (2008)
	Isopropanol	*C. acetobutylicum* *E. coli*	Glucose, anaerobic flask culture, batch fermentation	0.15–0.20	0.28–0.35	Jojima et al. (2008)
C4 chemicals	Butyric acid	*C. tyrobutyricum* *C. tyrobutyricum*	Glucose, fed-batch fermentation	0.38–0.46	0.24–1.1	Zhu et al. (2010), Liu et al. (2006), Jiang et al. (2011)
	Succinic Acid	*Engineered rumen bacteria* *E. coli*	Glucose, anaerobic fed-batch fermentation	0.76–1	0.7–2.8	Thakker et al. (2012)
	1-Butanol	*C. acetobutylicum* *E. coli*	Glucose, anaerobic batch fermentation, fed-batch fermentation	0.33–036	0.20–0.29	Dellomonaco et al. (2011), Shen et al. (2011), Atsumi et al. (2008)
	Isobutanol	*E. coli* *C. glutamicum*	Glucose, batch and fed-batch fermentation	0.20	0.15–0.33	Blombach et al. (2011)

19.3.1 Carboxylic Acid: Propionic Acid

Propionic acid is a C3 carboxylic acid with many industrial applications as a specialty chemical, and its calcium, potassium, and sodium salts are widely used as food and feed preservatives (Boyaval and Corre 1995). Currently, propionic acid is produced almost exclusively via petrochemical processes, with an annual production capacity of 400 million lbs in the USA. As the crude oil prices had surpassed US$100 per barrel, there have been increasing interests in propionic acid production from renewable bioresources by fermentation using propionibacteria (Zhang and Yang 2009; Wang et al. 2012; Zhu et al. 2012). Propionic acid bacteria have long been used in the dairy industry. These bacteria play important roles in the development of the characteristic flavor and eye production in Swiss-type cheeses. However, conventional propionic acid fermentation suffers from low productivity, low product concentration, and low yield due to strong end-product inhibition by acidic pH and the coproduction of other by-products, mainly acetic and succinic acids. Consequently, the conventional fermentation route for propionic acid production is inefficient and it competes with difficulty with petrochemical routes. Presently, only small amounts of propionate are produced by fermentation and are used as a natural product in foods for the labeling purpose. In order to make the fermentation route economically viable, it is necessary to develop novel fermentation processes that use highly efficient bioreactors and separations techniques. To lower the product cost, recent efforts have focused on using industrial wastes or by-products as low-cost renewable feedstocks (Fig. 19.1) for propionic acid fermentation (Feng et al. 2011; Zhu et al. 2012). Several studies have shown that glycerol can be a good carbon source for propionic acid fermentation with a higher propionic acid (Zhang and Yang 2009; Zhu et al. 2010). The fast growth of biodiesel production, a promising substitute for petroleum diesel, produce annually a large amount of crude glycerol, that present an economically feasible feedstock for industrial uses (da Silva et al. 2009). Glycerol has a high reduction degree, which favors the production of more reduced metabolites (Ito et al. 2005) but can cause redox imbalance in metabolism, leading to reduced cell growth and productivity, when used as the sole carbon source in fermentation (Zhang and Yang 2009). To overcome this problem, co-fermentation of glycerol with glucose has been proposed as an efficient process supporting both product formation and cell growth (Liu et al. 2012). In glycerol fermentation, propionic acid is mainly produced through succinate. A metabolically engineered *Propionibacterium acidipropionici* strain produced 106 g/L propionic acid with a 0.56 g/g yield and 0.035 g/L/h productivity from glycerol during fed-batch fermentation in a fibrous-bed bioreactor (Zhang and Yang 2009). However, many hurdles must be overcome (toxicity, by-product formation, low productivity, and redox imbalance) before propionic acid can be economically produced on a large scale.

19.3.2 Hydroxy Acids: Lactic and 3-Hydroxypropionic Acids

Lactic and 3-hydroxypropionic acids, both of which containing a hydroxyl group and a carboxyl group, are versatile block compounds that are particularly useful for the production of biodegradable polyesters, such as PLA and poly (3-hydroxypropionate). Lactic acid has a long history of uses for fermentation and preservation of human foodstuffs (Akerberg and Zacchi 2000). It was first discovered in sour milk by Scheele in 1780, who initially considered it a milk component. Lactic acid can be produced by either microbial fermentation or chemical synthesis. In the early 1960s, a method to synthesize lactic acid chemically was developed due to the need for heat-stable lactic acid in the baking industry (Datta et al. 1995). There are two optical isomers of lactic acid: L-(+)-lactic acid and D-(−)-lactic acid. At present, lactic acid is produced by the fermentation of carbohydrates; the worldwide lactic acid production volume in 2007 was estimated to be 150,000 ton (Sauer et al. 2008). Lactic acid is produced directly from pyruvate by the lactate dehydrogenase (Ldh) of various microorganisms, including *Carnobacterium, Enterococcus, Lactobacillus, Lactococcus, Leuconostoc, Oenococcus, Pediococcus, Streptococcus, Tetragenococcus, Vagococcus, and Weissella*. Lactic acid is now considered to be one of the most useful chemicals, used in the food industry as a preservative, acidulant, and flavoring, in the textile and pharmaceutical industries, and in the chemical industry as a raw material for the production of lactate ester, propylene glycol, 2,3-pentanedione, propionic acid, acrylic acid, acetaldehyde, and dilactide (Akerberg and Zacchi 2000). Recently, lactic acid consumption has increased considerably because of its role as a monomer in the production of biodegradable PLA, which is well known as a sustainable bioplastic material (Litchfield 1996). Compared to chemical synthesis, the biotechnological process for lactic acid production offers several advantages: low substrate costs, production temperature, and energy consumption (Datta and Henry 2006). Lactic acid-producing microorganisms use pyruvic acid as the precursor for lactic acid production. The conversion of pyruvic acid to lactic acid can be catalyzed by two types of enzymes: NAD-dependent L-lactate dehydrogenase and NAD-dependent D-lactate dehydrogenase (Garvie 1980). Because the optical purity of lactic acid is a crucial factor in lactic acid-based industries, numerous studies have investigated the biotechnological production of optically pure lactic acid (John et al. 2007; Okano et al. 2010; Zhao et al. 2010). There are two problems in the biotechnological production of optically pure lactic acid. The first problem is the substrate cost because of the addition of sugars as carbon sources. This problem can be resolved through fermentative production of lactic acid from cheap materials. Many cheap, renewable raw materials such as molasses, starch, lignocellulose, and wastes from agricultural and agro-industrial residues have been used as substrates for lactic acid fermentation. The second problem for lactic acid production is the operating cost related to sterilization of the media and the separation and purification processes after fermentation. There have been numerous investigations on the development of biotechnological processes for lactic acid production, with the ultimate objectives to enable the process to be more efficient and economical. L-, D-, or DL-lactic acid may be biologically produced by utilizing lactic acid dehydrogenases (Ldh) with different stereospecificities (Wu et al. 2011). Recently, an engineered sporolactobacillus produced 207 g/L D-lactate from 223 g/L of glucose with a productivity of 3.8 g/L/h and an optical purity of 99.3 % in a 30-L fed-batch fermentation supplemented with 40 g/L of peanut meal as a nitrogen source (Wang et al. 2011). Lactic acid may also be produced by metabolically engineered *E. coli* strains (Zhu et al. 2007). A fed-batch fermentation of a metabolically engineered *E. coli* strain showed a production of 138 g/L of D-lactic acid with an overall yield of 0.86 g/g of glucose and a productivity of 3.5 g/L/h (Zhu et al. 2007). Although bio-based production methods for lactic acid have advanced enough to be used at an industrial level, further improvements of their economy could be achieved by developing inexpensive carbon substrates and reducing the requirement of lactic acid bacteria for complex nitrogen sources. Klebsiella pneumoniae is a promising microorganism for 3-hydroxypropionic acid production from sugars such as glycerol (Ashok et al. 2011).

3-Hydroxypropionate can be produced from glycerol via 3-hydroxypropionic aldehyde (3-HPA). Recently, the metabolic pathway of K. pneumoniae was engineered to direct carbon flux from glycerol to 3-hydroxypropionic acid by the overexpression of endogenous glycerol dehydratase (dhaB) and g-glutamyl-g-aminobutyraldehyde dehydrogenase (PuuC). Additionally, the dhaT gene (encoding 1,3-propanediol oxidoreductase; dhaT) was deleted to eliminate a completing pathway. A fed-batch fermentation of this engineered strain produced 16.0 g/L 3-hydroxypropionic acid from glycerol with a productivity of 0.66 g/L/h (Ashok et al. 2011). However, a recombinant *E. coli* strain harboring a-ketoglutaric semialdehyde dehydrogenase (Sadh) from Azospirillum brasilense, along with dhaB and glycerol dehydratase reactivase from K. pneumoniae, produced 38.7 g/L of 3-hydroxypropionic acid with a yield of 0.34 g/g and productivity of 0.53 g/L/h in glycerol fed-batch fermentation (Rathnasingh et al. 2009). Unlike K. pneumoniae, *E. coli* lacks the coenzyme B12 metabolism required for dhaB activity; therefore, addition of coenzyme B12 to the culture medium could increase the cost in large-scale production. However, utilizing glycerol for the biological production of 3-hydroxypropionic acid could be a promising sustainable process because glycerol is a major by-product of biodiesel production.

19.3.3 Alcohols: 1-Propanol and Isopropanol

1-Propanol and isopropanol can be used in place of methanol as esterifying reagents, and the formed esters show reduced crystallization at low temperatures (Hanai et al. 2007). Chemically, 1-propanol is produced from ethane, carbon monoxide, and hydrogen, while isopropanol is produced by a hydration reaction between water and propene (Jain and Yan 2011). 2-Ketobutyrate, a component of the amino acid biosynthesis pathways, was used as a precursor of 1-propanol in a metabolically engineered *E. coli* strain (Atsumi et al. 2008) by sequential reactions of *Lactococcus lactis* 2-ketoacid decarboxylase (Kdc) and *S. cerevisiae* Adh2. The production of 1-propanol was further improved by increasing the precursor availability using evolved *Methanococcus jannaschii* citramalate synthase (CimA) to enhance the conversion of pyruvate to 2-ketobutyrate (Atsumi and Liao 2008). The engineered *E. coli* strain produced 3.9 g/L 1-propanol from glucose with a productivity of 0.04 g/L/h in flask culture (Atsumi and Liao 2008). Several microorganisms have been reported for isopropanol production, including wild-type *Clostridium beijerinckii* (George et al. 1983), engineered *E. coli* (George et al. 1983; Hanai et al. 2007). The strategy for the microbial production of isopropanol is derived from the *C. beijerinckii* pathway for converting acetyl CoA to isopropanol via acetone (George et al. 1983). However, due to the limited metabolic information and tools, the maximum titer from the native producer was only 1.8 g/L (George et al. 1983). More recently, a *C. acetobutylicum* ATCC 824 (pIPA3) strain harboring the secondary AdhI of *C. beijerinckii* NRRL B593 and a synthetic acetone operon (adc-ctfA-ctfB) of *C. acetobutylicum* produced 5.1 g/L of isopropanol in anaerobic flask culture. Metabolically engineered *E. coli* harboring four genes (thl-adc-ctfA-ctfB) involved in the acetone-producing pathway of *C. acetobutylicum* ATCC 824 and the gene-encoding AdhI of NRRL B593 achieved a higher titer of 13.6 g/L with a 0.15 g/g yield and 0.28 g/L/h productivity in glucose fed-batch fermentation (Jojima et al. 2008).

19.4 Bioproduction of C4 Chemicals

19.4.1 Carboxylic Acid: Butyric Acid

Butyric acid has wide range of industrial application in the field of food, cosmetic, pharmaceutical, and fine chemical manufacturing. Moreover, it can be used to produce fuels, such as butanol, ethyl butyrate, and butyl butyrate (Dwidar et al. 2012). In fact, butyric acid is a precursor of biobutanol. However, owing to its toxicity, fermentative production of butanol is problematic. Therefore, firstly, a biomass could be fermented to butyric acid and then it could be used as a starting material for butanol production (Dwidar et al. 2012). In the production of cellulose acetate butyrate plastics, the raw material of textile fiber manufacturing is the present major industrial application of butyric acid (Dwidar et al. 2012; Huang et al. 2011). Likewise, polyhydroxybutyrate (PHB), another industrially important polymer, could also be produced from butyric acid (Dwidar et al. 2012). Butyric acid is known to have anticancer property and a wide range of pharmaceutical applications (Dwidar et al. 2012; Huang et al. 2011). Similarly, butyric acid derivatives, such as its methyl, ethyl, and amyl esters, are used as flavor or aroma compounds (Dwidar et al. 2012).

Anaerobic bacteria, such as *clostridia* sp., can produce butyric acid by fermentation of a variety of substrates ranging from glucose to complex agro-industrial waste materials. Among different members of the genus Clostridium, *C. butyricum* and *C. tyrobutyricum* are the two mostly studied species for butyric acid production. Jiang et al. (2011) have reported the use of glucose as the major substrate for butyric acid production by *C. tyrobutyricum* using a repeated fed-batch process. According to the authors, the process was successful in reducing the product inhibition and final butyric acid concentration as high as 86.9 ± 2.17 g/L could be achieved (Jiang et al. 2011). Similar to glucose, xylose is the other substrate commonly investigated for butyric acid production (Zhu et al. 2002; Zhu and Yang 2003). For improved economic feasibility of the process, various industrial waste materials have also been evaluated for possible application in butyric acid production. Hwang and Hansen (1997) have used cheese-processing industry wastewater for butyric acid production using a continuously stirred tank reactor. The authors have used a constant inoculum system to reduce the experimental error and different physiological parameters were optimized by using response surface methodology. According to the authors, pH 7.26 at 36.2 °C was found to be optimum; however, the response was unpredictable and a small change of these parameters could cause significant variations in cumulative butyric acid production (Hwang and Hansen 1997). Likewise, butyrate production by fermentation of corn steep liquor supplemented with corn fiber hydrolysate has been reported by Zhu et al. (2002). The authors have demonstrated that using this crude substrate, butyrate yield as high as 0.47 g/g substrate as well as maximum productivity of 2.91 g/L/h could be achieved (Zhu et al. 2002). Huang et al. (2011) have evaluated Jerusalem artichoke as a potential, less expensive substrate for butyric acid production by *Clostridium tyrobutyricum*. According to the authors, in a fed-batch process, final butyric acid concentration as high as 60.4 g/L could be achieved by using acid-treated Jerusalem artichoke hydrolysate as the substrate (Huang et al. 2011).

Product inhibition is a concern in butyric acid bioproduction; however, a considerable progress has been made in resolving this problem. Actually, due to inhibitory effect of butyric acid on microbial growth, it is difficult to achieve a relatively higher final product concentration by employing a traditional fermentation process. In turn, low final product concentration can increase the expense of product recovery and reduce economic feasibility of the process. In this context, cell immobilization is one of the frequently used strategies for improved butyric acid production. Jiang et al. (2011) have reported that microbial cells could be immobilized in fibrous matrix to develop a fibrous-bed bioreactor for industrial application. According to the authors, relatively higher cell density could be maintained and improved productivity could be achieved by using this technology. Similarly, the immobilized microorganisms were found to quickly adapt to the acidic environment and were capable of developing a mutant community with high acid tolerance (Jiang et al. 2011). Similarly, Zhu and Yang (2003) have reported that *Clostridium tyrobutyricum* immobilized in fibrous matrix developed a mutant variety capable continuing butyric acid production at final product concentration as high as 30 g/L, using glucose and xylose as substrate (Zhu and Yang 2003). Based on kinetic behavior of biomass production profile as well as activity of the enzymes in the butyric acid formation pathway, such as phosphotransbutyrylase and butyrate kinase, the authors have concluded that highly acid-tolerant mutant variety emerged out of the immobilized *Clostridium tyrobutyricum* were physiologically superior to the original culture (Zhu and Yang 2003). Butyric acid production by immobilized *C. tyrobutyricum* has also been reported by Huang et al. (2011) and the authors have discussed possible commercial application of the process (Huang et al. 2011). Similar to cell immobilization, in situ product extraction is another strategy frequently applied to resolve the problem of product inhibition. Vandák et al. (1997) have demonstrated an extractive fermentation process for butyric acid production by *Clostridium butyricum* S21. The authors have evaluated nine different organic extractive agents in terms of cell growth and butyric acid production by the organism (Vandák et al. 1997). Among different extracting agents, based on biocompatibility as well as improvement in butyric acid production, Hostarex A327 (20 % w/w) in oleyl alcohol was reported to be suitable for in situ product recovery (Vandák et al. 1997).

A number of genetic engineering strategies for enhanced butyric acid production could be found in the literature. In general, these approaches involve elimination of multiple product formation pathways so that butyric acid could be the major fermentation end product, for example, butyric acid, acetic acid, hydrogen, as well as carbon dioxide are the major products of glucose fermentation by *Clostridium tyrobutyricum*. Among these products, acetic acid production has negative effect on butyric acid yield. In an approach by Liu et al. (2006), genes encoding two important enzymes (phosphotransacetylase and acetate kinase) of acetic acid biosynthetic pathway have been inactivated for improved butyric acid yield (Liu et al. 2006). Zhu et al. (2005) also have reported genetic manipulation of acetic acid formation pathway by integrational mutagenesis and 15 % improvement as well as 14 % reduction, respectively, in butyrate and acetate production by mutated *C. tyrobutyricum* (Zhu et al. 2005). Thus, it could be concluded that butyric acid is an important platform chemical and considerable success has been made in its bioproduction.

19.4.2 Dicarboxylic Acids: Succinic, Malic, and Fumaric Acids

Succinic acid is an important platform chemical and it can be used as the precursor for different specialized polyesters (http://en.wikipedia.org/wiki/Succinic_acid, accessed on 09/04/2013). Similarly, succinic acid could be used as raw material for the production of 1,4-butanediol, nylon precursor (adipic acid), ethylenediamine disuccinate, tetrahydrofuran, diethyl succinate, 2-pyrrolidinone, *N*-methyl pyrrolidinone, and gamma-butyrolactone (McKinlay et al. 2007; Song and Lee 2006). Likewise, it has many other applications in food and beverage, agriculture, as well as pharmaceutical industries. Succinic acid can be produced either by fermentative conversion of organic substrates or by chemical conversion of petroleum product.

Out of these two options, at present, succinic acid is commercially produced from liquefied petroleum gas or petroleum oil using chemical method (McKinlay et al. 2007; Song and Lee 2006). Being a sustainable process, succinic acid bioproduction is an attractive option; however, in its present form, it is not economically comparable to that of petrochemical-based technologies (McKinlay et al. 2007). There are certain technical constraints in succinic acid bioproduction and they need to be resolved prior to its efficient microbial production. Firstly, final concentration of succinic acid in fermented media should be sufficiently higher for cost-effective product recovery. Similarly, downstream processing strategies should be simplified (McKinlay et al. 2007). Further, it should be ensured that by-product formation is reduced so that high product yield could be achieved. Apart from the aforementioned approaches, selection of appropriate microbial culture as well as strain improvement through metabolic engineering is another possible option for profitable succinic acid bioproduction.

In a report by Liu et al. (2008), *Actinobacillus succinogenes* CGMCC1593 has been evaluated for succinic acid bioproduction using cane molasses. The authors have evaluated the use of sulfuric acid-treated cane molasses and reported that a maximum succinic acid concentration of 50.6 ± 0.9 g/L could be achieved by using a small-scale anaerobic bioreactor (Liu et al. 2008). Further, the authors have mentioned that by employing a 5-L stirred tank bioreactor, faster cell growth and product formation could be achieved. Moreover, by changing the process mode to fed-batch, substrate inhibition could be reduced and final succinic acid concentration could be further increased (Liu et al. 2008). Lee et al. (2002) have isolated a novel nonspore-forming, mesophilic, succinic acid-producing bacterium from bovine rumen, which has been named as *Mannheimia succiniciproducens* MBEL55E. According to the authors, the bacterium can grow at pH 6.0–7.5 and can produce succinic acid, acetic acid, and formic acid at a ratio of 2:1:1 (Lee et al. 2002). Vemuri et al. (2002) have demonstrated a dual-phase fermentation process for improved succinic acid production by *Escherichia coli* AFP111. This unique process has an initial aerobic phase for improve biomass production followed by an anaerobic phase for succinic acid synthesis. The authors have reported that using the optimized conditions, the process can achieve a final succinic acid concentration of 99.2 g/L (Vemuri et al. 2002). In general, application of less expensive organic waste materials as the substrate is an option for enhancing the financial compatibility of a fermentation process, and the same is true for succinic acid bioproduction. Kim et al. (2004) have used NaOH-treated wood hydrolysate as the substrate for succinic acid production by *Mannheimia succiniciproducens* MBEL55E. The authors have used batch as well as continuous fermentation process and reported that succinic acid productivity of 1.17 g/L/h could be achieved by a batch process, whereas, the same by a continuous process could be 3.19 g/L/h (Kim et al. 2004). Similarly, succinic acid production by *Actinobacillus succinogenes* CGMCC1593 using straw hydrolysates as the major substrate has been reported by Zheng et al. (2009). In another study, Lee et al. (2001) have demonstrated that glycerol could be used as the substrate for succinic acid production by *Anaerobiospirillum succiniciproducens*. The authors have reported that by feeding yeast extract as a supplementary nutrient, the final succinic acid to acetic acid ratio could be enhanced up to 31.7:1 (Lee et al. 2001). Thus, it could be concluded that considerable progress has been made in succinic acid bioproduction. However, it seems that genetic engineering strategies are still less explored in this particular field.

19.4.3 Butanol and Isobutanol

1-Butanol and isobutanol are 4-carbon alcohol platform chemicals with several large-volume derivatives and are used in many applications such as in plasticizers, amino resins, in paint industry, and as fuel additives among others. 1-Butanol can be used as precursor for the production of latex surface coatings, enamels, and lacquers (Kirschner 2006; Lee et al. 2008). Furthermore, 1-butanol could be used as a bio-based transportation fuel and is more fuel efficient than ethanol on a volume basis. While isobutanol ester derivatives, such as diisobutyl phthalate, can be used as plasticizer agents (Atsumi et al. 2008). Also, through standard chemistry, isobutanol can be used as an ingredient in nearly 40 % of traditional chemicals (such as butenes, toluenes, and xylenes) as well as many in transportation fuels. Used as a solvent, isobutanol appears in paints and cosmetics such as nail polish. The solvent, rubber, and fuel ingredient markets are each worth several billion dollars. Butanol production has been undertaken by chemical processes in which propylene is catalytically converted to 1-butanol (Jones and Woods 1986). More recently, the primary method of butanol production was carried out using biological processes by anaerobic fermentation with *Clostridium acetobutylicum* the acetone–butanol–ethanol (ABE) fermentation pathway (Jang et al. 2012a, b, c). In this context, clostridial species have advantages over other bacteria and yeasts in utilizing various carbon sources (Steen et al. 2008; Jang et al. 2012a, b, c). However, the butanol yields were low and the production cost was high. The US DOE's Small Business Technology Transfer Program is funding research to improve the bio-based route to butanol and make it cost competitive with the petroleum-based route. Researchers are attempting to increase butanol yield by using improved bacteria strains, employing advanced reactor technology, and separating the organic acid production and organic acid-to-alcohol phases into different vessels. Thus, metabolic engineering of Clostridium has been undertaken to improve 1-butanol production by reducing by-product formation. One notable example is the production of 16.7 g/L of 1-butanol with a productivity of 0.31 g/L/h in anaerobic batch fermentation of glucose, using a butyrate kinase-inactive mutant of *C. acetobutylicum* in which formation of the by-product, butyrate, was blocked (Harris et al. 2000). Recently, 1-butanol production of 14–15 g/L was achieved during glucose fermentations by engineered *E. coli* harboring a modified Clostridial pathway (Dellomonaco et al. 2011; Shen et al. 2011). Biological isobutanol production was carried out; reported using *E. coli* strain, more than 20 g/L of isobutanol was produced from glucose (Atsumi et al. 2008). Isobutanol production has also been implemented in several other organisms such as *C. glutamicum* (Smith et al. 2010; Blombach et al. 2011; Li et al. 2011). *C. glutamicum* is an attractive producer of isobutanol that was successfully engineered to produce 13 g/L of isobutanol with a yield of 0.20 g/g on glucose

(Blombach et al. 2011). Researchers have made significant progress in biological 1-butanol and isobutanol production using metabolic engineering of *E. coli*, *Clostridium acetobutylicum,* and *C. glutamicum*; the production efficiency still needs to be improved for it to become economically viable. Further work on engineering these strains and exploring their tolerances and the use of low-cost renewable substrate like biomass may allow researchers to increase the yields of butanol and isobutanol to an industrial scale.

19.5 Conclusions

Most of platform chemicals have been produced from fossil oil that is currently facing global crises, such as climate change and fossil-resource depletion. Thus, there is an increasing demand for sustainable production of bio-based platform chemicals using biomass sugars as a substrate fermented by microorganisms. Chemical products with a broad range of carbon lengths, including C3–C4 carboxylic acids, dicarboxylic acids, *hydroxy* acids, amino carboxylic acids, alcohols, and amines, can be produced by the sugar fermentations of microorganisms. Presently, bio-based lactic acid has already been commercialized, and succinic, isobutanol, are nearing large-scale commercialization as bio-based chemicals. However, propionic acid, 1-butanol, 3-hydroxypropionic acids, butyric acid, and 1-propanol are produced only in small amounts by fermentation because of the low efficiency and the high cost of the fermentation. In order to make the fermentation route economically viable, it is necessary to develop novel fermentation processes that use highly efficient bioreactors and separation techniques and low-cost renewable resources, namely, industrial wastes and by-products, as substrate for fermentation.

Acknowledgements The authors are sincerely thankful to the Natural Sciences and Engineering Research Council of Canada (Discovery Grants 355254) and INRS-ETE for financial support. The views or opinions expressed in this article are those of the authors.

References

Akerberg C, Zacchi G (2000) An economic evaluation of the fermentative production of lactic acid from wheat flour. Biores Technol 75:119–126

Ashok S, Raj SM, Rathnasingh C, Park S et al (2011) Development of recombinant *Klebsiella pneumoniae* dhaT strain for the co-production of 3-hydroxypropionic acid and 1,3-propanediol from glycerol. Appl Microbiol Biotechnol 90:1253–1265

Atsumi S, Cann AF, Connor MR, Shen CR, Smith KM, Brynildsen MP, Chou KJ, Hanai T, Liao JC et al (2008) Metabolic engineering of *Escherichia coli* for 1-butanol production. Metab Eng 10:305–311

Atsumi S, Liao JC (2008) Directed evolution of *Methanococcus jannaschii* citramalate synthase for biosynthesis of 1-propanol and 1-butanol by *Escherichia coli*. Appl Environ Microbiol 74:7802–7808

Bechthold I, Bretz K, Kabasci S, Kopitzky R, Springer A et al (2008) Succinic acid: a new platform chemical for biobased polymers from renewable resources. Chem Eng Technol 31:647–654

Blombach B, Riester T, Wieschalka S, Ziert C, Youn JW, Wendisch VF, Eikmanns BJ et al (2011) *Corynebacterium glutamicum* tailored for efficient isobutanol production. Appl Environ Microbiol 77:3300–3310

Boyaval P, Corre C (1995) Production of propionic acid. Lait 75:453–461

Carbonell P, Fichera D, Pandit SB, Faulon JL et al (2012) Enumerating metabolic pathways for the production of heterologous target chemicals in chassis organisms. BMC Syst Biol 6:10

Curran KA, Alper HS (2012) Expanding the chemical palate of cells by combining systems biology and metabolic engineering. Metab Eng 14:289–297

da Silva GP, Mack M, Contiero J et al (2009) A promising and abundant carbon source for industrial microbiology. Biotechnol Adv 27:30–39

Datta R, Tsai SP, Bonsignore P, Moon SH, Frank JR et al (1995) Technological and economic potential of poly(lactic acid) and lactic acid derivatives. FEMS Microbiol Rev 16:221–231

Datta R, Henry M (2006) Lactic acid: recent advances in products, processes and technologies – a review. J Chem Technol Biotechnol 81:1119–1129

Dellomonaco C, Clomburg JM, Miller EN, Gonzalez R et al (2011) Engineered reversal of the beta-oxidation cycle for the synthesis of fuels and chemicals. Nature 476:355–359

Dwidar M, Park JY, Mitchell RJ, Sang BI et al (2012) The future of butyric acid in industry. SciWorldJ 2012:471417

Feng X, Chen F, Xu H, Wu B, Li H, Li S, Ouyang P et al (2011) Green and economical production of propionic acid by *Propionibacterium freudenreichii* CCTCC M207015 in plant fibrous-bed bioreactor. Bioresour Technol 102:6141–6146

Garvie EI (1980) Bacterial lactate dehydrogenases. Microbiol Rev 44:106–139

George HA, Johnson JL, Moore WE, Holdeman LV, Chen JS et al (1983) Acetone, isopropanol, and butanol production by *Clostridium beijerinckii* (*syn. Clostridium butylicum*) and *Clostridium aurantibutyricum*. Appl Environ Microbiol 45:1160–1163

Hanai T, Atsumi S, Liao JC et al (2007) Engineered synthetic pathway for isopropanol production in *Escherichia coli*. Appl Environ Microbiol 73:7814–7818

Harris LM, Desai RP, Welker NE, Papoutsakis ET et al (2000) Characterization of recombinant strains of the *Clostridium acetobutylicum* butyrate kinase inactivation mutant: need for new phenomenological models for solventogenesis and butanol inhibition? Biotechnol Bioeng 67:1–11

Hatti-Kaul R, Tornvall U, Gustafsson L, Borjesson P et al (2007) Industrial biotechnology for the production of bio-based chemicals—a cradle-to-grave perspective. Trends Biotechnol 25:119–124

Hohenschutz H, Franz D, Buelow H, Dinkhauser G et al (1974). Production of propionic acid. United States patent US 3835185

Hong KK, Nielsen J (2012) Metabolic engineering of Saccharomyces cerevisiae: a key cell factory platform for future biorefineries. Cell Mol Life Sci 69:2671–2690. doi:10.1007/s00018-012-0945-1

Huang J, Cai J, Wang J, Zhu X, Huang L, Yang ST, Xu Z et al (2011) Efficient production of butyric acid from Jerusalem artichoke by immobilized *Clostridium tyrobutyricum* in a fibrous-bed bioreactor. Biores Technol 102:3923–3926

Hwang S, Hansen CL (1997) Modeling and optimization in anaerobic bioconversion of complex substrates to acetic and butyric acids. Biotechnol Bioeng 54:451–460

Ito T, Nakashimada Y, Senba K, Matsui K, Nishio N et al (2005) Hydrogen and ethanol production from glycerol-containing wastes discharged after Biodiesel manufacturing process. J Biosci Bioeng 100:260–265

Jäger G, Büchs J (2012) Biocatalytic conversion of lignocellulose to platform chemicals. Biotechnol J 7:1122–1136

Jain R, Yan Y (2011) Dehydratase mediated 1-propanol production in metabolically engineered *Escherichia coli*. Microb Cell Fact 10:97

Jang YS, Lee J, Malaviya A, Seung DY, Cho JH, Lee SY et al (2012a) Butanol production from renewable biomass: rediscovery of metabolic pathways and metabolic engineering. Biotechnol J 7:186–198

Jang YS, Park JM, Choi S, Choi YJ, Seung DY, Cho JH, Lee SY et al (2012b) Engineering of microorganisms for the production of biofuels and perspectives based on systems metabolic engineering approaches. Biotechnol Adv 30:989–1000. doi:10.1016/j.biotechadv.2011.08.015

Jang YS, Kim J, Shin JH, Choi YJ, Choi S, Song CW, Lee J, Park HG, Lee SY et al (2012c) Bio-based production of C2–C6 platform chemicals. Biotechnol Bioeng 109:2437–2459. doi:10.1002/bit.24599

Jiang L, Wang J, Liang S, Cai J, Xu Z, Cen P, Yang S, Li S et al (2011) Enhanced butyric acid tolerance and bioproduction by *Clostridium tyrobutyricum* immobilized in a fibrous bed bioreactor. Biotechnol Bioeng 108:31–40

John RP, Nampoothiri KM, Pandey A et al (2007) Fermentative production of lactic acid from biomass: an overview on process developments and future perspectives. Appl Microbiol Biotechnol 74:524–534

Jojima T, Inui M, Yukawa H et al (2008) Production of isopropanol by metabolically engineered *Escherichia coli*. Appl Microbiol Biotechnol 77:1219–1224

Jones DT, Woods DR (1986) Acetone–butanol fermentation revisited. Microbiol Rev 50:484–524

Kirschner M (2006) n-Butanol. Chem Mark Rep 269:42, January 30–February 5. ABI/INFORM Global

Kim DY, Yim SC, Lee PC, Lee WG, Lee SY, Chang HN et al (2004) Batch and continuous fermentation of succinic acid from wood hydrolysate by Mannheimia succiniciproducens MBEL55E. Enz Microbiol Technol 35:648–653

Lee PC, LeeWG LSY, Chang HN et al (2001) Succinic acid production with reduced by product formation in the fermentation of *Anaerobiospirillum succiniciproducens* using glycerol as a carbon source. Biotechnol Bioeng 72:41–48

Lee PC, Lee SY, Hong SH, Chang HN et al (2002) Isolation and characterization of a new succinic acid-producing bacterium, Mannheimia succiniciproducens MBE L55E, from bovine rumen. Appl Microbiol Biotechnol 58:663–668

Lee J, Jang YS, Choi SJ, Im JA, Song H, Cho JH, Seung do Y, Papoutsakis ET, Lee SY, Park JH, Jang SH, Nielsen LK, Kim J, Jung KS et al (2008) Fermentative butanol production by Clostridia. Biotechnol Bioeng 101:209–228

Li S, Wen J, Jia X et al (2011) Engineering Bacillus subtilis for isobutanol production by heterologous Ehrlich pathway construction and the biosynthetic 2-ketoisovalerate precursor pathway overexpression. Appl Microbiol Biotechnol 91:577–589

Litchfield JH (1996) Microbial production of lactic acid. Adv Appl Microbiol 42:45–95

Liu X, Zhu Y, Yang ST et al (2006) Butyric acid and hydrogen production by *Clostridium tyrobutyricum* ATCC 25755 and mutants. Enz Microbiol Technol 38:521–528

Liu YP, Zheng P, Sun ZH, Ni Y, Dong JJ, Zhu LL et al (2008) Economical succinic acid production from cane molasses by Actinobacillus succinogenes. Biores Technol 99:1736–1742

Liu L, Zhu Y, Li J, Wang M, Lee P, Du G, Chen J et al (2012) Microbial production of propionic acid from propionibacteria: current state, challenges and perspectives. Crit Rev Biotechnol 32:374–381. doi:10.3109/07388551.2011.651428

McKinlay JB, Vieille C, Zeikus JG et al (2007) Prospects for a bio-based succinate industry. Appl Microbiol Biotechnol 76:727–740

Murphy AC (2011) Metabolic engineering is key to a sustainable chemical industry. Nat Prod Rep 28:1406–1425

Odian G (2004) Principle of polymerization. Wiley, Hoboken, NJ

Okano K, Tanaka T, Ogino C, Fukuda H, Kondo A et al (2010) Biotechnological production of enantiomeric pure lactic acid from renewable resources: recent achievements, perspectives, and limits. Appl Microbiol Biotechnol 85:413–423

Rathnasingh C, Raj SM, Jo JE, Park S et al (2009) Development and evaluation of efficient recombinant *Escherichia coli* strains for the production of 3-hydroxypropionic acid from glycerol. Biotechnol Bioeng 104:729–739

Sauer M, Porro D, Mattanovich D, Branduardi P et al (2008) Microbial production of organic acids: expanding the markets. Trends Biotechnol 26:100–107

Sels BF, Dusselier M, Van Wouwe P, Dewaele A, Makshina E (2013) Lactic acid as platform chemical in the biobased economy: the role of chemocatalysis. Energy Environ Sci 6:1415–1442

Shen CR, Lan EI, Dekishima Y, Baez A, Cho KM, Liao JC et al (2011) Driving forces enable high-titer anaerobic 1-butanol synthesis in *Escherichia coli*. Appl Environ Microbiol 77:2905–2915

Smith KM, Cho KM, Liao JC et al (2010) Engineering *Corynebacterium glutamicum* for isobutanol production. Appl Microbiol Biotechnol 87:1045–1055

Song H, Lee SY (2006) Production of succinic acid by bacterial fermentation. Enz Microbiol Technol 39:352–361

Steen EJ, Chan R, Prasad N, Myers S, Petzold CJ, Redding A, Ouellet M, Keasling JD et al (2008) Metabolic engineering of *Saccharomyces cerevisiae* for the production of n-butanol. Microb Cell Fact 7:36

Thakker C, Martinez I, San KY, Bennett GN et al (2012) Succinate production in Escherichia coli. Biotechnol J 7:213–224

Wang L, Zhao B, Li F, Xu K, Ma C, Tao F et al (2011) Highly efficient production of D-lactate by *Sporolactobacillus sp*. CASD with simultaneous enzymatic hydrolysis of peanut meal. Appl Microbiol Biotechnol 89:1009–1017

Wang P, Wang Y, Su Z et al (2012) Microbial production of propionic acid with *Propionibacterium freudenreichii* using an anion exchanger-based in situ product recovery (ISPR) process with direct and indirect contact of cells. Appl Biochem Biotechnol 166:974–986

Wu X, Jiang S, Liu M, Pan L, Zheng Z, Luo S et al (2011) Production of L-lactic acid by *Rhizopus oryzae* using semicontinuous fermentation in bioreactor. J Ind Microbiol Biotechnol 38:565–571

Vandák D, Zigová J, Šturdík E, Schlosser Š et al (1997) Evaluation of solvent and pH for extractive fermentation of butyric acid. Proc Biochem 32:245–251

Vemuri G, Eiteman M, Altman E et al (2002) Succinate production in dual-phase *Escherichia coli* fermentations depends on the time of transition from aerobic to anaerobic conditions. J Ind Microbiol Biotechnol 28:325–332

Ye VM, Bhatia SK (2012) Metabolic engineering for the production of clinically important molecules: omega-3 fatty acids, artemisinin, and taxol. Biotechnol J 7:20–33

Zhang A, Yang ST (2009) Engineering Propionibacterium acidipropionici for enhanced propionic acid tolerance and fermentation. Biotechnol Bioeng 104:766–773

Zhao B, Wang L, Li F, Hua D, Ma C, Ma Y et al (2010) Kinetics of D-lactic acid production by *Sporolactobacillus* sp. strain CASD using repeated batch fermentation. Bioresour Technol 101:6499–6505

Zheng P, Dong JJ, Sun ZH, Ni Y, Fang L et al (2009) Fermentative production of succinic acid from straw hydrolysate by Actinobacillus succinogenes. Biores Technol 100:2425–2429

Zhu Y, Wu Z, Yang ST et al (2002) Butyric acid production from acid hydrolysate of corn fibre by *Clostridium tyrobutyricum* in a fibrous-bed bioreactor. Proc Biochem 38:657–666

Zhu Y, Yang ST (2003) Adaptation of *Clostridium tyrobutyricum* for enhanced tolerance to butyric acid in a fibrous-bed bioreactor. Biotechnol Prog 19:365–372

Zhu Y, Liu X, Yang ST et al (2005) Construction and characterization of pta gene-deleted mutant of *Clostridium tyrobutyricum* for enhanced butyric acid fermentation. Biotechnol Bioeng 90:154–166

Zhu Y, Lee YY, Elander RT et al (2007) Conversion of aqueous ammonia-treated corn stover to lactic acid by simultaneous saccharification and cofermentation. Appl Biochem Biotechnol 136–140:722–737

Zhu Y, Li J, Tan M, Liu L, Jiang L, Sun J, Lee P, Du G, Chen J et al (2010) Optimization and scale-up of propionic acid production by propionic acid-tolerant *Propionibacterium acidipropionici* with glycerol as the carbon source. Bioresour Technol 101:8902–8906

Zhu L, Wei P, Cai J, Zhu X, Wang Z, Huang L, Xu Z et al (2012) Improving the productivity of propionic acid with FBB-immobilized cells of an adapted acid-tolerant Propionibacterium acidipropionici. Bioresour Technol 112:248–253

About the Editors

Carlos Ricardo Soccol is the research group leader of *DEBB* (*Department of Bioprocess Engineering and Biotechnology*) at the Federal University of Paraná, Brazil, with 20 years of experience in biotechnological research and development of bioprocesses with industrial application. He is a graduate in Chemical Engineering (UFPR, 1979), Master in Food Technology (UFPR, 1986) and a Ph.D. in *Genie Enzymatique, Microbiologie et Bioconversion* (*Université de Technologie de Compiègne*,-France, 1992). He obtained his post-doctorate at Institut ORSTOM/IRD (Montpellier, 1994 and 1997) and at the *Université de Provence et de la Méditerranée* (Marseille, 2000). He is HDR Professor at *Ecole d'Ingénieurs Supériure* of *Luminy*, Marseille-France. He has experience in the areas of Science and Food Technology, with emphasis on Agro-industrial and Agroalimentary Biotechnology, acting in the following areas: bioprocess engineering and solid-state fermentation, submerged fermentation, bioseparations, industrial bioprocesses, enzyme technology, tissue culture, bio-industrial projects and bioproduction.

S.K. Brar et al. (eds.), *Biotransformation of Waste Biomass into High Value Biochemicals*, DOI 10.1007/978-1-4614-8005-1, © Springer Science+Business Media New York 2014

He is currently Coordinator of Master BIODEV-UNESCO, Associate Editor of five international journals and Editor in Chief of Brazilian Archives of Biology and Technology Journal. Professor Soccol received several national and international awards which include Science & Technology award of the Govt. of Paraná (1996), Scopus/Elsevier award (2009), Dr. Honoris Causa, University Blaise Pascal-France (2010), Outstanding Scientist—Fifth International Conference on Industrial Bioprocesses, Taipei, Taiwan (2012). He is a technical and scientific consultant of several companies, agencies, and scientific journals in Brazil and abroad. He has supervised and formed 90 master science students, 40 Ph.D. students and 12 post-doctorate Students. He has 946 publications/communications which include 16 books, 95 book chapters, 255 original research papers, 543 research communications in international and national conferences and has registered 37 patents. His research articles until the moment were cited 4,300 Times with Index $h = 34$.

Gurpreet Singh Dhillon earned his master's degree in Molecular Biology and Biochemistry from Guru Nanak Dev University, India, in 2005 and PhD. Water Sciences from the *Institut national de la recherche scientifique* (*INRS*), Centre for Water, Earth and Environment, University of Quebec, Canada. Dr. Dhillon started his research career working on the important domain of bioenergy at central Institute of Post-Harvest Engineering and Technology (CIPHET), India. Currently, Dhillon is pursuing his postdoctoral degree in the Department of Agricultural, Food and Nutritional Sciences (AFNS), University of Alberta, Edmonton, Canada.

Dr. Dhillon is a strong advocate for taking an integrated multidisciplinary approach for developing novel, eco-friendly and industrial processes. His research work is typically based on the concept of *biorefining* and involves waste management through value addition using multidisciplinary approach combining bioprocess technology, microbiology, biotechnology, and analytical chemistry. To date, his research activities have concentrated on biotransformation of biomass into multitude of products e.g. biofuels, platform chemicals, biocatalysts, biopolymers, bioactive compounds, nanoparticles and treatment of emerging contaminants.

Dr. Dhillon has received several international awards and scholarships. He is the recipient of prestigious: (1) Postdoctoral scholarship (Quebec-India, 2I) for foreign students (MELS) by Fonds de recherche du Quebec (FQRNT), Quebec, Canada (May 2013–April 2014); (2) Merit Scholarship for Doctoral Studies by Institut National de la Recherche Scientifique (INRS), Centre Eau, Terre & Environnement (ETE), University of Québec, Canada (Jan. 2010- Feb. 2013); (3) Graduate teaching assistantship, Department of Chemistry and Biochemistry, Miami University Oxford, Ohio, USA and; (4) Graduate research assistantship for doctorate studies, Department of Cellular & Molecular Biology (CEMB), University of Arkansas, Fayetteville, USA (2010–2014), among others. Currently, he is serving as the associate editor of the *International Journal of Life Sciences* and editorial board member of *Journal of Agricultural engineering and Biotechnology*. He has supervised several master's and bachelor's students working in different research areas. He has more than 55 publications, which include 25 research articles, 10 review articles, 2 books, 15 book chapters, and 8 research communications in international and national conferences and seminars.

Satinder Kaur Brar is associate professor at Institut National de la Recherche Scientifique (Eau, Terre et Environnement, INRS-ETE). She graduated with master's in Organic Chemistry from National Chemical Laboratory, Pune, India, followed by master's in Technology in Environmental Sciences and Engineering from Indian Institute of Technology, Bombay, Mumbai, India and obtained a Ph.D. in biochemical engineering from INRS, Quebec, Canada. She gained professional experience in between her master's and Ph.D. of research at Defence Research and Development Organization, Delhi, India where she worked on phytoremediation of explosive contaminated sites. Dr. Brar is a recipient of the *ASCE State-of-the-Art of*

Civil Engineering award (2007) for her article titled, "Bioremediation of Hazardous Wastes - A Review," which was published in the *Practice Periodical of Hazardous, Toxic & Radioactive Waste Management—Special issue on Bioremediation*. She has also received the Rudolf gold medal (2008) for her originality of the article published in *Practice Periodical of Hazardous, Toxic & Radioactive Waste Management*. Her research interests lie in the development of finished products (formulations) of wastewater and wastewater sludge-based value-added bioproducts, such as enzymes, organic acids, platform chemicals, biocontrol agents, biopesticides, butanol and biohydrogen. She is also interested in the fate of endocrine disrupter compounds, pharmaceuticals, nanoparticles and other toxic organic compounds during value addition of wastewater and wastewater sludge in turn finding suitable biological detoxification technologies. She is on the editorial board of Brazilian Archives of Biology and Technology Journal and associate editor of two international repute journals. She has won several accolades through her professional career through awards, such as outstanding young scientist in India and several others. She has supervised and co-supervised at least eight master's science students, ten Ph.D. students and four post-doctorate students. She has more than 162 research publications, which include 3 books, 30 book chapters, 80 original research papers, 50 research communications in international and national conferences and has registered 2 patents to her credit.

Index

S.K. Brar et al. (eds.), *Biotransformation of Waste Biomass into High Value Biochemicals*,
DOI 10.1007/978-1-4614-8005-1, © Springer Science+Business Media New York 2014

Printed by Printforce, the Netherlands